VOLUME ONE HUNDRED AND SEVENTY EIGHT

INTERNATIONAL REVIEW OF NEUROBIOLOGY

Advances in Alcohol Use Disorder Treatment

INTERNATIONAL REVIEW OF NEUROBIOLOGY

VOLUME 178

SERIES EDITOR

PATRICIA JANAK

Janak Lab, Department of Neuroscience,
Dunning Hall, John Hopkins University,
Baltimore, MD, USA

PETER JENNER

Division of Pharmacology and Therapeutics,
GKT School of Biomedical Sciences
King's College, London, UK

VOLUME ONE HUNDRED AND SEVENTY EIGHT

INTERNATIONAL REVIEW OF NEUROBIOLOGY

Advances in Alcohol Use Disorder Treatment

Edited by

ROSANA CAMARINI

Laboratory of Neurochemistry and Behavioural Pharmacology, Department of Pharmacology, Institute of Biomedical Sciences Universidade de São Paulo, São Paulo, SP, Brazil

FÁBIO CARDOSO CRUZ

Department of Pharmacology, Escola Paulista de Medicina, São Paulo Federal University, São Paulo, SP, Brazil

ELSEVIER

ACADEMIC PRESS

An imprint of Elsevier

Academic Press is an imprint of Elsevier
125 London Wall, London, EC2Y 5AS, United Kingdom
50 Hampshire Street, 5th Floor, Cambridge, MA 02139, United States
525 B Street, Suite 1650, San Diego, CA 92101, United States

First edition 2024

ISBN: 978-0-443-29638-3
ISSN: 0074-7742

For information on all Academic Press publications
visit our website at https://www.elsevier.com/books-and-journals

Publisher: Zoe Kruze
Acquisitions Editor: Mariana Kuhl
Editorial Project Manager: Palash Sharma
Production Project Manager: Abdulla Sait
Cover Designer: Gopalakrishnan Venkatraman

Typeset by MPS Limited, India

Contents

Contributors

Sofía Adasme-Reyes
Department of Pharmacological and Toxicological Chemistry, Faculty of Chemical Sciences and Pharmacy, University of Chile, Santiago, Chile

Alexia dos Anjos-Santos
Molecular and Behavioral Neuroscience Laboratory, Department of Pharmacology, Universidade Federal de São Paulo (UNIFESP), São Paulo, São Paulo, Brazil

Howard C. Becker
Department of Psychiatry and Behavioral Sciences, Charleston Alcohol Research Center; Department of Neuroscience, Charleston Alcohol Research Center, Medical University of South Carolina; RHJ Veterans Administartion Health Care System, Charleston, SC, United States

Laís F. Berro
Department of Health Sciences, Universidade Estadual de Santa Cruz, Ilhéus, BA, Brazil; Department of Psychiatry and Human Behavior, Center for Innovation and Discovery in Addictions, University of Mississippi Medical Center, Jackson, MS, United States

Pablo Berríos-Cárcamo
Center for Regenerative Medicine, Faculty of Medicine, Clínica Alemana-Universidad del Desarrollo, Santiago, Chile

Natalia Bonetti Bertagna
Molecular and Behavioral Neuroscience Laboratory, Pharmacology Department, Escola Paulista de Medicina; Molecular and Behavioral Neuroscience Laboratory, Department of Pharmacology, Universidade Federal de São Paulo (UNIFESP), São Paulo, São Paulo, Brazil

Rosana Camarini
Department of Pharmacology, Institute of Biomedical Sciences, Universidade de Sao Paulo, São Paulo, Brazil

Beatriz Yamada Costa
Department of Pharmacology, Institute of Biomedical Sciences, Universidade de Sao Paulo, São Paulo, Brazil

Giovanna Victória Lopes Costa
Molecular and Behavioral Neuroscience Laboratory, Department of Pharmacology, Universidade Federal de São Paulo (UNIFESP), São Paulo, São Paulo, Brazil

Fábio Cardoso Cruz
Molecular and Behavioral Neuroscience Laboratory, Department of Pharmacology, Universidade Federal de São Paulo (UNIFESP), São Paulo, São Paulo, Brazil

Talissa Dezanetti
Programa de Pós-Graduação em Neurociências, Universidade Federal de Santa Catarina (UFSC), Florianópolis, Brazil

Fernando Ezquer
Center for Regenerative Medicine, Faculty of Medicine, Clínica Alemana-Universidad del Desarrollo; Research Center for the Development of Novel Therapeutics Alternatives for Alcohol Use Disorders, Santiago, Chile

Cristiane Aparecida Favoretto
Molecular and Behavioral Neuroscience Laboratory, Department of Pharmacology, Universidade Federal de São Paulo (UNIFESP), São Paulo, São Paulo, Brazil; Department of Molecular Medicine, The Scripps Research Institute, La Jolla, CA, United States

Gustavo Juliate Damaceno Fernandes
Molecular and Behavioral Neuroscience Laboratory, Department of Pharmacology, Universidade Federal de São Paulo (UNIFESP), São Paulo, São Paulo, Brazil

Juan Fuentes
Department of Pharmacological and Toxicological Chemistry, Faculty of Chemical Sciences and Pharmacy, University of Chile, Santiago, Chile

Javiera Gallardo
Center for Regenerative Medicine, Faculty of Medicine, Clínica Alemana-Universidad del Desarrollo, Santiago, Chile

Julien Guiraud
Department of Psychiatry, Amsterdam Neuroscience, Amsterdam UMC, Academic Medical Center, University of Amsterdam, Amsterdam, The Netherlands; Vergio, Clichy, France

Ignacio Gutiérrez-Vega
Department of Pharmacological and Toxicological Chemistry, Faculty of Chemical Sciences and Pharmacy, University of Chile, Santiago, Chile

Mario Herrera-Marschitz
Program of Molecular and Clinical Pharmacology, Faculty of Medicine, University of Chile, Santiago, Chile

Eduardo Isla
Department of Pharmacological and Toxicological Chemistry, Faculty of Chemical Sciences and Pharmacy, University of Chile, Santiago, Chile

Nilma do Rocio Lara de Lima Janisset
Molecular and Behavioral Neuroscience Laboratory, Department of Pharmacology, Universidade Federal de São Paulo (UNIFESP), São Paulo, São Paulo, Brazil

Elisabet Jerlhag
Institute of Neuroscience and Physiology, Department of Pharmacology, The Sahlgrenska Academy at the University of Gothenburg, Gothenburg, Sweden

Caio Jovita-Farias
Molecular and Behavioral Neuroscience Laboratory, Department of Pharmacology, Universidade Federal de São Paulo (UNIFESP), São Paulo, São Paulo, Brazil

Rodrigo Molini Leão
Laboratory of Pharmacology, Biomedical Sciences Institute, Department of Pharmacology, Federal University of Uberlândia, Uberlândia, MG, Brazil

Marcelo F. López
Department of Psychiatry and Behavioral Sciences, Charleston Alcohol Research Center, Medical University of South Carolina, Charleston, SC, United States

Tânia Marcourakis
Laboratory of Neurotoxicology, Department of Clinical and Toxicological Analysis, School of Pharmaceutical Sciences, Universidade de São Paulo, São Paulo, SP, Brazil

Priscila Marianno
Department of Pharmacology, Institute of Biomedical Sciences, Universidade de Sao Paulo, São Paulo, Brazil

Eduardo A.V. Marinho
Department of Health Sciences, Universidade Estadual de Santa Cruz, Ilhéus, BA, Brazil

Júlia Dalfovo Marquez
Programa de Pós-Graduação em Neurociências, Universidade Federal de Santa Catarina (UFSC), Florianópolis, Brazil

Minna H. McFarland
Bowles Center for Alcohol Studies; Neuroscience Curriculum, University of North Carolina at Chapel Hill, School of Medicine, Chapel Hill, NC, United States

Tarciso Tadeu Miguel
Pharmacology Department, Universidade Federal de Uberlândia (Ufu), MG, Brazil

A. Leslie Morrow
Department of Psychiatry; Department of Pharmacology; Bowles Center for Alcohol Studies, University of North Carolina at Chapel Hill, School of Medicine, Chapel Hill, NC, United States

Ana Regina Noto
Department of Psychobiology, Universidade Federal de Sao Paulo, São Paulo, Brazil

Alexandre J. Oliveira-Lima
Department of Health Sciences, Universidade Estadual de Santa Cruz, Ilhéus, BA, Brazil

Todd K. O'Buckley
Bowles Center for Alcohol Studies, University of North Carolina at Chapel Hill, School of Medicine, Chapel Hill, NC, United States

Paola Palombo
Department of Psychobiology, Universidade Federal de Sao Paulo, São Paulo, Brazil

Donna M. Platt
Department of Psychiatry and Human Behavior, Center for Innovation and Discovery in Addictions, University of Mississippi Medical Center, Jackson, MS, United States

Carolina Ponce
Department of Pharmacological and Toxicological Chemistry, Faculty of Chemical Sciences and Pharmacy, University of Chile, Santiago, Chile

Vicente Pérez
Department of Pharmacological and Toxicological Chemistry, Faculty of Chemical Sciences and Pharmacy, University of Chile, Santiago, Chile

Isabel M.H. Quadros
Psychobiology Department, Escola Paulista de Medicina, Universidade Federal de São Paulo (Unifesp), SP, Brazil

María Elena Quilaqueo
Department of Pharmacological and Toxicological Chemistry, Faculty of Chemical Sciences and Pharmacy, University of Chile, Santiago, Chile

María Elena Quintanilla
Program of Molecular and Clinical Pharmacology, Faculty of Medicine, University of Chile, Santiago, Chile

Thamires Righi
Molecular and Behavioral Neuroscience Laboratory, Department of Pharmacology, Universidade Federal de São Paulo (UNIFESP), São Paulo, São Paulo, Brazil

Mario Rivera-Meza
Department of Pharmacological and Toxicological Chemistry, Faculty of Chemical Sciences and Pharmacy, University of Chile, Santiago, Chile

Donita L. Robinson
Department of Psychiatry; Bowles Center for Alcohol Studies, University of North Carolina at Chapel Hill, School of Medicine, Chapel Hill, NC, United States

Ben Tagami Rodolpho
Molecular and Behavioral Neuroscience Laboratory, Department of Pharmacology, Universidade Federal de São Paulo (UNIFESP), São Paulo, São Paulo, Brazil

Fernando Bezerra Romualdo da Silva
Laboratory of Experimental and Clinical Neuroscience, INSERM, Université de Poitiers, France

James K. Rowlett
Department of Psychiatry and Human Behavior, Center for Innovation and Discovery in Addictions, University of Mississippi Medical Center, Jackson, MS, United States

Sutton M. Scarboro
Department of Psychiatry and Behavioral Sciences, Charleston Alcohol Research Center; Department of Neuroscience, Charleston Alcohol Research Center, Medical University of South Carolina, Charleston, SC, United States

Yasmim A. Serra
Department of Health Sciences, Universidade Estadual de Santa Cruz, Ilhéus, BA, Brazil

David Vásquez
Department of Pharmacological and Toxicological Chemistry, Faculty of Chemical Sciences and Pharmacy, University of Chile, Santiago, Chile

Roger Walz
Programa de Pós-Graduação em Neurociências; Departamento de Ciencias Clinica Medica, Universidade Federal de Santa Catarina (UFSC), Florianopolis, Brazil

Cristiane Ribeiro de Carvalho
Programa de Pós-Graduação em Neurociências; Departamento de Patologia, Universidade Federal de Santa Catarina (UFSC), Florianopolis, Brazil

Wim van den Brink
Department of Psychiatry, Amsterdam Neuroscience, Amsterdam UMC, Academic Medical Center, University of Amsterdam, Amsterdam, The Netherlands

Preface

Alcohol Use Disorder (AUD) is a highly prevalent psychiatric condition that poses significant health challenges due to its complex nature and treatment difficulties. Treatment coverage for AUD remains low compared to other psychiatric disorders, with only about 7% of adults aged 18 and older receiving help. Gender differences also impact treatment-seeking behavior, as more men seek help than women. Furthermore, while individuals with severe AUD are more likely to pursue treatment, many with mild to moderate AUD do not, despite the benefits of early intervention. This book provides an overview of various approaches to treating AUD. It features chapters on both established and emerging treatments, including those initially approved for other conditions but now showing promise for AUD management. The book also explores preclinical models used in AUD research and discusses how environmental factors and complementary clinical interventions can influence alcohol use. Motivated by rapid advancements in the field and the lack of a unified resource covering these developments, this book brings together expert contributions to present innovative approaches and offer a practical guide to the current state of AUD treatment.

In Chapter 1, Camarini and Cruz explore the global impact of alcohol misuse, the most widely consumed psychoactive substance. The chapter explores the genetic factors involved in AUD and reviews the current pharmacological treatments – disulfiram, naltrexone, nalmefene and acamprosate, and baclofen, presenting their history, mechanisms and adverse effects.

In Chapter 2, Scarboro, López and Becker focus on preclinical and clinical investigations of oxytocin (OT) and vasopressin (VP), suggest that these neuropeptide systems hold significant promise as potential targets for pharmacological intervention. OT system activation through agonists has shown anxiolytic effects and reduced alcohol consumption. Conversely, VP system modulation, particularly through V1b receptor antagonists, has demonstrated a reduction in alcohol-seeking behaviors and stress-induced drinking, supporting the potential for VP antagonism as a treatment strategy. The study underscores the importance of considering sex differences, stress interactions, and the biochemical challenges associated with targeting these systems.

Chapter 3, by Morrow, McFarland, O'Buckley, and Robinson, shifts the focus to the potential of neuroactive pregnane steroids, including pregnenolone, progesterone, and allopregnanolone, in treating AUD. It presents a thorough review of preclinical and clinical evidence, detailing how these steroids might influence neurobiological pathways related to alcohol use, stress, and neurotransmitter systems. The chapter highlights the promising early clinical findings with pregnenolone, suggesting a need for further investigation into its therapeutic potential.

In Chapter 4, Favoretto, Bertagna, Miguel, and Quadros examine the role of the Corticotropin Releasing Factor (CRF) and Urocortins (Ucns) system in AUD. It provides an overview of CRF and its related peptides, including urocortins, and explores how CRF and its receptors contribute to excessive alcohol consumption. The chapter also discusses pharmacological manipulations of CRF/Ucns targets in preclinical models of AUD and addresses the challenges in developing effective treatments for stress and AUD.

Chapter 5, by Adasme-Reyes, Fuentes, Gutiérrez-Vega, Isla, Pérez et al., shifts focus to alcohol metabolism, highlighting the role of ADH and ALDH polymorphisms in AUD. It explains why individuals carrying the inactive ALDH2*2 variant are less likely to develop AUD, while also highlighting acetaldehyde as a central metabolite involved in the rewarding effects of ethanol. Additionally, it evaluates Alda-1, a drug that activates ALDH2, as a potential treatment for reducing alcohol consumption and protecting against alcohol-related tissue damage.

Gallardo, Berríos-Cárcamo, and Ezquer introduce mesenchymal stem cell (MSC) therapies as a novel approach to AUD in Chapter 6. The authors discuss how glutamatergic pathways and GLT1 expression are relate to ethanol consumption. The chapter then covers the roles of oxidative stress and neuroinflammation in AUD and evaluates anti-inflammatory and antioxidant therapies. It focuses on MSCs as a promising therapy, presenting preclinical evidence of their effectiveness in reducing ethanol consumption, and discusses the therapeutic potential of MSC-derived secretome as an advanced treatment for AUD.

In Chapter 7, Guiraud and van den Brink review the use of sodium oxybate (SMO), a sodium salt form of gamma-hydroxybutyrate (GHB), as a treatment for alcohol dependence (AD). The chapter evaluates clinical evidence supporting SMO's effectiveness in reducing withdrawal symptoms and maintaining abstinence, positioning it as a promising candidate for treating severe cases of AD. SMO has been clinically used in Italy and

Austria since the 1990s, approved for treating alcohol withdrawal syndrome and for maintaining abstinence in AD patients. Its therapeutic effects are related to its action on GABA receptors and its influence on neurotransmitter systems.

In Chapter 8, Marinho, Serra, Oliveira-Lima, Marcourakis, and Berro explore the emerging role of psychedelics in treating AUD, such as LSD, psilocybin, and ayahuasca. This chapter reviews both clinical and preclinical evidence suggesting that these compounds may reduce alcohol consumption and promote abstinence. The therapeutic potential of ayahuasca, an Amazonian brew, is examined in detail, with discussions on its mechanisms of action and its implications for AUD treatment.

Chapter 9, by Marquez, Dezanetti, Walz, and Carvalho, focuses on the endocannabinoid system, particularly the roles of cannabinoid receptors CB1 and CB2 in modulating the brain's reward pathways. It evaluates the potential of cannabis and cannabinoid compounds as treatments for AUD, weighing their benefits against risks, in light of increasing cannabis availability due to legalization movements.

In Chapter 10, Camarini, Marianno, Costa, Palombo, and Noto examine the impact of environmental factors and complementary clinical interventions on AUD. They explore how stress and environmental enrichment (EE) influence alcohol consumption and its rewarding effects, discussing potential mechanisms through which EE modulates ethanol's effects in preclinical models. The chapter also highlights the translational implications of EE as a therapeutic intervention, emphasizing activities such as physical exercise, cognitive behavioral therapy, and social interaction as strategies to mitigate the effects of chronic alcohol use.

Chapter 11, by Favoretto, Righi, Fernandes, Bertagna, Rodolpho et al., offers an overview of animal models utilized in AUD research. It covers a range of preclinical models, including conditioned place preference, locomotor sensitization, self-administration, and the alcohol vapor chamber. The chapter details how these models contribute to our understanding of the neurobiological mechanisms underlying alcohol use and how they are used to evaluate potential therapeutic interventions.

The GABAergic system, long implicated in the behavioral effects of alcohol, is the focus of Chapter 12. Berro, Rowlett, and Platt review the evidence supporting the use of GABAergic compounds as potential treatments for AUD, with a particular emphasis on the promising results of positive (PAMs) and negative (NAMs) allosteric modulators of $GABA_A$ and $GABA_B$ receptors.

Finally, Jerlhag (Chapter 13) introduces the gut-brain axis and its peptides, ghrelin and glucagon-like peptide-1 (GLP-1), discussing their roles in regulating appetite, reward mechanisms, and alcohol consumption. The chapter examines the role of these peptides in regulating alcohol intake and their potential as emerging targets in the treatment of AUD, supported by both preclinical and clinical data.

In conclusion, we hope that this volume serves as a resource for those dedicated to understanding and addressing AUD. By presenting various traditional and innovative treatment strategies, we aim to advance the efforts to improve outcomes for individuals affected by this disorder. It is our belief that the insights and knowledge shared in these chapters will inform and inspire further research, clinical practice, and policy-making, ultimately advancing the field and fostering more effective interventions for AUD.

Editors

Rosana Camarini
Laboratory of Neurochemistry and Behavioural Pharmacology
Department of Pharmacology, Institute of Biomedical Sciences Universidade de São Paulo,
São Paulo, SP, Brazil

Fábio C. Cruz
Department of Pharmacology, Escola Paulista de Medicina,
São Paulo Federal University, São Paulo, SP, Brazil

CHAPTER ONE

Introduction: Approved treatments for alcohol use disorder by regulatory agencies

Rosana Camarini[a,*] **and Fábio Cardoso Cruz**[b,*]
[a]Department of Pharmacology, Institute of Biomedical Sciences, Universidade de Sao Paulo, São Paulo, Brazil
[b]Molecular and Behavioral Neuroscience Laboratory, Department of Pharmacology, Universidade Federal de São Paulo (UNIFESP), São Paulo, São Paulo, Brazil
*Corresponding author. e-mail address: camarini@icb.usp.br; f.cruz@unifesp.br

Contents

Abstract

Alcohol, the most widely consumed substance globally, can lead to severe adverse effects for both users and those around them. Chronic ethanol consumption may lead to alcohol use disorder (AUD), a chronic relapsing condition characterized by compulsive drinking despite negative consequences. AUD is marked by a high relapse rate among individuals attempting abstinence. Currently, only a few medications, such as disulfiram, naltrexone, nalmefene, and acamprosate, are approved to treat AUD. Moreover, genetic factors and comorbid conditions can significantly influence both the development of AUD and the efficacy of its treatment. This chapter explores the genetic underpinnings of AUD and reviews the main pharmacological treatments available for managing this disorder.

1. Introduction

Alcohol (or ethanol) stands out as the most widely consumed globally, according to the Global status report on alcohol and health (GDS, 2018). Overall, 97.1% of GDS respondents reported lifetime alcohol use

International Review of Neurobiology, Volume 178
ISSN 0074-7742, https://doi.org/10.1016/bs.irn.2024.07.001

detected by the AUDIT (Alcohol Use Disorders Identification Test) (Winstock et al., 2021). The harmful use of alcohol is a causal factor in more than 200 conditions, including liver cirrhosis, pancreatitis, cardiovascular diseases, stroke, cancer, gastritis, psychiatric diseases, among others. Ethanol misuse can impair cognitive functions such as assessment and judgment, thereby increasing the likelihood of engaging in risky behaviors such as driving under the influence or participating in violent episodes. This behavior underscores the significance of alcohol consumption as one of the major noncommunicable disease risk factors, alongside tobacco use, unhealthy diet, and physical inactivity (WHO, 2023).

Alcohol Use Disorder (AUD) is a chronic relapsing condition characterized by compulsive alcohol consumption, loss of control over intake, despite of adverse outcomes; and it is influenced by genetic, biological, psychological, and environmental factors. Approximately 5.1% of the global burden of disease and injury can be attributed to alcohol consumption, quantified through Disability-Adjusted Life Years (DALYs) (GDS, 2018).

Alcohol exerts a broad influence on the central nervous system (CNS) as well as other organs and tissues. Its action can occur directly or indirectly. Various ionotropic receptors and ion channels possess multiple sites sensitive to ethanol, where it can produce either agonistic or antagonistic effects. These sites have been identified at various receptors and channels, including the γ-aminobutyric-acid type A (GABA-A) receptor, glycine receptor, serotonin 5-hydroxytryptamine type 1 and 3 receptors, nicotinic acetylcholine receptors (nAChRs), ionotropic purinergic P2X4 receptors, L-type calcium channels, and G-protein-activated inwardly rectifying potassium channels (Olsen & Liang, 2017; Popova et al., 2020; Spanagel, 2009; Trudell et al., 2014). For instance, depending on the dose and brain region, ethanol can act as a positive allosteric modulator of the GABA receptor or promote the presynaptic release of the neurotransmitter GABA (Olsen & Liang, 2017). In the same way, Lovinger, White, and Weight (1989) demonstrated that ethanol inhibited N-methyl-D-aspartate (NMDA) receptor function in a concentration-dependent manner, indicating antagonistic action at NMDA receptors.

Chronic alcohol consumption leads to neuroplastic changes in those neurotransmission systems in brain regions associated with executive function, cognition, motivated behaviors, and stress responses, including the limbic system, basal ganglia, and extended amygdala. These neuroplastic alterations are associated with both positive and negative reinforcement mechanisms that are critical in the development and maintenance of AUD (Koob & Volkow, 2016; Witkiewitz, Litten, & Leggio, 2019). This brief overview highlights the complexity of the neurobiology involved in AUD and indicates the wide range of systems that can be targeted in developing new pharmacological treatments.

2. Genetic variability and drug responses

Nearly 50% of the risk for AUD is determined by genetic factors, while environmental influences shape the other half. Understanding the genetic factors involved in AUD allows for early identification of individuals at higher risk of developing the disorder. Genetic testing can provide valuable insights into an individual's susceptibility to AUD, enabling healthcare professionals to implement preventive measures and interventions at an earlier stage (Reilly, Noronha, Goldman, & Koob, 2017).

Genes such as alcohol dehydrogenase 1B (ADH1B) and aldehyde dehydrogenase 2 (ALDH2) play significant roles in alcohol metabolism and influence susceptibility to AUD. Variants in these genes affect ethanol metabolism, leading to differences in acetaldehyde levels and subsequent reactions to alcohol consumption. The ADH1B⋆48His allele accelerates ethanol metabolism, while the ALDH2⋆504K allele is associated with reduced enzymatic activity, resulting in the accumulation of acetaldehyde following alcohol consumption. As a result, individuals - usually from East Asia - carrying one or both variants may experience adverse reactions such as facial flushing and nausea following ethanol intake. This unpleasant response serves as a deterrent to excessive alcohol consumption and may contribute to a decreased risk of developing AUD (Edenberg & McClintick, 2018).

A large genome-wide association (GWAS) study that used measures from the AUDIT, involving over 20,000 participants, identified genetic variants associated with AUD, including one near the ADH cluster. The study revealed specific genetic correlations, such as those with tobacco use and education level, alongside unexpected associations with body mass index (BMI) and obesity rates. The gene ADH1C contributes to ethanol oxidation and has been recognized as the most significant association. Another considerable association resides near the KCNJ9 (GIRK3) gene. This gene encodes one of the G protein-activated inwardly rectifying K + channels and has preclinical correlates related to alcohol-related behaviors in mice (Sanchez-Roige et al., 2019).

Other genes have been identified as candidates associated with AUD phenotypes, such as GABRA2, OPRM1, DRD2, and ANNK1, although robust signals for them were not found in the study by Sanchez-Roige et al. (2019). This suggests that while these genes may be linked to AUD, they might not necessarily be associated with AUDIT.

Several epidemiological and genetic studies have indicated variations in specific GABAA receptor (GABAA-R) subunits, notably α2 and γ1, are associated with AUD. The GABAA-R subunit genes are clustered on various chromosomes, with one notable cluster on chromosome 4. This cluster encompasses genes such as GABRB1, GABRA4, GABRA2, and GABRG1. Despite varying results across studies, GABRA2 emerges as a prominent genetic contributor to AUD, supported by clinical and genetic evidence (Borghese & Harris, 2012).

The relationship between the mRNA transcript expression of GABAA receptor subunit genes and AUD outcomes was explored using autopsy tissue samples. Higher expression of GABRA6 and GABRA2 transcripts was found in cirrhotic AUD cases compared to controls and non-cirrhotic cases, with variations observed between sexes, genotypes, and brain regions. These findings suggest that sex differences in individuals with GABRA6 and GABRA2 variants may influence susceptibility to AUD and alcohol-induced cirrhosis (Ashton et al., 2022).

The identification of a particular genetic variation (A118G single-nucleotide polymorphism or SNP) in the μ-opioid receptor (MOR) gene (OPRM1) holds significant implications for alcohol dependence treatment. This variation affects the receptor affinity for the endogenous opioid peptide beta-endorphin, potentially altering stress-responsivity and response to naltrexone treatment. A significant association between genotypes containing the 118G allele and AUD has been found, suggesting an increased risk for dependence among individuals with this variant. This genetic variation, as further discussed in item 3, may play a role in the development of alcohol dependence (Bart et al., 2005).

3. Approved treatments

Effective treatments for AUD typically combine psychotherapy and medication. Trained therapists play a crucial role in assisting individuals to develop coping strategies aimed at reducing or halting drinking, while medications are instrumental in cessation and relapse prevention (NIAAA).

These treatment approaches align with guidelines emphasizing a recovery-oriented system of care. Such guidelines advocate for personalized treatment plans that include the involvement of peers and recovery communities, the establishment of flexible recovery goals, comprehensive care for co-occurring conditions, and active engagement with diverse communities (Sheedy & Whitter, 2009).

Additionally, combatting the stigma surrounding alcohol problems in healthcare and support services is essential for promoting effective treatment, as stigma poses a significant barrier to seeking help and engaging in treatment. Recognizing this, the World Health Organization (WHO, 2023) advocates for actions to address social stigma and discrimination in accessing alcohol treatment. Service providers must carefully examine policies, practices, and attitudes contributing to stigma to ensure that individuals feel respected and not judged based on their alcohol use. Additionally, regular review of language used to describe alcohol problems is indispensable to prevent stigmatization, as stigma may lead individuals to minimize their alcohol problems.

The treatments approved for AUD by the main regulatory agencies worldwide are disulfiram, naltrexone (oral and long-acting injectable), and acamprosate. In addition, oral nalmefene has been approved in the European Union disorder since 2013, and baclofen was approved in France in 2018.

3.1 Disulfiram: history, mechanisms and adverse effects

Disulfiram (Fig. 1) has an exciting history intertwined with the treatment of AUD. It was first synthesized by M. Grodzki, a chemist based in Berlin. During the 1920s, disulfiram, also known as tetraethylthiuram disulfide (TETD), found application in the rubber industry for the process of vulcanization, which involves stiffening rubber.

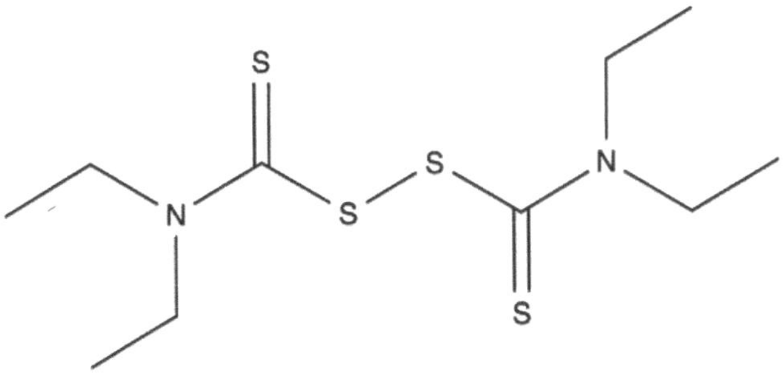

Fig. 1 Disulfiram molecular structure.

Notably, researchers observed that workers exposed to TETD experienced adverse reactions upon consuming alcohol (Kragh, 2008; Pal, Pattanayak, & Sagar, 2015). However, it wasn't until the 1940s that its clinical use was initiated when Dr. Erik Jacobsen, a Danish pharmacologist, investigated the potential of disulfiram to treat parasitic infections. In collaboration with Dr. Jens Hald, their investigation into the copper metabolism of intestinal parasites revealed that disulfiram forms chelates with copper, ultimately leading to the death of the organisms. Dr. Jacobsen's self-experimentation with disulfiram, prompted by promising results in rabbits, revealed its potent effect on altering alcohol effects, including adverse reactions following ethanol intake. Teaming up with Dr. Hald, they explored the potential use of disulfiram in treating individuals with alcohol-related problems. This serendipitous discovery led them to recognize disulfiram as a possible treatment to discourage alcohol consumption (Kragh, 2008; Pal et al., 2015).

In 1948, Jacobsen and Hald made a significant breakthrough by publishing the first clinical study showcasing the efficacy of disulfiram in curbing alcohol consumption. During their investigation, they closely monitored participants who were administered disulfiram along with alcohol, noting physiological alterations like facial vasodilation, tachycardia, and heightened pulmonary ventilation. It was when they detected a distinct odor of acetaldehyde in the laboratory that they conjectured about the possibility of elevated acetaldehyde levels being responsible for disulfiram's adverse effects when combined with alcohol (Kragh, 2008; Lanz et al., 2023). Subsequently to this finding, disulfiram was introduced as a treatment for AUD.

In recent years, research has revealed that TETD can function as an irreversible monoacylglycerol lipase (MAGL) inhibitor, thereby modulating the endocannabinoid system (ECS). This discovery suggests the potential repurposing of disulfiram for other therapeutic applications since the inhibition of MAGL activity would be beneficial for the treatment of a wide range of diseases, such as inflammation, neurodegeneration, metabolic disorders, and cancer (Omran, 2021).

Disulfiram was approved by the Danish agency in 1949 and by the U.S. Food and Drug Administration (FDA) for the treatment of AUD in 1951. Its approval marked a significant milestone in the pharmacological treatment of AUD (Pal et al., 2015).

Disulfiram works by inhibiting the enzyme aldehyde dehydrogenase, which metabolizes alcohol. When alcohol is consumed, disulfiram blocks the breakdown of acetaldehyde, a toxic byproduct of alcohol metabolism.

Accumulation of acetaldehyde results in unpleasant symptoms such as nausea, vomiting, flushing, headache, and palpitations, discouraging individuals from drinking alcohol (Jacobsen & Larsen, 1949; Lanz et al., 2023).

Despite the introduction of newer medications for AUD, disulfiram continues to be used in some instances, particularly for individuals who are highly motivated to abstain from alcohol and benefit from the aversive effects of disulfiram-alcohol interaction. Research also continues to explore the potential of disulfiram in combination with other treatments or for different indications, such as in the treatment of cocaine dependence (Lanz et al., 2023).

Common adverse effects include headache, drowsiness, halitosis, metallic taste, and fatigue. Severe adverse effects can occur if alcohol is consumed while taking disulfiram, including cardiovascular complications and respiratory depression. Serious side effects include hepatotoxicity, neurological events, such as psychosis and seizures, peripheral neuropathy, and dermatological reactions. Drug interactions, particularly with medications metabolized by the cytochrome P450 system, are also noted. Contraindications include significant coronary artery disease, heart failure, and psychosis, while caution is advised for patients with a history of liver disease or certain other conditions (Stokes & Abdijadid, 2024).

Overall, the history of disulfiram history reflects its evolution from a serendipitous discovery to an established pharmacological treatment for AUD.

3.2 Naltrexone: history, mechanisms and adverse effects

Naltrexone (Fig. 2), a medication commonly used in the treatment of AUD and opioid dependence, has a fascinating history and a varied response among patients. Naltrexone was initially synthesized in the early 1960s by Drs. Jack Fischman and Mozez Lewenstein from the Memorial Sloan Kettering Institute for Cancer Research. This development was inspired by works of Dr. Harold Blumberg, at Endo Laboratories in Long Island while researching opioid antagonists (Srivastava & Gold, 2018).

Since the 70's, researchers have shown that ethanol can alter the activity of opiate receptors (Davis & Walsh, 1970). Studies conducted by Dr. Joseph Volpicelli and his colleagues were determinant for the approval of naltrexone for AUD. In 1986, their research revealed the involvement of the endogenous opioid system in alcohol consumption by demonstrating that naltrexone blocked the increase in ethanol consumption induced by inescapable shock-induced stress in rats (Volpicelli et al., 1986). Preclinical research studies motivated the examination of naltrexone efficacy in treating patients. In a 12-week, double-blind, placebo-controlled trial,

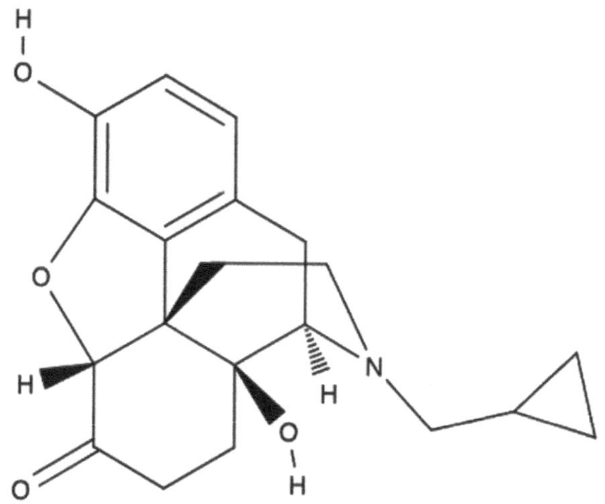

Fig. 2 Naltrexone molecular structure.

seventy male alcohol-dependent patients were treated with naltrexone hydrochloride following alcohol detoxification. The subjects reported reduced alcohol craving and fewer days of alcohol consumption compared to those on placebo. Notably, only 23% of naltrexone-treated patients relapsed, compared to 54.3% of those on placebo. The primary benefit of naltrexone was observed in patients who drank alcohol during outpatient treatment, with fewer naltrexone-treated patients meeting relapse criteria compared to those on placebo (Volpicelli et al., 1992).

The main mechanism of naltrexone is thought to involve blocking opioid receptors, thus inhibiting the heightened endogenous opioid activity induced by alcohol consumption. It's been proposed that inhibiting the μ-opioid receptor could suppress ethanol reinforcement by reducing ethanol-induced mesolimbic dopamine activity (Herz, 1997). This action blocks the rewarding effects of alcohol, ultimately reducing the craving to drink. This hypothesis is supported by studies showing that offspring of individuals with AUD exhibit increased levels of β-endorphin-related peptides in response to alcohol doses, indicating a dose-dependent association between alcohol intake and endogenous opioid release (Gianoulakis, Krishnan, & Thavundayil, 1996; Volpicelli, 1987). During the same period as Volpicelli's study, O'Malley and colleagues conducted a 12-week trial involving 97 patients. Their study found that naltrexone exhibited superiority over placebo across various drinking-related measures, including rates of abstinence, frequency of drinking days, and instances of relapse (O'Malley et al., 1992).

Advancements in neurobiology, along with findings from clinical trials, culminated in the FDA's approval of the opioid antagonist naltrexone for the treatment of AUD in 1994. Subsequently, in 2006, the FDA sanctioned an extended-release version of naltrexone. Its approval marked a significant advancement in pharmacological treatments for AUD.

Despite its approval and widespread use, not all patients with AUD experience beneficial effects from naltrexone; pharmacogenetic effects related to genetic variations in opioid receptor genes, particularly OPRM1, OPRK1, and OPRD1, may influence the response to naltrexone treatment. Variations in genes associated with opioid receptors are thought to potentially affect the efficacy of naltrexone in treating AUD patients. The μ-opioid receptor gene (OPRM1) stands as one of the most investigated genetic factors in humans and macaques associated with naltrexone treatment response, with polymorphisms within this gene, notably A118G (Asn40Asp), being linked to the reinforcing effects of alcohol (Barr et al., 2007; Chamorro et al., 2012). Most individuals across populations exhibit homozygosity for the A allele of this SNP, encoding the amino acid asparagine in the receptor protein. In contrast, a minority carries a variant that results in the substitution of asparagine with aspartate in the receptor protein (G allele). The allelic substitution is linked to receptor activity for β-endorphin, with individuals carrying the G allele demonstrating heightened affinity for the receptor (Bond et al., 1998), elevated levels of alcohol cue-induced craving, heightened subjective sensations of intoxication, stimulation, and sedation, in comparison to those homozygous for the A allele (Ray & Hutchison, 2004). In a Swedish study, a higher prevalence of the G allele variant was observed among adults with alcohol dependence compared to non-alcohol dependent controls (Bart et al., 2005). Despite these findings, other studies present conflicting results (Bergen et al., 1997; Beyer, Koch, Schröder, Schulz, & Höllt, 2004; Zhang, Wang, Johnson, Papp, & Sadée, 2005). In human autopsy brain tissues and CHO cells, a decrease in μ-opioid receptor function has been linked to reduced mRNA and OPRM1 protein levels associated with OPRM1-G118. Significantly, the OPRM1-Asn40 (A118) variant exhibited ten times more binding sites than the OPRM1-Asp40 (G118) variant (Zhang et al., 2005).

Among rhesus macaques, an association was observed between the G allele of the C77G OPRM1 polymorphism, increased β-endorphin binding affinity, and higher alcohol preference and consumption (Barr et al., 2007). Following naltrexone administration, these macaques exhibited more pronounced reductions in preference and alcohol consumption (Barr et al., 2010).

While this discovery has not been consistently validated (Gelernter et al., 2007), individuals possessing the G allele might exhibit improved responsiveness to naltrexone treatment for AUD. In the Combined Pharmacotherapies and Behavioral Interventions for Alcohol Dependence (COMBINE) study, conducted across eleven academic sites, Anton et al. (2008) found that the functionally significant OPRM1 Asp40 allele predicts the response to naltrexone treatment in individuals with AUD, highlighting the importance of pharmacogenomics.

Naltrexone demonstrates a high affinity for μ opioid receptors, but also binds to δ and κ opioid receptors (Takemori & Portoghese, 1992; Takemori et al., 1988). Specific κ receptor SNP and δ receptor SNP moderate the effects of naltrexone-induced changes in alcohol-induced sedation and stimulation/craving responses, respectively. These findings imply that genetic differences near these sites may contribute to individual differences in response to naltrexone (Ashenhurst, Bujarski, & Ray, 2012).

In addition to genetic variations near opioid receptors affecting naltrexone treatment response for AUD, it is essential to consider other factors influencing its efficacy and tolerability. Among these are pharmacokinetic properties and adverse effects, impacting patient adherence and overall treatment outcomes.

Oral administration of naltrexone is a standard treatment for AUD, although its efficacy is limited due to extensive hepatic metabolism and low bioavailability (Lobmaier, Gossop, Waal, & Bramness, 2010; Yin, Akalam, & Taylor, 2002). Additionally, frequent dosing and adverse effects reduce patient adherence. In a study comparing the effects of nefazodone, a serotonergic antidepressant, with the opioid antagonist naltrexone, and an inactive placebo in 183 alcohol-dependent subjects, naltrexone treatment yielded higher adverse effects, leading to decreased medication adherence and an increased rate of treatment discontinuation compared to the placebo (Kranzler, Modesto-Lowe, & Van Kirk, 2000). The study by Oncken, Van Kirk, and Kranzler (2001) investigated the relationship between adverse events, medication compliance, and study retention in individuals with AUD receiving naltrexone for relapse prevention. While neuropsychiatric adverse events (tiredness, sleepiness, drowsiness) had minimal impact on medication compliance but directly decreased study retention, gastrointestinal adverse events (abdominal pain, nausea, dry mouth) significantly affected medication compliance without direct impact on study retention.

Extended-release preparations have been developed to address these challenges, offering more predictable pharmacokinetics and bypassing first-pass metabolism (Dunbar et al., 2006), such as the injectable naltrexone microsphere formulation, administered via intramuscular injection once a month. There is a growing interest in developing delivery systems that allow for alternative routes of administration and provide a sustained release to maintain consistent plasma levels. Microemulsions capable of transitioning into gels with hexagonal structures in vivo have been characterized and optimized (Santos et al., 2020).

In summary, naltrexone has a rich history in the treatment of AUD, but its effectiveness can vary among patients due to genetic factors. Understanding these factors and tailoring treatment accordingly may be crucial for maximizing the benefits of naltrexone therapy.

3.3 Nalmefene: history, mechanisms, and adverse effects

Nalmefene hydrochloride (Fig. 3) is an antagonist at the μ and δ receptors and a partial k-opioid receptor agonist (Bart et al., 2005; DeHaven-Hudkins et al., 1990; Emmerson, Liu, Woods, & Medzihradsky, 1994).

The interplay among the opioid receptors plays a crucial role in regulating the rewarding and reinforcing aspects of alcohol consumption, ultimately influencing drinking behavior. It was primarily employed for the treatment of AUD in the European Union and has received approval for clinical use in Japan and several other countries.

Preclinical data showed that nalmefene reduced relapse-like alcohol consumption in male rats, especially those with higher pre-treatment

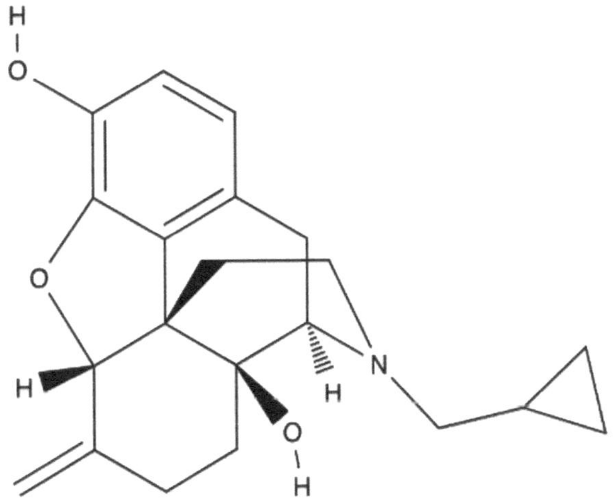

Fig. 3 Nalmefeme molecular structure.

alcohol intake (Foo et al., 2019). Meta-analyses have affirmed the efficacy of nalmefene, particularly at a dosage of 20 mg, in reducing both heavy drinking days and daily total alcohol consumption. These findings underscore the potential of reducing alcohol intake as a feasible treatment approach for individuals who may not be prepared for complete abstinence. Several meta-analyses confirmed its effectiveness in reducing heavy drinking days and daily total alcohol consumption, particularly at a dose of 20 mg (Mann et al., 2016).

A study conducted by Miyata et al. (2019) examined the effectiveness of nalmefene at doses of 20 and 10 mg, compared to a placebo, alongside psychosocial support, in alcohol-dependent Japanese patients with a high or very high drinking risk level (DRL). Both doses of nalmefene led to significant reductions in heavy drinking days and total alcohol consumption at week 12 compared to placebo. Adverse events were mostly mild or moderate in severity, with nalmefene being well-tolerated.

Overall, nalmefene offers an alternative approach to reducing alcohol consumption, potentially motivating more patients to treatment and addressing a significant public health concern.

3.4 Acamprosate: history, mechanisms, and adverse effects

Acamprosate (Fig. 4), chemically known as calcium 3-acetamido-1-propane-sulfonate or calcium-bis-acetyl-homotaurinate, has a chemical structure similar to the amino acids neurotransmitter γ-aminobutyric acid (GABA) and neuromodulator taurine (Chabenat et al., 1988).

The history of acamprosate in treating AUD traces back to the 1980s when researchers began exploring its potential to reduce voluntary alcohol consumption in rats (Boismare et al., 1984). This initial study paved the way for clinical trials to evaluate its effectiveness in humans. In double-blind studies, individuals with severe AUD who were successfully weaned off alcohol were treated with acamprosate or placebo as outpatients over three months. Abstinence at the end of the study was the primary measure of

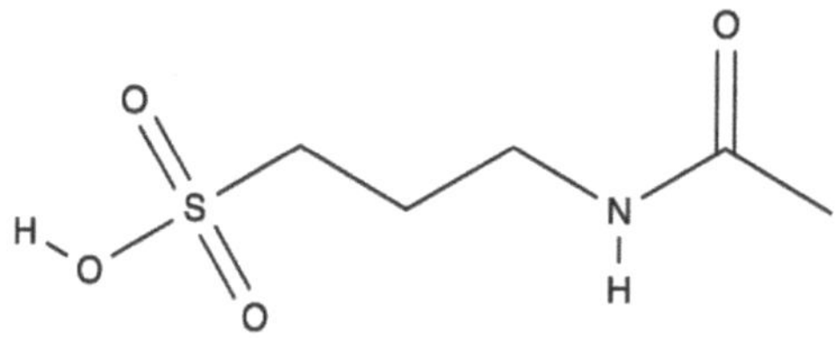

Fig. 4 Acamprosate molecular structure.

success. The patients on acamprosate showed a significantly higher rate of abstinence compared to those on placebo (Lhuintre et al., 1985, 1990). Subsequent investigations revealed similar results, including reduced rates of discontinuation from treatment (Sass, Soyka, Mann, & Zieglgänsberger, 1996). Pioneering preclinical studies investigated the impact of acamprosate on alcohol withdrawal symptoms in male Wistar rats (Spanagel, Hölter, Allingham, Landgraf, & Zieglgänsberger, 1996; Spanagel, Putzke, Stefferl, Schöbitz, & Zieglgänsberger, 1996). Using a radio-telemetric system, they monitored various parameters during alcohol withdrawal, including body temperature, locomotor activity, and food and water intake patterns. Acamprosate reduced hyperlocomotion and food intake in alcohol withdrawal animals, although it did not affect withdrawal-induced hyperthermia.

Interestingly, acamprosate induced transient psychomotor stimulant effects when administered to alcohol-naive animals (Spanagel, Hölter, Allingham, Landgraf, & Zieglgänsberger, 1996). The efficacy of acamprosate was also evaluated in a rat model simulating long-term alcohol consumption, mirroring relapse behavior observed in human alcoholics. After continuous alcohol access followed by deprivation phases, alcohol intake increased upon the reintroduction of alcohol solutions. However, administration of acamprosate twice daily decreased alcohol intake dose-dependently, with the highest dose even reducing intake below baseline levels (Spanagel et al., 1996).

Since then, extensive clinical trials have thoroughly assessed the therapeutic effectiveness and safety of acamprosate for treating AUD, particularly in maintaining abstinence among alcohol-dependent patients post-withdrawal (Bahji et al., 2022; Chick, Howlett, Morgan, & Ritson, 2000; Garbutt, West, Carey, Lohr, & Crews, 1999; Swift, 1999; Wilde & Wagstaff, 1997).

In a multicenter prospective study spanning five European countries, individuals with AUD were treated with acamprosate and various psychosocial support techniques over 24 weeks as part of standard patient care. A total of 1289 patients were enrolled, with 1230 receiving at least one dose of acamprosate and providing follow-up data. Factors such as physical and psychiatric comorbidities, history of drug addiction, previous detoxification episodes, employment status, living arrangements, and age influenced treatment outcomes. During the study, the overall proportion of cumulative abstinence days (CADP), defined as the ratio between the number of days of abstinence divided by the whole duration of the trial, was 0.48 (Pelc et al., 2002).

Acamprosate has been accessible for AUD treatment in Europe since 1996. After several clinical trials demonstrated acamprosate effectiveness in reducing alcohol cravings and promoting abstinence in individuals with alcohol dependence, the U.S. Food and Drug Administration (FDA) approved acamprosate calcium for the treatment of alcohol dependence in 2004. Acamprosate is typically prescribed as part of a comprehensive treatment plan that includes psychosocial interventions such as counseling, support groups, and behavioral therapies. It is most used in individuals who have already achieved abstinence from alcohol and are seeking to maintain sobriety. It is hypothesized that acamprosate helps maintain abstinence by reducing the negative emotional and physical symptoms associated with alcohol withdrawal, such as anxiety, dysphoria, and insomnia.

Although numerous trials have demonstrated its significant efficacy compared to placebo in reducing drinking days, enhancing abstinence, and prolonging time to relapse, the Combining Medications and Behavioral Interventions (COMBINE) study did not ascertain acamprosate to be more effective than placebo (Anton et al., 2006). Conducted among 1383 individuals with alcohol abstinence and diagnosed with primary alcohol dependence, the trial involved various treatment groups receiving naltrexone, acamprosate, or placebos, with or without combined behavioral intervention (CBI), over 16 weeks. Patients receiving naltrexone, CBI, or both alongside medical management exhibited higher rates of abstinence compared to those receiving placebos and medical management only. Naltrexone reduced the risk of heavy drinking days, especially when combined with medical management. Conversely, acamprosate did not show significant efficacy in reducing drinking. Patients who received CBI without pills or medical management showed lower rates of abstinence than those who received placebos alongside medical management, underscoring the potential positive influence of placebos and the interaction with healthcare professionals in promoting abstinence among patients.

A discrimination study was conducted in male Wistar rats trained to discriminate between ethanol and saline injections in a two-lever operant drug discrimination paradigm to elucidate further potential mechanisms behind acamprosate ability to reduce ethanol intake. Acamprosate was administered before testing in two ways: (i) generalization test and (ii) antagonism test. Acamprosate was not a substitute for the ethanol cue, nor did it alter the discrimination between ethanol and saline. It did not affect response latencies or act via the N-methyl-D-aspartic acid (NMDA) channel. The study suggested that acamprosate suppressed ethanol intake through a mechanism independent of NMDA receptors or substituting ethanol (Spanagel et al., 1996).

One of the mechanisms proposed to acamprosate has been the ability to modulate glutamatergic neurotransmission in the brain, restoring the balance between excitatory and inhibitory neurotransmitters disrupted by chronic alcohol exposure (Zeise, Kasparov, Capogna, & Zieglgänsberger, 1993).

A study conducted by Naassila, Hammoumi, Legrand, Durbin, and Daoust (1998) aimed to characterize [3H]acamprosate binding and its potential relation to NMDA receptor complex sites, revealed specific binding of [3H]acamprosate to rat brain membranes, sensitive to spermidine inhibition and modulated by arcane and glutamate. Acamprosate displaced [14C]spermidine binding, indicating an allosteric interaction with a spermidine binding site. Acamprosate altered [3H]dizocilpine (an NMDA antagonist) binding on the NMDA receptor in a concentration-dependent manner. In conjunction with the research conducted by Al Qatari, Bouchenafa, and Littleton (1998), which indicated similar findings, their work suggested that acamprosate functions as a partial co-agonist at the NMDA receptor. This implies that at low concentrations, it augments activation when receptor activity is low. In contrast, at high concentrations, it inhibits activation when receptor activity is high, potentially explaining its clinical effects in alcohol-dependent patients during abstinence.

While this evidence suggested acamprosate involvement in the glutamate system, its molecular mechanism remains unclear. A seminal study revealed that acamprosate does not interact with proposed glutamate receptor mechanisms, including NMDA receptors or metabotropic glutamate receptor group I. Preclinical animal models demonstrated that *N*-acetylhomotaurinate alone lacks psychotropic activity, with only calcium salts exhibiting acamprosate-like effects. High plasma calcium levels due to acamprosate treatment correlated with better efficacy parameters in patients, suggesting calcium as the active component of acamprosate. This finding challenged previous attributions of acamprosate effects solely to *N*-acetylhomotaurinate and underscored the role of calcium in its therapeutic action (Spanagel et al., 2014).

Following up, the effects of oral calcium administration on craving, withdrawal, and relapse risk were investigated in alcohol-dependent patients. Using a single-blind, randomized trial design, alcohol-dependent subjects received either calcium carbonate or sodium bicarbonate daily during a 14-day inpatient alcohol-withdrawal treatment. Calcium carbonate treatment led to faster attenuation of withdrawal intensity and significantly reduced alcohol craving compared to sodium bicarbonate, supporting the role of calcium in reducing alcohol withdrawal symptoms and craving (Schuster et al., 2021).

Regarding the potential adverse effects of calcium acamprosate, although calcium compounds, such as calcium carbonate, cause constipation and bloating, the most prevalent adverse effect of calcium acamprosate is diarrhea, which can be severe and persistent in certain patients. Other adverse effects may include headaches, dizziness, pruritus, and kidney problems.

3.5 Other drugs

In 2018, baclofen, which is an agonist of the γ-aminobutyric-acid (GABA) B receptor, was approved for treating AUD in France (Witkiewitz et al., 2019). Meta-analysis studies have shown that baclofen can lead to higher abstinence rates compared to placebo. However, it does not significantly increase the number of days individuals remain abstinent or significantly reduce the frequency of heavy drinking (Rose & Jones, 2018). Moreover, its effectiveness appears more pronounced in heavy drinkers (Pierce, Sutterland, Beraha, Morley, & van den Brink, 2018).

In laboratory animals and non-human primates, baclofen treatment consistently reduced multiple alcohol-related behaviors, including alcohol-induced locomotor stimulation, binge and relapse drinking, oral alcohol self-administration, alcohol seeking behavior, and reinstatement of alcohol seeking (Colombo & Gessa, 2018).

Besides baclofen, other drugs, such as topiramate, ondansetron, gabapentin, and varenicline, have been used off-label, and they have shown a promisor effect for being used in the AUD treatment (Witkiewitz et al., 2019).

In the subsequent chapters, we will present and discuss the efficacy of new pharmacological targets and drugs evaluated as promising for treating AUD.

4. Final considerations

Most pharmacotherapies have limitations, with a high relapse rate within six months to 1 year following treatment initiation (Brandon, Vidrine, & Litvin, 2007; Nguyen et al., 2020).

The National Institute of Health and Care Excellence (National Collaborating Centre for Mental Health NCCMH Alcohol-use Disorders: Diagnosis, Assessment and Management of Harmful Drinking and Alcohol Dependence, 2011) guidelines recommend the use of naltrexone and acamprosate as first-line options for relapse prevention in combination with psychological intervention. Disulfiram is considered alongside with psychological assistance for individuals for whom naltrexone and acamprosate are not suitable or accessible. Although current pharmacological interventions for AUD may yield positive outcomes for specific individuals, their efficacy is inconsistent across all patients, and treatment response varies significantly among individuals. Not all individuals diagnosed with AUD exhibit positive responses to existing medications, with some encountering severe adverse effects or challenges in maintaining treatment adherence. Off-label use of other medications to treat AUD still faces several challenges, highlighting the necessity of exploring additional medications.

There is growing recognition of the importance of individualized treatment approaches, taking into account the characteristics and circumstances of the patient, such as genetics, social and cultural factors, in order to enhance precision in prognosis, treatment choices, and treatment outcomes.

Many individuals with AUD experience concurrent mental health disorders like depression, anxiety, or post-traumatic stress disorder (PTSD) and medications targeting both AUD and these coexisting mental health conditions could enhance treatment effectiveness and lower relapse rates.

Advances in neuroscience and pharmacology have led to the identification of novel therapeutic targets for AUD, including specific neurotransmitter systems, neural circuits, and molecular pathways involved in addiction and reward processing. Research into these targets may lead to developing new medications with improved efficacy, tolerability, and safety profiles for the treatment of AUD.

Integrated approaches that combine pharmacological interventions with behavioral therapies, counseling, and support services may enhance treatment effectiveness by addressing biological aspects of AUD and the complex interaction with psychological and environmental factors.

In summary, while pharmacological treatments are available for AUD, there remains a need for continued research and innovation to develop additional medications, refine existing treatments, and tailor interventions to patients' individual needs. Studying the effects of the environment on AUD and adopting a holistic approach to therapy are essential for improving results and reducing the burden of AUD on individuals, families, and society.

References

Al Qatari, M., Bouchenafa, O., & Littleton, J. (1998). Mechanism of action of acamprosate. Part II. Ethanol dependence modifies effects of acamprosate on NMDA receptor binding in membranes from rat cerebral cortex. *Alcohol, Clinical and Experimental Research, 22*, 810–814.

Anton, R. F., O'Malley, S. S., Ciraulo, D. A., Cisler, R. A., Couper, D., Donovan, D. M., ... COMBINE Study Research Group (2006). Combined pharmacotherapies and behavioral interventions for alcohol dependence: The COMBINE study: A randomized controlled trial. *Journal of the American Medical Association, 295*, 2003–2017.

Anton, R. F., Oroszi, G., O'Malley, S., Couper, D., Swift, R., Pettinati, H., & Goldman, D. (2008). An evaluation of mu-opioid receptor (OPRM1) as a predictor of naltrexone response in the treatment of alcohol dependence: Results from the Combined Pharmacotherapies and Behavioral Interventions for Alcohol Dependence (COMBINE) study. *Archives of General Psychiatry, 65*, 135–144.

Ashenhurst, J. R., Bujarski, S., & Ray, L. A. (2012). Delta and kappa opioid receptor polymorphisms influence the effects of naltrexone on subjective responses to alcohol. *Pharmacology Biochemistry & Behavior, 103*, 253–259.

Ashton, M. K., Rueda, A. V. L., Ho, A. M., Noor Aizin, N. A. B. M., Sharma, H., Dodd, P. R., ... Camarini, R. (2022). Sex differences in GABAA receptor subunit transcript expression are mediated by genotype in subjects with alcohol-related cirrhosis of the liver. *Genes, Brain, and Behavior, 21*, e12785.

Bahji, A., Bach, P., Danilewitz, M., Crockford, D., Devoe, D. J., El-Guebaly, N., & Saitz, R. (2022). Pharmacotherapies for adults with alcohol use disorders: A systematic review and network meta-analysis. *Journal of Addiction Medicine, 16*, 630–638.

Barr, C. S., Chen, S. A., Schwandt, M. L., Lindell, S. G., Sun, H., Suomi, S. J., & Heilig, M. (2010). Suppression of alcohol preference by naltrexone in the rhesus macaque: A critical role of genetic variation at the μ-opioid receptor gene locus. *Biological Psychiatry, 67*, 78–80.

Barr, C. S., Schwandt, M., Lindell, S. G., Chen, S. A., Goldman, D., Suomi, S. J., ... Heilig, M. (2007). Association of a functional polymorphism in the μ-opioid receptor gene with alcohol response and consumption in male rhesus macaques. *Archives of General Psychiatry, 64*, 369–376.

Bart, G., Kreek, M. J., Ott, J., LaForge, K. S., Proudnikov, D., Pollak, L., & Heilig, M. (2005). Increased attributable risk related to a functional μ-opioid receptor gene polymorphism in association with alcohol dependence in central Sweden. *Neuropsychopharmacology: Official Publication of the American College of Neuropsychopharmacology, 30*, 417–422.

Bergen, A. W., Kokoszka, J., Peterson, R., Long, J. C., Virkkunen, M., Linnoila, M., & Goldman, D. (1997). opioid receptor gene variants: Lack of association with alcohol dependence. *Molecular Psychiatry, 2*, 490–494.

Beyer, A., Koch, T., Schröder, H., Schulz, S., & Höllt, V. (2004). Effect of the A118G polymorphism on binding affinity and agonist-mediated endocytosis, desensitization, and resensitization of the human μ-opioid receptor. *Journal of Neurochemistry, 89*, 553–560.

Boismare, F., Daoust, M., Moore, N., Saligaut, C., Lhuintre, J. P., Chretien, P., & Durlach, J. (1984). A homotaurine derivative reduces the voluntary intake of ethanol by rats: Are cerebral GABA receptors involved? *Pharmacology Biochemistry & Behavior, 21*, 787–789.

Bond, C., LaForge, K. S., Tian, M., Melia, D., Zhang, S., Borg, L., ... Yu, L. (1998). Single-nucleotide polymorphism in the human mu opioid receptor gene alters beta-endorphin binding and activity: Possible implications for opiate addiction. *Proceedings of the National Academy of Sciences, 95*, 9608–9613.

Borghese, C. M., & Harris, R. A. (2012). Alcohol dependence and genes encoding α2 and γ1 GABAA receptor subunits: Insights from humans and mice. *Alcohol Research, 34*, 345–353.

Brandon, T. H., Vidrine, J. I., & Litvin, E. B. (2007). Relapse and relapse prevention. *Annual Review of Clinical Psychology, 3*, 257–284.

Chabenat, C., Chretien, P., Daoust, M., Moore, N., Andre, D., Lhuintre, J. P., ... Boismare, F. (1988). Physicochemical, pharmacological and pharmacokinetic study of a new GABAergic compound, calcium acetylhomotaurinate. *Methods & Findings in Experimental & Clinical Pharmacology, 10*, 311–317.

Chamorro, A. J., Marcos, M., Mirón-Canelo, J. A., Pastor, I., González-Sarmiento, R., & Laso, F. J. (2012). Association of μ-opioid receptor (OPRM1) gene polymorphism with response to naltrexone in alcohol dependence: A systematic review and meta-analysis. *Addiction Biology, 17*, 505–512.

Chick, J., Howlett, H., Morgan, M. Y., & Ritson, B. (2000). United Kingdom Multicentre Acamprosate Study (UKMAS): A 6-month prospective study of acamprosate versus placebo in preventing relapse after withdrawal from alcohol. *Alcohol and Alcoholism (Oxford, Oxfordshire), 35*, 176–187.

Colombo, G., & Gessa, G. L. (2018). Suppressing effect of baclofen on multiple alcohol-related behaviors in laboratory animals. *Frontiers in Psychiatry, 9*, 475.

Davis, V. E., & Walsh, M. D. (1970). Alcohol, amines, and alkaloids: A possible basis for alcohol addiction. *Science (New York, N. Y.), 167*, 1005–1007.

DeHaven-Hudkins, D. L., Brostrom, P. A., Allen, J. T., Lesko, L. J., Ferkany, J. W., Kaplita, P. V., ... Steranka, L. R. (1990). Pharmacologic profile of NPC 168 (naltrexone phenyl oxime), a novel compound with activity at opioid receptors. *Pharmacololgy, Biochemistry and Behavior, 37*, 497–504.

Dunbar, J. L., Turncliff, R. Z., Dong, Q., Silverman, B. L., Ehrich, E. W., & Lasseter, K. C. (2006). Single- and multiple-dose pharmacokinetics of long-acting injectable naltrexone. *Alcohol and Clinical Experimental Research, 30*, 480–490.

Edenberg, H. J., & McClintick, J. N. (2018). Alcohol dehydrogenases, aldehyde dehydrogenases, and alcohol use disorders: A critical review. *Alcohol, Clinical and Experimental Research, 42*, 2281–2297.

Emmerson, P. J., Liu, M. R., Woods, J. H., & Medzihradsky, F. (1994). Binding affinity and selectivity of opioids at mu, delta, and kappa receptors in monkey brain membranes. *Journal of Pharmacology and Experimental Therapeutics, 271*, 1630–1637.

Foo, J. C., Vengeliene, V., Noori, H. R., Yamaguchi, I., Morita, K., Nakamura, T., ... Spanagel, R. (2019). Drinking levels and profiles of alcohol-addicted rats predict response to nalmefene. *Frontiers in Pharmacology, 10*, 471.

Garbutt, J. C., West, S. L., Carey, T. S., Lohr, K. N., & Crews, F. T. (1999). Pharmacological treatment of alcohol dependence: A review of the evidence. *Journal of the American Medical Association, 281*, 1318–1325.

GDS. (2018). *Global status report on alcohol and health 2018*. Geneva: World Health Organization. Licence: CC BY-NC-SA 3.0 IGO.

Gelernter, J., Gueorguieva, R., Kranzler, H. R., Zhang, H., Cramer, J., Rosenheck, R., ... Cooperative Study #425 Study Group, V. A. (2007). Opioid receptor gene (OPRM1, OPRK1, and OPRD1) variants and response to naltrexone treatment for alcohol dependence: Results from the VA Cooperative Study. *Alcohol, Clinical and Experimental Research, 31*, 555–563.

Gianoulakis, C., Krishnan, B., & Thavundayil, J. (1996). Enhanced sensitivity of pituitary β-endorphin to ethanol in subjects at high risk of alcoholism. *Archives of General Psychiatry, 53*, 250–257.

Herz, A. (1997). Endogenous opioid systems and alcohol addiction. *Psychopharmacology (Berl), 129*, 99–111.

Jacobsen, E., & Larsen, V. (1949). Site of the formation of acetaldehyde after ingestion of antabuse (tetraethylthiuramdisulphide) and alcohol. *Acta Pharmacologica et Toxicologica (Copenh), 5*, 285–291.

Koob, G. F., & Volkow, N. D. (2016). Neurobiology of addiction: A neurocircuitry analysis. *Lancet Psychiatry, 3*, 760–773.

Kragh, H. (2008). From disulfiram to antabuse: The invention of a drug. *Bulletin for the History of Chemistry, 33*, 7.

Kranzler, H. R., Modesto-Lowe, V., & Van Kirk, J. (2000). Naltrexone vs. nefazodone for treatment of alcohol dependence. A placebo-controlled trial. *Neuropsychopharmacology: Official Publication of the American College of Neuropsychopharmacology, 22*, 493–503.

Lanz, J., Biniaz-Harris, N., Kuvaldina, M., Jain, S., Lewis, K., & Fallon, B. A. (2023). Disulfiram: Mechanisms, applications, and challenges. *Antibiotics (Basel), 12*, 524.

Lhuintre, J. P., Daoust, M., Moore, N. D., Chretien, P., Saligaut, C., Tran, G., ... Hillemand, B. (1985). Ability of calcium bis acetyl homotaurinate, a GABA agonist, to prevent relapse in weaned alcoholics. *Lancet, 1*(8436), 1014–1016.

Lhuintre, J. P., Moore, N., Tran, G., Steru, L., Langrenon, S., Daoust, M., ... Hillemand, B. (1990). Acamprosate appears to decrease alcohol intake in weaned alcoholics. *Alcohol and Alcoholism (Oxford, Oxfordshire), 25*, 613–622.

Lobmaier, P., Gossop, M., Waal, H., & Bramness, J. (2010). The pharmacological treatment of opioid addiction—A clinical perspective. *European Journal of Clinical Pharmacology, 66*, 537–545.

Lovinger, D. M., White, G., & Weight, F. F. (1989). Ethanol inhibits NMDA-activated ion current in hippocampal neurons. *Science (New York, N. Y.), 243*, 1721–1724.

Mann, K., Torup, L., Sørensen, P., Gual, A., Swift, R., Walker, B., & van den Brink, W. (2016). Nalmefene for the management of alcohol dependence: Review on its pharmacology, mechanism of action and meta-analysis on its clinical efficacy. *European Neuropsychopharmacology, 26*, 1941 -149.

Miyata, H., Takahashi, M., Murai, Y., Tsuneyoshi, K., Hayashi, T., Meulien, D., ... Higuchi, S. (2019). Nalmefene in alcohol-dependent patients with a high drinking risk: Randomized controlled trial. *Psychiatry and Clinical Neurosciences, 73*, 697–706.

Naassila, M., Hammoumi, S., Legrand, E., Durbin, P., & Daoust, M. (1998). Mechanism of action of acamprosate. Part I Characterization of spermidine-sensitive acamprosate binding site in rat brain. *Alcohol and Clinical Experimental Research, 22*, 802–809.

National Collaborating Centre for Mental Health (NCCMH) Alcohol-use Disorders: Diagnosis, Assessment and Management of Harmful Drinking and Alcohol Dependence. (2011). Leicester and London: The British Psychological Society and the Royal College of Psychiatrists.

Nguyen, L. C., Durazzo, T. C., Dwyer, C. L., Rauch, A. A., Humphreys, K., Williams, L. M., & Padula, C. B. (2020). Predicting relapse after alcohol use disorder treatment in a high-risk cohort: The roles of anhedonia and smoking. *Journal of Psychiatric Research, 126*, 1–7.

O'Malley, S. S., Jaffe, A. J., Chang, G., Schottenfeld, R. S., Meyer, R. E., & Rounsaville, B. (1992). Naltrexone and coping skills therapy for alcohol dependence: A controlled study. *Archives of General Psychiatry, 49*, 881–887.

Olsen, R. W., & Liang, J. (2017). Role of GABAA receptors in alcohol use disorders suggested by chronic intermittent ethanol (CIE) rodent model. *Molecular Brain, 10*, 45.

Omran, Z. (2021). New disulfiram derivatives as MAGL-selective inhibitors. *Molecules (Basel, Switzerland), 26*, 3296.

Oncken, C., Van Kirk, J., & Kranzler, H. R. (2001). Adverse effects of oral naltrexone: Analysis of data from two clinical trials. *Psychopharmacology (Berl), 154*, 397–402.

Pal, A., Pattanayak, R. D., & Sagar, R. (2015). Tracing the journey of disulfiram. From an unintended discovery to a treatment option for alcoholism. *Journal of. Mental Health Human Behaviour, 20*, 41–43.

Pelc, I., Ansoms, C., Lehert, P., Fischer, F., Fuchs, W. J., Landron, F., ... Morgan, M. Y. (2002). The European NEAT program: An integrated approach using acamprosate and psychosocial support for the prevention of relapse in alcohol-dependent patients with a statistical modeling of therapy success prediction. *Alcohol, Clinical and Experimental Research, 26*, 1529–1538.

Pierce, M., Sutterland, A., Beraha, E. M., Morley, K., & van den Brink, W. (2018). Efficacy, tolerability, and safety of low-dose and high-dose baclofen in the treatment of alcohol dependence: A systematic review and meta-analysis. *European Neuropsychopharmacology, 28*, 795–806.

Popova, N. K., Ilchibaeva, T. V., Antonov, E. V., Pershina, A. V., Bazovkina, D. V., & Naumenko, V. S. (2020). On the interaction between BDNF and serotonin systems: The effects of long-term ethanol consumption in mice. *Alcohol (Fayetteville, N. Y.), 87*, 1–15.

Ray, L. A., & Hutchison, K. E. (2004). A polymorphism of the mu opioid receptor gene and sensitivity to the effects of alcohol in humans. *Alcohol, Clinical and Experimental Research, 28*, 1789–1795.

Reilly, M. T., Noronha, A., Goldman, D., & Koob, G. F. (2017). Genetic studies of alcohol dependence in the context of the addiction cycle. *Neuropharmacology, 122*, 3–21.

Rose, A. K., & Jones, A. (2018). Baclofen: Its effectiveness in reducing harmful drinking, craving, and negative mood. A meta-analysis. *Addiction (Abingdon, England), 113*, 1396–1406.

Sanchez-Roige, S., Fontanillas, P., Elson, S. L., 23andMe Research Team, Gray, J. C., de Wit, H., ... Palmer, A. A. (2019). Genome-wide association study of alcohol use disorder identification test (AUDIT) scores in 20 328 research participants of European ancestry. *Addiction Biology, 24*, 121–131.

Santos, R. A., Rae, M., Dartora, V. F. M. C., Matos, J. K. R., Camarini, R., & Lopes, L. B. (2020). Bioresponsive nanostructured systems for sustained naltrexone release and treatment of alcohol use disorder: Development and biological evaluation. *International Journal of Pharmaceutics, 585*, 119474.

Sass, H., Soyka, M., Mann, K., & Zieglgänsberger, W. (1996). Relapse prevention by acamprosate. Results from a placebo-controlled study on alcohol dependence. Archives of General Psychiatry, 53, 673-680. *Erratum in: (1996) Archives of General Psychiatry, 53*, 1097.

Schuster, R., Winkler, M., Koopmann, A., Bach, P., Hoffmann, S., Reinhard, I., ... Kiefer, F. (2021). Calcium carbonate attenuates withdrawal and reduces craving: A randomized controlled trial in alcohol-dependent patients. *European Addiction Research, 27*, 332–340.

Sheedy, C. K., & Whitter, M. (2009). Guiding principles and elements of recovery-oriented systems of care: What do we know from the research? HHS Publication No. (SMA) 09–4439. Rockville, MD: Center for Substance Abuse Treatment, Substance Abuse and Mental Health Services Administration.

Spanagel, R. (2009). Alcoholism: A systems approach from molecular physiology to addictive behavior. *Physiological Reviews, 89*, 649–705.

Spanagel, R., Hölter, S. M., Allingham, K., Landgraf, R., & Zieglgänsberger, W. (1996). Acamprosate and alcohol: I. Effects on alcohol intake following alcohol deprivation in the rat. *European Journal of Pharmacology, 305*, 39–44.

Spanagel, R., Putzke, J., Stefferl, A., Schöbitz, B., & Zieglgänsberger, W. (1996). Acamprosate and alcohol: II. Effects on alcohol withdrawal in the rat. *European Journal of Pharmacology, 305*, 45–50.

Spanagel, R., Vengeliene, V., Jandeleit, B., Fischer, W. N., Grindstaff, K., Zhang, X., ... Kiefer, F. (2014). Acamprosate produces its anti-relapse effects via calcium. *Neuropsychopharmacology: Official Publication of the American College of Neuropsychopharmacology, 39*, 783–791.

Spanagel, R., Zieglgänsberger, W., & Hundt, W. (1996). Acamprosate and alcohol: III. Effects on alcohol discrimination in the rat. *European Journal of Pharmacology, 305*, 51–56.

Srivastava, A. B., & Gold, M. S. (2018). Naltrexone: A history and future directions. *Cerebrum, 2018, vcer-*, 13–18.

Stokes, M., & Abdijadid, S. (2024). *Disulfiram. StatPearls [Internet]*. Treasure Island, FL: StatPearls Publishing.

Swift, R. M. (1999). Drug therapy for alcohol dependence. *New England Journal of Medicine, 340*, 1482–1487.

Takemori, A. E., Ho, B. Y., Naeseth, J. S., & Portoghese, P. S. (1988). Nor-binaltorphimine, a highly selective kappa-opioid antagonist in analgesic and receptor binding assays. *Journal of Pharmacology and Experimental Therapeutics, 246*, 255–258.

Takemori, A. E., & Portoghese, P. S. (1992). Selective naltrexone-derived opioid receptor antagonists. *Annual Review of Pharmacology and Toxicology, 32*, 239–269.

Trudell, J. R., Messing, R. O., Mayfield, J., & Harris, R. A. (2014). Alcohol dependence: molecular and behavioral evidence. *Trends in Pharmacological Sciences, 35*, 317–323.

Volpicelli, J. R. (1987). Uncontrollable events and alcohol drinking. *British Journal of Addiction, 82*, 385–396.

Volpicelli, J. R., Alterman, A. I., Hayashida, M., & O'Brien, C. P. (1992). Naltrexone in the treatment of alcohol dependence. *Archives of General Psychiatry, 49*, 876–880.

Volpicelli, J. R., Davis, M. A., & Olgin, J. E. (1986). Naltrexone blocks the post-shock increase of ethanol consumption. *Life Sciences, 38*, 841–847.

WHO (2023). *Alcohol action plan 2022–2030.* Geneva: World Health Organization. Licence: CC BY-NC-SA 3.0 IGO.

Wilde, M. I., & Wagstaff, A. J. (1997). Acamprosate. A review of its pharmacologyand clinical potential in the management of alcohol dependence afterdetoxification. *Drugs, 53*, 1038–1053.

Winstock, A. R., Maier, L. J., Zhuparris, A., Davies, E., Puljevic, C., Kuypers, K. P. C., & Barratt, M. J. (2021). Global Drug Survey (GDS) 2021 Key findings report.

Witkiewitz, K., Litten, R. Z., & Leggio, L. (2019). Advances in the science and treatment of alcohol use disorder. *Science Advances, 5*, eaax4043.

Yin, W., Akalam, E. O., & Taylor, R. E. (2002). Design of naltrexone-loaded hydrolyzable crosslinked nanoparticles. *International Journal of Pharmacology, 244*, 9–19.

Zeise, M. L., Kasparov, S., Capogna, M., & Zieglgänsberger, W. (1993). Acamprosate (calciumacetylhomotaurinate) decreases postsynaptic potentials in the rat neocortex: Possible involvement of excitatory amino acid receptors. *European Journal of Pharmacology, 231*, 47–52.

Zhang, Y., Wang, D., Johnson, A. D., Papp, A. C., & Sadée, W. (2005). Allelic expression imbalance of human mu opioid receptor (OPRM1) caused by variant A118G. *Journal of Biological Chemistry, 280*, 32618–32624.

CHAPTER TWO

Role of oxytocin and vasopressin in alcohol use disorder

Sutton M. Scarboro[a,b], Marcelo F. López[a], and Howard C. Becker[a,b,c,*]

[a]Department of Psychiatry and Behavioral Sciences, Charleston Alcohol Research Center, Medical University of South Carolina, Charleston, SC, United States
[b]Department of Neuroscience, Charleston Alcohol Research Center, Medical University of South Carolina, Charleston, SC, United States
[c]RHJ Veterans Administartion Health Care System, Charleston, SC, United States
*Corresponding author. e-mail address: beckerh@musc.edu

Contents

Abstract

Published works highlight the role of neuropeptides in both the development and treatment of AUD. Closely related hypothalamic neuropeptides, oxytocin (OT) and vasopressin (VP), initially recognized for their physiological hormone effects, are increasingly acknowledged for their behavioral influences. Studies consistently demonstrate that OT and VP impact alcohol consumption and related behaviors, implicating them in the neurobiology of addiction. Moreover, stress is a pivotal risk factor for alcohol use and relapse, with OT and VP playing an integral role in the body's stress response system. While previous work has explored the interaction of OT and VP with other substances of abuse, this review focuses on their roles in alcohol-associated behaviors specifically to better understand the role of OT and VP in AUD. Here we synthesize recent preclinical and clinical literature examining changes in OT and VP protein and receptor expression in response to alcohol, as well as research

International Review of Neurobiology, Volume 178
ISSN 0074-7742, https://doi.org/10.1016/bs.irn.2024.08.006

investigating the effects of modulating these systems on alcohol-related behaviors. This review aims to deepen the understanding of OT and VP in the context of AUD with the goal of facilitating future research and enhancing treatment outcomes.

1. Introduction

Alcohol use disorder (AUD) represents a significant global health challenge with substantial consequences for public health and individual well-being. Despite extensive research efforts, the precise etiology of AUD remains unclear, hindering the development of optimal treatment strategies. Available therapeutic options exhibit varying success rates, emphasizing the need for developing additional and more effective interventions to improve treatment outcomes for individuals with AUD. Elucidating the underlying biological mechanisms that drive this disorder will not only help address the complexity of AUD, but also provide valuable information with broad implications for personalized medicine, targeted interventions and overall improvement of treatment outcomes.

Existing literature underlines the involvement of neuropeptides in various aspects of AUD. The neuropeptide oxytocin (OT), classically acknowledged for its role as a hormone in maternal physiology (e.g., parturition and uterine contractions) has more recently been identified as a positive regulator of social behaviors including pair-bonding and social reward processing (Baskerville & Douglas, 2010; Burkett & Young, 2012; Gimpl & Fahrenholz, 2001; Neumann & Landgraf, 2012). The structurally related hypothalamic peptide vasopressin (VP) was identified as a key component of the endocrine system, serving a major homeostatic function (e.g., water balance and blood pressure) (Bilezikjian & Vale, 1987; Rigney et al., 2022; Stoop, 2012). As with OT, VP was later recognized for behavioral regulation of pair-bonding and social recognition. Interestingly, in contrast to OT, the VP system also has been shown to facilitate aggressive behavior (Johnson & Young, 2017; Nair & Young, 2006; Rigney et al., 2022; Walum & Young, 2018; Williams Avram & Cymerblit-Sabba, 2017). Although oppositely, both peptides influence the hypothalamic-pituitary-adrenal (HPA) axis, a pivotal component of the body's stress regulation system, and are increasingly recognized for their contribution to behaviors often associated with neuropsychiatric disorders like depression and anxiety (Bilezikjian & Vale, 1987; King et al., 2020; Lee et al., 2009; Rigney et al., 2022; Stoop, 2012). Research indicates that OT can suppress HPA-axis activity and as such ameliorate anxiety-like behavior (King et al., 2020).

Conversely, VP stimulates HPA-axis reactivity resulting in enhanced anxiety (Rotondo et al., 2016). There is also evidence supporting the interaction of OT and VP with drug-related reward (King et al., 2020; Wronikowska-Denysiuk et al., 2023). Expanding from physiological functions to social behavior, growing acknowledgment of the roles of OT and VP in influencing the brain's reward system and stress-related behaviors positions the peptides as promising candidates for the treatment of AUD, considering that stress is a critical risk factor for alcohol use and relapse. Existing literature demonstrates changes in the OT system in response to alcohol and provides evidence that OT affects alcohol-related behaviors in both preclinical and clinical studies. While less extensive than the body of work surrounding OT interactions with alcohol, research indicates that the vasopressin system is also sensitive to alcohol.

The relationship between both OT and VP with other substances of abuse has been researched; however, this review concentrates specifically on the association of these peptides with alcohol-related behaviors. Here we summarize the current landscape of preclinical and clinical literature investigating alterations in expression of the neuropeptides and their receptors induced by alcohol exposure and outline results from studies examining the use of OT and VP as potential therapeutic agents to reduce alcohol consumption and the modulating effects of stress on alcohol-related behaviors. This review synthesizes the present understanding of the relationship between the OT and VP systems and alcohol and identifies areas where further research is needed to advance our comprehension of these neuropeptides in the context of AUD.

2. The oxytocin system

Acting as a hormone and neuropeptide, oxytocin (OT) is responsible for various behavioral and physiological effects (e.g., parturition, lactation, maternal behavior, pro-social behavior, anxiolytic effects). Due to its responsiveness to both stress and alcohol, OT has gained interest as a potential therapeutic in the treatment of AUD.

2.1 Oxytocin (OT) neurons

OT is a nine amino acid peptide that is primarily synthesized within three hypothalamic brain regions: the paraventricular nucleus (PVN), the supraoptic nuclei (SON) and the accessory nuclei (AN) (Busnelli & Chini, 2018; Grinevich & Neumann, 2021). Projections from these neurons to the posterior pituitary constitute the hypothalamic-neurohypophysial system, where the peptide is released into circulation to mediate its varied endocrine functions. OT producing neurons are defined as magnocellular (magnOT) or parvocellular (parvOT), which are characterized by size, shape, location, and projections. magnOT neurons exhibit both peripheral and central functions. Depolarization of posterior pituitary-projecting magnOT neurons results in secretion of OT into the bloodstream via the neurohypophysis, a posterior pituitary region of fenestrated capillaries that permit the exocytosis of OT-containing dense core vesicles (Brown et al., 2013; Grinevich & Neumann, 2021). In addition to pituitary projections, magnOT neurons project centrally to cortical regions and subcortical structures, including striatal (e.g., nucleus acucumbens; NAc) and limbic areas (e.g., hippocampus (HPC), amygdala (AMY), bed nucleus of stria terminalis (BNST), septum). Central OT release is induced by reproductive stimuli, stress (physical and emotional), and social stimuli. OT release may occur through various mechanisms (e.g., synaptic or non-synaptic) and can stimulate its own release via a peptide feedback loop (Grinevich & Neumann, 2021; Mitre et al., 2018). These oxytocinergic pathways are thought to play a significant role in modulating stress-related and prosocial behaviors (Busnelli & Chini, 2018; Grinevich & Ludwig, 2021; Grinevich & Neumann, 2021). Smaller in size than magnOT neurons, parvOT neurons are primarily located within the PVN and these neurons project to the midbrain, brainstem, and spinal cord. As such, parvOT neurons mediate various autonomic functions (e.g., breathing, cardiovascular responses, somatosensory processing and inhibition of pain transmission). parvOT neurons have no direct projection to the posterior pituitary but may indirectly modulate peripheral OT release by synapsing onto magnOT neurons (Grinevich & Ludwig, 2021; Grinevich & Neumann, 2021).

2.2 The OT receptor (OTR)

To date, only one OT receptor (OTR) has been identified. The OTR is a member of the class-A/rhodopsin-type G-protein coupled receptor (GPCR) family. The OTR is desensitized and internalized upon activation and is expressed by both neurons and glial cells (Grinevich & Neumann, 2021). The OTR can couple to both excitatory (G_q/G_{11}) and inhibitory (G_i/G_o) G-proteins, depending on brain region, cell type, local OT concentration, local OTR expression, and local G-protein expression (Busnelli & Chini, 2018; Menon & Neumann, 2023). OTR-$G_{q/11}$ signaling stimulates the phospholipase-C-β (PLC-β) pathway, resulting in the phosphorylation of proteins that regulate reproductive (e.g., parturition and lactation) and anxiolytic effects of OT (Busnelli & Chini, 2018; Grinevich & Neumann, 2021). G_q/G_{11} activation also initiates the mitogen-activated protein kinase (MAPK) pathway. Activation of the G_i/G_o pathway modulates ion channel function via stimulation of phosphatidylinositol-4,5-bisphosphate 3-kinase (PI3K) MAPK pathways (Busnelli & Chini, 2018). Ubiquitous expression of the $G_{\alpha q}$ subunit in the brain favors G_q-OTR coupling as the widespread outcome; however, G_i/G_o-OTR coupling can occur (Busnelli & Chini, 2018; Busnelli et al., 2012). In addition to ligand availability, OTR expression, and G-protein availability, OTRs have been found to be regulated by epigenetic mechanisms, estrogen receptor activation, and acute or chronic stressors (Grinevich & Neumann, 2021; Mitre et al., 2018). In summary, the effects of OTR activation are highly regulated, resulting in brain-region specific and diverse effects of OT-OTR signaling.

OTRs are highly expressed in the HPC (ventral, CA2), BNST, central nucleus of the amygdala (CeA), and ventromedial hypothalamic nucleus (VM). OTR expression has also been identified in regions implicated in addiction-related phenotypes (e.g., emotional processing and reward) including the brainstem, ventral pallidum (VP), NAc, caudate putamen (CPu), medial preoptic area (MPOA), lateral septum (LS), ventral tegmental area (VTA), and prefrontal cortex (PFC) (Schimmer et al., 2023). While not detailed in this review, it is important to note that there are species, sex, and developmental differences in OTR expression (Dumais & Veenema, 2016; Grinevich & Neumann, 2021; Menon & Neumann, 2023; Mitre et al., 2018).

2.3 OT—alcohol interactions

2.3.1 Effects of alcohol on the OT system

Preclinical studies illustrate a complex relationship between alcohol exposure and the OT system. Animal studies have shown a reduction in OT immunoreactive magnocellular neurons in the hypothalamus following chronic alcohol exposure in male Wistar rats and prairie voles (Silva et al., 2002; Stevenson, Young, et al., 2017). Notably, there was no change in the number of OT positive neurons during withdrawal (Silva et al., 2002). In contrast, OTR expression was increased in frontal and striatal regions of male Wistar rats following chronic intermittent ethanol vapor exposure. Specifically, after 3 weeks of abstinence, OTR mRNA and protein levels were increased in the PFC, striatum, AMY and HPC, but decreased protein expression was found in the hypothalamus (Hansson et al., 2018). Interestingly, these changes were not observed in female rats following similar chronic alcohol exposure (Hansson & Spanagel, 2021). Thus, chronic alcohol exposure alters expression of OT and OTR in the brain, with the latter effect varying across different brain regions, and possibly sex.

Clinical investigations also demonstrate a relationship between alcohol and the OT system. For instance, immunoreactivity studies in post-mortem tissue of male individuals with AUD showed decreased OT expression in the SON and increased OT expression in the PVN (Sivukhina et al., 2006). More recently, analysis of post-mortem tissue from male individuals diagnosed with AUD indicated reduced expression of the peptide in the hypothalamus and elevated levels of OTR mRNA in frontal and striatal regions compared to controls (Hansson et al., 2018; Lee et al., 2017). As was the case for changes reported in rats, a follow-up study showed no changes in OT or OTR expression in tissue from female subjects (Hansson & Spanagel, 2021). Analysis of male subjects revealed a positive correlation between OT mRNA fold change and both daily and weekly alcohol intake (Lee et al., 2017). Identifying peripheral biomarkers associated with AUD is critical to less-invasively predict risk for harmful alcohol drinking. To investigate OTR's usefulness as an AUD biomarker, Muhle and colleagues examined blood OTR concentrations in males and females diagnosed with AUD. In males and females, OTR concentrations were not different between patients diagnosed with AUD and controls; however, in males previously diagnosed with AUD, increased blood OTR expression predicted hospital readmission (Muhle et al., 2022). It is unclear to what extent plasma levels reflect central levels of OT. Nevertheless, studies involving analysis of postmortem tissue derived from individuals diagnosed with AUD show a history of heavy alcohol use produces significant alterations in the OT system, and this effect may differ in males and females.

2.3.2 Effect of OT on alcohol tolerance and withdrawal

There are physiological symptoms associated with alcohol use, including tolerance and withdrawal. Tolerance is observed when the same dose of alcohol results in less of an effect and/or when higher doses of alcohol are required to reach the same effect. This leads to higher levels of alcohol consumption associated with AUD. Existing literature suggests that OT plays a role in the physiological effects of alcohol tolerance, including sedation, ataxia, and hypothermic effects (Jodogne et al., 1991; Puciklowski et al., 1985; Szabo, Kovacs, Balaspiri, et al., 1987; Szabo, Kovacs, Szekeli, et al., 1987; Szabo et al., 1985, 1989; Tirelli et al., 1992). The development of hypothermic, hypnotic, and ataxic manifestations of alcohol tolerance was prevented by multiple intraperitoneal (ip.) injections of OT (Jodogne et al., 1991; Szabo et al., 1985, 1989). Negative emotional state associated with withdrawal is also a driver of alcohol drinking. Subcutaneous (sc.) administration of OT prior to alcohol exposure reduces the intensity of alcohol withdrawal symptoms (seizures) (Szabo, Kovacs, Balaspiri, et al., 1987) and central administration of OT was more efficacious at producing this effect than systemic administration of OT (Szabo et al., 1989). However, more information is needed regarding the effect of OT on other effects of alcohol withdrawal that play a significant role in alcohol drinking, such as nociception and anxiety.

2.3.3 Effects of OT on alcohol consummatory behavior

Just as alcohol has effects on the endogenous OT system, manipulating the OT system has been shown to alter alcohol consumption. The following describes some of the models employed to examine the influence of OT on motivational effects of alcohol.

2.3.4 Preclinical studies

2.3.4.1 2-Bottle choice drinking

OT effects on alcohol consumption have been examined under conditions in which the choice to consume alcohol is presented continuously or intermittently (e.g., every other day), the latter of which engenders escalated intake (Becker & Lopez, 2024). Examining alcohol consumption in the presence of an alternative liquid (typically water) provides insight into the potential for OT to alter alcohol preference and general fluid consumption. Systemic (ip.) administration of OT, as well as direct (intracerebroventricular; icv.) administration of OT, reduced voluntary alcohol consumption in male C57BL/6N mice (Peters et al., 2013). Similar results were obtained with icv. administration of OT in male Wistar rats with no change in water consumption, thereby reducing alcohol preference in an intermittent access paradigm (Peters et al., 2017). In another study, systemic administration of OT (1 mg/kg ip.) in adolescent male Wistar rats resulted in reduced alcohol (beer) intake with no change in water consumption during early adulthood (Bowen et al., 2011). This same investigative group showed that a single injection of OT (1 mg/kg) produced a long-lasting reduction in consumption of a sweetened alcoholic beverage compared to a sucrose solution in alcohol-preferring P rats (McGregor & Bowen, 2012). These findings suggest a long-acting effect of OT, although most studies have found the effects of OT on alcohol drinking to last only a few hours. In male C57BL/6J mice, OT (1 mg/kg; ip.) reduced alcohol intake under continuous (24-h/day) free-choice conditions. The decrease in alcohol consumption was accompanied by an increase in water intake, resulting in reduced alcohol preference and no change in total fluid intake (King et al., 2017). In another study, group-housed C57BL/6J mice given 24-h continuous free-choice access, repeated daily OT injections (3 mg/kg; ip.) reduced alcohol intake with minimal effects on water intake in males and females. The effects of OT were relatively short-lived (about 2–4 h) (Caruso et al., 2021). Notably, female OTR knockout (KO) mice showed a significant increase in alcohol intake and preference (24-hr access) that was not observed in male subjects (Rodriguez et al., 2020). OTR deletion did not alter quinine or saccharin solutions intake (Rodriguez et al., 2020). Overexpression of the OTR in the NAc of male C57BL/6 mice resulted in reduced anxiety (elevated plus maze, open field, and marble burying testing) and reduced voluntary alcohol intake and preference (24-hr access) without altering the intake or preference of saccharin or quinine solutions (Bahi et al., 2016). Similar results were obtained in group-housed prairie voles, except OT (3 mg/kg ip.) also reduced water intake (Walcott & Ryabinin, 2021). In an earlier study, OT (1–10 mg/kg; ip.) reduced alcohol intake and preference in male and female prairie voles under continuous or intermittent access situations (Stevenson, Wenner, et al., 2017). Thus, across three rodent species (rats, mice, prairie voles) OT treatment was found to reduce voluntary alcohol consumption. These effects were typically (but not always) evident over a few hours following systemic administration of the peptide. *See* Table 1.

2.3.5 Binge-like drinking

In contrast to free-choice 24-h/day alcohol access (whether continuous or intermittent), restricting rodent alcohol access to a shorter period of time during the dark phase of the circadian cycle produces high levels of intake with corresponding elevation in blood alcohol levels. For example, the "drinking in the dark" (DID) paradigm is a preclinical model of binge-like alcohol consumption, wherein alcohol intake in a relatively short period of time produces blood alcohol levels that exceed the legal limit of intoxication (0.08 g/dL) (Becker & Lopez, 2024). Using this model, systemic administration of OT (0.3–10 mg/kg; ip.) was found to significantly decrease binge-like alcohol consumption and accompanying elevated blood alcohol levels in male C57BL/6J mice in a dose-dependent manner,

Table 1 Oxytocin in preclinical studies.

Authors	Year	Subjects	Model	Treatment	Results
Bowen et al.	2011	Male Wistar rats, Adolescent	2-Bottle Choice	1 mg/kg OT ip. x 10 days	↓ alcohol intake
McGregor & Bowen	2012	Male & Female P-rats	2-Bottle Choice	0.3 or 1 mg/kg OT ip.	↓ alcohol preference in males & females
Peters et al.	2013	Male C57BL/6J	2-Bottle Choice following CSC	10 mg/kg OT ip., 0.5 μg/ 2 μL OT icv.	10 mg/kg OT ip.: ↓ alcohol intake in CSC and CTL animals 0.5 μg/ 2 μL OT icv.: ineffective
MacFadyen et al.	2016	Male Sprague-Dawley rats	Limited Access, 3-bottle choice	0.1–0.5 mg/kg OT ip.	↓ alcohol-saccharin intake but not saccharin intake alone
MacFadyen et al.	2016	Male Sprague-Dawley rats	Operant Alcohol Self-Administration	0.3 mg/kg OT ip.	↓ alcohol responding
Bahi, Mansouri & Maamari et al.	2016	Male C57BL/6J mice	2-bottle choice & other assessments: EPM, OF, MBT, Wire-hanging, LORR,	Nucleus Accumbens *OxtR* lentiviral overexpression	↓ anxiety-like behaviors (EPM, OF, MBT) ↓ ethanol consumption & preference (no effect on saccharin nor quinine intake/preference) Reversed ethanol-induced motor and hypnotic responses (Wire-hanging, LORR)

(continued)

Table 1 Oxytocin in preclinical studies. *(cont'd)*

Authors	Year	Subjects	Model	Treatment	Results
Peters et al.	2017	Male Wistar rats	2-Bottle Choice	1 μg/5 μL OT icv.	↓ alcohol intake and preference.
King et al.	2017	Male C57BL/6J mice	Continuous Access	1 mg/kg OT ip.	↓ alcohol intake and preference
Stevenson et al.	2017	Male & Female prairie voles	Continuous or Intermittent Access	1–10 mg/kg OT ip.	↓ alcohol intake and preference
King et al.	2017	Male C57BL/6J mice	Drinking in the Dark	0.3–10 mg/kg ip.	↓ alcohol intake and BALs dose-dependently, reversed by OTR antagonist (ip.)
King et al.	2017	Male C57BL/6J mice	Operant Alcohol Self-Administration	0.1–1.0 mg/kg OT ip .0.3 mg/kg OT ip.	**A.** ↓ alcohol responding and intake dose-dependently **B.** ↓ breakpoint response values for alcohol but not sucrose
Hansson et al.	2018	Male Wistar rats	Operant Alcohol Self-Administration & CIE	10 nM OT icv.	↓ active lever responding during cue-induced reinstatement in alcohol-dependent rats only

Tunstall et al.	2019	Male Wistar rats	Operant Alcohol Self-Administration following CIE	0.125–1.0 mg/kg OT ip. or 0.25–1.0 mg/kg OT in.OTR agonist PF−06655075 icv.	**A.** ↓ alcohol self-administration and break point response values dose-dependently in alcohol-dependent rats **B.** ↓ alcohol self-administration
King & Becker	2019	Male & Female C57 BL/6J mice	Operant Alcohol Self-Administration	0.1–1.0 mg/kg OT Iip.	↓ stress-induced reinstatement dose-dependently
Rodriguez, Smith & Caldwell	2020	Male & Female *Oxtr* -/- and wild-type mice	2-Bottle Choice & FST	—	↑ alcohol consumption in Oxtr -/- females pre- and post-FST. No significant difference in male mice.
Caruso et al.	2021	Male & Female C57BL/6J mice	Continuous Access	3 mg/kg OT ip., Repeated	↓ alcohol intake
Walcott & Ryabinin.	2021	Male & Female prairie voles	Continuous Access	3 mg/kg OT ip.	↓ alcohol and water intake

(continued)

Table 1 Oxytocin in preclinical studies. *(cont'd)*

Authors	Year	Subjects	Model	Treatment	Results
King et al.	2021	Male *Oxt-IRES-Cre* mice	Drinking in the Dark	Chemogenetic activation of PVN OT neurons	↓ alcohol intake, reversed by central, but not peripheral OTR antagonist (ip.)
Ballas et al.	2021	Male Sprague-Dawley rats	Operant Alcohol Self-Administration	0.5 μg/0.5 μL OT Intra-CeA	↓ stress-induced reinstatement
Becker et al.	2023	Male & Female C57Bl/6J mice	PTSD-AUD Comorbidity Model	1 mg/kg OT ip.	↓ sensitized stress-induced reinstatement
Wilfur et al.	2023	Female Sprague-Dawley rats	Operant Alcohol Self-Administration	0.5 μg/0.5 μL OT Intra-CeA	↓ stress-induced reinstatement

Subordinate Colony Housing (CSC), Elevated Plus Maze (EPM), Open-field Test (OF), Marble-Burying Test (MBT), Loss of Righting Reflex Test (LORR), Forced Swim Test (FST).

and this effect was blocked by pretreatment with an OT receptor antagonist (L-368,889; 10 mg/kg; ip.). Importantly, lower doses of OT (0.3, 1.0 mg/kg) that significantly reduced alcohol intake did not alter general fluid consumption and had minimal effects on general open-field locomotor activity (King et al., 2017). In a follow-up study, the effects of directly engaging the endogenous OT system on alcohol consumption (as opposed to exogenous OT administration) were examined. Selective chemogenetic activation of hypothalamic OT neurons, using male Oxy-IRES-Cre mice, produced a robust reduction in binge-like alcohol intake (King et al., 2021). Further, OT receptor antagonists were used to pharmacologically determine whether alcohol consumption reduction through endogenous OT activation is attributed to peripheral versus central effects. Reduced alcohol consumption produced by chemogenetic activation of hypothalamic OT neurons was reversed by pretreatment with the brain-penetrant OTR antagonist L-368–899 (10 mg/kg; ip.) but not by the peripherally restricted OTR antagonist Atosiban (1 mg/kg; ip.). Neither antagonist given alone altered alcohol intake. These results suggest that OT decreases binge-like alcohol consumption and this effect appears to be mediated by signaling at OTRs in the brain (King et al., 2021). Similar results on limited access (2-hr/day) alcohol intake were reported in a study using male Sprague-Dawley rats. In a 3-bottle choice paradigm, systemic administration of OT (0.1–0.5 mg/kg; ip.) reduced 10% or 15% alcohol mixed with saccharin but did not alter intake of 0.5% saccharin alone (MacFadyen et al., 2016). *See* Table 1.

2.3.6 Operant alcohol self-administration and reinstatement

Operant conditioning procedures used in studies on alcohol self-administration allow for independent evaluation of alcohol 'seeking' and consumption components, as well as alcohol relapse-like behavior. Studies in mice and rats have demonstrated OT treatment to significantly reduce alcohol responding and intake in operant alcohol self-administration models. For example, OT treatment (0.1–1.0 mg/kg; ip.) reduced alcohol responding (active lever responses) and intake (g/kg) in a dose-related manner in male C57BL/6J mice (King et al., 2017). In a separate cohort of mice, doses of OT that reduced alcohol (12% v/v, unsweetened) self-administration did not alter responding or intake for sucrose (5% w/v). Further, using a progressive-ratio schedule (increasing the amount of work required to attain reinforcement), OT (0.3 mg/kg; ip.) significantly reduced breakpoint response values for alcohol but did not alter this metric for sucrose (King et al., 2017). Similarly, in male Sprague-Dawley rats, OT (0.3 mg/kg ip.) significantly reduced operant responding for alcohol (10%) containing gelatin while having a smaller effect in rats responding for control (alcohol-free) gelatin (MacFadyen et al., 2016). In another study, effects of OT delivered by different routes of administration on alcohol self-administration was compared in dependent (via chronic intermittent alcohol vapor exposure) and nondependent male Wistar rats. OT given by injection (0.125–1.0 mg/kg; ip.) or intranasally (0.25–1.0 mg/kg; in.) reduced alcohol self-administration in a dose-dependent manner in alcohol-dependent but not nondependent rats (Tunstall et al., 2019). Using a progressive-ratio schedule, dependent rats showed the expected higher breakpoint values compared to nondependent rats, and this effect was attenuated by OT (ip. or in.) treatment. Interestingly, injected (ip.) OT at the highest dose tested also reduced general locomotor activity and consumption of a saccharin solution, but these effects were not observed following in. administration of the neuropeptide (Tunstall et al., 2019). These investigators also showed that direct (icv.) infusion of the OT receptor agonist PF-06655075 reduced alcohol self-administration but systemic administration of this agonist (which does not cross the blood-brain barrier) did not produce this effect.

Several studies have also shown OT treatment to attenuate the ability of stress to trigger relapse-like responding. In these studies, once stable operant oral alcohol self-administration behavior is established, animals are tested under extinction conditions (responding no longer results in delivery of reinforcement). Once responding is substantially decreased below baseline levels, animals are challenged with stress and reinstatement of alcohol responding ('seeking') is then assessed. Systemic administration of OT (0.1–1.0 mg/kg; ip.) reduced this reinstatement (relapse) responding in male and female C57BL/6J mice following exposure to a pharmacological stressor yohimbine (YOH, alpha-2 adrenoreceptor antagonist yohimbine) or the predator odor 2,3,5-trimethyl-3-thiazoline (TMT) (King & Becker, 2019). The effect was dose-related in both sexes, although there was a trend for female mice to exhibit greater sensitivity to OT treatment. In a model of PTSD-AUD comorbidity, sensitized stress (TMT)-induced alcohol relapse-like behavior in mice with a history of chronic multimodal stress exposure was blocked by injection of OT (1 mg/kg; ip.) prior to the reinstatement test session (Becker et al., 2023). Recently, it was shown that direct infusion of OT into the CeA reduced YOH-induced reinstatement of alcohol 'seeking' behavior in male (Ballas et al., 2021) and female

(Wilfur et al., 2023) Sprague-Dawley rats. Collectively, these results indicate a role for OT in the regulation of alcohol self-administration and, owing to its stress-buffering effects, susceptibility to stress-induced relapse-like behavior. There is also evidence to suggest that these effects are mediated, at least in part, by OT signaling in the CeA. *See* Table 1.

2.3.7 Dependence models

Alcohol-dependent individuals experience negative physiological and psychological side effects upon attenuation of alcohol consumption, a phenomenon known as withdrawal. This aversive experience often results in relapse and continuation of the addiction cycle (Becker & Lopez, 2024; Koob & Volkow, 2016). Animal models of alcohol dependence often include intermittent periods of withdrawal that promote increasingly higher levels of voluntary alcohol consumption. In rat models, operant self-administration is often combined with ethanol vapor exposure to assess alcohol-seeking behavior after intermittent high alcohol exposure. In alcohol-dependent rats exclusively, alcohol seeking, consumption (fixed-ratio) and motivation (progressive-ratio) for alcohol were attenuated by systemic (0.25, 0.50, and 1 mg/kg ip.), intranasal (1 mg/kg, in.), and central (icv.) administration of OT. It is important to note that there was a slight decrease in locomotion in response to ip. OT administration only (Tunstall et al., 2019). Additionally, OT administration (10 nM icv.) blunted cue-induced reinstatement of alcohol seeking only in alcohol-dependent male Wistar rats (Hansson et al., 2018). As few studies have explored the interaction between alcohol dependence and the OT system, this research area requires further development. *See* Table 1.

The above research supports the involvement of the OT system in multiple models of preclinical AUD paradigms including binge-like drinking, alcohol preference, operant alcohol self-administration, and alcohol dependence. These findings indicate OT as a promising therapeutic option for AUD. However, these studies predominately examine male animals and including female subjects is necessary to identify potential sex differences.

2.3.8 Clinical studies

OT has gained attention in clinical trials as a pharmacological intervention for reducing multiple facets of harmful alcohol use. This section aims to synthesize current clinical research on OT's role in mitigating factors that may promote excessive alcohol use and/or relapse.

In exploring potential treatments for AUD, researchers have evaluated the impact of OT on neural response reactivity to alcohol cues using functional magnetic resonance imaging (fMRI). In a pilot study, adult males characterized as heavy social drinkers received a single dose of OT (24 International Units (IU), in.), and neural activity was assessed via fMRI. This treatment reduced alcohol cue reactivity in the insular cortex (IC), hippocampal/parahippocampal formation, anterior cingulate cortex (Cg), inferior and medial frontal gyrus, and visual motor regions (Hansson et al., 2018). Although self-reported changes in alcohol craving were not measured, the study indicates that OT may modulate neural activity in response to alcohol cues. Bach et al. (2019) found that a single dose of OT (24 IU, in.) reduced NAc connectivity in male social drinkers and this correlated with alcohol craving scores (Bach et al., 2019). To further understand how OT may influence alcohol drinking, the same research group evaluated the effect of OT on neural responses to negative social-emotional stimuli, correlating these responses with changes in alcohol craving and consumption. Following a single administration of OT (24 IU, in.), fMRI data showed reduced activation of the AMY, which was associated with decreased alcohol craving and fewer heavy drinking days. This dose also reduced activation of the frontal gyri and parietal lobe during the task. Based on these results, the authors propose that OT may influence drinking behaviors by reducing anxiety (Bach et al., 2021).

The previously described experiments were conducted in non-alcohol dependent individuals.

AUD involves various stages, each contributing to the persistent challenge of achieving and maintaining abstinence. One critical stage is withdrawal, during which OT emerges as a potential therapeutic tool to alleviate aversive symptoms experienced. When given twice a day for three days, OT (24IU, in.) reduced the amount of lorazepam needed for detoxification and decreased patient scores on various clinical assessments (Clinical Institute Withdrawal Assessment for Alcohol (CIWA)) and Alcohol Withdrawal Symptom Checklist (AWSC), Alcohol Craving Visual Analog Scale (ACVAS) and the Profile of Mood States (POM) (Pedersen et al., 2013). These results imply a reduction in alcohol craving post-OT treatment, aligning with previous findings (Bach et al., 2021; Hansson et al., 2018). This evidence supports OT as a potential target for preventing relapse driven by aversive symptoms during withdrawal and suggests that OT could serve as an alternative treatment for benzodiazepines during assisted withdrawal/detoxification. Another symptom observed in alcohol-dependent individuals is emotional dysregulation, which may act as a catalyst for alcohol drinking and relapse. A clinical tool used to evaluate emotional stimuli processing, the reading the mind in the eyes test (RMET), is a common assessment for emotional dysregulation. Markedly, individuals with AUD perform worse on the RMET (Onuoha et al., 2016). Despite previous data indicating that OT can enhance RMET scores in alcohol-dependent individuals, OT had no significant effect on RMET scores of alcohol-dependent individuals during withdrawal. However, preliminary data posits that OT may enhance performance on the RMET negative emotion subscale during withdrawal. In other words, OT may improve social cognition in the context of negative emotions (Melby et al., 2022). The same group evaluated the effects of self-administered OT in alcohol-dependent patients post-detoxification. 4 weeks of patient-administered OT (8IU, in. up to 3 times/day) had no effect on alcohol intake, number of days to relapse, or proportion of subjects relapsing; however, results indicated that OT reduced self-reported nervousness scores (Melby et al., 2021). These results convey that this dose of OT may not be efficient in preventing relapse but does seem to modulate negative affect.

Post-traumatic stress disorder (PTSD) and AUD are highly comorbid, and OT is a prospective treatment. Flanagan and colleagues (Flanagan et al., 2019) examined the effect of OT on stress-reactivity (cortisol) in male patients diagnosed with co-occurring PTSD/AUD using the Trier Social Stress Task (TSST). Results indicated that a single dose of OT (40 IU, in.) slightly reduced cortisol reactivity following the stress task. Interestingly, this reduction was greater in individuals with higher baseline cortisol levels. However, this dosing regimen had no effect on subjective ratings of alcohol craving (Flanagan et al., 2019). Further, individuals diagnosed with both PTSD and AUD showed no change in hyperarousal or stress reactivity after receiving a single administration of OT (20IU, in.), as measured by startle reactivity and surprisingly, 40IU in. enhanced hyperarousal (Stauffer et al., 2022). Stress, alcohol craving, and alcohol withdrawal symptoms are significant contributors to alcohol use and/or relapse. Current literature indicates that OT holds promise as a therapeutic intervention for reducing these symptoms. While some studies yielded unexpected results, they enrich existing knowledge and aid in discerning effective OT dosing. This prompts further exploration into potential variations in treatment efficacy among individuals characterized as heavy social drinkers and those meeting diagnostic criteria for AUD, as well as individuals experiencing anxiety and negative affect. *See* Table 2.

Table 2 Oxytocin in clinical studies.

Authors	Year	Subjects	Outcome Measure	Treatment	Results
Pederson et al.	2013	Treatment seeking alcohol-dependent patients	CIWA, AWSC, PACS, ACVAS, POMS	2/day x 3days of OT (24 IU, in.) during withdrawal	↓ symptoms of alcohol withdrawal
Hansson et al.	2018	Non-treatment seeking heavy social drinking males	Alcohol cue-induced neural activation (fMRI)	Single administration of OT (24 IU, in.)	↓ neural activity to alcohol-related cues in multiple brain regions
Bach et al.	2019	Male Social Drinkers	Alcohol cue-reactivity task (fMRI)	Single administration of OT (24 IU, in.)	↓ NAc connectivity related to subjective alcohol craving scores
Flanagan et al.	2019	Males diagnosed with co-occurring PTSD and AUD	Cortisol changes in response to TSST	Single administration of OT (40IU, in.)	↓ cortisol in response to TSST, greater reduction in individuals with higher baseline cortisol levels.No effect alcohol craving
Bach et al.	2021	Non-treatment seeking heavy social drinking males	Face-matching task induced neural activation (fMRI)	Single administration of OT (24IU, in.)	↓ activation in the bilateral amygdala associated with lower alcohol craving and lower percentage of heavy drinking days.↓ activation in the frontal gyri and parietal lobe.

Melby et al.	2021	Alcohol dependent males and females post-detoxification	*Primary:* alcohol intake, number of days to relapse, proportion of subjects relapsing *Secondary:* self-reported alcohol craving, sleep and mental distress	OT (8IU, in.) self-administered as needed up to 3/day x 4weeks	No difference in daily alcohol intake, number of drinking days or proportion of subjects relapsing.Larger reduction in self-reported nervousness
Melby et al.	2022	Treatment seeking alcohol-dependent males and females	RMET	2/day x 3days of OT (24 IU, in.) during withdrawal	No effect on RMET scores. Preliminary trend toward enhancing RMET negative emotion subset scores.
Stauffer et al.	2022	Males and Females diagnosed with co-occurring PTSD-AUD	Startle reactivity	Single administration of OT (20IU or 40IU, in.)	20IU had no effect on startle reactivity, 40IU ↑ startle reactivity

International Unit (IU), Clinical Institute Withdrawal Assessment (CIWA), assisted, Alcohol Withdrawal Symptoms Checklist (AWSC), Penn Alcohol Craving Scale (PACS), Alcohol Craving Visual Analog Scale (ACVAS), Profile of Mood States (POM), Reading the Mind in the Eyes Test (RMET), Trier Social Stress Task (TSST).

3. The vasopressin (VP) system

Vasopressin (VP) was initially recognized as a key component of the endocrine system, regulating osmotic balance, body temperature, blood pressure and HPA axis activity (Bilezikjian & Vale, 1987; Rigney et al., 2022; Stoop, 2012). More recently, VP has been implicated in many social behaviors, including pair-bonding, social recognition, aggression, and parental behaviors (Johnson & Young, 2017; Nair & Young, 2006; Rigney et al., 2022; Walum & Young, 2018; Williams Avram & Cymerblit-Sabba, 2017). In parallel with OT, targeting the VP system represents a growing field of research in the treatment of AUD.

3.1 Vasopressin (VP) neurons

Much like OT neurons, VP-producing neurons are largely found in the hypothalamus (PVN, SON, SCN, AN) and are of the magnocellular or parvocellular type with projections that extend centrally and/or to the pituitary gland. To a lesser extent, VP neurons are also found in the BNST, medial AMY, anterior hypothalamus, and preoptic area (Harper et al., 2018; Rigney et al., 2022; Williams Avram & Cymerblit-Sabba, 2017). Neurosecretory magnocellular VP neurons release VP from large dense core vesicles into the bloodstream via fenestrated capillaries in the posterior pituitary. Parvocellular VP neurons project to the anterior pituitary and modulate the release of other hormones (e.g., adrenocorticotropin). These non-neurosecretory parvocellular neurons also project to the spinal cord (Harper et al., 2018; Rigney et al., 2022). Additionally, PVN VP neurons have major projections to the cortex, thalamus, striatum, HPC, and AMY (Rigney et al., 2022). Central VP is often released in response to physiological stressors (e.g., changes in plasma osmolality, changes blood pressure/volume, and pain) and behavioral stressors (Douglas, 2005; Garrahy & Thompson, 2019).

3.2 The VP receptor (VPR)

VP exerts its effects through three identified G-protein coupled vasopressin receptors (VPRs): vasopressin receptor 1a (V1a), vasopressin receptor 1b (V1b) and vasopressin receptor 2 (V2). The V2 receptor subtype is involved in maintaining osmotic homeostasis, while the V1 receptor subtypes regulate the behavioral effects of VP (Rigney et al., 2022). Both V1a and V1b receptor subtypes (V1aR, V1bR) activate the PI3K pathway and are expressed in the frontal cortex, piriform cortex, lateral septum, HPC,

PVN, BNST, NAc, and AMY. In addition to overlapping regions of expression, V1aRs are present in the anterior olfactory nucleus, SCN, ventromedial hypothalamic nucleus, arcuate nucleus, lateral habenula, cerebellum, ventral pallidum, and ventrolateral hypothalamus and V1bRs are present in the cingulate cortex, entorhinal cortex, parietal cortex, caudate putamen, and lower brainstem (Harper et al., 2018; Williams Avram & Cymerblit-Sabba, 2017). Compared to the broad expression of V1Rs, V2Rs are primarily found in the cerebellum and periphery where activation stimulates the cyclic adenosine 3,5-monophosphate (cAMP) pathway (Harper et al., 2018). Preclinical and clinical evidence primarily support a relationship between alcohol and the V1bR subtype (discussed below). VP mediates both endocrine regulation and social behaviors via diverse neural circuits. Exploration of these mechanisms will enhance our understanding of VP's contribution to both normal physiology and pathological conditions, including its involvement in alcohol drinking behavior.

3.3 VP—alcohol interactions

3.3.1 Effects of alcohol on the VP system

Research suggests that genetic differences in VP can predict high alcohol preference. For example, an early study demonstrated that the alcohol-preferring (P) rat strain and high alcohol drinking (HAD) strain expressed higher levels of VP mRNA in the PVN compared to controls, with no significant change observed in the SON (Hwang et al., 1998). A more recent study examined VP mRNA expression in both alcohol naïve and alcohol exposed P rats, as well as alcohol non-preferring (NP) rats. Higher VP mRNA expression levels were detected in the AMY and HPC of alcohol naïve P rats compared to NP rats (Zhou et al., 2011). To expand upon this, Zhou and colleagues designed a study to determine if voluntary alcohol exposure (2-bottle choice, 17-days) differentially influences VP mRNA expression in preferring versus non-preferring rats. Compared to alcohol naïve P rats, alcohol exposure reduced AMY and HPC VP expression. Interestingly, no difference in VP expression was detected between alcohol naïve and alcohol-exposed NP rats, suggesting a neurobiological adaptation specific to animals with a genetic predisposition to consume alcohol (Zhou et al., 2011). VP expression has also been investigated in alcohol-preferring C57BL mice. A reduction in hypothalamic (PVN, SON, lateral hypothalamic area, anterior hypothalamic area) and extrahypothalamic (BNST, MPOA) VP expression was detected following a 7-day alcohol exposure paradigm shown to produce dependence in mice (Gulya et al., 1991, 1993). In a more recent study, C57BL/6J mice categorized as "high drinkers" showed elevated VP mRNA expression in the AMY; however, 10 days of binge alcohol exposure (20% alcohol in water, intragastric gavage) showed no effect on AMY VP expression (Nelson et al., 2018). These data indicate elevated VP expression in various brain regions of rodents genetically predisposed to prefer alcohol. In rats, this is followed by a reduction in expression. Notably, this is consistent with clinical literature. In humans, plasma VP (pVP) measurements are often used as a proxy for central VP expression. Early clinical studies indicate that alcohol initially decreases pVP levels, followed by an increase in pVP after cessation of use (Eisenhofer & Johnson, 1982; Gill et al., 1982; Harper et al., 2018; Helderman et al., 1978; Mander et al., 1989). It is important to acknowledge that these changes could be associated with hydration state and plasma measurements of neuropeptides do not directly correlate to central changes.

The preceding preclinical and clinical studies focus on VP expression; however, it is crucial to also consider the impact of alcohol on VP receptor expression. As such, Edwards, and colleagues (2012) examined the interaction between alcohol dependence and V1bR expression in the AMY of male Wistar rats. Following acquisition of operant alcohol self-administration, animals were exposed to ethanol vapor to induce dependence. An initial reduction in V1bR expression following exposure to alcohol was reversed upon establishment of alcohol dependence, highlighting the possible involvement of V1bR in the progression toward alcohol dependence (Edwards et al., 2012). Furthermore, in Sprague-Dawley rats, adolescent intermittent ethanol (AIE) exposure produced sex-specific changes in adult V1bR expression. In males, alcohol consumption in a social setting enhanced hypothalamic V1bR expression compared to controls, independent of adolescent exposure. In adult females, alcohol drinking increased hypothalamic VP and V1bR expression, but only in animals with no adolescent alcohol exposure (Towner et al., 2022). Interestingly, in another study AIE increased V1bR neuronal surface expression. This effect was specific to males (Dannenhoffer et al., 2018). These data suggest sex-specific effects of AIE on hypothalamic V1bR expression. In summary, evidence suggests that alcohol exposure modulates V1bR expression, yet further research is warranted.

Several studies have shown that VP signaling perturbation modulates tolerance to several effects of chronic alcohol exposure. Studies have shown that alcohol tolerance is prolonged by systemic (Hoffman et al., 1978; Hoffman & Tabakoff, 1984) and icv. (Hung et al., 1984) administration of VP in male C57BL mice. Similar results were obtained by injections (sc.) of a VP analog (DGAVP) in Swiss male mice (Rigter et al., 1980) and central (icv.) administration of a V1bR agonist in C57BL/6NCR mice (Szabo et al., 1988). Interestingly, there was no difference in ethanol induced motor coordination or in the hypothermic effects of alcohol In VPR1a/b knockout mice (Caldwell et al., 2006). Another study demonstrated that icv. administration of VP blocked the acquisition of tolerance to the sedative and hypothermic effects of alcohol in male C57BL mice; an effect that could be related to learned associations between repeated alcohol injections (ip.) and environmental cues (Mannix et al., 1986). *See* Table 3.

3.3.2 Effects of VP on alcohol consummatory behavior

3.3.2.1 Preclinical studies

3.3.2.1.1 2-Bottle choice drinking. Modulating the VP system facilitates changes in alcohol consummatory behavior as demonstrated by studies utilizing the selective V1bR antagonist, SSR149415 (Nelivaptan). In a 2-bottle choice paradigm (alcohol 10% v/v vs. water), SSR149415 was administered (3, 10 and 30 mg/kg ip.) to determine the effects on alcohol consumption in male Sardinian alcohol preferring (sP) rats. Although treatment decreased both alcohol consumption (SSR149415 30 mg/kg) and alcohol preference ratio (SSR149415 10 and 30 mg/kg), increased water intake was noted which could suggest a homeostatic mechanism due to the role of VP in water balance (Zhou et al., 2011). The same group conducted a similar study in male and female C57BL/6J mice. Mice underwent chronic intermittent alcohol exposure characterized by 24-h access to ethanol (7.5%, 15% or 30% v/v) every other day for 3-weeks, followed by administration of SSR149415 (1, 3, 10, 30 mg/kg). In both males and females, SSR149415 decreased alcohol (15% v/v) intake and preference in a dose-dependent manner. An increase in water consumption was observed following administration of 10 mg/kg SSR149415. No effect of SSR149415 (10 mg/kg) on alcohol at concentrations of 7.5% or 30% v/v were found. There was also no effect on sucrose or saccharin intake or preference (Zhou et al., 2018). Despite promising pharmacological data, Caldwell and colleagues (2006) reported no change in voluntary alcohol consumption or acute intoxication effects of ethanol (e.g., motor coordination, hypothermia, mood) following targeted disruption of V1aR or V1bR in male and female mice compared to wild-type littermates (Caldwell et al., 2006). *See* Table 3.

Table 3 Vasopressin in preclinical studies.

Authors	Year	Species	Drinking model	Treatment	Effect on alcohol-related behavior
Caldwell et al.	2006	Male and Female VPR knockout mice (Avpr1a-/-; Avpr1b -/-)	2-Bottle Choice	—	No effect on alcohol consumption
Zhou et al.	2011	Male P Rats	2-Bottle Choice	3–30 mg/kg SSR149415 ip.	↓ alcohol intake (30 mg/kg) and preference (10 & 30 mg/kg)
Edwards et al.	2012	Male Wistar Rats	Operant Alcohol Self-Administration +/− Ethanol Vapor Exposure	10, 20 & 30 mg/kg SSR149415 ip.	↓ alcohol responding in alcohol-dependent rats, dose-dependently
Zhou et al.	2018	Male and Female C57BL/6J Mice	Chronic Intermittent Access	1–30 mg/kg SSR149415 ip.	↓ alcohol intake and preference dose-dependently
Harper et al.	2019	Male Wistar or Sprague-Dawley Rats	Chronic Intermittent Ethanol Exposure (CIE)	**A.** 1.5 ng SSR149415 intra-CeA **B.** 4.34 μg/.5 μL or 0.434 μg/.5 μL VP intra-CeA	**A.** ↓ anxiety-like behavior during withdrawal **B.** ↑ anxiety-like behavior during withdrawal

3.3.3 *Operant alcohol self-administration*

The effects of V1bR antagonism on operant alcohol self-administration are consistent with findings from 2-bottle choice drinking studies. Edwards et al. (2012) investigated the effects of SSR149415 on operant oral alcohol self-administration (alcohol 10% v/v, FR1, 30 min/day) in male Wistar rats. In alcohol-dependent animals, SSR149415 (10, 20 and 30 mg/kg ip.) reduced alcohol seeking in a dose-dependent manner reaching statistical significance at 30 mg/kg ip. There was no effect of the compound on alcohol self-administration in non-dependent animals, which could indicate sensitization in alcohol-dependent animals. The authors observed no change in water consumption in either group of animals (Edwards et al., 2012). *See* Table 3.

The above pharmacological data implicates V1bR signaling in alcohol seeking and consumption behaviors, but it is also important to consider how modulating the VP system may affect withdrawal. As mentioned previously, anxiety is a risk factor for relapse, especially during withdrawal from chronic alcohol use, and the CeA VP system has been implicated in anxiety-like behaviors during withdrawal (Dannenhoffer et al., 2018; Hernandez et al., 2016; Hernandez-Perez et al., 2018). Harper et al. (2019) examined the effects of VP antagonism in the CeA on anxiety like behaviors during CIE-induced withdrawal and demonstrated that intra-CeA.

infusion of SSR149415 (1.5 ng) suppressed anxiety-related phenotypes in male rats, as measured by the social interaction test. Expectedly, intra-CeA AVP infusion induced anxiety in this model (Harper et al., 2019). As such, V1bR signaling appears to influence both alcohol consummatory and stress-related behavior, suggesting that pharmacological antagonism of this receptor may be an effective treatment to reduce alcohol consumption and prevent relapse. However, further research is warranted to determine the precise mechanism of SSR149415. For instance, this compound showed minimal brain penetrance in a positron emission tomography (PET) study (Schonberger et al., 2010), which could suggest a peripheral function of V1bRs in the behavioral studies discussed. Furthermore, the specificity of this compound must be examined, as there is evidence indicating that SSR-149415 may act as an OTR antagonist (Griffante et al., 2005). Additionally, many of these studies were conducted in male rodents. Further investigation of V1bR signaling in the context of alcohol-related behaviors is necessary to differentiate between central and peripheral effects, ascertain the specificity of V1bR antagonism, and identify any sex-differences in treatment outcome. *See* Table 3.

3.3.4 Clinical studies

Given the involvement of V1bR signaling in HPA axis activity, pharmacological antagonism of the V1bR is considered a prospective therapeutic strategy in clinical trials regarding stress-related behaviors. Akin to preclinical studies, SSR149415 has been utilized in clinical trials investigating generalized anxiety disorder and major depressive disorder; however, to the best of our knowledge, there have been no clinical trials evaluating the effectiveness of SSR149415 in the treatment of AUD. Nevertheless, a novel, selective, and orally active V1bR antagonist, ABT-436, was developed for use in humans. This compound was shown to reduce HPA activity in humans and had no negative interactions with a single dose of alcohol, indicating its potential for continued exploration in AUD clinical trials (Katz, Liu, et al., 2016; Katz, Locke, et al., 2016). A separate clinical trial including male and female patients diagnosed with alcohol dependence demonstrated that 12-weeks of ABT-436 significantly increased the percentage of days abstinent compared to placebo. A reduction in the percentage of heavy drinking days was observed; however, this outcome was not statistically significant. Strikingly, individuals reporting higher stress levels at the outset responded better compared to placebo, suggesting that this compound may be more effective at reducing heavy drinking in individuals with heightened stress levels (Ryan et al., 2017).

The development of ABT-436 was discontinued for reasons nondisclosed, nonetheless, this study provides evidence supporting V1bR antagonism as a promising therapeutic target for AUD. Another novel V1bR antagonist, TS-121 (THY1773), is currently under clinical investigation for the treatment of major depressive disorder (Inatani et al., 2021; Kamiya et al., 2020). While this compound has not yet been utilized in alcohol-related clinical trials, the development of novel V1bR antagonists may aid in the advancement of AUD therapeutics. The clinical evidence discussed suggests the involvement of V1bR signaling in behaviors related to AUD and highlights V1bR antagonism as a promising pharmacological intervention for AUD.

4. Discussion

The profound public health repercussions of AUD and the varying success rates of existing treatments emphasize the need to identify more effective therapies for the disorder. Here we have presented results from a range of preclinical and clinical studies suggesting that the OT and VP systems hold promise as potential targets for pharmacological intervention in treating AUD. Although nuanced, activating the OT system and suppressing the VP system appears to yield beneficial outcomes on alcohol-related behaviors.

Historically recognized as a mediator of maternal physiology, recent research highlights the anxiolytic properties of oxytocin (OT) and shows that alcohol exposure alters OT and OTR expression in a brain region-specific, and possibly species-specific, manner. In turn, OT also influences alcohol tolerance, withdrawal-related symptoms, and alcohol consumption. These findings suggest a significant relationship between the OT system and alcohol, supporting OTRs as a potential therapeutic target for AUD.

Vasopressin (VP) is structurally similar to OT, but functionally mediates opposing effects. There appears to be a positive correlation between VP expression and alcohol preference. Further, alcohol exposure has been shown to reduce hypothalamic VP expression. Interestingly, this reduction in VP expression is reversed upon the development of alcohol dependence, indicating a distinction between moderate alcohol exposure and heavy chronic alcohol exposure on the VP system. There is evidence indicating that pharmacological antagonism of the V1bR reduces alcohol consumption and stress-related alcohol behaviors in preclinical models. However, gaining a deeper understanding of peripheral vs. central effects of this modulation is needed. While V1bR antagonism in clinical trials resulted in variable results, its ability to modulate HPA axis activity, promote abstinence, and marginally reduce heavy drinking days shows the potential for this VP receptor subtype as a promising therapeutic for treating AUD.

Although pharmacological manipulation of the OT and VP systems exhibit potential for treating AUD, several fundamental questions remain to be addressed. Despite indication of sexual dimorphisms in protein and receptor expression for both systems (Dumais & Veenema, 2016), there is a lack of research specifically exploring sex differences in treatment outcomes. Furthermore, the OT and VP clinical trials discussed in this review suggest differences in treatment effectiveness dependent on baseline stress levels of the subjects. In developing effective treatments for AUD, biological and environmental factors should be considered to better tailor treatment and enhance overall effectiveness. Therefore, further research into sex differences, and stress interactions will be necessary to fully understand the interaction between these peptide systems and alcohol-related behaviors. Additionally, the reviewed literature highlights numerous considerations often encountered when investigating novel pharmacological treatments, including peptide biochemistry and detection methods. The structural similarity between OT and VP facilitates crosstalk between these systems (Chini & Manning, 2007; Menon & Neumann, 2023; Song & Albers, 2018), thereby reinforcing the importance of considering their joint influence on individual experiments. Current clinical detection methods for peptides include blood/plasma levels; however, it has been established that peripheral levels of peptides do not always correlate to central levels (Grinevich & Neumann, 2021). Additionally, lacking a standardized detection method results in decreased comparability between studies. As such, it will be necessary to generate novel and standardized peptide detection methods (Tabak et al., 2023). Brain penetrance represents an additional challenge when considering the use of OT as a therapeutic agent. Rather than crossing the blood-brain barrier in amounts necessary to elicit a behavioral response, OT was traditionally believed to stimulate endogenous OT production through a feedforward mechanism (King et al., 2020). Several studies have investigated this important issue. Central infusion of OT has been shown to influence behavior and peripheral administration activates OT neurons in the PVN. However, detectable concentrations of central OT following systemic administration are variable. While research supports increased central OT levels following peripheral administration, the process by which this occurs has yet to be determined (King et al., 2020).

Overall, there is a convergent body of evidence that suggests the OT and VP neuropeptide systems contribute to behaviors associated with alcohol use disorder. This has spurred increased attention and investigation into their potential as therapeutic targets for AUD. Further studies, both preclinical and clinical, are needed to further define the optimal conditions in which drug development targeting these peptide systems will be most effective in treating AUD.

Acknowledgments

Supported by NIH grants P50 AA010761 (HCB), R01 AA026536 (HCB), U01 AA014095 (HCB), U24 AA029968 (MFL), NIAAA T32 AA007474 (SMS), and a grant from the Department of Veterans Affairs (BLRD BX000813) (HCB).

References

Bach, P., Koopmann, A., Bumb, J. M., Zimmermann, S., Buhler, S., Reinhard, I., ... Kiefer, F. (2021). Oxytocin attenuates neural response to emotional faces in social drinkers: An fMRI study. *European Archives of Psychiatry and Clinical Neuroscience, 271*(5), 873–882. https://doi.org/10.1007/s00406-020-01115-0.

Bach, P., Reinhard, I., Buhler, S., Vollstadt-Klein, S., Kiefer, F., & Koopmann, A. (2019). Oxytocin modulates alcohol-cue induced functional connectivity in the nucleus accumbens of social drinkers. *Psychoneuroendocrinology, 109*, 104385. https://doi.org/10.1016/j.psyneuen.2019.104385.

Bahi, A., Al Mansouri, S., & Al Maamari, E. (2016). Nucleus accumbens lentiviral-mediated gain of function of the oxytocin receptor regulates anxiety- and ethanol-related behaviors in adult mice. *Physiology & Behavior, 164*(Pt A), 249–258. https://doi.org/10.1016/j.physbeh.2016.06.009.

Ballas, H. S., Wilfur, S. M., Freker, N. A., & Leong, K. C. (2021). Oxytocin attenuates the stress-induced reinstatement of alcohol-seeking in male rats: Role of the central amygdala. *Biomedicines, 9*(12), https://doi.org/10.3390/biomedicines9121919.

Baskerville, T. A., & Douglas, A. J. (2010). Dopamine and oxytocin interactions underlying behaviors: Potential contributions to behavioral disorders. *CNS Neuroscience & Therapeutics, 16*(3), e92–e123. https://doi.org/10.1111/j.1755-5949.2010.00154.x.

Becker, H. C., & Lopez, M. F. (2024). Animal models of excessive alcohol consumption in rodents. *Current Topics in Behavioral Neurosciences.* https://doi.org/10.1007/7854_2024_461.

Becker, H. C., Lopez, M. F., King, C. E., & Griffin, W. C. (2023). Oxytocin reduces sensitized stress-induced alcohol relapse in a model of posttraumatic stress disorder and alcohol use disorder comorbidity. *Biological Psychiatry, 94*(3), 215–225. https://doi.org/10.1016/j.biopsych.2022.12.003.

Bilezikjian, L. M., & Vale, W. W. (1987). Regulation of ACTH secretion from corticotrophs: The interaction of vasopressin and CRF. *Annals of the New York Academy of Sciences, 512*, 85–96. https://doi.org/10.1111/j.1749-6632.1987.tb24952.x.

Bowen, M. T., Carson, D. S., Spiro, A., Arnold, J. C., & McGregor, I. S. (2011). Adolescent oxytocin exposure causes persistent reductions in anxiety and alcohol consumption and enhances sociability in rats. *PLoS One, 6*(11), e27237. https://doi.org/10.1371/journal.pone.0027237.

Brown, C. H., Bains, J. S., Ludwig, M., & Stern, J. E. (2013). Physiological regulation of magnocellular neurosecretory cell activity: Integration of intrinsic, local and afferent mechanisms. *Journal of Neuroendocrinology, 25*(8), 678–710. https://doi.org/10.1111/jne.12051.

Burkett, J. P., & Young, L. J. (2012). The behavioral, anatomical and pharmacological parallels between social attachment, love and addiction. *Psychopharmacology (Berl), 224*(1), 1–26. https://doi.org/10.1007/s00213-012-2794-x.

Busnelli, M., & Chini, B. (2018). Molecular basis of oxytocin receptor signalling in the brain: What we know and what we need to know. *Current Topics in Behavioral Neurosciences, 35*, 3–29. https://doi.org/10.1007/7854_2017_6.

Busnelli, M., Sauliere, A., Manning, M., Bouvier, M., Gales, C., & Chini, B. (2012). Functional selective oxytocin-derived agonists discriminate between individual G protein family subtypes. *The Journal of Biological Chemistry, 287*(6), 3617–3629. https://doi.org/10.1074/jbc.M111.277178.

Caldwell, H. K., Stewart, J., Wiedholz, L. M., Millstein, R. A., Iacangelo, A., Holmes, A., ... Wersinger, S. R. (2006). The acute intoxicating effects of ethanol are not dependent on the vasopressin 1a or 1b receptors. *Neuropeptides, 40*(5), 325–337. https://doi.org/10.1016/j.npep.2006.08.001.

Caruso, M. A., Robins, M. T., Fulenwider, H. D., & Ryabinin, A. E. (2021). Temporal analysis of individual ethanol consumption in socially housed mice and the effects of oxytocin. *Psychopharmacology (Berl), 238*(3), 899–911. https://doi.org/10.1007/s00213-020-05741-3.

Chini, B., & Manning, M. (2007). Agonist selectivity in the oxytocin/vasopressin receptor family: New insights and challenges. *Biochemical Society Transactions, 35*(Pt 4), 737–741. https://doi.org/10.1042/BST0350737.

Dannenhoffer, C. A., Kim, E. U., Saalfield, J., Werner, D. F., Varlinskaya, E. I., & Spear, L. P. (2018). Oxytocin and vasopressin modulation of social anxiety following adolescent intermittent ethanol exposure. *Psychopharmacology (Berl), 235*(10), 3065–3077. https://doi.org/10.1007/s00213-018-5003-8.

Douglas, A. J. (2005). Chapter 2.5—Vasopressin and oxytocin. In T. Steckler, N. H. Kalin, & J. M. H. M. Reul (Vol. Eds.), *Techniques in the behavioral and neural sciences: Vol. 15*, (pp. 205–229). (pp. 205) Elsevier. https://doi.org/10.1016/S0921-0709(05)80014-8.

Dumais, K. M., & Veenema, A. H. (2016). Vasopressin and oxytocin receptor systems in the brain: Sex differences and sex-specific regulation of social behavior. *Frontiers in Neuroendocrinology, 40*, 1–23. https://doi.org/10.1016/j.yfrne.2015.04.003.

Edwards, S., Guerrero, M., Ghoneim, O. M., Roberts, E., & Koob, G. F. (2012). Evidence that vasopressin V1b receptors mediate the transition to excessive drinking in ethanol-dependent rats. *Addiction Biology, 17*(1), 76–85. https://doi.org/10.1111/j.1369-1600.2010.00291.x.

Eisenhofer, G., & Johnson, R. H. (1982). Effect of ethanol ingestion on plasma vasopressin and water balance in humans. *The American Journal of Physiology, 242*(5), R522–R527. https://doi.org/10.1152/ajpregu.1982.242.5.R522.

Flanagan, J. C., Allan, N. P., Calhoun, C. D., Badour, C. L., Moran-Santa Maria, M., Brady, K. T., & Back, S. E. (2019). Effects of oxytocin on stress reactivity and craving in veterans with co-occurring PTSD and alcohol use disorder. *Experimental and Clinical Psychopharmacology, 27*(1), 45–54. https://doi.org/10.1037/pha0000232.

Garrahy, A., & Thompson, C. J. (2019). Vasopressin. In I. Huhtaniemi, & L. Martini (Eds.). *Encyclopedia of endocrine diseases* (pp. 29–35)(2nd ed.). Academic Press. https://doi.org/10.1016/B978-0-12-801238-3.65216-9.

Gill, G. V., Baylis, P. H., Flear, C. T., Skillen, A. W., & Diggle, P. H. (1982). Acute biochemical responses to moderate beer drinking. *British Medical Journal (Clinical Research Edition), 285*(6357), 1770–1773. https://doi.org/10.1136/bmj.285.6357.1770.

Gimpl, G., & Fahrenholz, F. (2001). The oxytocin receptor system: Structure, function, and regulation. *Physiological Reviews, 81*(2), 629–683. https://doi.org/10.1152/physrev.2001.81.2.629.

Griffante, C., Green, A., Curcuruto, O., Haslam, C. P., Dickinson, B. A., & Arban, R. (2005). Selectivity of d[Cha4]AVP and SSR149415 at human vasopressin and oxytocin receptors: Evidence that SSR149415 is a mixed vasopressin V1b/oxytocin receptor antagonist. *British Journal of Pharmacology, 146*(5), 744–751. https://doi.org/10.1038/sj.bjp.0706383.

Grinevich, V., & Ludwig, M. (2021). The multiple faces of the oxytocin and vasopressin systems in the brain. *Journal of Neuroendocrinology, 33*(11), e13004. https://doi.org/10.1111/jne.13004.

Grinevich, V., & Neumann, I. D. (2021). Brain oxytocin: How puzzle stones from animal studies translate into psychiatry. *Molecular Psychiatry, 26*(1), 265–279. https://doi.org/10.1038/s41380-020-0802-9.

Gulya, K., Dave, J. R., & Hoffman, P. L. (1991). Chronic ethanol ingestion decreases vasopressin mRNA in hypothalamic and extrahypothalamic nuclei of mouse brain. *Brain Research, 557*(1–2), 129–135. https://www.ncbi.nlm.nih.gov/pubmed/1747746.

Gulya, K., Orpana, A. K., Sikela, J. M., & Hoffman, P. L. (1993). Prodynorphin and vasopressin mRNA levels are differentially affected by chronic ethanol ingestion in the mouse. *Brain Research. Molecular Brain Research, 20*(1–2), 1–8. https://doi.org/10.1016/0169-328x(93)90105-x.

Hansson, A. C., Koopmann, A., Uhrig, S., Buhler, S., Domi, E., Kiessling, E., ... Spanagel, R. (2018). Oxytocin reduces alcohol cue-reactivity in alcohol-dependent rats and humans. *Neuropsychopharmacology: Official Publication of the American College of Neuropsychopharmacology, 43*(6), 1235–1246. https://doi.org/10.1038/npp.2017.257.

Hansson, A. C., & Spanagel, R. (2021). No changes in the oxytocin system in alcohol-dependent female rodents and humans: Towards a sex-specific psychopharmacology in alcoholism. *Addiction Biology, 26*(2), e12945. https://doi.org/10.1111/adb.12945.

Harper, K. M., Knapp, D. J., Butler, R. K., Cook, C. A., Criswell, H. E., Stuber, G. D., & Breese, G. R. (2019). Amygdala arginine vasopressin modulates chronic ethanol withdrawal anxiety-like behavior in the social interaction task. *Alcoholism, Clinical and Experimental Research, 43*(10), 2134–2143. https://doi.org/10.1111/acer.14163.

Harper, K. M., Knapp, D. J., Criswell, H. E., & Breese, G. R. (2018). Vasopressin and alcohol: A multifaceted relationship. *Psychopharmacology (Berl), 235*(12), 3363–3379. https://doi.org/10.1007/s00213-018-5099-x.

Helderman, J. H., Vestal, R. E., Rowe, J. W., Tobin, J. D., Andres, R., & Robertson, G. L. (1978). The response of arginine vasopressin to intravenous ethanol and hypertonic saline in man: The impact of aging. *Journal of Gerontology, 33*(1), 39–47. https://doi.org/10.1093/geronj/33.1.39.

Hernandez, V. S., Hernandez, O. R., Perez De La Mora, M., Gomora, M. J., Fuxe, K., Eiden, L. E., & Zhang, L. (2016). Hypothalamic vasopressinergic projections innervate central amygdala GABAergic neurons: Implications for anxiety and stress coping. *Frontiers in Neural Circuits, 10*, 92. https://doi.org/10.3389/fncir.2016.00092.

Hernandez-Perez, O. R., Crespo-Ramirez, M., Cuza-Ferrer, Y., Anias-Calderon, J., Zhang, L., Roldan-Roldan, G., ... Perez de la Mora, M. (2018). Differential activation of arginine-vasopressin receptor subtypes in the amygdaloid modulation of anxiety in the rat by arginine-vasopressin. *Psychopharmacology (Berl), 235*(4), 1015–1027. https://doi.org/10.1007/s00213-017-4817-0.

Hoffman, P. L., Ritzmann, R. F., Walter, R., & Tabakoff, B. (1978). Arginine vasopressin maintains ethanol tolerance. *Nature, 276*(5688), 614–616. https://doi.org/10.1038/276614a0.

Hoffman, P. L., & Tabakoff, B. (1984). Neurohypophyseal peptides maintain tolerance to the incoordinating effects of ethanol. *Pharmacology, Biochemistry, and Behavior, 21*(4), 535–543. https://www.ncbi.nlm.nih.gov/pubmed/6504951.

Hung, C. R., Tabakoff, B., Melchior, C. L., & Hoffman, P. L. (1984). Intraventricular arginine vasopressin maintains ethanol tolerance. *European Journal of Pharmacology, 106*(3), 645–648. https://doi.org/10.1016/0014-2999(84)90072-4.

Hwang, B. H., Froehlich, J. C., Hwang, W. S., Lumeng, L., & Li, T. K. (1998). More vasopressin mRNA in the paraventricular hypothalamic nucleus of alcohol-preferring rats and high alcohol-drinking rats selectively bred for high alcohol preference. *Alcoholism, Clinical and Experimental Research, 22*(3), 664–669. https://doi.org/10.1111/j.1530-0277.1998.tb04309.x.

Inatani, S., Mizuno-Yasuhira, A., Kamiya, M., Nishino, I., Sabia, H. D., & Endo, H. (2021). Prediction of a clinically effective dose of THY1773, a novel V(1B) receptor antagonist, based on preclinical data. *Biopharmaceutics & Drug Disposition, 42*(5), 204–217. https://doi.org/10.1002/bdd.2273.

Jodogne, C., Tirelli, E., Klingbiel, P., & Legros, J. J. (1991). Oxytocin attenuates tolerance not only to the hypothermic but also to the myorelaxant and akinesic effects of ethanol in mice. *Pharmacology, Biochemistry, and Behavior, 40*(2), 261–265. https://doi.org/10.1016/0091-3057(91)90549-h.

Johnson, Z. V., & Young, L. J. (2017). Oxytocin and vasopressin neural networks: Implications for social behavioral diversity and translational neuroscience. *Neuroscience and Biobehavioral Reviews, 76*(Pt A), 87–98. https://doi.org/10.1016/j.neubiorev.2017.01.034.

Kamiya, M., Sabia, H. D., Marella, J., Fava, M., Nemeroff, C. B., Umeuchi, H., ... Nishino, I. (2020). Efficacy and safety of TS-121, a novel vasopressin V(1B) receptor antagonist, as adjunctive treatment for patients with major depressive disorder: A randomized, double-blind, placebo-controlled study. *Journal of Psychiatric Research, 128*, 43–51. https://doi.org/10.1016/j.jpsychires.2020.05.017.

Katz, D. A., Liu, W., Locke, C., Dutta, S., & Tracy, K. A. (2016). Clinical safety and hypothalamic-pituitary-adrenal axis effects of the arginine vasopressin type 1B receptor antagonist ABT-436. *Psychopharmacology (Berl), 233*(1), 71–81. https://doi.org/10.1007/s00213-015-4089-5.

Katz, D. A., Locke, C., Liu, W., Zhang, J., Achari, R., Wesnes, K. A., & Tracy, K. A. (2016). Single-dose interaction study of the arginine vasopressin type 1B receptor antagonist ABT-436 and alcohol in moderate alcohol drinkers. *Alcoholism, Clinical and Experimental Research, 40*(4), 838–845. https://doi.org/10.1111/acer.12996.

King, C. E., & Becker, H. C. (2019). Oxytocin attenuates stress-induced reinstatement of alcohol seeking behavior in male and female mice. *Psychopharmacology (Berl), 236*(9), 2613–2622. https://doi.org/10.1007/s00213-019-05233-z.

King, C. E., Gano, A., & Becker, H. C. (2020). The role of oxytocin in alcohol and drug abuse. *Brain Research, 1736*, 146761. https://doi.org/10.1016/j.brainres.2020.146761.

King, C. E., Griffin, W. C., Lopez, M. F., & Becker, H. C. (2021). Activation of hypothalamic oxytocin neurons reduces binge-like alcohol drinking through signaling at central oxytocin receptors. *Neuropsychopharmacology: Official Publication of the American College of Neuropsychopharmacology, 46*(11), 1950–1957. https://doi.org/10.1038/s41386-021-01046-x.

King, C. E., Griffin, W. C., Luderman, L. N., Kates, M. M., McGinty, J. F., & Becker, H. C. (2017). Oxytocin reduces ethanol self-administration in mice. *Alcoholism, Clinical and Experimental Research, 41*(5), 955–964. https://doi.org/10.1111/acer.13359.

Koob, G. F., & Volkow, N. D. (2016). Neurobiology of addiction: A neurocircuitry analysis. *Lancet Psychiatry, 3*(8), 760–773. https://doi.org/10.1016/s2215-0366(16)00104-8.

Lee, H.-J., Macbeth, A. H., Pagani, J. H., & Young, W. S., 3rd (2009). Oxytocin: The great facilitator of life. *Progress in Neurobiology, 88*(2), 127–151.

Lee, M. R., Schwandt, M. L., Sankar, V., Suchankova, P., Sun, H., & Leggio, L. (2017). Effect of alcohol use disorder on oxytocin peptide and receptor mRNA expression in human brain: A post-mortem case-control study. *Psychoneuroendocrinology, 85*, 14–19. https://doi.org/10.1016/j.psyneuen.2017.07.481.

MacFadyen, K., Loveless, R., DeLucca, B., Wardley, K., Deogan, S., Thomas, C., & Peris, J. (2016). Peripheral oxytocin administration reduces ethanol consumption in rats. *Pharmacology, Biochemistry, and Behavior, 140*, 27–32. https://doi.org/10.1016/j.pbb.2015.10.014.

Mander, A. J., Young, A., Merrick, M. V., & Morton, J. J. (1989). Fluid balance, vasopressin and withdrawal symptoms during detoxification from alcohol. *Drug and Alcohol Dependence, 24*(3), 233–237. https://doi.org/10.1016/0376-8716(89)90060-4.

Mannix, S. A., Hoffman, P. L., & Melchior, C. L. (1986). Intraventricular arginine vasopressin blocks the acquisition of ethanol tolerance in mice. *European Journal of Pharmacology, 128*(1–2), 137–141. https://doi.org/10.1016/0014-2999(86)90568-6.

McGregor, I. S., & Bowen, M. T. (2012). Breaking the loop: Oxytocin as a potential treatment for drug addiction. *Hormones and Behavior, 61*(3), 331–339. https://doi.org/10.1016/j.yhbeh.2011.12.001.

Melby, K., Grawe, R. W., Aamo, T. O., Skovlund, E., & Spigset, O. (2021). Efficacy of self-administered intranasal oxytocin on alcohol use and craving after detoxification in patients with alcohol dependence. A double-blind placebo-controlled trial. *Alcohol and Alcoholism (Oxford, Oxfordshire), 56*(5), 565–572. https://doi.org/10.1093/alcalc/agaa133.

Melby, K., Spigset, O., Grawe, R. W., Aamo, T. O., & Quintana, D. S. (2022). The effect of intranasal oxytocin on processing emotional stimuli during alcohol withdrawal: A randomized placebo-controlled double-blind clinical trial. *Hormones and Behavior, 146*, 105268. https://doi.org/10.1016/j.yhbeh.2022.105268.

Menon, R., & Neumann, I. D. (2023). Detection, processing and reinforcement of social cues: Regulation by the oxytocin system. *Nature Reviews. Neuroscience, 24*(12), 761–777. https://doi.org/10.1038/s41583-023-00759-w.

Mitre, M., Minder, J., Morina, E. X., Chao, M. V., & Froemke, R. C. (2018). Oxytocin modulation of neural circuits. *Current Topics in Behavioral Neurosciences, 35*, 31–53. https://doi.org/10.1007/7854_2017_7.

Muhle, C., Mazza, M., Weinland, C., von Zimmermann, C., Bach, P., Kiefer, F., ... Lenz, B. (2022). Elevated oxytocin receptor blood concentrations predict higher risk for, more, and earlier 24-month hospital readmissions after in-patient detoxification in males with alcohol use disorder. *International Journal of Molecular Sciences, 23*(17), https://doi.org/10.3390/ijms23179940.

Nair, H. P., & Young, L. J. (2006). Vasopressin and pair-bond formation: genes to brain to behavior. *Physiology (Bethesda), 21*, 146–152. https://doi.org/10.1152/physiol.00049.2005.

Nelson, B. S., Sequeira, M. K., & Schank, J. R. (2018). Bidirectional relationship between alcohol intake and sensitivity to social defeat: Association with Tacr1 and Avp expression. *Addiction Biology, 23*(1), 142–153. https://doi.org/10.1111/adb.12494.

Neumann, I. D., & Landgraf, R. (2012). Balance of brain oxytocin and vasopressin: Implications for anxiety, depression, and social behaviors. *Trends in Neurosciences, 35*(11), 649–659. https://doi.org/10.1016/j.tins.2012.08.004.

Onuoha, R. C., Quintana, D. S., Lyvers, M., & Guastella, A. J. (2016). A meta-analysis of theory of mind in alcohol use disorders. *Alcohol and Alcoholism (Oxford, Oxfordshire), 51*(4), 410–415. https://doi.org/10.1093/alcalc/agv137.

Pedersen, C. A., Smedley, K. L., Leserman, J., Jarskog, L. F., Rau, S. W., Kampov-Polevoi, A., ... Garbutt, J. C. (2013). Intranasal oxytocin blocks alcohol withdrawal in human subjects. *Alcoholism, Clinical and Experimental Research, 37*(3), 484–489. https://doi.org/10.1111/j.1530-0277.2012.01958.x.

Peters, S., Slattery, D. A., Flor, P. J., Neumann, I. D., & Reber, S. O. (2013). Differential effects of baclofen and oxytocin on the increased ethanol consumption following chronic psychosocial stress in mice. *Addiction Biology, 18*(1), 66–77. https://doi.org/10.1111/adb.12001.

Peters, S. T., Bowen, M. T., Bohrer, K., McGregor, I. S., & Neumann, I. D. (2017). Oxytocin inhibits ethanol consumption and ethanol-induced dopamine release in the nucleus accumbens. *Addiction Biology, 22*(3), 702–711. https://doi.org/10.1111/adb.12362.

Puciklowski, O., Kostowski, W., & Trzaskowska, E. (1985). The effect of oxytocin and fragment (MIF-I) on the development of tolerance to hypothermic and hypnotic action of ethanol in the rat. *Peptides, 6*(1), 7–10. https://doi.org/10.1016/0196-9781(85)90067-1.

Rigney, N., De Vries, G. J., Petrulis, A., & Young, L. J. (2022). Oxytocin, vasopressin, and social behavior: From neural circuits to clinical opportunities. *Endocrinology, 163*(9), https://doi.org/10.1210/endocr/bqac111.

Rigter, H., Rijk, H., & Crabbe, J. C. (1980). Tolerance to ethanol and severity of withdrawal in mice are enhanced by a vasopressin fragment. *European Journal of Pharmacology, 64*(1), 53–68. https://doi.org/10.1016/0014-2999(80)90369-6.

Rodriguez, K. M., Smith, B. L., & Caldwell, H. K. (2020). Voluntary alcohol consumption is increased in female, but not male, oxytocin receptor knockout mice. *Brain and Behavior, 10*(9), e01749. https://doi.org/10.1002/brb3.1749.

Rotondo, F., Butz, H., Syro, L. V., Yousef, G. M., Di Ieva, A., Restrepo, L. M., ... Kovacs, K. (2016). Arginine vasopressin (AVP): A review of its historical perspectives, current research and multifunctional role in the hypothalamo-hypophysial system. *Pituitary, 19*(4), 345–355. https://doi.org/10.1007/s11102-015-0703-0.

Ryan, M. L., Falk, D. E., Fertig, J. B., Rendenbach-Mueller, B., Katz, D. A., Tracy, K. A., ... Litten, R. Z. (2017). A phase 2, double-blind, placebo-controlled randomized trial assessing the efficacy of ABT-436, a novel V1b receptor antagonist, for alcohol dependence. *Neuropsychopharmacology: Official Publication of the American College of Neuropsychopharmacology, 42*(5), 1012–1023. https://doi.org/10.1038/npp.2016.214.

Schimmer, J., Patwell, R., Küppers, S., & Grinevich, V. (2023). The relationship between oxytocin and alcohol dependence. *Current Topics in Behavioral Neurosciences*. https://doi.org/10.1007/7854_2023_444.

Schonberger, M., Leggett, C., Kim, S. W., & Hooker, J. M. (2010). Synthesis of [11C] SSR149415 and preliminary imaging studies using positron emission tomography. *Bioorganic & Medicinal Chemistry Letters, 20*(10), 3103–3106. https://doi.org/10.1016/j.bmcl.2010.03.108.

Silva, S. M., Madeira, M. D., Ruela, C., & Paula-Barbosa, M. M. (2002). Prolonged alcohol intake leads to irreversible loss of vasopressin and oxytocin neurons in the paraventricular nucleus of the hypothalamus. *Brain Research, 925*(1), 76–88. https://doi.org/10.1016/s0006-8993(01)03261-9.

Sivukhina, E. V., Dolzhikov, A. A., Morozov Iu, E., Jirikowski, G. F., & Grinevich, V. (2006). Effects of chronic alcoholic disease on magnocellular and parvocellular hypothalamic neurons in men. *Hormone and Metabolic Research. Hormon- und Stoffwechselforschung. Hormones et Metabolisme, 38*(6), 382–390. https://doi.org/10.1055/s-2006-944522.

Song, Z., & Albers, H. E. (2018). Cross-talk among oxytocin and arginine-vasopressin receptors: Relevance for basic and clinical studies of the brain and periphery. *Frontiers in Neuroendocrinology, 51*, 14–24. https://doi.org/10.1016/j.yfrne.2017.10.004.

Stauffer, C. S., Morrison, T. E., Meinzer, N. K., Leung, D., Buffington, J., Sheh, E. G., ... Woolley, J. D. (2022). Effects of oxytocin administration on fear-potentiated acoustic startle in co-occurring PTSD and alcohol use disorder: A randomized clinical trial. *Psychiatry Research, 308*, 114340. https://doi.org/10.1016/j.psychres.2021.114340.

Stevenson, J. R., Wenner, S. M., Freestone, D. M., Romaine, C. C., Parian, M. C., Christian, S. M., ... O'Kane, C. M. (2017). Oxytocin reduces alcohol consumption in prairie voles. *Physiology & Behavior, 179*, 411–421. https://doi.org/10.1016/j.physbeh.2017.07.021.

Stevenson, J. R., Young, K. A., Bohidar, A. E., Francomacaro, L. M., Fasold, T. R., Buirkle, J. M., ... Christian, S. C. (2017). Alcohol consumption decreases oxytocin neurons in the anterior paraventricular nucleus of the hypothalamus in prairie voles. *Alcoholism, Clinical and Experimental Research, 41*(8), 1444–1451. https://doi.org/10.1111/acer.13430.

Stoop, R. (2012). Neuromodulation by oxytocin and vasopressin. *Neuron, 76*(1), 142–159. https://doi.org/10.1016/j.neuron.2012.09.025.

Szabo, G., Kovacs, G. L., Balaspiri, L., & Telegdy, G. (1987). D-pipecolyl-leucyl-glycinamide, a substituted tripeptide analogue of the C-terminal part of oxytocin, influences tolerance to and dependence on ethanol in mice. *Alcohol and Drug Research, 7*(2), 99–105. https://www.ncbi.nlm.nih.gov/pubmed/3778581.

Szabo, G., Kovacs, G. L., Szekeli, S., Balaspiri, L., & Telegdy, G. (1987). C-terminal fragments of oxytocin (prolyl-leucyl-glycinamide and Z-prolyl-D-leucine) attenuate the development of tolerance to ethanol. *Acta Physiologica Hungarica, 69*(1), 115–122. https://www.ncbi.nlm.nih.gov/pubmed/2884803.

Szabo, G., Kovacs, G. L., Szekeli, S., & Telegdy, G. (1985). The effects of neurohypophyseal hormones on tolerance to the hypothermic effect of ethanol. *Alcohol (Fayetteville, N. Y.), 2*(4), 567–574. https://doi.org/10.1016/0741-8329(85)90082-5.

Szabo, G., Kovacs, G. L., & Telegdy, G. (1989). Intraventricular administration of neurohypophyseal hormones interferes with the development of tolerance to ethanol. *Acta Physiologica Hungarica, 73*(1), 97–103. https://www.ncbi.nlm.nih.gov/pubmed/2711843.

Szabo, G., Tabakoff, B., & Hoffman, P. L. (1988). Receptors with V1 characteristics mediate the maintenance of ethanol tolerance by vasopressin. *The Journal of Pharmacology and Experimental Therapeutics, 247*(2), 536–541. https://www.ncbi.nlm.nih.gov/pubmed/2972830.

Tabak, B. A., Leng, G., Szeto, A., Parker, K. J., Verbalis, J. G., Ziegler, T. E., ... Mendez, A. J. (2023). Advances in human oxytocin measurement: Challenges and proposed solutions. *Molecular Psychiatry, 28*(1), 127–140. https://doi.org/10.1038/s41380-022-01719-z.

Tirelli, E., Jodogne, C., & Legros, J. J. (1992). Oxytocin blocks the environmentally conditioned compensatory response present after tolerance to ethanol-induced hypothermia in mice. *Pharmacology, Biochemistry, and Behavior, 43*(4), 1263–1267. https://doi.org/10.1016/0091-3057(92)90512-e.

Towner, T. T., Papastrat, K. M., Spear, L. P., Varlinskaya, E. I., & Werner, D. F. (2022). Impact of adolescent intermittent ethanol exposure in male and female rats on social drinking and neuropeptide gene expression. *Alcoholism, Clinical and Experimental Research, 46*(6), 979–993. https://doi.org/10.1111/acer.14847.

Tunstall, B. J., Kirson, D., Zallar, L. J., McConnell, S. A., Vendruscolo, J. C. M., Ho, C. P., ... Vendruscolo, L. F. (2019). Oxytocin blocks enhanced motivation for alcohol in alcohol dependence and blocks alcohol effects on GABAergic transmission in the central amygdala. *PLoS Biology, 17*(4), e2006421. https://doi.org/10.1371/journal.pbio.2006421.

Walcott, A. T., & Ryabinin, A. E. (2021). Assessing effects of oxytocin on alcohol consumption in socially housed prairie voles using radio frequency tracking. *Addiction Biology, 26*(2), e12893. https://doi.org/10.1111/adb.12893.

Walum, H., & Young, L. J. (2018). The neural mechanisms and circuitry of the pair bond. *Nature Reviews. Neuroscience, 19*(11), 643–654. https://doi.org/10.1038/s41583-018-0072-6.

Wilfur, S. M., McNeely, E. C., Lackan, A. A., Bowers, C. P., & Leong, K. C. (2023). Oxytocin attenuates yohimbine-induced reinstatement of alcohol-seeking in female rats via the central amygdala. *Behavioral Sciences (Basel), 13*(7), https://doi.org/10.3390/bs13070556.

Williams Avram, S. K., & Cymerblit-Sabba, A. (2017). 3.14—Vasopressin: Roles in modulating social behaviors. In D. W. Pfaff, & M. Joëls (Eds.). *Hormones, brain and behavior* (pp. 279–304)(3rd ed.). Academic Press. https://doi.org/10.1016/B978-0-12-803592-4.00057-2.

Wronikowska-Denysiuk, O., Mrozek, W., & Budzyńska, B. (2023). The role of oxytocin and vasopressin in drug-induced reward-implications for social and non-social factors. *Biomolecules, 13*(3), https://doi.org/10.3390/biom13030405.

Zhou, Y., Colombo, G., Carai, M. A., Ho, A., Gessa, G. L., & Kreek, M. J. (2011). Involvement of arginine vasopressin and V1b receptor in alcohol drinking in Sardinian alcohol-preferring rats. *Alcoholism, Clinical and Experimental Research, 35*(10), 1876–1883. https://doi.org/10.1111/j.1530-0277.2011.01532.x.

Zhou, Y., Rubinstein, M., Low, M. J., & Kreek, M. J. (2018). V1b receptor antagonist SSR149415 and Naltrexone synergistically decrease excessive alcohol drinking in male and female mice. *Alcoholism, Clinical and Experimental Research, 42*(1), 195–205. https://doi.org/10.1111/acer.13544.

CHAPTER THREE

Emerging evidence for pregnane steroid therapeutics for alcohol use disorders

A. Leslie Morrow[a,b,c,*], Minna H. McFarland[c,d], Todd K. O'Buckley[c], and Donita L. Robinson[a,c]

[a]Department of Psychiatry, University of North Carolina at Chapel Hill, School of Medicine, Chapel Hill, NC, United States
[b]Department of Pharmacology, University of North Carolina at Chapel Hill, School of Medicine, Chapel Hill, NC, United States
[c]Bowles Center for Alcohol Studies, University of North Carolina at Chapel Hill, School of Medicine, Chapel Hill, NC, United States
[d]Neuroscience Curriculum, University of North Carolina at Chapel Hill, School of Medicine, Chapel Hill, NC, United States
*Corresponding author. e-mail address: leslie_morrow@med.unc.edu

Contents

International Review of Neurobiology, Volume 178
ISSN 0074-7742, https://doi.org/10.1016/bs.irn.2024.07.004

Abstract

Many lines of research have suggested that the neuroactive pregnane steroids, including pregnenolone, progesterone, and allopregnanolone ([3α,5α]-3-hydroxypregnan-20-one, 3α,5α-THP), have therapeutic potential for treatment of alcohol use disorders (AUDs). In this chapter, we systematically address the preclinical and clinical evidence that supports this approach for AUD treatment, describe the underlying neurobiology of AUDs that are targeted by these treatments, and delineate how pregnane steroids may address various components of the disease. This review updates the theoretical framework for understanding how endogenous steroids that modulate the effects of alcohol, stress, excitatory/inhibitory and dopamine transmission, and the innate immune system appear to play a key role in the prevention and mitigation of AUDs. We further discuss newly discovered limitations of pregnane steroid therapies as well as the challenges that are inherent to development of endogenous compounds for therapeutics. We argue that overcoming these challenges presents the opportunity to help millions who suffer from AUDs across the world.

1. Introduction

Alcohol addiction is a complex alcohol use disorder (AUD) resulting from dysregulation of brain circuits, neurochemistry, stress responses, and inflammatory immune signaling (see (Morrow, Boero, & Porcu, 2020), for review). Only three drugs are approved to treat AUD by the US Food and Drug Administration (disulfiram, acamprosate, and naltrexone) but they are only moderately effective at a population level (Garbutt, West, Carey, Lohr, & Crews, 1999; Ray et al., 2019) and they are under-prescribed. Treatments for AUDs typically address one modality of alcohol pathology or another, and this may partially explain why the success of those approaches has been limited.

This review addresses the therapeutic potential of endogenous pregnane steroids to simultaneously address multiple pathologies associated with alcohol use disorders and addiction. We will review the evidence that excessive alcohol intake leads to dysregulation of multiple brain circuits to disrupt excitatory/inhibitory balance in the central nervous system (CNS) by altering brain neurochemistry, including inflammatory pathways and the regulation of stress responses. The endogenous pregnane steroids regulate each of these domains through pleiotropic actions that appear to be favorable for restoration of homeostasis and mental health. We argue that simultaneous treatment of each of these pathological domains of AUD holds great promise for recovery, a thesis that is borne out by preclinical and early clinical studies. In addition, we present evidence that pregnane neuroactive steroids moderate many of the co-occurring symptoms of alcohol addiction, including anxiety, depression, seizures, sleep disturbance, and pain, that likely contribute to the spiral of addiction to alcohol.

This article updates previous work suggesting that the pleiotropic actions of allopregnanolone make it an attractive candidate for the treatment of AUDs (Morrow et al., 2020). In this chapter, we further argue that modulation of dopamine neurotransmission is an important component of AUD pathology to address; that pregnenolone, progesterone and allopregnanolone may all share the ability to modulate various aspects of AUD pathology; and that these pregnane steroids represent promising treatments for the disorder.

2. Pregnane neurosteroid biosynthesis and modes of action

Pregnane neuroactive steroids are a class of endogenous steroids synthesized de novo within the CNS, independent of the endocrine gland steroidogenesis (Cook, Werner, et al., 2014; Lloyd-Evans & Waller-Evans, 2020). The enzymes necessary for the synthesis of neurosteroids are found in several brain regions (Agis-Balboa, Graziano et al., 2006; Corpechot et al., 1993; Lloyd-Evans & Waller-Evans, 2020; Mellon & Deschepper, 1993; Mensah-Nyagan et al., 1999; Purdy, Morrow, Blinn, & Paul, 1990; Purdy, Moore, Morrow, & Paul, 1991; Reddy, 2010). In the brain, neurons are the primary producers of pregnenolone, progesterone, and allopregnanolone (Agis-Balboa, Guidotti, & Pinna, 2014; Agis-Balboa, Pinna et al., 2006; Baulieu, Robel, & Schumacher, 2001; Cook, Dumitru, O'Buckley, & Morrow, 2014; Tsutsui, Ukena, Takase, Kohchi, & Lea, 1999; Tsutsui, 2008). Neurosteroidogenesis begins with cholesterol or other steroidal precursors. The precursors are synthesized in the brain and/or derived from systemic steroids that cross the blood-brain barrier. Steroidogenesis is initiated by the transport of cholesterol to the inner mitochondrial membrane (Papadopoulos, Liu, & Culty, 2007). Here, cholesterol is converted to pregnenolone, the precursor of all steroids and a key intermediate for neurosteroid biosynthesis. Pregnenolone is converted to progesterone, then allopregnanolone and pregnanolone via progressive A-ring reductions (Purdy et al., 1990; Purdy, Moore, et al., 1991). Exogenous pregnenolone and progesterone are also converted to allopregnanolone and pregnanolone in vivo (Marx et al., 2009; Milivojevic, Fox, Sofuoglu, Covault, & Sinha, 2016; Porcu et al., 2009), so administration of either precursor also leads to enhancement of circulating and brain levels of allopregnanolone.

The pregnane neurosteroids have multiple modes of action that include both genomic and non-genomic actions. Pregnenolone undergoes metabolic conversion to conventional steroids, such as progesterone, estrogen, androgen, glucocorticoid, and mineralocorticoid, which are then mediated by their respective nuclear receptors (Rupprecht, 2003; Slater, Hesse, & Beato, 1994). It is also converted by sulfatases to pregnenolone sulfate which has direct effects as an NMDA receptor agonist (Irwin et al., 1992; Wu, Gibbs, & Farb, 1991) as well as modulating synaptic GABA release (Mtchedlishvili & Kapur, 2003). Furthermore, pregnenolone also has the capacity to directly interact with microtubule-associated protein 2 and the cannabinoid receptor type 1 (CB1) (Murakami, Fellous, Baulieu, & Robel, 2000; Vallée et al., 2014). Hence pregnenolone has direct actions that impact brain function besides its conversion to progesterone and allopregnanolone.

The pregnane neuroactive steroids that are 3α-reduced derivatives of progesterone, including allopregnanolone and pregnanolone, are potent positive allosteric modulators of $GABA_A$ receptors (Belelli, Casula, Ling, & Lambert, 2002; Majewska, Harrison, Schwartz, Barker, & Paul, 1986; Morrow, Suzdak, & Paul, 1987; Puia, Vicini, Seeburg, & Costa, 1991). Allopregnanolone exhibits high potency at most subtypes of recombinant $GABA_A$ receptors (Puia et al., 1990; Puia, Ducic, Vicini, & Costa, 1993), although it is more potent at extrasynaptic $GABA_A$ receptors containing the δ subunit (Modgil et al., 2017; Stell, Brickley, Tang, Farrant, & Mody, 2003). Neuroactive steroid binding sites are located at the interface of the α and β subunits of synaptic GABA receptors (Hosie, Wilkins, da Silva, & Smart, 2006), however, the presence of the δ subunit, rather than the γ subunit, enhances the sensitivity of $GABA_A$ receptors to allopregnanolone and (3α,5α)-3,21-dihydroxypregnan-20-one (THDOC). In fact, δ subunit knockout mice have blunted tonic currents, along with reduced sensitivity to neuroactive steroid modulation of $GABA_A$ receptors (Mihalek et al., 1999; Spigelman et al., 2003; Stell et al., 2003). Thus, the GABAergic neurosteroids modulate multiple subtypes of $GABA_A$ receptors, including inhibition that is insensitive to benzodiazepines, such as the δ subunit containing receptors. The GABAergic actions of allopregnanolone may be critical in the treatment of AUDs since many studies have shown that AUD involves the dysregulation of inhibitory $GABA_A$ receptor expression (Kumar et al., 2009) and inhibitory pathways in the brain, as well as excessive excitatory drive (see below).

Recent studies have revealed that allopregnanolone has novel direct effects on expression of multiple proteins that regulate the hypothalamic-pituitary-adrenal (HPA) axis during stress, including the glucocorticoid receptor, corticotropin releasing factor (CRF), CRF binding protein, the CRF1 receptor, and pro-opiomelanocortin. These effects were found in male and female rats after restraint stress and forced swim stress, but they differed by sex and the type of stressor (Boero et al., 2022). In females, allopregnanolone inhibited the effects of restraint stress on corticosterone, CRF, and pro-opiomelanocortin, while it inhibited the effects of swim stress on CRF, CRF1 receptor, CRF binding protein, and pro-opiomelanocortin. In males, allopregnanolone inhibited the effects of restraint stress and swim stress on all the aforementioned stress markers except corticosterone. These data indicate that allopregnanolone targets multiple sites to restore the animal to homeostasis following stress in both males and females. The ability of allopregnanolone to modulate the stress response may be critical in AUD since many studies have shown that HPA axis stress reactivity is dysregulated in those who suffer from it (see below).

By enhancing GABA transmission, pregnane neurosteroids can indirectly regulate a variety of neurotransmitters, including the catecholamine dopamine. Dopamine is involved in many aspects of voluntary behavior and motivation, and increases can signal salient stimuli such as rewards or stressors, and accumulating evidence shows that GABAergic neurosteroids modulate dopamine transmission. For example, while stress increased tissue content of dopamine in prefrontal cortex in male rats, intraventricular THDOC blunted that accumulation (Grobin, Roth, & Deutch, 1992). Similarly, intraventricular allopregnanolone modulated tissue content of dopamine in the nucleus accumbens of female rats, although the direction of change was dependent on estrous cycle stage (Laconi, Reggiani, Penissi, Yunes, & Cabrera, 2007). We recently conducted direct measurements of dopamine release and observed that systemic allopregnanolone reduced dopamine release in the nucleus accumbens of male and female rats, either evoked by electrical stimulation in anesthetized animals (Dornellas et al., 2020) or spontaneously occurring in awake animals (McFarland et al., unpublished). These findings are consistent with the ability of other drugs that act as GABA agonists to reduce dopamine transmission (e.g., Brodnik, Batra, Oleson, & Espana, 2019; Sanchez et al., 2022; Smolders, De Klippel, Sarre, Ebinger, & Michotte, 1995), although allopregnanolone reduced dopamine release at doses (15 mg/kg) that did not induce locomotor sedation (Boero et al., 2023). The ability of allopregnanolone to reduce stimulated dopamine release may be helpful in AUD treatment since previous studies have shown that dopamine release is linked to the rewarding properties of alcohol (Wightman & Robinson, 2002).

In 2019, Morrow's group and others found that the pregnane neurosteroids inhibit inflammatory signals mediated by toll-like receptors (TLR) in macrophages and the brain (Balan, Beattie, O'Buckley, Aurelian, & Morrow, 2019). Inflammation is caused by pathogen activation of TLRs that initiate a signaling cascade producing specific transcription factors that regulate the production of inflammatory cytokines and chemokines. Allopregnanolone inhibits inflammatory signaling by blocking protein-protein interactions that are required for TLR activation and consequently blocks the transcription factors and the cytokine production. These steroids block TLR2, TLR4, and TLR7 pro-inflammatory signal activation and the production of cytokines/chemokines through their ability to block TLR2-MyD88 binding, TLR4-MD2 binding, TLR4-MyD88 binding, and TLR7-MyD88 binding in the mouse macrophage RAW264.7 cell line and the alcohol-preferring (P) rat brain (Balan et al., 2019; Balan et al., 2021). Pregnenolone inhibited the TLR4 pathway, but was not studied in the other TLR pathways. Other studies found that pregnenolone, progesterone, and allopregnanolone each inhibit inflammation in macrophage and microglial cell lines by promoting ubiquitination and degradation of TLR2 and its accessory proteins and by preventing TLR4- and TLR2-mediated production of TNF-α and IL-6 (Murugan, Jakka, Namani, Mujumdar, & Radhakrishnan, 2019). These effects of pregnane steroids, including pregnenolone, progesterone, and allopregnanolone, are likely to be key to the treatment of AUDs, since both systemic and brain inflammation are prominent symptoms of the disorder and may prove to be causative (See Section 3.4).

3. Neurobiological basis of AUDs

Many approaches to AUD seek to identify the underling etiology of the disorder to find an appropriate intervention. While this approach has great utility, AUD is a very complex disorder with multiple apparent etiologies that likely contribute to the condition. Furthermore, AUD is often co-morbid with other conditions that may exacerbate symptoms, such as chronic anxiety, depression, sleep disruptions, and pain.

Therefore, effective treatment of AUD may require that treatment can address multiple etiologies of the disorder as well as symptoms that are difficult to distinguish from co-occurring conditions. The complexity of AUD is likely to necessitate multiple approaches to address various components of the disease. The following sections cover the major hypotheses of the neurobiology of AUD that may lead to effective treatments.

3.1 Disruptions in excitatory/inhibitory balance in the CNS

Ethanol induces a wide range of effects in the brain, with initial stimulatory effects at low doses and predominately sedative effects at higher doses. Moreover, neuroadaptations to the acute inhibition associated with repeated and chronic alcohol use result in tolerance to the sedative and GABA-mimetic effects of alcohol, resulting in a shift toward excitation during withdrawal and abstinence from alcohol (for review, see (Morrow et al., 2020; Olsen & Liang, 2017)). Indeed, AUD in humans and chronic ethanol exposure in rodents results in acute and persistent hyperexcitability after cessation of alcohol that, along with long-term negative affect, characterizes alcohol withdrawal syndrome (Koob, 2024). The hyperexcitability associated with AUD is often conceptualized as a shift in excitatory/inhibitory (E/I) balance and can be experimentally observed at the level of cells and circuits (for reviews, see (Becker & Mulholland, 2014; Olsen & Liang, 2017)).

Some therapeutic approaches to AUD treatment aim to restore optimal E/I balance, including glutamatergic antagonists such as acamprosate and topiramate. For example, acute withdrawal of AUD in humans is associated with higher levels of excitatory neurotransmitters (glutamate, glutamine) in the nucleus accumbens and anterior cingulate cortex as measured by proton magnetic resonance spectroscopy, as compared to healthy controls (Umhau et al., 2010). However, treatment with acamprosate reduced magnetic resonance spectroscopy (MRS)-measured glutamate in the anterior cingulate as compared to untreated or placebo-treated subjects, and this reduction correlated with less craving (Bauer et al., 2013). A complimentary approach to the treatment of alcohol use disorder is to enhance GABAergic function. Indeed, benzodiazepines are commonly used to increase inhibition during acute alcohol withdrawal (Airagnes, Ducoutumany, Laffy-Beaufils, Le Faou, & Limosin, 2019; Long, Long, & Koyfman, 2017). However, as may be expected, there can be notable side effects from benzodiazepines and pharmacological cross-tolerance to alcohol that prevent their use beyond acute alcohol detoxication.

3.2 Disruptions in stress homeostasis including excessive CRF signaling

Stress plays a major role in alcohol addiction, starting from the initial activation of the HPA axis through enhancement of hypothalamic CRF by acute ethanol exposure to the subsequent phases of binge intake, withdrawal, and relapse/reinstatement (Koob, 2013) that involves the extra-hypothalamic CRF system in the extended amygdala (Zorrilla, Logrip, & Koob, 2014). CRF is a 41 amino acid polypeptide widely distributed throughout the brain. The highest densities of CRF neurons are found in the paraventricular nucleus (PVN) of the hypothalamus, where CRF is the main effector of the response to stress mediated by HPA activation (Rivier & Vale, 1983). CRF-positive cells are also abundant in extrahypothalamic structures of the extended amygdala, in particular the central nucleus of the amygdala, and the bed nucleus of the stria terminalis (BNST) (Peng et al., 2017), which mediate the behavioral and emotional response to stress (Schreiber & Gilpin, 2018). CRF binds to two types of Gs protein-coupled receptors: CRF receptor type 1 (CRFR1) and type 2 (CRFR2). CRFR1s are widely expressed throughout the brain and trigger the HPA response to stress as well as anxiogenic behavior. The role of CRFR2s, whom CRF binds with lower affinity, is less clear and may involve maintenance of homeostasis (Henckens, Deussing, & Chen, 2016).

Ethanol activates the HPA response to stress during acute withdrawal via hypothalamic CRF and CRFR1 activation that results in elevated adrenocorticotropic hormone (ACTH) and corticosteroids; within hours it also activates the extrahypothalamic CRF response, with subsequent increased CRF release from the amygdala, which increases GABA release in the CeA through CRFR1. Likewise, binge alcohol drinking also affects hypothalamic and extrahypothalamic CRF signaling in different ways depending on species, age, and experimental procedures employed. Chronic alcohol exposure increases hypothalamic and extrahypothalamic CRF gene expression immediately after consumption. However, during withdrawal hypothalamic CRF and CRFR1 expression is reduced, with subsequent blunting of the HPA response, whereas extrahypothalamic CRF release is increased within the CeA and BNST (see (Agoglia & Herman, 2018; Roberto, Spierling, Kirson, & Zorrilla, 2017) and (Schreiber & Gilpin, 2018) for review). Moreover, it has been hypothesized that extrahypothalamic upregulation of CRF signaling may contribute to negative affect during withdrawal that motivates the escalation of alcohol drinking (Zorrilla et al., 2014).

The majority of these studies have been conducted in male rodents. However, preclinical studies have demonstrated cellular and molecular sex differences in stress response systems, including CRFR1 coupling, signaling, and trafficking (Bangasser & Valentino, 2014; Logrip, Milivojevic, Bertholomey, & Torregrossa, 2018; Oyola & Handa, 2017). A dysregulation of the extrahypothalamic CRF systems is hypothesized to play a role in the negative symptoms of withdrawal and subsequent relapse. In fact, the extrahypothalamic CRF signaling, activated by multiple withdrawal cycles, sensitizes the brain to such repeated cycles, thus contributing to negative emotional states (i.e. anxiety-like behavior) that trigger relapse and maintenance of drug taking (Zorrilla et al., 2014). In line with this interpretation, it has recently been shown that optogenetic inactivation of CRF neurons in the CeA reduced alcohol consumption and withdrawal symptoms in dependent male rats (de Guglielmo et al., 2019). CRFR1 is also involved in these processes; in fact, withdrawal-induced drinking is reduced in CRFR1 knockout mice or can be blocked by CRFR1 antagonists in alcohol dependent animals (Chu, Koob, Cole, Zorrilla, & Roberts, 2007). Further, CRFR1 antagonists attenuate stress and cue-induced reinstatement of alcohol seeking behavior, likely by inhibition of extrahypothalamic CRF, given that it is not influenced by adrenalectomy. In contrast, hypothalamic CRF appears to mediate cue-induced reinstatement because metyrapone, an inhibitor of glucocorticoid synthesis, mimics the reduction in reinstatement induced by CRFR1 antagonists (Schreiber & Gilpin, 2018).

Importantly, clinical studies also show that acute alcohol intoxication activates the HPA axis in healthy non-dependent subjects (Mendelson & Stein, 1966), but alcohol dependence leads to adaptations in the HPA/CRF signaling that result in a blunted cortisol and ACTH response to stress (Adinoff et al., 1990; Adinoff, Junghanns, Kiefer, & Krishnan-Sarin, 2005; Adinoff, Krebaum, et al., 2005a, 2005b). Further, blunted stress responses are accompanied by increased alcohol craving, in binge/heavy drinkers as well as in moderate social drinkers, and these measures are predictive of higher risk for relapse after treatment (Blaine, Nautiyal, Hart, Guarnaccia, & Sinha, 2019; Sinha et al., 2011). This work strongly indicates that CRF signaling represents a critical therapeutic target for multiple aspects of AUD pathology.

3.3 Disruptions in dopamine signaling

The neuromodulator dopamine is involved in many neurobiological systems, including reward and reinforcement processing, voluntary movement, cognition, and mood. Most addictive drugs share the common mechanism of action to increase dopamine transmission, either directly or indirectly, and this effect is considered an important aspect of the addiction liability of a substance. Ethanol can acutely increase activity of dopamine neurons (Brodie, Pesold, & Appel, 1999) as well as dopamine release (Mereu, Fadda, & Gessa, 1984), although the resulting dopamine release is modest compared to that induced by stimulants such as cocaine and amphetamines. Nevertheless, alcohol-induced dopamine release is thought to contribute to the activating, rewarding, and reinforcing aspects of alcohol use (Gonzales, Job, & Doyon, 2004). Moreover, dopamine is a critical learning signal, and a vital role of dopamine transmission is to signal salient and predictive cues in the environment (for reviews, see (Schultz, 1998; Wightman & Robinson, 2002)). Thus, dopamine is also released to cues that predict alcohol availability (Fiorenza et al., 2018; Shnitko & Robinson, 2015) and is thought to thereby trigger alcohol seeking, craving, and relapse (Domi, Domi, Adermark, Heilig, & Augier, 2021).

Due to the ubiquitous involvement of dopamine in voluntary movement and motivation, it is unsurprising that dopamine transmission is dysregulated in many psychiatric disorders, including AUD. For example, Spitta, Garbusow, Buchert, and Heinz (2023) recently reviewed studies that indirectly measured dopamine transmission by using positron emission tomography (PET) and related function imaging techniques in human drinkers. Many studies to date found that alcohol acutely increases dopamine release in both social and heavy drinking humans. Interestingly, one study disentangled the pharmacological effect of alcohol and the sensory (tastant) cues to reveal that both aspects increased dopamine release in heavy drinking males (Oberlin et al., 2015). Other studies measured dopamine release evoked by psychostimulants to probe dopamine transmission apart from alcohol. The studies generally reported blunted evoked dopamine release in people with AUD compared to healthy controls, providing evidence of dysregulated dopamine release in AUD. However, dopamine release measured via PET is also impacted by subject age and smoking status, both of which impact dopamine signaling. Sex may also impact the effect of acute and chronic alcohol on dopamine systems, although data from females are sparse (Verplaetse, Cosgrove, Tanabe, & McKee, 2021). While one PET study reported more dopamine release to alcohol in young male social drinkers than females (Urban et al., 2010), another found no sex difference in dopamine release after acute alcohol in either people with AUD or healthy controls (Kegeles et al., 2018). As the number of women engaging in risky drinking and developing AUD increases, it will be important to understand how dopamine systems change in women with AUD in order to effectively treat the disorder.

3.4 Excessive inflammatory signaling in the CNS

Several groups have suggested that excessive neuroimmune signaling in the brain drives alcohol addiction, based on studies in animal models as well as analysis of post-mortem brains of humans with alcohol use disorder (Alfonso-Loeches, Pascual-Lucas, Blanco, Sanchez-Vera, & Guerri, 2010;

Aurelian & Balan, 2019; Crews & Vetreno, 2016; Crews, Lawrimore, Walter, & Coleman, 2017; Mayfield & Harris, 2017; Vetreno, Qin, Coleman, & Crews, 2021). Indeed, alcohol activates multiple TLR signaling pathways in many brain regions (Kane et al., 2014; Leclercq et al., 2012; Pascual, Balino, Aragon, & Guerri, 2015), enhances the microglial response to peripheral inflammation, and increases the expression of pro-inflammatory cytokines and chemokines (Coleman, Zou, & Crews, 2017; Coleman, Zou, Qin, & Crews, 2018; Crews et al., 2017; Fernandez-Lizarbe, Montesinos, & Guerri, 2013) in neurons and glial cells. Furthermore, alcohol alters the balance of pro- and anti-inflammatory cytokines, thereby potentially contributing to increased risk from AUDs (Aurelian & Balan, 2019; Qin et al., 2008).

The innate immune cells in the brain, microglia, as well as neurons and other glial cells signal via TLRs to promote innate immune gene expression and produce pro-inflammatory cytokines and chemokines in a progressive and feed-forward fashion (Pavlov & Tracey, 2017) that is particularly long lasting in the brain (Crews et al., 2017). The inflammatory pathways involved in AUD are primarily the MyD88-dependent TLR4 and TLR7 complexes that result in increased levels of nuclear transcription factor nuclear factor-kappa-light chain enhancer of activated B cells (NFκB) and the induction of TLR agonist high mobility group box 1 protein (HMGB1) (Vetreno et al., 2021). These effects of ethanol have been directly linked to neurodegeneration (Qin et al., 2021), cognitive and behavioral inflexibility (Vetreno & Crews, 2012) and excessive drinking in animal models, as well as lifetime alcohol consumption in humans (Crews, Qin, Sheedy, Vetreno, & Zou, 2013; Vetreno et al., 2021). Clearly, the excessive activation and expression of these TLR pathways contributes to the pathology of AUDs, and the presence of inflammation both systemically and in the brain needs to be addressed in its treatment (Morrow et al., 2020).

3.5 Loss of endogenous neurosteroid tone

AUD patients, both male and female, have reduced circulating levels of progesterone, allopregnanolone, and THDOC during acute alcohol withdrawal, which normalized upon recovery (Hill et al., 2005; Romeo et al., 1996). Moreover, male alcohol-dependent subjects exhibited a delayed deoxycorticosterone response to CRF challenge (Porcu, O'Buckley, Morrow, & Adinoff, 2008), a likely consequence of the blunted HPA axis response in AUD patients (Porcu & Morrow, 2014).

In animal models of AUD, chronic ethanol consumption also depletes neuroactive steroid concentrations, but there are species, sex, and regional differences across the brain. In cynomolgus monkeys, there appears to be a pronounced sex difference in serum allopregnanolone levels following prolonged voluntary ethanol self-administration with males showing a depletion of allopregnanolone (Beattie et al., 2017), but female monkeys showing no change (Dozier et al., 2019). However, allopregnanolone immunoreactivity in the lateral and basolateral amygdala was reduced following 12 months of voluntary alcohol self-administration in the male monkey study, and the decrease was inversely related to daily ethanol intake and more pronounced in heavy drinkers (Beattie et al., 2017). However, in the same subjects, allopregnanolone immunoreactivity in the hippocampus was slightly enhanced (Beattie et al., 2018), suggesting that chronic alcohol may selectively dysregulate GABAergic neuroactive steroids across brain regions. In support of this hypothesis, increased allopregnanolone immunoreactivity was reported in the VTA and substantia nigra pars medialis of male human AUD subjects (Hasirci, Maldonado-Devincci, Beattie, O'Buckley, & Morrow, 2017).

Ethanol-dependent male rats, but not females, also have decreased basal cerebrocortical and hippocampal allopregnanolone levels (Cagetti, Pinna, Guidotti, Baicy, & Olsen, 2004; Janis, Devaud, Mitsuyama, & Morrow, 1998; Morrow, Porcu, Boyd, & Grant, 2006), and there is a blunted elevation of cerebrocortical allopregnanolone, and of plasma and brain deoxycorticosterone levels in male rats following an acute ethanol challenge (Khisti, Boyd, Kumar, & Morrow, 2005; Morrow et al., 2006), suggesting tolerance to ethanol-induced increases in neuroactive steroid levels. In mice, chronic alcohol exposure differentially affects GABAergic neuroactive steroids, depending on strain, sex, and brain regions examined (Finn, Beckley, Kaufman, & Ford, 2010; Porcu & Morrow, 2014; Snelling et al., 2014). Increased allopregnanolone levels were reported in whole brain from male, but not female, C57BL/6J mice, following voluntary ethanol consumption (Finn et al., 2010). By contrast, chronic intermittent ethanol exposure and subsequent withdrawal in male C57BL/6J mice decreased allopregnanolone immunoreactivity in the lateral amygdala and VTA (Maldonado-Devincci et al., 2014).

Overall, it appears that allopregnanolone is depleted in alcohol-dependent subjects, and this is particularly clear in human studies. Loss of endogenous neurosteroid tone may contribute to multiple etiologies of the disorder, such as E/I imbalance, dysregulation of the HPA axis, dysregulation of stimulated dopamine release, and excessive inflammatory signaling. Furthermore, the deficit in endogenous neurosteroid levels may exacerbate common symptoms of AUD such as anxiety, sleep disturbance, depression, irritability, seizures, and pain. Each of these common symptoms of AUDs have previously shown to be ameliorated by allopregnanolone or its precursors, pregnenolone and progesterone (Morrow et al., 2020).

4. Pregnane steroids mitigate neurobiological adaptations in AUD

4.1 Effects of allopregnanolone on E/I balance

The GABAergic effects of allopregnanolone have been well established since the late 1980's. Furthermore, these GABAergic effects are also known to mitigate anxiety, HPA axis activation, seizure susceptibility (see (Morrow et al., 2020) for review), and likely contribute to amelioration of depression and anhedonia as well (Antonoudiou et al., 2022). As an endogenous positive allosteric modulator of the $GABA_A$ receptor, allopregnanolone acutely shifts E/I balance toward inhibition.

E/I ratio may also be influenced by neuroimmune signaling, since inflammatory cytokines such as TNF-α can have effects on neurotransmission through decreasing $GABA_A$ receptor expression and increasing NMDA receptor expression (Stellwagen, Beattie, Seo, & Malenka, 2005). Therefore, the ability of allopregnanolone to increase neuronal inhibition is mediated through a combination of its effects on $GABA_A$ receptors as well as CRF and neuroimmune pathways in the brain. These mechanisms all enhance inhibitory transmission and, therefore, reduce the E/I ratio. Since E/I ratio is elevated in patients with AUD as well as animal models of AUD (vide supra), this factor may contribute to its therapeutic potential for treatment.

4.2 Allopregnanolone regulation of stress responses and stress sensitivity

The activation of the HPA axis stress response, leads to the synthesis and release of allopregnanolone. Allopregnanolone increases following stress in rats, in both plasma and brain (Barbaccia, Concas, Serra, & Biggio, 1998; Purdy, Morrow, Moore, & Paul, 1991), and in human serum following CRF or ACTH stimulation (Genazzani et al., 1998). Pretreatment with allopregnanolone and other neurosteroids, such as THDOC (another positive allosteric modulator of $GABA_A$ receptors) or progesterone, attenuates stress-induced circulating ACTH and corticosterone (Owens, Ritchie, & Nemeroff, 1992; Patchev, Hassan, Holsboer, & Almeida, 1996). Moreover, allopregnanolone treatment decreased baseline levels of hypothalamic CRF mRNA expression (Boero et al., 2021; Patchev, Shoaib, Holsboer, & Almeida, 1994) as well as CRF peptides in hippocampus, VTA, and CeA of males rats, while females rats only exhibited CRF peptide reductions in VTA (Boero et al., 2021). However, these studies did not consider other components of the HPA axis or examine different types of stressors.

Boero et al. (2022) recently showed that allopregnanolone administration prior to stress alters the HPA axis response at multiples levels, all of which serve to restore homeostasis, but differ by sex. They examined the regulation of stress-induced hypothalamic CRF, CRFR1, CRFBP, POMC, and GR by western blot and circulating CORT by ELISA in male and female Sprague Dawley rats. In females, allopregnanolone inhibited the effects of restraint stress on corticosterone, CRF, and pro-opiomelanocortin, while it inhibited the effects of swim stress on CRF, CRF1 receptor, CRF binding protein, and pro-opiomelanocortin. In males, allopregnanolone inhibited the effects of restraint stress and swim stress on all the stress markers except corticosterone. These findings indicate that the type of stressor makes a difference, but allopregnanolone inhibits the stress response through some mechanism for both restraint stress and forced swim stress. It is important to note that all rats showed a normal corticosterone response to stress in the absence of allopregnanolone. These data suggest that allopregnanolone plays an essential regulatory role on the neuroendocrine modifications induced by the stress response.

In human laboratory studies, progesterone and pregnenolone have been shown to normalize stress dysregulation in both cocaine users and heavy drinking subjects (Fox, Sofuoglu, Morgan, Tuit, & Sinha, 2013; Milivojevic et al., 2016; Milivojevic, Charron, Fogelman, Hermes, & Sinha, 2022; Milivojevic et al., 2023). Therefore, elevations of allopregnanolone levels by replacement therapy, selective neurosteroidogenic drugs, or steroid precursor administration all appear to be effective for the treatment of stress-dysregulation in general and particularly associated with cocaine and alcohol addiction.

4.3 Allopregnanolone regulation of dopamine signaling

Several studies have investigated the effect of allopregnanolone on dopaminergic transmission in vitro and in vivo and have found broad and varied effects. In one in vitro study, allopregnanolone acutely and dose-dependently induced the production of norepinephrine and dopamine measured by radioimmunoassay in a rat pheochromocytoma cell line (PC12) (Charalampopoulos et al., 2005). Moreover, allopregnanolone increased dopamine release evoked by N-methyl-D-aspartic acid (NMDA) in striatal slices from female rats. This effect was found to be estrous cycle dependent, as the observed increase was specific to slices from rats in estrus, but not ovariectomized or diestrus rats (Cabrera, Bregonzio, Laconi, & Mampel, 2002). In vivo studies recapitulated this in male rats, as intracerebroventricular 50 and 100 pmol allopregnanolone dose-dependently increased tonic dopamine release in the nucleus accumbens, as measured by microdialysis (Rouge-Pont et al., 2002). In contrast, however, another study found that intracerebroventricular allopregnanolone dose-dependently inhibited basal dopamine in the prefrontal cortex and nucleus accumbens of male rats. Furthermore, it blunted a foot-shock-induced increase in extracellular dopamine in the nucleus accumbens (Motzo et al., 1996). In support of this inverse correlation, reducing allopregnanolone levels via finasteride (a 5α-reductase inhibitor) increased tonic dopamine release in response to pharmacologically-induced stress (Dazzi et al., 2002). It is worth noting that Motzo et al. (1996) used 10 and 15 μg allopregnanolone, doses far greater in concentration than those used by Rouge-Pont and colleagues. Thus, the divergent effects of allopregnanolone, both increasing and reducing dopamine release, may be dose responsive. As allopregnanolone is known to produce effects on $GABA_A$ receptors at nanomolar concentrations, it may be that physiologically relevant doses elicit variable responses to those that are substantially higher (for review, see Lambert, Belelli, Hill-Venning, & Peters, 1995; Paul, Pinna, & Guidotti, 2020). In further support of studies reporting allopregnanolone-induced reductions in dopamine, Laconi and colleagues found that intracerebroventricular injection of 6 μM allopregnanolone decreased tissue levels of dopamine levels in striatum of female rats in estrus, using high performance liquid chromatography. Yet this same investigation found allopregnanolone increased striatal dopamine concentrations in ovariectomized rats primed with estrogen and progesterone (Laconi et al., 2007). These findings suggest there exist divergent effects of allopregnanolone based on dose and estrous cycle stage, highlighting the complexity of its modulation of dopamine transmission.

The above studies reflected tissue content of dopamine and tonic dopamine, representing the slow changes in dopamine concentrations over time, whereas phasic, sub-second fluctuations in dopamine release reflect salient stimuli, reward prediction error, and associative phases of reward (Schultz, 2016; Wightman & Robinson, 2002). A study from the Robinson lab found that systemically injected allopregnanolone dose-dependently reduced electrically-evoked phasic dopamine release in the nucleus accumbens of anesthetized male and female rats. This effect was shown to be sex- and estrous-cycle-stage-dependent, as females were less sensitive to allopregnanolone relative to males, and females in proestrus were less sensitive relative to non-proestrus females (Dornellas et al., 2020). To reconcile the potential confound of the compounding effect of anesthesia and allopregnanolone, the same group assessed the effect of allopregnanolone on spontaneous dopamine transients in awake and freely-moving male and female rats. Allopregnanolone significantly decreased the frequency and marginally reduced the amplitude of spontaneous dopamine transients in female, but not male rats (McFarland, Morrow, & Robinson, 2024b). The apparent sex difference remains to be clarified; however, prior research may suggest some involvement of estrous cycle stage in the differential responses to allopregnanolone (Dornellas et al., 2020).

This divergence between the impact of allopregnanolone on tonic and phasic dopamine release is also seen with other GABA agonists, like ethanol. Notably, ethanol reliably and dose-dependently increased tonic dopamine concentrations in the nucleus accumbens of awake male rats, whereas it both increased and decreased phasic dopamine release in a site-specific manner (Robinson, Howard, McConnell, Gonzales, & Wightman, 2009). As dopamine is involved in reward and motivation, one might speculate whether the observed reductions in dopamine release produced by allopregnanolone are aversive. In fact, allopregnanolone produced a positive place preference in male mice (Finn, Phillips, Okorn, Chester, & Cunningham, 1997) and in male and female rats (McFarland et al., 2024b). Notably, the latter study used the same dose (15 mg/kg) and timing (15 min post-injection) as the study that described reductions in phasic dopamine release (Dornellas et al., 2020; McFarland et al., 2024b). Moreover, in multiple studies, allopregnanolone did not alter motivation for a non-drug (sucrose) reward in male and female rats (Anker, Zlebnik, & Carroll, 2010; Janak & Michael Gill, 2003; McFarland et al., 2024b). Thus, while allopregnanolone can alter spontaneous dopamine release, it leaves intact the reward-associated behaviors that are regulated in part by dopamine neurotransmission.

The effect of allopregnanolone on dopamine release in pathological conditions and drug states remains to be determined. However, research on dopamine modulation by other compounds suggests a potential therapeutic role for allopregnanolone in treating AUD. Interestingly, CVT-10216 is a selective, reversible inhibitor of ALDH2 that suppressed self-administration and reinstatement of alcohol seeking in heavy drinking male rats (Arolfo et al., 2009). Moreover, it attenuated the increase in tonic dopamine release in the nucleus accumbens induced by binge sucrose consumption (Bocarsly et al., 2014) and intraperitoneal injection of alcohol (1 g/kg) (Arolfo et al., 2009). CVT-10216 also demonstrated anxiolytic effects during social interaction in rats subjected to repeated alcohol withdrawal (Overstreet, Knapp, Breese, & Diamond, 2009). While one effect of CVT-10216 is an aversive accumulation of acetaldehyde that likely contributes to the reduction of alcohol consumption, reductions in dopamine release also likely contribute. Similarly, one might investigate whether allopregnanolone blunts alcohol-induced increases in dopamine release, either in naive or alcohol-dependent subjects, as allopregnanolone has been shown to reduce electrically-evoked dopamine release (Dornellas et al., 2020), spontaneous dopamine transient frequency and amplitude (trend) in females, (McFarland et al., 2024), and tonic dopamine release (Motzo et al., 1996). It is plausible that allopregnanolone may normalize dysregulated dopamine signaling induced by alcohol dependence and AUD by mimicking alcohol actions in a limited way and reducing the incentive salience of the drug reward. In the cocaine literature, Anker and colleagues demonstrated that allopregnanolone reduced the escalation of cocaine-self administration and reinstatement in female rats (Anker, Holtz, Zlebnik, & Carroll, 2009; Anker et al., 2010). Thus, it is possible that reductions in dopamine transmission (frequency or amplitude) may correspond to these behaviors and this hypothesis warrants future investigation. Probing allopregnanolone effects across various experimental models is essential for understanding their neurobiological implications and evaluating allopregnanolone's therapeutic efficacy as it relates to dopaminergic transmission. Considering the dopamine-modulating effects of compounds like CVT-10216, we can hypothesize that allopregnanolone might also be effective as a sole or adjunct treatment for AUD through similar dopaminergic mechanisms.

4.4 Pregnane steroid inhibition of inflammatory signaling

Inflammation in AUD is caused by activation of TLRs, which induce multiple nuclear transcription factors that regulate the production of pro-inflammatory cytokines and chemokines. These inflammatory pathways are activated in blood macrophages, peripheral organs, and the brain. Both allopregnanolone and pregnenolone inhibit inflammatory signaling by blocking protein-protein interactions, such as TLR4-MD2 binding and TLR4-MyD88 binding, that are required for TLR4 activation (Balan et al., 2019). Allopregnanolone also blocks TLR2 and TLR7 pro-inflammatory signal activation and the production of cytokines/chemokines through their ability to block TLR2-MyD88 binding, and TLR7-MyD88 binding in the mouse macrophage RAW264.7 cell line and the alcohol-preferring (P) rat brain (Balan et al., 2021). More recent studies suggest that allopregnanolone inhibition of TLR4-TIRAP binding in the brain of alcohol-preferring P rats may also be involved in the blockade of TLR4 pathway activation (Lopez, O'Buckley, Balan, & Morrow, 2023). Neurosteroid inhibition of TLRs is specific for MyD88-dependent pathways and does not impact the TIR-domain-containing adapter-inducing interferon-β dependent TLR3 pathway. Furthermore, these effects are independent of $GABA_A$ receptor activity and could not be mimicked by the GABAergic steroid THDOC, although THDOC shared some effects on downstream chemokines (Balan et al., 2019). No sex differences in the responses to allopregnanolone were observed in the brain of P rats, in which TLR4 is innately activated (Balan et al., 2021).

The ability of allopregnanolone to inhibit TLR signal activation was confirmed in human studies of cultured monocyte-derived macrophages, where inhibition of TLR4 activation and its inflammatory pathway members tumor necrosis factor α (TNF-α), monocyte chemoattractant protein (MCP-1), phospho-CREB, and phospho-signal transducer and activator of transcription factor 1 (pSTAT1) were observed in cells from both male and female subjects (Balan et al., 2022). TLR7 activation by imiquimod was inhibited by allopregnanolone in human macrophages, blocking IL1-β, IL-6, pSTAT1, and pIRF7 elevations, but this effect was only observed in cells from in female donors. These data suggest that neurosteroid actions on TLR7 receptors may differ in male and female subjects, and this area deserves much more focus and investigation in future studies.

Other studies have shown that pregnenolone, progesterone, and allopregnanolone inhibit inflammation in macrophages and microglial cell lines by promoting the ubiquitination and degradation of TIRAP and TLR2 and thereby prevent TLR4- and TLR2-mediated production of TNF-α and IL-6 (Murugan et al., 2019). These studies show equivalent activities of all three steroids which share structural homology of the C and D rings, pointing to the importance of this structural feature of the molecules for inhibition of TLR signaling.

Furthermore, TLR4 mediates anti-inflammatory as well as pro-inflammatory pathways through distinct protein interactions (Siegemund & Sauer, 2012). Specifically, TLR4 engages the sorting adaptor TRAM and the signaling adaptor TRIF, inducing anti-inflammatory modulators such as IL-10 and brain derived neurotrophic factor (BDNF). We recently discovered that allopregnanolone can promote the TLR4-TRAM-TRIF pathway, likely in endosomes, to increase IL-10 and BDNF in P rat amygdala and nucleus accumbens of male, but not female, rats (Balan, Grusca, O'Buckley, & Morrow, 2023). Allopregnanolone administration to P rats elevated both IL-10 and BDNF levels in conjunction with the elevation of phosphorylated/activated pTRAM indicating activation of the endosomal TRIF-dependent TLR4 signals. The TRAM-dependent TLR4 activation was accompanied by elevation of the transcription factor specificity protein-1 (SP1) and the p110δ isoform of phosphatidylinositol-3-OH kinase (PI(3)K), as well as a reduction of TIRAP. Moreover, allopregnanolone facilitated TLR4, but not TLR3, accumulation in endosomes and promoted the transition between early endosomes (EEA1 levels were unchanged) and late endosomes (Rab7 levels were increased by 35%). Together, the data supports a role for allopregnanolone in the facilitation of anti-inflammatory pathways by elevations of IL-10 and BDNF. The mechanism of sex specificity in the effects of allopregnanolone on IL-10 and BDNF are unknown and may involve sex-dependent allopregnanolone doses. Furthermore, no studies of the structural requirements for this activity have been undertaken to date, so the direct actions of pregnenolone and progesterone are yet unknown. Further studies are clearly warranted.

In contrast, allopregnanolone induces a distinct anti-inflammatory pathway in brain of female rats (Balan et al., 2024). Allopregnanolone administration to P rats elevates fracktalkine (CX3CL1) levels by 16% in the NAc, 32% in the amygdala, 22% in the midbrain, and 23% in the hypothalamus of female P rats, with no significant changes observed in males. Additionally, female P rats treated with allopregnanolone also exhibited a 48% increase in the fracktalkine receptor (CX3CR1) and 24% increase in growth factor beta 1 levels, along with heightened activation (phosphorylation) of STAT1 (85%), indicating enhancement of the entire anti-inflammatory pathway in the NAc. No corresponding alterations were observed in male P rats. Furthermore, allopregnanolone decreased GFAP levels by 19% in both female and male P rat NAc, without affecting microglial markers IBA1 and TMEM119.

Anti-inflammatory modulators play an important role in cell function and health. In cases of neuronal damage, the anti-inflammatory chemokine fracktalkine induces the production of the anti-inflammatory cytokine IL-10, which helps repair damaged brain tissue and exhibits neuroprotective effects (Wang et al., 2021). IL-10 production in lung tissue protects against lethal oxidant injury (Zhang, Zhang et al., 2013). BDNF has trophic or regenerative actions in brain that alleviate depressive behaviors (Ge et al., 2015; Sun et al., 2021), restores $GABA_A$ receptors losses after ethanol exposure in cerebellar neurons (Ericson, Haythornthwaite, Yeh, & Yeh, 2003), and has protective effects against cerebral ischemia (Zhang, Zhao et al., 2013).

There is a constant interplay between endogenous pregnane steroids and neuroimmune pathways in both the periphery and the brain (Morrow, Boero, & Balan, 2024). Pregnenolone, progesterone, and allopregnanolone are present in circulation and in the brain (Purdy, Morrow, et al., 1991) along with endotoxin, cytokines, and other TLR activators (Akhter et al., 2019; Walter, Vetreno, & Crews, 2017; Zhang et al., 2008). The concurrent production of the pregnane steroids serves to prevent the activation of TLR signaling and thus prevents the production of pro-inflammatory molecules that promote inflammatory conditions. This is a healthy response to environmental challenges, and it likely protects against these disorders. Under conditions where neuroactive steroids are depleted, such as chronic excessive alcohol intake, chronic stress, or exposure to other toxins, inflammation and neuroinflammation are unchecked by the endogenous steroid TLR inhibitors and the production of inflammatory modulators feeds-forward to promote, exacerbate, or prolong inflammation. The ability of the endogenous pregnane steroids to inhibit the innate inflammatory pathways and enhance the innate anti-inflammatory pathways make them particularly attractive for treatment of AUDs, as well as other inflammatory brain conditions.

5. Potential benefits of pregnane steroid replacement in AUD

Our primary argument in this review is that the pregnane neurosteroids address multiple pathologies associated with AUDs, including the deficits in GABAergic transmission leading to altered E/I balance in the CNS, dysregulation of the HPA stress axis, dysregulation of dopamine signaling, and the inflammatory syndrome associated with AUDs (Fig. 1). However, it may be equally important to recognize that these steroids can also address many of the most common symptoms of AUDs, including anxiety, anhedonia, sleep disturbances, seizures, and pain. This argument was previously reviewed in Morrow et al. (2020), but will also be summarized here.

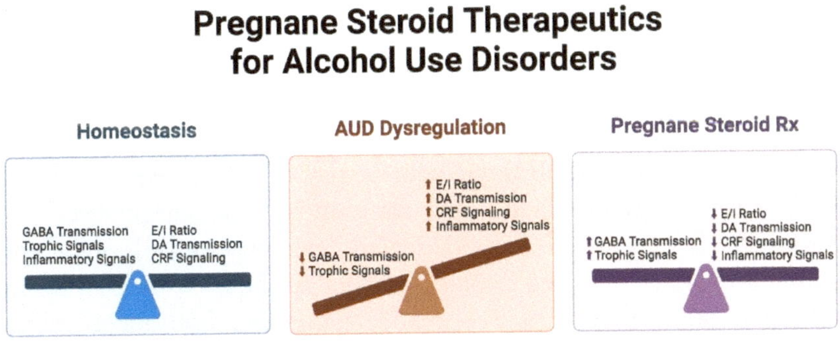

Fig. 1 Pleiotropic actions of pregnane steroids. Schematic showing the hypothesis that homeostasis in healthy individuals involves a balance in glutamate and GABA transmission that underlie the E/I ratio, dopamine (DA) transmission, CRF signaling, as well as inflammatory and trophic pathways. These crucial elements are dysregulated in AUD such that E/I ratio, stimulated DA transmission, CRF and inflammatory signaling are elevated while GABAergic and trophic pathways are inhibited. Pregnane steroids, particularly allopregnanolone, act to reverse the adaptations caused by chronic ethanol exposure and restore homeostasis by reducing E/I ratio, stimulated DA transmission, CRF activation and inflammatory signaling while enhancing GABAergic and trophic activity. Therefore, pregnane steroids are promising therapeutic candidates for the treatment of AUDs.

Allopregnanolone exerts anxiolytic-like and antidepressant-like effects in rodents (Bitran, Hilvers, & Kellogg, 1991; Crawley, Glowa, Majewska, & Paul, 1986; Khisti, Chopde, & Jain, 2000), which may be important for prevention and recovery from AUDs. Similar effects are reported in women treated for post-partum depression, where intravenous infusion of brexanolone, the proprietary formulation of allopregnanolone,exerts a rapid effect in counteracting postpartum anxiety and depression (Kanes et al., 2017; Meltzer-Brody et al., 2018). In addition, endogenous neuroactive steroids also exert sedative-hypnotic actions (Lancel et al., 1997; Mendelson et al., 1987), similar to other exogenous GABAergic modulators such as THIP, which act upon extrasynaptic $\alpha4\beta\delta$ $GABA_A$ receptors (Belelli & Lambert, 2005). Sleep disturbances are reported in AUDs, and alcohol's chronic effect on $GABA_A$ receptors is thought to contribute to sleep loss and increased wakefulness (Koob & Colrain, 2019; Kumar et al., 2009). Thus, administration of neuroactive steroids might ameliorate sleep quality in AUDs by restoring GABAergic tone. Allopregnanolone also has potent anticonvulsant effects in several animal models (Reddy & Rogawski, 2012). Moreover, allopregnanolone has potent anticonvulsant effects on seizures induced during ethanol withdrawal (Devaud, Purdy, & Morrow, 1995;

Devaud, Fritschy, & Morrow, 1998; Finn, Roberts, & Crabbe, 1995). Ethanol-dependent male rats are sensitized to the anticonvulsant effects of allopregnanolone and their $GABA_A$ receptors exhibit enhanced sensitivity to allopregnanolone as well (Cagetti et al., 2004; Devaud, Purdy, Finn, & Morrow, 1996). Thus, GABAergic neuroactive steroids may be therapeutic during ethanol withdrawal, and their use may have advantages over benzodiazepine therapy since benzodiazepines exhibit cross-tolerance with ethanol. Moreover, allopregnanolone and its synthetic 3β-methyl analog ganaxolone both proved effective in rodent models of treatment-resistant status epilepticus (Rogawski, Loya, Reddy, Zolkowska, & Lossin, 2013; Zolkowska, Wu, & Rogawski, 2018) and intravenous allopregnanolone proved therapeutic for status epilepticus in a number of individuals (Broomall et al., 2014). Preclinical studies suggest that neuroactive steroids play an important role in both inflammatory and neuropathic pain (Gonzalez et al., 2019; Kavaliers & Wiebe, 1987; Ocvirk, Pearson Murphy, Franklin, & Abbott, 2008). Thus, administration of neuroactive steroids might be beneficial to ameliorate pain symptoms. Indeed, a recent clinical study in 92 veterans showed that pregnenolone administration for 4 weeks reduced chronic pain symptoms by 20% (Naylor et al., 2020). Thus, there is a strong argument for therapeutic actions of allopregnanolone or its precursors for in the treatment of anxiety, anhedonia, sleep disturbances, seizures, and pain in AUD patients.

6. Potential limitations of pregnane steroid therapeutics

6.1 Allopregnanolone dependence risk and mitigation strategies

With the considerable and growing interest in the use of pregnane steroids for the treatment of psychiatric and neurological disorders, it becomes essential to examine the potential limitations and challenges associated with their use. Allopregnanolone produces robust anxiolytic, anticonvulsant, and sedative effects reminiscent of benzodiazepines and other $GABA_A$ receptor agonists. Consequently, brexanolone (intravenous allopregnanolone) was classified as a Schedule IV controlled substance by the United States Food and Drug Administration. As such, there are concerns about the development of tolerance, dependence, and withdrawal symptoms as observed with benzodiazepine cessation. Animal studies have established the risk of dependence upon allopregnanolone in female rats that exhibited anxiety and $GABA_A$ receptor dysfunction following withdrawal (Costa, Spence, Smith, & ffRench-Mullen, 1995; Moran & Smith, 1998; Moran, Goldberg, & Smith, 1998; Smith et al., 1998). However, no signs of dependence have been observed in patients treated with the intravenous formulation of allopregnanolone, brexanolone, for post-partum depression (Kanes et al., 2017; Meltzer-Brody et al., 2018). This work indicates that short term brexanolone therapy with gradual tapering of dose may prevent the possibility of dependence or withdrawal. More importantly, brexanolone therapy for post-partum depression has proven to be long-lasting with favorable outcomes as long as nine months after treatment (Patterson, Krohn, Richardson, Kimmel, & Meltzer-Brody, 2022). The long-lasting anti-depressant effects of brexanolone may be related to its anti-inflammatory actions, since brexanolone inhibition of inflammatory signals predicted the therapeutic outcome (reduction in the Hamilton Depression Ratings or HAM-D scores) in follow-up studies (Balan, Patterson, et al., 2023). Thus, the available data to date suggest that brexanolone therapy is safe, effective, and does not induce dependence with short term administration. Furthermore, short term therapy may be adequate to produce meaningful clinical responses in AUD patients. Progesterone and pregnenolone have also been used in clinical populations, posing no threat of dependence risk in available studies (Milivojevic et al., 2016; Milivojevic et al., 2022).

7. Potential interactions with other drugs of abuse

Potential interactions with ethanol or benzodiazepines are an obvious concern, particularly since relapse is a common feature in AUD. This possibility could be mitigated by treatment with brexanolone in a medically supervised setting.

Surprisingly, an untoward interaction between allopregnanolone and opiates was recently detected in studies of forced swim stress in rats (Boero et al., 2023). Allopregnanolone administration produced deleterious and potentially lethal submersion behavior during forced swim stress that could be attributed to stress-induced increases in serum β-endorphin as well as hypothalamic Mu-opiate receptors. Pre-treatment with the mu-opiate receptor antagonist D-Phe-Cys-Tyr-D-Trp-Arg-Thr-Pen-Thr-NH2 (CTAP) prevented the submersion behavior, indicating that the interaction involved the opiate response to swim stress in the rats. This interaction between allopregnanolone and the opioid system in the context of forced swim stress highlights a potentially lethal interaction between allopregnanolone and opiates that was not previously recognized or predicted by the known mechanisms of allopregnanolone action.

To avoid these potential issues, AUD treatment with pregnenolone or progesterone should be given more serious consideration. Since both steroids are converted in vivo to allopregnanolone, some GABAergic activity would be inherent to treatment, however, the anti-inflammatory effects of pregnenolone and progesterone do not require the conversion to allopregnanolone and these effects may prove therapeutic in their own right. Early clinical studies with pregnenolone for AUD are promising (Milivojevic et al., 2023) and this line of research deserves further exploration.

8. Challenges in the development of pregnane steroid therapeutics

8.1 Endogenous compounds preclude patent protection

The fact that pregnenolone, progesterone, and allopregnanolone are endogenous compounds presents complications in drug development since the cost cannot be subsidized by patent protection for the drug development industry. Obviously, analogs of these steroids could be developed to mimic some of their actions, but it may be challenging to mimic all of the many therapeutic actions of allopregnanolone for AUDs. While multiple steroids have been developed that mimic the GABAergic actions of allopregnanolone, including ganaxolone and zuranolone, it is unknown if these steroids mimic the effects of allopregnanolone on the modulation of the HPA axis, or its anti-inflammatory or trophic actions. Further studies on the structural requirements for each of the pleiotropic actions of allopregnanolone are needed.

9. Conclusions

Pregnane steroids hold great potential for the treatment of AUDs. Clinical studies are needed to establish the efficacy of these compounds in various stages of AUD severity across the lifespan. Efficacy is likely to differ by sex of the patient, as sex differences in the many effects of these steroids have been noted and most of the preclinical studies on these steroids were conducted in male animals. Since patent protection is not available for the endogenous pregnane steroids, government or private funding will be needed to address this goal. Industry development of novel compounds for AUD can also build upon the advances described in this review. Considering the huge magnitude of AUD diagnoses and the cost of AUDs to society, it seems the investment would present a considerable gain for all involved.

References

Adinoff, B., Junghanns, K., Kiefer, F., & Krishnan-Sarin, S. (2005). Suppression of the HPA axis stress-response: Implications for relapse. *Alcoholism: Clinical and Experimental Research, 29*(7), 1351–1355. https://doi.org/10.1097/01.alc.0000176356.97620.84.

Adinoff, B., Krebaum, S. R., Chandler, P. A., Ye, W., Brown, M. B., & Williams, M. J. (2005a). Dissection of hypothalamic-pituitary-adrenal axis pathology in 1-month-abstinent alcohol-dependent men, part 1: Adrenocortical and pituitary glucocorticoid responsiveness. *Alcoholism: Clinical and Experimental Research, 29*(4), 517–527. https://doi.org/10.1097/01.alc.0000158940.05529.0a.

Adinoff, B., Krebaum, S. R., Chandler, P. A., Ye, W., Brown, M. B., & Williams, M. J. (2005b). Dissection of hypothalamic-pituitary-adrenal axis pathology in 1-month-abstinent alcohol-dependent men, part 2: Response to ovine corticotropin-releasing factor and naloxone. *Alcoholism: Clinical and Experimental Research, 29*(4), 528–537. https://doi.org/10.1097/01.alc.0000158939.25531.ee.

Adinoff, B., Martin, P. R., Bone, G. H., Eckardt, M. J., Roehrich, L., George, D. T., ... Gold, P. W. (1990). Hypothalamic-pituitary-adrenal axis functioning and cerebrospinal fluid corticotropin releasing hormone and corticotropin levels in alcoholics after recent and long-term abstinence. *Archives of General Psychiatry, 47*(4), 325–330. http://www.ncbi.nlm.nih.gov/entrez/query.fcgi?cmd=Retrieve&db=PubMed&dopt=Citation&list_uids=2157379.

Agis-Balboa, R., Pinna, G., Zhubi, A., Veldic, M., Costa, E., & Guidotti, A. (2006a). Location and expression of brain enzymes catalyzing neurosteroid biosynthesis. *Proceedings of the National Academy of Sciences of the United States of America, 103*, 14602–14607.

Agis-Balboa, R. C., Graziano, P., Zhubi, A., Maloku, E., Veldic, M., Costa, E., & Guidotti, A. (2006b). Characterization of brain neurons that express enzymes mediating neurosteroid biosynthesis. *Proceedings of the National Academy of Sciences of the United States of America, 103*, 14602–14607.

Agis-Balboa, R. C., Guidotti, A., & Pinna, G. (2014). 5α-reductase type i expression is down-regulated in the prefrontal cortex/Brodmann's area 9 (BA9) of depressed patients [Article. *Psychopharmacology, 231*(17), 3569–3580. https://doi.org/10.1007/s00213-014-3567-5.

Agoglia, A. E., & Herman, M. A. (2018). The center of the emotional universe: Alcohol, stress, and CRF1 amygdala circuitry. *Alcohol (Fayetteville, N. Y.), 72*, 61–73. https://doi.org/10.1016/j.alcohol.2018.03.009.

Airagnes, G., Ducoutumany, G., Laffy-Beaufils, B., Le Faou, A. L., & Limosin, F. (2019). Alcohol withdrawal syndrome management: Is there anything new? *La Revue de Medecine Interne/Fondee ... par la Societe Nationale Francaise de Medecine Interne, 40*(6), 373–379. https://doi.org/10.1016/j.revmed.2019.02.001.

Akhter, N., Madhoun, A., Arefanian, H., Wilson, A., Kochumon, S., Thomas, R., ... Sindhu, S. (2019). Oxidative stress induces expression of the toll-like receptors (TLRs) 2 and 4 in the human peripheral blood mononuclear cells: Implications for metabolic inflammation. *Cellular Physiology and Biochemistry, 53*(1), 1–18. https://doi.org/10.33594/000000117.

Alfonso-Loeches, S., Pascual-Lucas, M., Blanco, A. M., Sanchez-Vera, I., & Guerri, C. (2010). Pivotal role of TLR4 receptors in alcohol-induced neuroinflammation and brain damage. *Journal of Neuroscience, 30*(24), 8285–8295. https://doi.org/10.1523/JNEUROSCI.0976-10.2010.

Anker, J. J., Holtz, N. A., Zlebnik, N., & Carroll, M. E. (2009). Effects of allopregnanolone on the reinstatement of cocaine-seeking behavior in male and female rats. *Psychopharmacology (Berl), 203*(1), 63–72. https://doi.org/10.1007/s00213-008-1371-9.

Anker, J. J., Zlebnik, N. E., & Carroll, M. E. (2010). Differential effects of allopregnanolone on the escalation of cocaine self-administration and sucrose intake in female rats. *Psychopharmacology (Berl), 212*(3), 419–429. https://doi.org/10.1007/s00213-010-1968-7.

Antonoudiou, P., Colmers, P. L. W., Walton, N. L., Weiss, G. L., Smith, A. C., Nguyen, D. P., ... Maguire, J. L. (2022). Allopregnanolone mediates affective switching through modulation of oscillatory states in the basolateral amygdala. *Biological Psychiatry, 91*(3), 283–293. https://doi.org/10.1016/j.biopsych.2021.07.017.

Arolfo, M. P., Overstreet, D. H., Yao, L., Fan, P., Lawrence, A. J., Tao, G., ... Diamond, I. (2009). Suppression of heavy drinking and alcohol seeking by a selective ALDH-2 inhibitor. *Alcoholism, Clinical and Experimental Research, 33*(11), 1935–1944. https://doi.org/10.1111/j.1530-0277.2009.01031.x.

Aurelian, L., & Balan, I. (2019). $GABA_AR$ α2-activated neuroimmune signal controls binge drinking and impulsivity through regulation of the CCL2/CX3CL1 balance. *Psychopharmacology (Berl), 236*(10), 3023–3043. https://doi.org/10.1007/s00213-019-05220-4.

Balan, I., Aurelian, L., Schleicher, R., Boero, G., O'Buckley, T., & Morrow, A. L. (2021). Neurosteroid allopregnanolone (3alpha,5alpha-THP) inhibits inflammatory signals induced by activated MyD88-dependent toll-like receptors. *Translational Psychiatry, 11*(1), 145. https://doi.org/10.1038/s41398-021-01266-1.

Balan, I., Aurelian, L., Williams, K. S., Campbell, B., Meeker, R. B., & Morrow, A. L. (2022). Inhibition of human macrophage activation via pregnane neurosteroid interactions with toll-like receptors: Sex differences and structural requirements. *Frontiers in Immunology, 13*, 940095. https://doi.org/10.3389/fimmu.2022.940095.

Balan, I., Beattie, M. C., O'Buckley, T. K., Aurelian, L., & Morrow, A. L. (2019). Endogenous neurosteroid (3α,5α)3-hydroxypregnan-20-one inhibits toll-like-4 receptor activation and pro-inflammatory signaling in macrophages and brain. *Scientific Reports, 9*(1), 1220. https://doi.org/10.1038/s41598-018-37409-6.

Balan, I., Grusca, A., Chery, S. L., Materia, B. R., O'Buckley, T. K., & Morrow, A. L. (2024). Neurosteroid [3α,5α]-3-hydroxy-pregnan-20-one enhances the CX3CL1-CX3CR1 pathway in the brain of alcohol-preferring rats with sex-specificity. *Life, 14*, 860. https://doi.org/10.3390/life14070860.

Balan, I., Grusca, A., O'Buckley, T. K., & Morrow, A. L. (2023). Neurosteroid [3α,5α]-3-hydroxy-pregnan-20-one enhances IL-10 production via endosomal TRIF-dependent TLR4 signaling pathway [Original Research]. *Frontiers in Endocrinology, 14*. https://doi.org/10.3389/fendo.2023.1299420.

Balan, I., Patterson, R., Boero, G., Krohn, H., O'Buckley, T. K., Meltzer-Brody, S., & Morrow, A. L. (2023). Brexanolone therapeutics in post-partum depression involves inhibition of systemic inflammatory pathways. *EBioMedicine, 89*, 104473. https://doi.org/10.1016/j.ebiom.2023.104473.

Bangasser, D. A., & Valentino, R. J. (2014). Sex differences in stress-related psychiatric disorders: Neurobiological perspectives. *Frontiers in Neuroendocrinology, 35*(3), 303–319. https://doi.org/10.1016/j.yfrne.2014.03.008.

Barbaccia, M. L., Concas, A., Serra, M., & Biggio, G. (1998). Stress and neurosteroids in adult and aged rats. *Experimental Gerontology, 33*(7-8), 697–712. http://www.ncbi.nlm.nih.gov/entrez/query.fcgi?cmd=Retrieve&db=PubMed&dopt=Citation&list_uids=9951617.

Bauer, J., Pedersen, A., Scherbaum, N., Bening, J., Patschke, J., Kugel, H., ... Ohrmann, P. (2013). Craving in alcohol-dependent patients after detoxification is related to glutamatergic dysfunction in the nucleus accumbens and the anterior cingulate cortex. *Neuropsychopharmacology: Official Publication of the American College of Neuropsychopharmacology, 38*(8), 1401–1408. https://doi.org/10.1038/npp.2013.45.

Baulieu, E. E., Robel, P., & Schumacher, M. (2001). Neurosteroids: Beginning of the story. *International Review of Neurobiology, 46*, 1–32. https://doi.org/10.1016/s0074-7742(01)46057-0.

Beattie, M. C., Maldonado-Devincci, A. M., Porcu, P., O'Buckley, T. K., Daunais, J. B., Grant, K. A., & Morrow, A. L. (2017). Voluntary ethanol consumption reduces GABAergic neuroactive steroid (3α,5α)3-hydroxypregnan-20-one (3α,5α-THP) in the amygdala of the cynomolgus monkey. *Addiction Biology, 22*(2), 318–330. https://doi.org/10.1111/adb.12326.

Beattie, M. C., Reguyal, C. S., Porcu, P., Daunais, J. B., Grant, K. A., & Morrow, A. L. (2018). Neuroactive steroid (3α,5α)3-hydroxypregnan-20-one (3α,5α-THP) and pro-inflammatory cytokine MCP-1 levels in hippocampus CA1 are correlated with voluntary ethanol consumption in cynomolgus monkey. *Alcoholism: Clinical and Experimental Research, 42*(1), 12–20. https://doi.org/10.1111/acer.13545.

Becker, H. C., & Mulholland, P. J. (2014). Neurochemical mechanisms of alcohol withdrawal. *Handbook of Clinical Neurology, 125*, 133–156. https://doi.org/10.1016/B978-0-444-62619-6.00009-4.

Belelli, D., Casula, A., Ling, A., & Lambert, J. J. (2002). The influence of subunit composition on the interaction of neurosteroids with GABA(A) receptors. *Neuropharmacology, 43*(4), 651–661. https://doi.org/10.1016/s0028-3908(02)00172-7.

Belelli, D., & Lambert, J. J. (2005). Neurosteroids: Endogenous regulators of the GABA(A) receptor. *Nature Reviews. Neuroscience, 6*(7), 565–575. https://doi.org/10.1038/nrn1703.

Bitran, D., Hilvers, R. J., & Kellogg, C. K. (1991). Anxiolytic effects of 3 alpha-hydroxy-5 alpha[beta]-pregnan-20-one: Endogenous metabolites of progesterone that are active at the GABAA receptor. *Brain Research, 561*, 157–161. https://doi.org/10.1016/0006-8993(91)90761-j.

Blaine, S. K., Nautiyal, N., Hart, R., Guarnaccia, J. B., & Sinha, R. (2019). Craving, cortisol and behavioral alcohol motivation responses to stress and alcohol cue contexts and discrete cues in binge and non-binge drinkers. *Addiction Biology, 24*(5), 1096–1108. https://doi.org/10.1111/adb.12665.

Bocarsly, M. E., Hoebel, B. G., Paredes, D., von Loga, I., Murray, S. M., Wang, M., ... Avena, N. M. (2014). GS 455534 selectively suppresses binge eating of palatable food and attenuates dopamine release in the accumbens of sugar-bingeing rats. *Behavioral Pharmacology, 25*(2), 147–157. https://doi.org/10.1097/FBP.0000000000000029.

Boero, G., McFarland, M. H., Tyler, R. E., O'Buckley, T. K., Chery, S. L., Robinson, D. L., ... Morrow, A. L. (2023). Deleterious interaction between the neurosteroid (3alpha,5alpha)3-hydroxypregnan-20-one (3alpha,5alpha-THP) and the mu-opioid system activation during forced swim stress in rats. *Biomolecules, 13*(8), https://doi.org/10.3390/biom13081205.

Boero, G., Tyler, R. E., O'Buckley, T. K., Balan, I., Besheer, J., & Morrow, A. L. (2022). 3alpha,5alpha)3-Hydroxypregnan-20-one (3alpha,5alpha-THP) regulation of the HPA axis in the context of different stressors and sex. *Biomolecules, 12*(8), https://doi.org/10.3390/biom12081134.

Boero, G., Tyler, R. E., Todd, C. A., O'Buckley, T. K., Balan, I., Besheer, J., & Morrow, A. L. (2021). (3alpha,5alpha)3-hydroxypregnan-20-one (3alpha,5alpha-THP) regulation of hypothalamic and extrahypothalamic corticotropin releasing factor (CRF): Sexual dimorphism and brain region specificity in Sprague Dawley rats. *Neuropharmacology, 186*, 108463. https://doi.org/10.1016/j.neuropharm.2021.108463.

Brodie, M. S., Pesold, C., & Appel, S. B. (1999). Ethanol directly excites dopaminergic ventral tegmental area reward neurons. *Alcoholism: Clinical and Experimental Research, 23*(11), 1848–1852.

Brodnik, Z. D., Batra, A., Oleson, E. B., & Espana, R. A. (2019). Local GABA(A) receptor-mediated suppression of dopamine release within the nucleus accumbens. *ACS Chemical Neuroscience, 10*(4), 1978–1985. https://doi.org/10.1021/acschemneuro.8b00268.

Broomall, E., Natale, J. E., Grimason, M., Goldstein, J., Smith, C. M., Chang, C., ... Wainwright, M. S. (2014). Pediatric super-refractory status epilepticus treated with allopregnanolone. *Annals of Neurology, 76*(6), 911–915. https://doi.org/10.1002/ana.24295.

Cabrera, R. J., Bregonzio, C., Laconi, M., & Mampel, A. (2002). Allopregnanolone increase in striatal N-methyl-D-aspartic acid evoked [3H]dopamine release is estrogen and progesterone dependent. *Cellular and Molecular Neurobiology, 22*(4), 445–454. https://doi.org/10.1023/a:1021015705597.

Cagetti, E., Pinna, G., Guidotti, A., Baicy, K., & Olsen, R. W. (2004). Chronic intermittent ethanol (CIE) administration in rats decreases levels of neurosteroids in hippocampus, accompanied by altered behavioral responses to neurosteroids and memory function. *Neuropharmacology, 46*(4), 570–579. https://doi.org/10.1016/j.neuropharm.2003.10.001.

Charalampopoulos, I., Dermitzaki, E., Vardouli, L., Tsatsanis, C., Stournaras, C., Margioris, A. N., & Gravanis, A. (2005). Dehydroepiandrosterone sulfate and allopregnanolone directly stimulate catecholamine production via induction of tyrosine hydroxylase and secretion by affecting actin polymerization. *Endocrinology, 146*(8), 3309–3318. https://doi.org/10.1210/en.2005-0263.

Chu, K., Koob, G. F., Cole, M., Zorrilla, E. P., & Roberts, A. J. (2007). Dependence-induced increases in ethanol self-administration in mice are blocked by the CRF1 receptor antagonist antalarmin and by CRF1 receptor knockout. *Pharmacology, Biochemistry, and Behavior, 86*(4), 813–821. https://doi.org/10.1016/j.pbb.2007.03.009.

Coleman, L. G., Jr., Zou, J., & Crews, F. T. (2017). Microglial-derived miRNA let-7 and HMGB1 contribute to ethanol-induced neurotoxicity via TLR7. *Journal of Neuroinflammation, 14*(1), 22. https://doi.org/10.1186/s12974-017-0799-4.

Coleman, L. G., Jr., Zou, J., Qin, L., & Crews, F. T. (2018). HMGB1/IL-1beta complexes regulate neuroimmune responses in alcoholism. *Brain, Behavior, and Immunity, 72*, 61–77. https://doi.org/10.1016/j.bbi.2017.10.027.

Cook, J. B., Dumitru, A. M., O'Buckley, T. K., & Morrow, A. L. (2014). Ethanol administration produces divergent changes in GABAergic neuroactive steroid immunohistochemistry in the rat brain. *Alcoholism: Clinical and Experimental Research, 38*(1), 90–99. https://doi.org/10.1111/acer.12223.

Cook, J. B., Werner, D. F., Maldonado-Devincci, A. M., Leonard, M. N., Fisher, K. R., O'Buckley, T. K., ... Morrow, A. L. (2014). Overexpression of the steroidogenic enzyme cytochrome P450 side chain cleavage in the ventral tegmental area increases 3alpha,5alpha-THP and reduces long-term operant ethanol self-administration. *Journal of Neuroscience, 34*(17), 5824–5834. https://doi.org/10.1523/JNEUROSCI.4733-13.2014.

Corpechot, C., Young, J., Calvel, M., Wehrey, C., Veltz, J. N., Touyer, G., ... Robel, P. (1993). Neurosteroids: 3a-hydroxy-5a-pregnan-20-one and its precursors in the brain, plasma, and steroidogenic glands of male and female rats. *Endocrinology, 133*, 1003–1009.

Costa, A. M., Spence, K. T., Smith, S. S., & ffRench-Mullen, J. (1995). Withdrawal from the endogenous steroid progesterone results in GABAA currents insensitive to benzodiazepine modulation in rat CA1 hippocampus. *Journal of Neurophysiology, 74*, 464–469. https://doi.org/10.1152/jn.1995.74.1.464.

Crawley, J. N., Glowa, J. R., Majewska, M. D., & Paul, S. M. (1986). Anxiolytic activity of an endogenous adrenal steroid. *Brain Research, 398*, 382–385. https://doi.org/10.1016/0006-8993(86)91500-3.

Crews, F. T., Lawrimore, C. J., Walter, T. J., & Coleman, L. G., Jr. (2017). The role of neuroimmune signaling in alcoholism. *Neuropharmacology, 122*, 56–73. https://doi.org/10.1016/j.neuropharm.2017.01.031.

Crews, F. T., Qin, L., Sheedy, D., Vetreno, R. P., & Zou, J. (2013). High mobility group box 1/Toll-like receptor danger signaling increases brain neuroimmune activation in alcohol dependence. *Biological Psychiatry, 73*(7), 602–612. https://doi.org/10.1016/j.biopsych.2012.09.030.

Crews, F. T., & Vetreno, R. P. (2016). Mechanisms of neuroimmune gene induction in alcoholism. *Psychopharmacology (Berl), 233*(9), 1543–1557. https://doi.org/10.1007/s00213-015-3906-1.

Dazzi, L., Serra, M., Vacca, G., Ladu, S., Latrofa, A., Trapani, G., & Biggio, G. (2002). Depletion of cortical allopregnanolone potentiates stress-induced increase in cortical dopamine output. *Brain Research, 932*(1-2), 135–139. https://doi.org/10.1016/s0006-8993(02)02290-4.

de Guglielmo, G., Kallupi, M., Pomrenze, M. B., Crawford, E., Simpson, S., Schweitzer, P., ... George, O. (2019). Inactivation of a CRF-dependent amygdalofugal pathway reverses addiction-like behaviors in alcohol-dependent rats. *Nature Communications, 10*(1), 1238. https://doi.org/10.1038/s41467-019-09183-0.

Devaud, L. L., Fritschy, J.-M., & Morrow, A. L. (1998). Influence of gender on chronic ethanol-induced alterations in GABAA receptors in rats. *Brain Research, 796*, 222–230. https://doi.org/10.1016/s0006-8993(98)00357-6.

Devaud, L. L., Purdy, R. H., Finn, D. A., & Morrow, A. L. (1996). Sensitization of gamma-aminobutyric acidA receptors to neuroactive steroids in rats during ethanol withdrawal. *Journal of Pharmacology and Experimental Therapeutics, 278*(2), 510–517. ⟨https://www.ncbi.nlm.nih.gov/pubmed/8768698⟩.

Devaud, L. L., Purdy, R. H., & Morrow, A. L. (1995). The neurosteroid, 3α-hydroxy-5α-pregnan-20-one, protects against bicuculline-induced seizures during ethanol withdrawal in rats. *Alcoholism: Clinical and Experimental Research, 19*, 350–355. https://doi.org/10.1111/j.1530-0277.1995.tb01514.x.

Domi, E., Domi, A., Adermark, L., Heilig, M., & Augier, E. (2021). Neurobiology of alcohol seeking behavior. *Journal of Neurochemistry, 157*(5), 1585–1614. https://doi.org/10.1111/jnc.15343.

Dornellas, A. P. S., Macedo, G. C., McFarland, M. H., Gomez, A. A., O'Buckley, T. K., Da Cunha, C., ... Robinson, D. L. (2020). Allopregnanolone decreases evoked dopamine release differently in rats by sex and estrous stage. *Frontiers in Pharmacology, 11*, 608887. https://doi.org/10.3389/fphar.2020.608887.

Dozier, B. L., Stull, C. A., Baker, E. J., Ford, M. M., Jensen, J. P., Finn, D. A., & Grant, K. A. (2019). Chronic ethanol drinking increases during the luteal menstrual cycle phase in rhesus monkeys: Implication of progesterone and related neurosteroids. *Psychopharmacology (Berl), 236*(6), 1817–1828. https://doi.org/10.1007/s00213-019-5168-9.

Ericson, M., Haythornthwaite, A. R., Yeh, P. W., & Yeh, H. H. (2003). Brain-derived neurotrophic factor mitigates chronic ethanol-induced attenuation of gamma-aminobutyric acid responses in cultured cerebellar granule cells. *Journal of Neuroscience Research, 73*(5), 722–730. http://www.ncbi.nlm.nih.gov/entrez/query.fcgi?cmd=Retrieve&db=PubMed&dopt=Citation&list_uids=12929140.

Fernandez-Lizarbe, S., Montesinos, J., & Guerri, C. (2013). Ethanol induces TLR4/TLR2 association, triggering an inflammatory response in microglial cells. *Journal of Neurochemistry, 126*(2), 261–273. https://doi.org/10.1111/jnc.12276.

Finn, D. A., Beckley, E. H., Kaufman, K. R., & Ford, M. M. (2010). Manipulation of GABAergic steroids: Sex differences in the effects on alcohol drinking- and withdrawal-related behaviors. *Hormones and Behavior, 57*(1), 12–22. https://doi.org/10.1016/j.yhbeh.2009.07.002 S0018-506X(09)00143-3 [pii].

Finn, D. A., Phillips, T. J., Okorn, D. M., Chester, J. A., & Cunningham, C. L. (1997). Rewarding effect of the neuroactive steroid 3a-hydroxy-5a-pregnan-20-one in mice. *Pharmacology, Biochemistry, and Behavior, 56*, 261–264.

Finn, D. A., Roberts, A. J., & Crabbe, J. C. (1995). Neuroactive steroid sensitivity in withdrawal seizure-prone and -resistant mice. *Alcoholism: Clinical and Experimental Research, 18*, 410–415.

Fiorenza, A. M., Shnitko, T. A., Sullivan, K. M., Vemuru, S. R., Gomez, A. A., Esaki, J. Y., ... Robinson, D. L. (2018). Ethanol exposure history and alcoholic reward differentially alter dopamine release in the nucleus accumbens to a reward-predictive cue. *Alcoholism, Clinical and Experimental Research, 42*(6), 1051–1061. https://doi.org/10.1111/acer.13636.

Fox, H. C., Sofuoglu, M., Morgan, P. T., Tuit, K. L., & Sinha, R. (2013). The effects of exogenous progesterone on drug craving and stress arousal in cocaine dependence: Impact of gender and cue type. *Psychoneuroendocrinology, 38*(9), 1532–1544. https://doi.org/10.1016/j.psyneuen.2012.12.022.

Garbutt, J. C., West, S. L., Carey, T. S., Lohr, K. N., & Crews, F. T. (1999). Pharmacological treatment of ethanol dependence: A review of the evidence. *JAMA: The Journal of the American Medical Association, 281*, 1318–1325.

Ge, L., Liu, L., Liu, H., Liu, S., Xue, H., Wang, X., ... Liu, D. (2015). Resveratrol abrogates lipopolysaccharide-induced depressive-like behavior, neuroinflammatory response, and CREB/BDNF signaling in mice. *European Journal of Pharmacology, 768*, 49–57. https://doi.org/10.1016/j.ejphar.2015.10.026.

Genazzani, A. R., Petraglia, F., Bernardi, F., Casarosa, E., Salvestroni, C., Tonetti, A., ... Luisi, M. (1998). Circulating levels of allopregnanolone in humans: Gender age, and endocrine influences. *The Journal of Clinical Endocrinology and Metabolism, 83*(6), 2099–2103.

Gonzales, R. A., Job, M. O., & Doyon, W. M. (2004). The role of mesolimbic dopamine in the development and maintenance of ethanol reinforcement. *Pharmacology & Therapeutics, 103*(2), 121–146. https://doi.org/10.1016/j.pharmthera.2004.06.002 S0163-7258(04)00095-6 [pii].

Gonzalez, S. L., Meyer, L., Raggio, M. C., Taleb, O., Coronel, M. F., Patte-Mensah, C., & Mensah-Nyagan, A. G. (2019). Allopregnanolone and progesterone in experimental neuropathic pain: Former and new insights with a translational perspective. *Cellular and Molecular Neurobiology, 39*(4), 523–537. https://doi.org/10.1007/s10571-018-0618-1.

Grobin, A. C., Roth, R. H., & Deutch, A. Y. (1992). Regulation of the prefrontal cortical dopamine system by the neuroactive steroid 3a,21-dihydroxy-5a-pregnane-20-one. *Brain Research, 578*, 351–356.

Hasirci, A. S., Maldonado-Devincci, A. M., Beattie, M. C., O'Buckley, T. K., & Morrow, A. L. (2017). Cellular GABAergic neuroactive steroid (3α,5α)-3-hydroxy-pregnan-20-One (3α,5α-THP) immunostaining levels are increased in the ventral tegmental area of human alcohol use disorder patients: A postmortem study. *Alcoholism: Clinical and Experimental Research, 41*(2), 299–311. https://doi.org/10.1111/acer.13300.

Henckens, M. J., Deussing, J. M., & Chen, A. (2016). Region-specific roles of the corticotropin-releasing factor-urocortin system in stress. *Nature Reviews. Neuroscience, 17*(10), 636–651. https://doi.org/10.1038/nrn.2016.94.

Hill, M., Popov, P., Havlikova, H., Kancheva, L., Vrbikova, J., Meloun, M., ... Starka, L. (2005). Reinstatement of serum pregnanolone isomers and progesterone during alcohol detoxification therapy in premenopausal women. *Alcoholism: Clinical and Experimental Research, 29*(6), 1010–1017. http://www.ncbi.nlm.nih.gov/entrez/query.fcgi?cmd=Retrieve&db=PubMed&dopt=Citation&list_uids=15976527.

Hosie, A. M., Wilkins, M. E., da Silva, H. M., & Smart, T. G. (2006). Endogenous neurosteroids regulate GABAA receptors through two discrete transmembrane sites. *Nature, 444*, 486–489. https://doi.org/10.1038/nature05324.

Irwin, R. P., Maragakis, N. J., Rogawski, M. A., Purdy, R. H., Farb, D. H., & Paul, S. M. (1992). Pregnenolone sulfate augments NMDA receptor mediated increases in intracellular Ca^{2+} in cultured rat hippocampal neurons. *Neuroscience Letters, 141*, 30–34.

Janak, P. H., & Michael Gill, T. (2003). Comparison of the effects of allopregnanolone with direct GABAergic agonists on ethanol self-administration with and without concurrently available sucrose. *Alcohol (Fayetteville, N. Y.), 30*(1), 1–7.

Janis, G. C., Devaud, L. L., Mitsuyama, H., & Morrow, A. L. (1998). Effects of chronic ethanol consumption and withdrawal on the neuroactive steroid 3α-hydroxy-5α-pregnan-20-one in male and female rats. *Alcoholism: Clinical and Experimental Research, 22*, 2055–2061.

Kane, C. J., Phelan, K. D., Douglas, J. C., Wagoner, G., Johnson, J. W., Xu, J., ... Drew, P. D. (2014). Effects of ethanol on immune response in the brain: Region-specific changes in adolescent versus adult mice. *Alcoholism: Clinical and Experimental Research, 38*(2), 384–391. https://doi.org/10.1111/acer.12244.

Kanes, S., Colquhoun, H., Gunduz-Bruce, H., Raines, S., Arnold, R., Schacterle, A., ... Meltzer-Brody, S. (2017). Brexanolone (SAGE-547 injection) in post-partum depression: A randomised controlled trial. *Lancet, 390*(10093), 480–489. https://doi.org/10.1016/S0140-6736(17)31264-3.

Kavaliers, M., & Wiebe, J. P. (1987). Analgesic effects of the progesterone metabolite, 3 alpha-hydroxy-5 alpha-pregnan-20-one, and possible modes of action in mice. *Brain Research, 415*(2), 393–398. https://doi.org/10.1016/0006-8993(87)90228-9.

Kegeles, L. S., Horga, G., Ghazzaoui, R., Rosengard, R., Ojeil, N., Xu, X., ... Abi-Dargham, A. (2018). Enhanced striatal dopamine release to expectation of alcohol: A potential risk factor for alcohol use disorder. *Biological Psychiatry: Cognitive Neuroscience and Neuroimaging, 3*(7), 591–598. https://doi.org/10.1016/j.bpsc.2018.03.018.

Khisti, R. T., Boyd, K. N., Kumar, S., & Morrow, A. L. (2005). Systemic ethanol administration elevates deoxycorticosterone levels and chronic ethanol exposure attenuates this response. *Brain Research, 1049*(1), 104–111. http://www.ncbi.nlm.nih.gov/entrez/query.fcgi?cmd=Retrieve&db=PubMed&dopt=Citation&list_uids=15936001.

Khisti, R. T., Chopde, C. T., & Jain, S. P. (2000). Antidepressant-like effect of the neurosteroid 3a-hydroxy-5a-pregnan-20-one in mice forced swim test. *Pharmacology, Biochemistry, and Behavior, 67*(1), 137–143. https://doi.org/10.1016/s0091-3057(00)00300-2.

Koob, G. F. (2013). Theoretical frameworks and mechanistic aspects of alcohol addiction: Alcohol addiction as a reward deficit disorder. *Current Topics in Behavioral Neurosciences, 13*, 3–30. https://doi.org/10.1007/7854_2011_129.

Koob, G. F. (2024). Alcohol use disorder treatment: Problems and solutions. *Annual Review of Pharmacology and Toxicology, 64*, 255–275. https://doi.org/10.1146/annurev-pharmtox-031323-115847.

Koob, G. F., & Colrain, I. M. (2019). Alcohol use disorder and sleep disturbances: A feed-forward allostatic framework. *Neuropsychopharmacology: Official Publication of the American College of Neuropsychopharmacology*. https://doi.org/10.1038/s41386-019-0446-0.

Kumar, S., Porcu, P., Werner, D. F., Matthews, D. B., Diaz-Granados, J. L., Helfand, R. S., & Morrow, A. L. (2009). The role of GABA(A) receptors in the acute and chronic effects of ethanol: A decade of progress. *Psychopharmacology (Berl), 205*, 529–564. http://www.ncbi.nlm.nih.gov/entrez/query.fcgi?cmd=Retrieve&db=PubMed&dopt=Citation&list_uids=19455309.

Laconi, M. R., Reggiani, P. C., Penissi, A., Yunes, R., & Cabrera, R. J. (2007). Allopregnanolone modulates striatal dopamingergic activity of rats under different gonadal hormones conditions. *Neurological Research, 29*(6), 622–627. https://doi.org/10.1179/016164107X166281.

Lambert, J. J., Belelli, D., Hill-Venning, C., & Peters, J. A. (1995). Neurosteroids and GABAA receptor function. *Trends in Pharmacological Sciences, 16*, 295–303.
Lancel, M., Faulhaber, J., Schiffelholz, T., Romeo, E., di Michele, F., Holsboer, F., & Rupprecht, R. (1997). Allopregnanolone affects sleep in a benzodiazepine-like fashion. *Journal of Pharmacology and Experimental Therapeutics, 282*, 1213–1218.
Leclercq, S., Cani, P. D., Neyrinck, A. M., Starkel, P., Jamar, F., Mikolajczak, M., ... de Timary, P. (2012). Role of intestinal permeability and inflammation in the biological and behavioral control of alcohol-dependent subjects. *Brain, Behavior, and Immunity, 26*(6), 911–918. https://doi.org/10.1016/j.bbi.2012.04.001.
Lloyd-Evans, E., & Waller-Evans, H. (2020). Biosynthesis and signalling functions of central and peripheral nervous system neurosteroids in health and disease. *Essays in Biochemistry, 64*(3), 591–606. https://doi.org/10.1042/ebc20200043.
Logrip, M. L., Milivojevic, V., Bertholomey, M. L., & Torregrossa, M. M. (2018). Sexual dimorphism in the neural impact of stress and alcohol. *Alcohol (Fayetteville, N. Y.), 72*, 49–59. https://doi.org/10.1016/j.alcohol.2018.02.002.
Long, D., Long, B., & Koyfman, A. (2017). The emergency medicine management of severe alcohol withdrawal. *The American Journal of Emergency Medicine, 35*(7), 1005–1011. https://doi.org/10.1016/j.ajem.2017.02.002.
Lopez, A. G., O'Buckley, T. K., Balan, I., & Morrow, A. L. (2023). Neuroimmune actions of (3a,5a)3-hydroxypregnan-20-one in alcohol preferring (P) rat brain. *Alcoholism, Clinical and Experimental Research.*
Majewska, M. D., Harrison, N. L., Schwartz, R. D., Barker, J. L., & Paul, S. M. (1986). Steroid hormone metabolites are barbiturate-like modulators of the GABA receptor. *Science (New York, N. Y.), 232*, 1004–1007.
Maldonado-Devincci, A. M., Cook, J. B., O' Buckley, T. K., Morrow, D. H., McKinley, R. E., Lopez, M. F., ... Morrow, A. L. (2014). Chronic intermittent ethanol exposure and withdrawal alters (3alpha,5alpha)-3-hydroxy-pregnan-20-one immunostaining in cortical and limbic brain regions of C57BL/6J mice. *Alcoholism: Clinical and Experimental Research, 38*(10), 2561–2571. https://doi.org/10.1111/acer.12530.
Marx, C. E., Keefe, R. S., Buchanan, R. W., Hamer, R. M., Kilts, J. D., Bradford, D. W., ... Shampine, L. J. (2009). Proof-of-concept trial with the neurosteroid pregnenolone targeting cognitive and negative symptoms in schizophrenia. *Neuropsychopharmacology: Official Publication of the American College of Neuropsychopharmacology, 34*(8), 1885–1903. https://doi.org/10.1038/npp.2009.26.
Mayfield, J., & Harris, R. A. (2017). The neuroimmune basis of excessive alcohol consumption. *Neuropsychopharmacology: Official Publication of the American College of Neuropsychopharmacology, 42*(1), 376. https://doi.org/10.1038/npp.2016.177.
McFarland, M. H., Machado, M. M. F., Sansbury, G. M., Musselman, K. C., Boero, G., O'Buckley, T. K., ... Robinson, D. L. (2024a). Acute, but not repeated, cocaine exposure alters allopregnanolone levels in the midbrain of male and female rats. *Psychopharmacology (Berl), 241*(5), 1011–1025. https://doi.org/10.1007/s00213-024-06534-8.
McFarland, M.H., Morrow, A.L., Robinson, D.L. (2024b) Allopregnanolone regulation of phasic dopamine release and motivated behavior. Submitted for publication.
Mellon, S. H., & Deschepper, C. F. (1993). Neurosteroid biosynthesis: Genes for adrenal steroidogenic enzymes are expressed in the brain. *Brain Research, 629*, 283–292.
Meltzer-Brody, S., Colquhoun, H., Riesenberg, R., Epperson, C. N., Deligiannidis, K. M., Rubinow, D. R., ... Kanes, S. (2018). Brexanolone injection in post-partum depression: Two multicentre, double-blind, randomised, placebo-controlled, phase 3 trials. *Lancet, 392*(10152), 1058–1070. https://doi.org/10.1016/S0140-6736(18)31551-4.
Mendelson, J. H., & Stein, S. (1966). Serum cortisol levels in alcoholic and nonalcoholic subjects during experimentally induced ethanol intoxication. *Psychosomatic Medicine, 28*(4), 616–626.

Mendelson, W. B., Martin, J. V., Perlis, M., Wagner, R., Majewska, M. D., & Paul, S. M. (1987). Sleep induction by an adrenal steroid in the rat. *Psychopharmacology, 93*, 226–229. https://doi.org/10.1007/bf00179939.

Mensah-Nyagan, A. G., Do-Rego, J. L., Beaujean, D., Luu-The, V., Pelletier, G., & Vaudry, H. (1999). Neurosteroids: Expression of steroidogenic enzymes and regulation of steroid biosynthesis in the central nervous system. *Pharmacological Reviews, 51*(1), 63–81.

Mereu, G., Fadda, F., & Gessa, G. L. (1984). Ethanol stimulates the firing rate of nigral dopaminergic neurons in unanesthetized rats. *Brain Research, 292*(1), 63–69. https://doi.org/10.1016/0006-8993(84)90890-4.

Mihalek, R. M., Banerjee, P. K., Korpi, E. R., Quinlan, J. J., Firestone, L. L., Mi, Z. P., ... Homanics, G. E. (1999). Attenuated sensitivity to neuroactive steroids in gamma-aminobutyrate type A receptor delta subunit knockout mice. *Proceedings of the National Academy of Sciences of the United States of America, 96*(22), 12905–12910. https://doi.org/10.1073/pnas.96.22.12905.

Milivojevic, V., Charron, L., Fogelman, N., Hermes, G., & Sinha, R. (2022). Pregnenolone reduces stress-induced craving, anxiety, and autonomic arousal in individuals with cocaine use disorder. *Biomolecules, 12*(11), https://doi.org/10.3390/biom12111593.

Milivojevic, V., Fox, H. C., Sofuoglu, M., Covault, J., & Sinha, R. (2016). Effects of progesterone stimulated allopregnanolone on craving and stress response in cocaine dependent men and women. *Psychoneuroendocrinology, 65*, 44–53. https://doi.org/10.1016/j.psyneuen.2015.12.008.

Milivojevic, V., Sullivan, L., Tiber, J., Fogelman, N., Simpson, C., Hermes, G., & Sinha, R. (2023). Pregnenolone effects on provoked alcohol craving, anxiety, HPA axis, and autonomic arousal in individuals with alcohol use disorder. *Psychopharmacology (Berl), 240*(1), 101–114. https://doi.org/10.1007/s00213-022-06278-3.

Modgil, A., Parakala, M. L., Ackley, M. A., Doherty, J. J., Moss, S. J., & Davies, P. A. (2017). Endogenous and synthetic neuroactive steroids evoke sustained increases in the efficacy of GABAergic inhibition via a protein kinase C-dependent mechanism. *Neuropharmacology, 113*(Pt A), 314–322. https://doi.org/10.1016/j.neuropharm.2016.10.010.

Moran, M. H., Goldberg, N., & Smith, S. S. (1998). Progesterone withdrawal II: Insensitivity to the sedative effects of a benzodiazepine. *Brain Research, 807*, 91–100. https://doi.org/10.1016/s0006-8993(98)00781-1.

Moran, M. H., & Smith, S. S. (1998). Progesterone withdrawal I: Pro-convulsant effects. *Brain Research, 807*, 84–90. https://doi.org/10.1016/s0006-8993(98)00782-3.

Morrow, A. L., Boero, G., & Balan, I. (2024). Emerging evidence for endogenous neurosteroid modulation of pro-inflammatory and anti-inflammatory pathways that impact neuropsychiatric disease. *Neuroscience & Biobehavioral Reviews* 105558. https://doi.org/10.1016/j.neubiorev.2024.105558.

Morrow, A. L., Boero, G., & Porcu, P. (2020). A rationale for allopregnanolone treatment of alcohol use disorders: Basic and clinical studies. *Alcoholism: Clinical and Experimental Research, 44*(2), 320–339. https://doi.org/10.1111/acer.14253.

Morrow, A. L., Porcu, P., Boyd, K. N., & Grant, K. A. (2006). Hypothalamic-pituitary-adrenal axis modulation of GABAergic neuroactive steroids influences ethanol sensitivity and drinking behavior. *Dialogues in Clinical Neuroscience, 8*, 463–477.

Morrow, A. L., Suzdak, P. D., & Paul, S. M. (1987). Steroid hormone metabolites potentiate GABA receptor-mediated chloride ion flux with nanomolar potency. *European Journal of Pharmacology, 142*, 483–485. https://doi.org/10.1016/0014-2999(87)90094-X.

Motzo, C., Porceddu, M. L., Maira, G., Flore, G., Concas, A., Dazzi, L., & Biggio, G. (1996). Inhibition of basal and stress-induced dopamine release in the cerebral cortex and nucleus accumbens of freely moving rats by the neurosteroid allopregnanolone. *Journal of Psychopharmacology, 10*, 266–272.

Mtchedlishvili, Z., & Kapur, J. (2003). A presynaptic action of the neurosteroid pregnenolone sulfate on GABAergic synaptic transmission. *Molecular Pharmacology, 64*, 857–864.

Murakami, K., Fellous, A., Baulieu, E. E., & Robel, P. (2000). Pregnenolone binds to microtubule-associated protein 2 and stimulates microtubule assembly. *Proceedings of the National Academy of Sciences of the United States of America, 97*(7), 3579–3584. http://www.ncbi.nlm.nih.gov/entrez/query.fcgi?cmd=Retrieve&db=PubMed&dopt=Citation&list_uids=10737804.

Murugan, S., Jakka, P., Namani, S., Mujumdar, V., & Radhakrishnan, G. (2019). The neurosteroid pregnenolone promotes degradation of key proteins in the innate immune signaling to suppress inflammation. *Journal of Biological Chemistry, 294*(12), 4596–4607. https://doi.org/10.1074/jbc.RA118.005543.

Naylor, J. C., Kilts, J. D., Shampine, L. J., Parke, G. J., Wagner, H. R., Szabo, S. T., ... Marx, C. E. (2020). Effect of pregnenolone vs placebo on self-reported chronic low back pain among US Military Veterans: A randomized clinical trial. -e200287 *JAMA Network Open, 3*(3), e200287. https://doi.org/10.1001/jamanetworkopen.2020.0287.

Oberlin, B. G., Dzemidzic, M., Tran, S. M., Soeurt, C. M., O'Connor, S. J., Yoder, K. K., & Kareken, D. A. (2015). Beer self-administration provokes lateralized nucleus accumbens dopamine release in male heavy drinkers. *Psychopharmacology (Berl), 232*(5), 861–870. https://doi.org/10.1007/s00213-014-3720-1.

Ocvirk, R., Pearson Murphy, B. E., Franklin, K. B., & Abbott, F. V. (2008). Antinociceptive profile of ring A-reduced progesterone metabolites in the formalin test. *Pain, 138*(2), 402–409. https://doi.org/10.1016/j.pain.2008.01.019.

Olsen, R. W., & Liang, J. (2017). Role of GABAA receptors in alcohol use disorders suggested by chronic intermittent ethanol (CIE) rodent model. *Molecular Brain, 10*(1), 45. https://doi.org/10.1186/s13041-017-0325-8.

Overstreet, D. H., Knapp, D. J., Breese, G. R., & Diamond, I. (2009). A selective ALDH-2 inhibitor reduces anxiety in rats. *Pharmacology, Biochemistry, and Behavior, 94*(2), 255–261. https://doi.org/10.1016/j.pbb.2009.09.004.

Owens, M. J., Ritchie, J. C., & Nemeroff, C. B. (1992). 5 alpha-pregnane-3 alpha, 21-diol-20-one (THDOC) attenuates mild stress-induced increases in plasma corticosterone via a non-glucocorticoid mechanism: Comparison with alprazolam. *Brain Research, 573*, 353–355. https://doi.org/10.1016/0006-8993(92)90788-B.

Oyola, M. G., & Handa, R. J. (2017). Hypothalamic-pituitary-adrenal and hypothalamic-pituitary-gonadal axes: Sex differences in regulation of stress responsivity. *Stress (Amsterdam, Netherlands), 20*(5), 476–494. https://doi.org/10.1080/10253890.2017.1369523.

Papadopoulos, V., Liu, J., & Culty, M. (2007). Is there a mitochondrial signaling complex facilitating cholesterol import? *Molecular and Cellular Endocrinology, 265-266*, 59–64.

Pascual, M., Balino, P., Aragon, C. M., & Guerri, C. (2015). Cytokines and chemokines as biomarkers of ethanol-induced neuroinflammation and anxiety-related behavior: Role of TLR4 and TLR2. *Neuropharmacology, 89*, 352–359. https://doi.org/10.1016/j.neuropharm.2014.10.014.

Patchev, V. K., Hassan, A. H. S., Holsboer, F., & Almeida, O. F. X. (1996). The neurosteroid tetrahydroprogesterone attenuates the endocrine response to stress and exerts glucocorticoid-like effects on vasopressin gene transcription in the rat hypothalamus. *Neuropsychopharmacology: Official Publication of the American College of Neuropsychopharmacology, 15*, 533–540.

Patchev, V. K., Shoaib, M., Holsboer, F., & Almeida, O. F. X. (1994). The neurosteroid tetrahydroprogesterone counteracts corticotropin-releasing hormone-induced anxiety and alters the release and gene expression of corticotropin-releasing hormone in the rat hypothalamus. *Neuroscience, 62*, 265–271. https://doi.org/10.1016/0306-4522(94)90330-1.

Patterson, R., Krohn, H., Richardson, E., Kimmel, M., & Meltzer-Brody, S. (2022). A brexanolone treatment program at an academic medical center: Patient selection, 90-day posttreatment outcomes, and lessons learned. *Journal of the Academy of Consultation-Liaison Psychiatry, 63*(1), 14–22. https://doi.org/10.1016/j.jaclp.2021.08.001.

Paul, S. M., Pinna, G., & Guidotti, A. (2020). Allopregnanolone: From molecular pathophysiology to therapeutics. A historical perspective. *Neurobiol Stress, 12*, 100215. https://doi.org/10.1016/j.ynstr.2020.100215.

Pavlov, V. A., & Tracey, K. J. (2017). Neural regulation of immunity: Molecular mechanisms and clinical translation. *Nature Neuroscience, 20*(2), 156–166. https://doi.org/10.1038/nn.4477.

Peng, J., Long, B., Yuan, J., Peng, X., Ni, H., Li, X., ... Li, A. (2017). A quantitative analysis of the distribution of CRH neurons in whole mouse brain. *Frontiers in Neuroanatomy, 11*, 63. https://doi.org/10.3389/fnana.2017.00063.

Porcu, P., & Morrow, A. L. (2014). Divergent neuroactive steroid responses to stress and ethanol in rat and mouse strains: Relevance for human studies. *Psychopharmacology (Berl), 231*(17), 3257–3272. https://doi.org/10.1007/s00213-014-3564-8.

Porcu, P., O'Buckley, T. K., Alward, S. E., Marx, C. E., Shampine, L. J., Girdler, S. S., & Morrow, A. L. (2009). Simultaneous quantification of GABAergic 3alpha,5alpha/3alpha,5beta neuroactive steroids in human and rat serum. *Steroids, 74*, 463–473. https://doi.org/10.1016/j.steroids.2008.12.015.

Porcu, P., O'Buckley, T. K., Morrow, A. L., & Adinoff, B. (2008). Differential hypothalamic-pituitary-adrenal activation of the neuroactive steroids pregnenolone sulfate and deoxycorticosterone in healthy controls and alcohol-dependent subjects. *Psychoneuroendocrinology, 33*(2), 214–226. https://doi.org/10.1016/j.psyneuen.2007.11.003.

Puia, G., Ducic, I., Vicini, S., & Costa, E. (1993). Does neurosteroid modulatory efficacy depend on GABAA receptor subunit composition? *Receptors and Channels, 1*, 135–142.

Puia, G., Santi, M., Vicini, S., Pritchett, D. B., Purdy, R. H., Paul, S. M., ... Costa, E. (1990). Neurosteroids act on recombinant human GABAA receptors. *Neuron, 4*, 759–765. https://doi.org/10.1016/0896-6273(90)90202-q.

Puia, G., Vicini, S., Seeburg, P. H., & Costa, E. (1991). Influence of recombinant gamma-aminobutyric acid-A receptor subunit composition on the action of allosteric modulators of gamma-aminobutyric acid-gated Cl- currents. *Molecular Pharmacology, 39*, 691–696.

Purdy, R. H., Moore, P. H., Morrow, A. L., & Paul, S. M. (1991). The 3a-hydroxy ring-A-reduced metabolites of progesterone and deoxycorticosterone: Natural ligands of central $GABA_A$ receptors. In E. Costa, & S. M. Paul (Eds.). *Neurosteroids and brain function* (pp. 95–102) Raven Press.

Purdy, R. H., Morrow, A. L., Blinn, J. R., & Paul, S. M. (1990). Synthesis, metabolism, and pharmacological activity of 3 alpha-hydroxy steroids which potentiate GABA-receptor-mediated chloride ion uptake in rat cerebral cortical synaptoneurosomes. *Journal of Medicinal Chemistry, 33*, 1572–1581. https://doi.org/10.1021/jm00168a008.

Purdy, R. H., Morrow, A. L., Moore, P. H., Jr., & Paul, S. M. (1991). Stress-induced elevations of gamma-aminobutyric acid type A receptor-active steroids in the rat brain. *Proceedings of the National Academy of Sciences of the United States of America, 88*, 4553–4557. https://doi.org/10.1073/pnas.88.10.4553.

Qin, L., He, J., Hanes, R. N., Pluzarev, O., Hong, J. S., & Crews, F. T. (2008). Increased systemic and brain cytokine production and neuroinflammation by endotoxin following ethanol treatment. *Journal of Neuroinflammation, 5*, 10. https://doi.org/10.1186/1742-2094-5-10.

Qin, L., Zou, J., Barnett, A., Vetreno, R. P., Crews, F. T., & Coleman, L. G., Jr. (2021). TRAIL mediates neuronal death in AUD: A link between neuroinflammation and neurodegeneration. *International Journal of Molecular Science, 22*(5). https://doi.org/10.3390/ijms22052547.

Ray, L. A. P., Bujarski, S. P., Grodin, E. P., Hartwell, E. P., Green, R. M., Venegas, A. B., ... Miotto, K. M. (2019). State-of-the-art behavioral and pharmacological treatments for alcohol use disorder. *The American Journal of Drug and Alcohol Abuse, 45*(2), 124–140. https://doi.org/10.1080/00952990.2018.1528265.

Reddy, D. S. (2010). Neurosteroids: Endogenous role in the human brain and therapeutic potentials. *Progress in Brain Research, 186,* 113–137. https://doi.org/10.1016/B978-0-444-53630-3.00008-7.

Reddy, D. S., & Rogawski, M. A. (2012). In J. L. Noebels, M. Avoli, M. A. Rogawski, R. W. Olsen, & A. V. Delgado-Escueta (Eds.). *Neurosteroids - Endogenous regulators of seizure susceptibility and role in the treatment of epilepsy*Jasper's Basic Mechanisms of the Epilepsieshttps://www.ncbi.nlm.nih.gov/pubmed/22787590\.

Rivier, C., & Vale, W. (1983). Modulation of stress-induced ACTH release by corticotropin-releasing factor, catecholamines and vasopressin. *Nature, 305*(5932), 325–327. https://doi.org/10.1038/305325a0.

Roberto, M., Spierling, S. R., Kirson, D., & Zorrilla, E. P. (2017). Corticotropin-releasing factor (CRF) and addictive behaviors. *International Review of Neurobiology 136,* 5–51. https://doi.org/10.1016/bs.irn.2017.06.004.

Robinson, D. L., Howard, E. C., McConnell, S., Gonzales, R. A., & Wightman, R. M. (2009). Disparity between tonic and phasic ethanol-induced dopamine increases in the nucleus accumbens of rats. *Alcoholism: Clinical and Experimental Research, 33*(7), 1187–1196. https://doi.org/10.1111/j.1530-0277.2009.00942.x ACER942 [pii].

Rogawski, M. A., Loya, C. M., Reddy, K., Zolkowska, D., & Lossin, C. (2013). Neuroactive steroids for the treatment of status epilepticus. *Epilepsia, 54*(*Suppl 6*), 93–98. https://doi.org/10.1111/epi.12289.

Romeo, E., Brancati, A., De Lorenzo, A., Fucci, P., Furnari, C., Pompili, E., ... Pasini, A. (1996). Marked decrease of plasma neuroactive steroids during alcohol withdrawal. *Clinical Neuropharmacology, 19*(4), 366–369. http://www.ncbi.nlm.nih.gov/pubmed/8829001.

Rouge-Pont, F., Mayo, W., Marinelli, M., Gingras, M., Le Moal, M., & Piazza, P. V. (2002). The neurosteroid allopregnanolone increases dopamine release and dopaminergic response to morphine in the rat nucleus accumbens. *European Journal of Neuroscience, 16*(1), 169–173. https://doi.org/10.1046/j.1460-9568.2002.02084.x[pii].

Rupprecht, R. (2003). Neuroactive steroids: Mechanisms of action and neuropsychopharmacological properties. *Psychoneuroendocrinology, 28*(2), 139–168 https://doi.org/10.1016/s0306-4530(02)00064-1 [pii].

Sanchez, W. N., Pochapski, J. A., Jessen, L. F., Ellenberger, M., Schwarting, R. K., Robinson, D. L., ... Da Cunha, C. (2022). Diazepam attenuates the effects of cocaine on locomotion, 50-kHz ultrasonic vocalizations and phasic dopamine in the nucleus accumbens of rats. *British Journal of Pharmacology, 179*(8), 1565–1577. https://doi.org/10.1111/bph.15658.

Schreiber, A. L., & Gilpin, N. W. (2018). *Corticotropin-releasing factor (CRF) neurocircuitry and neuropharmacology in alcohol drinking. Handbook of Experimental Pharmacology, 248,* Springer435–471. https://doi.org/10.1007/164_2017_86.

Schultz, W. (1998). Predictive reward signal of dopamine neurons. *Journal of Neurophysiology, 80*(1), 1–27. http://www.ncbi.nlm.nih.gov/pubmed/9658025.

Schultz, W. (2016). Dopamine reward prediction-error signalling: A two-component response. *Nature Reviews. Neuroscience, 17*(3), 183–195. https://doi.org/10.1038/nrn.2015.26.

Shnitko, T. A., & Robinson, D. L. (2015). Regional variation in phasic dopamine release during alcohol and sucrose self-administration in rats. *ACS Chemical Neuroscience, 6*(1), 147–154. https://doi.org/10.1021/cn500251j.

Siegemund, S., & Sauer, K. (2012). Balancing pro- and anti-inflammatory TLR4 signaling. *Nature Immunology, 13*(11), 1031–1033. https://doi.org/10.1038/ni.2452.

Sinha, R., Fox, H. C., Hong, K. I., Hansen, J., Tuit, K., & Kreek, M. J. (2011). Effects of adrenal sensitivity, stress- and cue-induced craving, and anxiety on subsequent alcohol relapse and treatment outcomes. *Archives of General Psychiatry, 68*(9), 942–952. https://doi.org/10.1001/archgenpsychiatry.2011.49.

Slater, E. P., Hesse, H., & Beato, M. (1994). Regulation of transcription by steroid hormones. *Annals of the New York Academy of Sciences, 733*, 103–112. https://doi.org/10.1111/j.1749-6632.1994.tb17260.x.

Smith, S. S., Gong, Q. H., Hsu, F.-C., Markowitz, R. S., Ffrench-Mullen, J. M. H., & Li, X. (1998). GABA(A) receptor alpha4 subunit suppression prevents withdrawal properties of an endogenous steroid. *Nature, 392*, 926–930. https://doi.org/10.1038/31948.

Smolders, I., De Klippel, N., Sarre, S., Ebinger, G., & Michotte, Y. (1995). Tonic GABA-ergic modulation of striatal dopamine release studied by in vivo microdialysis in the freely moving rat. *European Journal of Pharmacology, 284*(1-2), 83–91. https://doi.org/10.1016/0014-2999(95)00369-v.

Snelling, C., Tanchuck-Nipper, M. A., Ford, M. M., Jensen, J. P., Cozzoli, D. K., Ramaker, M. J., ... Finn, D. A. (2014). Quantification of ten neuroactive steroids in plasma in Withdrawal Seizure-Prone and -Resistant mice during chronic ethanol withdrawal. *Psychopharmacology (Berl), 231*(17), 3401–3414. https://doi.org/10.1007/s00213-014-3618-y.

Spigelman, I., Li, Z., Liang, J., Cagetti, E., Samzadeh, S., Mihalek, R. M., ... Olsen, R. W. (2003). Reduced inhibition and sensitivity to neurosteroids in hippocampus of mice lacking the GABA(A) receptor delta subunit. *Journal of Neurophysiology, 90*(2), 903–910. https://doi.org/10.1152/jn.01022.2002.

Spitta, G., Garbusow, M., Buchert, R., & Heinz, A. (2023). Dopamine and alcohol: A review of in vivo PET and SPECT studies. *Neuropsychobiology, 82*(6), 319–345. https://doi.org/10.1159/000534620.

Stell, B. M., Brickley, S. G., Tang, C. Y., Farrant, M., & Mody, I. (2003). Neuroactive steroids reduce neuronal excitability by selectively enhancing tonic inhibition mediated by delta subunit-containing GABAA receptors. *Proceedings of the National Academy of Sciences of the United States of America, 100*(24), 14439–14444. https://doi.org/10.1073/pnas.2435457100.

Stellwagen, D., Beattie, E. C., Seo, J. Y., & Malenka, R. C. (2005). Differential regulation of AMPA receptor and GABA receptor trafficking by tumor necrosis factor-alpha. *Journal of Neuroscience, 25*(12), 3219–3228. https://doi.org/10.1523/JNEUROSCI.4486-04.2005.

Sun, Y., Zhang, H., Wu, Z., Yu, X., Yin, Y., Qian, S., ... Chen, G. (2021). Quercitrin rapidly alleviated depression-like behaviors in lipopolysaccharide-treated mice: The involvement of PI3K/AKT/NF-kappaB signaling suppression and CREB/BDNF signaling restoration in the hippocampus. *ACS Chemical Neuroscience, 12*(18), 3387–3396. https://doi.org/10.1021/acschemneuro.1c00371.

Tsutsui, K. (2008). Neurosteroids in the Purkinje cell: Biosynthesis, mode of action and functional significance. *Molecular Neurobiology, 37*(2-3), 116–125. https://doi.org/10.1007/s12035-008-8024-1.

Tsutsui, K., Ukena, K., Takase, M., Kohchi, C., & Lea, R. W. (1999). Neurosteroid biosynthesis in vertebrate brains. *Comparative Biochemistry and Physiology - Part C: Toxicology, 124*(2), 121–129.

Umhau, J. C., Momenan, R., Schwandt, M. L., Singley, E., Lifshitz, M., Doty, L., ... Heilig, M. (2010). Effect of acamprosate on magnetic resonance spectroscopy measures of central glutamate in detoxified alcohol-dependent individuals: A randomized controlled experimental medicine study. *Archives of General Psychiatry, 67*(10), 1069–1077. https://doi.org/10.1001/archgenpsychiatry.2010.125.

Urban, N. B., Kegeles, L. S., Slifstein, M., Xu, X., Martinez, D., Sakr, E., ... Abi-Dargham, A. (2010). Sex differences in striatal dopamine release in young adults after oral alcohol challenge: A positron emission tomography imaging study with [(1)(1)C]raclopride. *Biological Psychiatry, 68*(8), 689–696. https://doi.org/10.1016/j.biopsych.2010.06.005.

Vallée, M., Vitiello, S., Bellocchio, L., Hébert-Chatelain, E., Monlezun, S., Martin-Garcia, E., ... Piazza, P. V. (2014). Pregnenolone can protect the brain from cannabis intoxication. *Science (New York, N. Y.), 343*(6166), 94–98. https://doi.org/10.1126/science.1243985.

Verplaetse, T. L., Cosgrove, K. P., Tanabe, J., & McKee, S. A. (2021). Sex/gender differences in brain function and structure in alcohol use: A narrative review of neuroimaging findings over the last 10 years. *Journal of Neuroscience Research, 99*(1), 309–323. https://doi.org/10.1002/jnr.24625.

Vetreno, R. P., & Crews, F. T. (2012). Adolescent binge drinking increases expression of the danger signal receptor agonist HMGB1 and Toll-like receptors in the adult prefrontal cortex. *Neuroscience, 226*, 475–488. https://doi.org/10.1016/j.neuroscience.2012.08.046.

Vetreno, R. P., Qin, L., Coleman, L. G., Jr., & Crews, F. T. (2021). Increased toll-like receptor-myD88-NFkappaB-proinflammatory neuroimmune signaling in the orbitofrontal cortex of humans with alcohol use disorder. *Alcoholism, Clinical and Experimental Research, 45*(9), 1747–1761. https://doi.org/10.1111/acer.14669.

Walter, T. J., Vetreno, R. P., & Crews, F. T. (2017). Alcohol and stress activation of microglia and neurons: Brain regional effects. *Alcoholism, Clinical and Experimental Research, 41*(12), 2066–2081. https://doi.org/10.1111/acer.13511.

Wang, J., Pan, H., Lin, Z., Xiong, C., Wei, C., Li, H., ... Dong, X. (2021). Neuroprotective effect of fractalkine on radiation-induced brain injury through promoting the M2 polarization of microglia. *Molecular Neurobiology, 58*(3), 1074–1087. https://doi.org/10.1007/s12035-020-02138-3.

Wightman, R. M., & Robinson, D. L. (2002). Transient changes in mesolimbic dopamine and their association with 'reward'. *Journal of Neurochemistry, 82*(4), 721–735. https://doi.org/10.1046/j.1471-4159.2002.01005.x.

Wu, F. S., Gibbs, T. T., & Farb, D. H. (1991). Pregnenolone sulfate: A positive allosteric modulator at the *N*-methyl-D-aspartate receptor. *Molecular Pharmacology, 40*, 333–336.

Zhang, Y., Zhang, X., Shan, P., Hunt, C. R., Pandita, T. K., & Lee, P. J. (2013). A protective Hsp70-TLR4 pathway in lethal oxidant lung injury. *Journal of Immunology, 191*(3), 1393–1403. https://doi.org/10.4049/jimmunol.1300052.

Zhang, L., Zhao, H., Zhang, X., Chen, L., Zhao, X., Bai, X., & Zhang, J. (2013). Nobiletin protects against cerebral ischemia via activating the p-Akt, p-CREB, BDNF and Bcl-2 pathway and ameliorating BBB permeability in rat. *Brain Research Bulletin, 96*, 45–53. https://doi.org/10.1016/j.brainresbull.2013.04.009.

Zhang, Y., Woodruff, M., Zhang, Y., Miao, J., Hanley, G., Stuart, C., ... Yin, D. (2008). Toll-like receptor 4 mediates chronic restraint stress-induced immune suppression. *Journal of Neuroimmunology, 194*(1-2), 115–122. https://doi.org/10.1016/j.jneuroim.2007.12.002.

Zolkowska, D., Wu, C. Y., & Rogawski, M. A. (2018). Intramuscular allopregnanolone and ganaxolone in a mouse model of treatment-resistant status epilepticus. *Epilepsia, 59*(*Suppl 2*), 220–227. https://doi.org/10.1111/epi.13999.

Zorrilla, E. P., Logrip, M. L., & Koob, G. F. (2014). Corticotropin releasing factor: A key role in the neurobiology of addiction. *Frontiers in Neuroendocrinology, 35*(2), 234–244. https://doi.org/10.1016/j.yfrne.2014.01.001.

CHAPTER FOUR

The CRF/Urocortin systems as therapeutic targets for alcohol use disorders

Cristiane Aparecida Favoretto[a,b,1], Natalia Bonetti Bertagna[a,1], Tarciso Tadeu Miguel[c], and Isabel M.H. Quadros[d,*]

[a]Molecular and Behavioral Neuroscience Laboratory, Pharmacology Department, Escola Paulista de Medicina, Universidade Federal de São Paulo (Unifesp), SP, Brazil
[b]Department of Molecular Medicine, The Scripps Research Institute, La Jolla, CA, United States
[c]Pharmacology Department, Universidade Federal de Uberlândia (Ufu), MG, Brazil
[d]Psychobiology Department, Escola Paulista de Medicina, Universidade Federal de São Paulo (Unifesp), SP, Brazil
*Corresponding author. e-mail address: isabel.quadros@unifesp.br

Contents

Abstract

Development and maintenance of alcohol use disorders have been proposed to recruit critical mechanisms involving Corticotropin Releasing Factor and Urocortins (CRF/Ucns). The CRF/Ucns system is comprised of a family of peptides (CRF, Ucn 1, Ucn 2, Ucn 3) which act upon two receptor subtypes, CRFR1 and CRFR2, each with different affinity profiles to the endogenous peptides and differential brain distribution. Activity of CRF/Ucn system is further modulated by CRF binding protein (CRF-BP), which regulates availability of CRF and Ucns to exert their actions. Extensive evidence in preclinical models support the involvement of CRF/Ucn targets in escalated alcohol drinking, as well as point to changes in CRF/Ucn brain function as a result of chronic alcohol exposure and/or withdrawal. It highlights the role of CRF and CRFR1-mediated signaling in conditions of excessive alcohol taking and seeking, including during

[1] *Shared first authorship.*

International Review of Neurobiology, Volume 178
ISSN 0074-7742, https://doi.org/10.1016/bs.irn.2024.08.002

various stages of withdrawal and relapse to alcohol. Besides its role in the hypothalamic-pituitary-adrenal (HPA) axis, the importance of extra-hypothalamic CRF pathways, especially in the extended amygdala, in the neurobiology of alcohol abuse and dependence is emphasized. Emerging roles for other targets of the CRF/Ucn system, such as CRF2 receptors, CRF-BP and Ucns in escalated alcohol drinking is also discussed. Finally, the limited translational value of CRF/Ucn interventions in stress-related and alcohol use disorders is discussed. So far, CRFR1 antagonists have shown little or no efficacy in human clinical trials, although a range of unexplored conditions and possibilities remain to be explored.

1. Overview

Drug and alcohol use disorders can be viewed as "stress surfeit disorders", with dysregulation of brain stress systems playing a crucial role in the shift from reward-motivated drug taking to a behavior driven by dysphoria and negative reinforcement. The corticotropin-releasing factor (CRF), a key player in mediating brain stress responses, has emerged as a critical target in mechanisms contributing to alcohol dependence. Preclinical studies have highlighted the significant involvement of CRF and related peptides, Urocortins, in the escalation of alcohol consumption and other alcohol-related behaviors. This chapter will provide a glance into the CRF-related family of peptides, including CRF and Urocortins (Ucns) 1, 2 and 3, its receptors, modulators and signaling mechanisms. The role of CRF/Ucns systems in modulating neuroendocrine stress responses and extra-hypothalamic regions and pathways will be briefly presented.

We will discuss a number of preclinical studies showing how alcohol exposure, as well as withdrawal from alcohol, can impact CRF/Ucn signaling and plasticity, contributing to neuroadaptive changes in the brain, using animal models of alcohol use and dependence. Changes in CRF/Ucn systems are presented in stress- and reward-related brain regions, according to different molecular targets that participate in CRF/Ucn transmission, from gene expression to peptide/protein levels and electrophysiological properties of CRF/Ucn neurons.

In non-human animal models, pharmacological interventions targeting CRF/Ucn receptors show consistent effects in reducing excessive alcohol consumption and alleviating withdrawal-related drinking. We will discuss the effects of different CRF receptor antagonists on alcohol drinking models and protocols in rodents. Furthermore, studies with CRF receptor antagonists (and a few agonists) administered directly into specific brain regions shed light into the neurobiology of alcohol use disorders. Finally, we discuss a limited number of experimental and clinical human studies which assessed the effects of CRFR1 antagonists in stress-related disorders and alcohol- and drug-use disorders.

We refer readers to Chapter 1 of this book for a general overview on the importance, characterization of alcohol use disorders and current treatment options (*Introduction: treatments endorsed by regulatory agencies for alcohol use disorder*), which will not be addressed in this chapter. Further, Chapter 11 can be a source for a thorough discussion of preclinical models of alcohol-reward and alcohol-dependence, pharmacology and neurobiology of alcohol's effects (*Animal models and translational challenges*).

This chapter did not include studies in genetically modified animals or polymorphisms (see review by Phillips, Reed, & Pastor, 2015). We also did not include human studies on genetic polymorphisms in CRF-related genes associated with alcohol use disorders, which have been covered in different reviews (see Phillips, Reed, & Pastor, 2015; Quadros et al., 2016).

2. Corticotropin-releasing factor (CRF) and CRF family peptides: an overview

Corticotropin-releasing factor (CRF), or hormone (CRH), is a peptide with 41 amino acids, originally associated with neuroendocrine, immune, cardiovascular and behavioral responses to stress mediated by the Hypothalamus-Pituitary-Adrenal (HPA) axis (or HHA axis, hypothalamus-hypophysis-adrenal axis). More recently, studies have demonstrated that brain CRF function far exceeds its pivotal role in the HPA axis, exerting major modulation in several physiologic and pathophysiologic processes (Vale et al., 1981; Inda et al., 2017).

CRF is synthesized from the CRF gene, located on the long arm of chromosome 8, location 8q13.1. Chromosome 8 has around 700 genes that encode a variety of different proteins with different functions in the organism. Specifically, the CRF gene encodes the transcription and translation of an inactive precursor peptide of 196 aminoacids named preproCRF, which is cleaved by proteases to the final and functional peptide, CRF. CRF is identical in humans, rats and mice and has some differences in sheep and other animals (84% identity sheep x human/rat CRF), revealing functional differences (Castro et al., 1995; Myers et al., 1998; King & Nicholson, 2007). CRF action is terminated by proteolytic cleavage by three exopeptidases. ACE (angiotensin converting enzyme), involved in angiotensin 2 synthesis from angiotensin 1, cleaves CRF in four fragments (CRF1–27 the largest). Neutral endopeptidase (NEP) also cleaves CRF, but in seven fragments (CRF1–26 is the largest). Finally, N-aminopeptidase (AP-N) also can cleave CRF. All fragments originated are devoid of function.

Several factors and conditions may activate HPA axis, with stress serving as a major factor. Psychological and physical stressors such as excessive cold, infections, social stress and others, promote the activity of parvocellular neurons in the paraventricular nucleus of the hypothalamus that produce CRF and store it in large dense vesicles. With increased neuronal activity, action potentials induce calcium-dependent CRF exocytosis. Once released, CRF will reach the hypothalamic-pituitary portal vascular system, reaching the adenohypophysis gland (anterior hypophysis) and activating production and secretion of adrenocorticotropic hormone (ACTH) by specialized hormone-producing cells. The neuropeptide arginine-vasopressin (AVP), also produced in PVN neurons, has synergistic actions with CRF in the activation of the HPA axis, mediated by activation of V1b receptors in the anterior hypophysis (Rivier et al., 1984; Papadimitriou & Priftis, 2009; Quadros et al., 2016).

Once released into systemic circulation, ACTH reaches the adrenal glands, where it stimulates the production and release of glucocorticoids (cortisol in humans, corticosterone in rats and mice). Cortisol and corticosterone hormones have diverse functions, including participation in the metabolism of carbohydrates, lipids and proteins, as well as anti-inflammatory and immunosuppressant actions. HPA function is carefully regulated by negative feedback mechanisms which inhibit the expression and release of hypothalamic CRF in the presence of ACTH (short negative feedback loop) and glucocorticoids (long negative feedback loop). Dysregulation of CRF/ACTH/Cortisol levels can trigger severe health consequences (Vale et al., 1981; Papadimitriou & Priftis, 2009; Inda et al., 2017).

CRF, ACTH and cortisol levels can be elevated in patients with stress-related disorders including anxiety, PTSD (Post-Traumatic Stress Disorder) and depression. Investigation of mood disorders often include measurements of CRF/ACTH/cortisol levels, and HPA axis hyperactivity is a proposed mechanism in the neuroendocrine hypothesis of depression. Cortisol activates nuclear glucocorticoid receptors (GR), activating gene transcription of factors that promote neuronal death and provoking gene repression of factors related to hippocampal and cortical neurogenesis. This causes an unfavorable balance between neuronal death and neurogenesis and, may contribute to neuroanatomical changes observed in patients with depression. In these patients, a reduction in hippocampal metabolism and volume is often observed, with progressive hippocampal atrophy, coincident with the time course of the worsening of depressive symptoms (Sapolsky, 2000; Lee et al., 2002).

Besides the hypothalamus, a relevant number of brain regions have neurons that produce CRF and/or present CRF receptors, such as the amygdala, bed nucleus of the stria terminalis (BNST), nucleus accumbens, periaqueductal gray (PAG), hippocampus, cerebral cortex, locus coeruleus and other sites. In these areas, CRF is involved with diverse actions and changes in CRF activity are involved in the pathophysiology of different disorders such as epilepsy, anxiety, depression, drug and alcohol use disorders (Steckler & Holsboer, 1999; Van Pett et al., 2000; Henckens et al., 2016; Inda et al., 2017).

CRF binds as agonist to two types of receptors: CRF type 1 receptor (CRFR1), with 415 amino acids encoded by the CRFR1 gene present on human chromosome 17, and the CRF type 2 receptor (CRFR2), encoded by the CRFR2 gene present on chromosome 7. Despite the high homology between these two receptors (around 70%), there are different splice variants for both receptors, the CRFR1α and CRFR1β subtypes, as well as CRFR2α (411 amino acids), CRFR2β (431 amino acids) and CRFR2γ (397 amino acids). Both CRFR1 and CRFR2, as well as their functional variants, are coupled to the stimulatory G protein signaling, which when activated leads to the activation of guanylyl cyclase converting ATP to cAMP, which, in turn, activates protein kinase A (PKA; Lovenberg et al., 1995; McCarthy et al., 1999; Markovic & Grammatopoulos, 2010; Henckens et al., 2016; Vandael & Gounko, 2019). The activation of this pathway leads to the phosphorylation of several substrates such as enzymes, ion channels and transcription factors such as CREB (cAMP Response Element Binding Protein), an important transcription factor associated with cAMP and PKA.

Although CRF actions through cAMP/PKA signaling are widely studied, findings show distinct pathways also triggered by CRF receptor activation. Some studies characterized a signaling cascade mediated by the activation of MAPK (Mitogen Activated Protein Kinase), which also results in the phosphorylation and activation of CREB. Other cascades have been proposed in different studies, with CRF actions being mediated by PKC through the activation of Gqα protein, with activation of the Beta-Gamma subunit of the G protein. The participation of the Beta-Gamma subunit has also been proposed to occur for CRF-induced phosphorylation of CREB. Moreover, CRF receptors may also interact with other G-protein systems, including Gqα, Gi, Go, Gi1/2 and Gz. Thus, depending on their location and cellular context, CRF receptors can recruit different cellular signaling cascades, in a concentration-dependent manner. These findings demonstrate the complexity of CRF actions, which can occur through the classic and most studied pathway (cAMP/PKA), but also through lesser-known routes, and for this reason, require further studies (Blank et al., 2003; Hauger et al., 2006; Henckens et al., 2016).

The distribution of CRFR1 and CRFR2 in the rodent brain is heterogeneous, with high densities of CRFR1 in the anterior/adenohypophysis, cerebral cortex, cerebellum, amygdala, hippocampus and striatum, while CRFR2 have a more restricted distribution in the mesencephalon, raphé nuclei, lateral septum, amygdala and hypothalamus (Chalmers et al., 1995; Lovenberg et al., 1995). CRF affinity to CRFR1 is about 10 times higher than to CRFR2 (Vandael & Gounko, 2019). While most of the known modulation of CRF in different physiological and pathological processes involves actions in CRFR1, specific roles for CRFR2 have emerged and are still being explored (Henckens et al., 2016; Quadros et al., 2016). For example, a study showed that CRFR1 antagonism, but not CRFR2 antagonism, attenuated anxiety-like behaviors induced by CRF injection into the periaqueductal gray of mice (Miguel & Nunes-de-Souza, 2011). Studies in rats also showed the same pattern of response in other brain region involved in anxiety-like behavior, the bed nucleus of stria terminalis (BNST, (Sahuque et al., 2006).

A second important factor in the complexity of CRF function is the fact that other endogenous peptides can also bind to CRF receptors, the Urocortin (Ucn) peptides, which also belong to the CRF-family of peptides. There are three types of Urocortins: Urocortin 1 (Ucn1), composed of 40 amino acids encoded by the UCN gene present on chromosome 2, Urocortin 2 (Ucn2), also known as stresscopin-peptide related, composed of 38 amino acids encoded by the UCN2 gene present on chromosome 3 and, finally, Urocortin 3 (Ucn3), also known as stresscopin. Ucn3 has two isoforms originating from a post-translational precursor with 161 amino acids. The more prevalent form is composed of 38 amino acids, and the other isoform with 41 amino acids, less prevalent, both encoded by the UCN3 gene present on chromosome 10. In addition to these three peptides present in mammals, two other related peptides are found in fish (Urotensin) and in amphibians (Sauvagine) (Skelton et al., 2000; Vandael & Gounko, 2019).

Peptides in the CRF family show differences in binding and affinity with CRF receptors. CRF binds, as mentioned above, with high affinity to CRFR1, but it also binds to CRFR2 mainly when in high concentrations. Ucn1 binds with similar affinity to both receptors, while Ucn2 and Ucn3 bind preferentially to CRFR2 (Henckens et al., 2016; Vandael & Gounko, 2019). Despite a more discrete distribution relative to CRF, urocortins are expressed both in peripheral and central tissues. In the rodent brain, Ucn1 is present in the hypothalamus, especially in the Supraoptic Nucleus, in the PAG, in the nucleus ambiguous, Edinger-Westphal nucleus, among other nuclei. Regarding Ucn2, this peptide is also present in the hypothalamus, in the arcuate, paraventricular and supraoptic nuclei, in the *locus coeruleus* and other areas. Finally, Ucn3, like the other two, is also expressed in the hypothalamus, but in different subnuclei as the medial preoptic nucleus, perifornical nucleus, besides BNST and the medial amygdala (Henckens et al., 2016).

Like CRF, in general, Ucns are also involved in stress responses and implicated in the modulation of anxiety disorders. Despite common effects, each subtype of urocortin presents some specificity in its actions. Ucn1, for example, is associated with anorexigenic effects, and also participates in the regulation of anxiety, learning, memory, body temperature, and may have neuroprotective effects. Ucn2 is also involved in anxiety and anorexigenic effects, besides hypotension and changes in cardiovascular function and central modulation of insulin metabolism. Lastly, Ucn3 is also involved in anxiety and depression, presents anorexigenic effects and may promote beneficial effects on metabolic disorders, such as obesity, diabetes, and cardiovascular disease. Both Ucn3 and Ucn2 participate in the regulation of insulin secretion. However, knowledge of specific mechanisms and regulation of urocortins function still require more investigation (Henckens et al., 2016; Vandael & Gounko, 2019; Kavalakatt et al., 2021).

Regulation and availability of CRF and some Ucns are also modulated by the CRF binding protein (CRF-BP). The human gene CRHBP, present in the Chromossome 5, location 5q13.3 encodes this protein of 322 amino acids. CRF-BP binds to CRF and Ucn1 with notable affinity, to Ucn2 with low affinity, and it does not bind to Ucn3 (Behan et al., 1995a). When bound to CRF-BP, CRF and Ucns are temporarily sequestered and cannot exert their effects until available in free form. For example, CRF-BP inhibits CRF effects on ACTH production and release in vitro and, in the periphery, CRF-BP forms a dimer after binding to CRF and clears this peptide from blood (Behan et al., 1989; Potter et al., 1991). In the brain, CRF-BP is usually membrane-bound and is predominantly expressed in the cerebral cortex and subcortical limbic structures. In some brain regions, CRF-BP colocalizes with CRF receptors. In the hypophysis, CRF-BP is present in the secretory ACTH cells where it modulates CRF-induced ACTH release. Thus, expression of the CRF-BP, specifically in the HPA axis, results in lower levels of free CRF and decreases CRF-R1 activation and downstream signaling (Potter et al., 1992; Behan et al., 1995a; Vandael & Gounko, 2019).

Pharmacological manipulations of CRF-BP have emerged as interventions that can affect CRF/Urocortin function. For instance, CRF-BP binds with high affinity to a CRF fragment (CRF6–33), leading to increased levels of free CRF which can exert its actions of CRF receptors (Behan et al., 1995b). In fact, treatment with CRF6–33 restored brain CRF levels and induced improvements in memory and cognition in an animal model for Alzheimer's Disease (Behan et al., 1995b).

Increased complexity for CRF/Urocortin regulation and function was further added with the discovery of a rodent-specific splice variant α-isoform of CRFR2 (sCRFRα2, with 153 amino acids), which encodes a soluble ligand-binding domain of CRFR2 (Chen et al., 2005). Soluble CRFRα2 binds to CRF and Ucn1, but not to Ucn2 or Ucn3, and it is expressed in the olfactory bulb, cortex, midbrain, and the hypophysis. Its function is yet to be determined, but as sCRFRα2 binds to both CRF and Ucn1 with high affinity, it is possible that it acts as a regulator of the activity of both agonists. By binding to CRF and Ucn1, sCRFRα2 may be able to sequester these peptides and reduce their post-synaptic activity (Chen et al., 2005; Vandael & Gounko, 2019).

This chapter will focus on the involvement of the CRF-family of peptides and signaling in alcohol use disorders. While a large body of studies focus on CRF and actions mediated by CRFR1, there is growing interest towards Urocortin peptides, CRFR2 and even CRF-BP. Evidence of CRF inputs to VTA dopamine neurons, as well as expression of CRFR1, CRFR2 and CRF-BP in brain reward regions indicate the involvement of CRF/Ucn systems in both reward and stress circuits associated with drug and alcohol use disorders, as discussed in the following sections (e.g., Van Pett et al., 2000; Ungless et al., 2003; Sparta et al., 2008; Boyson et al., 2011; Hwa et al., 2013; Albrechet-Souza et al., 2015; Quadros et al., 2016).

3. Alcohol exposure impacts CRF/Urocortins signaling

Alcohol exposure has a significant impact on CRFergic transmission by altering transcripts and protein expression levels and binding dynamics of CRF, CRFR1, and CRFR2 receptors, corticotropin-releasing factor-binding protein (CRF-BP), and urocortins (Ucns). Moreover, alcohol may induce changes in excitability and other electrophysiological properties of cells expressing CRF, CRFR1, and CRFR2 at baseline or in response to the application of other substances. Such alcohol-induced neuroadaptations are possibly associated with behavioral changes induced by drug exposure and/or withdrawal. This section will gather and discuss studies reporting alcohol-produced alterations in different components of the CRF/Urocortins system in different brain regions. Most studies on alcohol effects via CRF/Ucn signaling will focus on amygdala, hypothalamus, pituitary, bed nucleus of stria terminalis (BNST), prefrontal cortex (PFC), nucleus accumbens (NAc), ventral tegmental area (VTA), hippocampus, and subregions. However, other brain areas will also be briefly approached—due to limited studies in specific literature. Importantly, we want to direct the reader's attention to the fact that the outcome seems dependent on protocol specificities and characteristics of experimental subjects, such as species, strain, sex, and age.

3.1 CRF

Firstly, we report studies on how alcohol exposure (single or repeated/chronic) and withdrawal (acute or protracted) affect CRF peptide or gene expression in different brain regions. Alcohol consequences on electrophysiological properties of CRF-expressing cells in the hypothalamus and BNST will also be discussed.

Following repeated alcohol exposure and withdrawal, studies reported increases in CRF peptide (Läck et al., 2005) or prepro-CRF mRNA levels (Lowery-Gionta et al., 2012) in the central nucleus of the amygdala (CeA). However, long-term alcohol intake did not alter CRF mRNA levels in alcohol-preferring rats' amygdala (Alhaddad et al., 2020). Interestingly, Caputi et al. (2021) reported that alcohol consumption reduced CRF mRNA levels in the amygdala of alcohol-preferring marchigian sardinian (msP) but not Wistar rats, suggesting that strain may influence ethanol-induced changes in CRFergic signaling.

Furthermore, repeated alcohol exposure also appears to increase activity levels of CeA CRFergic neurons in a sex-dependent manner. For example, male but not female rats consuming an alcohol liquid diet presented increased sustained neuronal activation of CeA CRF neurons, evaluated as the count of cells double-labeled for CRF and ΔFosB proteins (Retson et al., 2015).

Although some studies reported that single exposure to alcohol may also elevate CRF levels (Lam & Gianoulakis, 2011), and electrophysiological properties of CRF-containing neurons (Silberman et al., 2015) in the CeA, others failed to detect alterations in CeA CRF concentration after acute exposure (Eisenhardt et al., 2015). In the same way, while several studies found upregulated CRF expression (peptide or mRNA) following acute or protracted alcohol withdrawal in the amygdala (Pich et al., 1995) or the CeA (Sommer et al., 2008; Roberto et al., 2010; Eisenhardt et al., 2015; Li et al., 2019; Kreifeldt et al., 2022), Ketchesin et al. (2016) did not observe altered CeA CRF gene expression during acute abstinence. In the BLA, reduced preproCRF mRNA levels were found during protracted withdrawal after repeated alcohol exposure (Falco et al., 2009).

Adolescent alcohol exposure also seems to produce upregulations in CeA CRF levels and CRFergic cell activation, assessed later in life (Boutros et al., 2016; de Guglielmo et al., 2019; Sampedro-Piquero et al., 2022). On the other hand, considering the entire amygdala region, and not only CeA, adolescent intermittent alcohol exposure from postnatal day (PND) 31 to 55 via ip administration did not affect CRF gene expression at PND 70 (Sanchez-Marin et al., 2017). Interestingly, another work from Sánchez-Marín et al. (2022a) found that adolescent intermittent alcohol exposure via intragastric (ig) gavage, from PND 39–63, reduced amygdala CRF gene expression at PND 77. Therefore, the route of alcohol administration and/or the exact period of drug exposure might influence CRF-related neuroadaptations observed in the amygdala.

In the hypothalamus, a single oral alcohol administration enhanced CRF peptide content or levels 30–60 min later in rats (László et al., 2001). Also, although acute alcohol exposure via oral gavage or ip injection did not affect CRF mRNA levels in the rat hypothalamus (Rivier & Lee, 1996; Zhou et al., 2000), single ip and intracerebroventricular (icv) injection increased heteronuclear CRF (hnCRF) expression shortly after administration, in parvocellular paraventricular nucleus (PVN) of rats (Rivier & Lee, 1996; Lee et al., 2004).

Some studies found enhanced CRF mRNA or peptide levels in the hypothalamus after repeated alcohol exposure (Todkar et al., 2016; Simon et al., 2023) and during acute or protracted alcohol withdrawal (Marty et al., 2020; Seo et al., 2021; Simon et al., 2023). Contrarily, other studies did not detect changes in CRF peptide or gene expression in the hypothalamus following repeated exposure to alcohol consumption (Lowery-Gionta et al., 2012; Odeon et al., 2017; Alhaddad et al., 2020), oral gavage administration (Zhou et al., 2000), or high doses of alcohol vapor (Rivier et al., 1984).

It is important to note that the amount of time elapsed from the final exposure to alcohol, or withdrawal period, seems to be very important for the verification of possible drug-induced CRF-related neuroadaptations. For example, CRF immunoreactivity in the PVN was significantly higher in rats repeatedly-exposed to alcohol, compared to alcohol-naive rats ten days into withdrawal, but not at one day of withdrawal or following an alcohol binge session (Brancato et al., 2021).

Early alcohol exposure may lead to hypothalamic CRF-related neuroadaptations later in life, possibly associated with the onset and development of harmful alcohol-related behaviors. Prenatal alcohol exposure enhanced hypothalamic CRF peptide levels of adolescent mice (Caldwell et al., 2014, 2015). However, prenatal exposure to alcohol did not impact CRF mRNA levels in the hypothalamus of neonatal mice (Ruffaner-Hanson et al., 2023), nor in the parvocellular PVN of adult male and female rats (Glavas et al., 2007).

In addition, adolescent alcohol exposure seems to induce contrasting effects on hypothalamus CRF peptide or transcript levels. Adolescent alcohol drinking enhanced CRF density in the PVN during adulthood in mice (Sampedro-Piquero et al., 2022). Contrarily, Allen et al. (2011) showed that adolescent alcohol exposure via vapor inhalation decreased CRF mRNA levels in response to acute alcohol challenge via ig catheter during young adult life. However, in another study, the same authors did not find alterations in CRF mRNA expression in the PVN of adult rats exposed to vapor during adolescence, following a challenge with ig alcohol (Allen et al., 2016). Logrip et al. (2013) also showed that adolescent exposure to alcohol vapor failed to modify CRF mRNA levels in the rat PVN following alcohol or saline challenge in adulthood.

In the hypothalamus, particularly the PVN, alcohol-induced adaptations in the cell counting and electrophysiological properties of CRF-positive cells, seem to vary in a sex-dependent fashion. For example, during acute withdrawal from long-term alcohol intake, hyperexcitability of PVN CRF-expressing neurons was detected in female, but not male, mice (Neira et al., 2023). On the other hand, during acute abstinence, male, but not female, mice presented reduced CRF-positive cell counting relative to water-exposed animals (Neira et al., 2023). Moreover, the study by Munier et al. (2022) suggested that, during protracted withdrawal from chronic intermittent alcohol (CIA) exposure, CRF parvocellular neuroendocrine cells in the PVN of male rats, but not females, presented more depolarized resting membrane potential, reduction in the evoked spike firing rate and decreased intrinsic excitability. Also, noradrenaline application produced excitatory and inhibitory effects on CRF parvocellular neuroendocrine cells of the PVN in depending on alcohol exposure and in a sex-dependent manner (Munier et al., 2022).

In the BNST, studies have shown increased CRF levels (Jiao et al., 2021; Olive et al., 2002) or excitability of CRF neurons (Pati et al., 2020) during alcohol withdrawal. Of note, while alcohol drinking reduced CRF gene expression in BNST of Wistar rats, it enhanced CRF mRNA levels in msP rats (Caputi et al., 2021). Snyder et al. (2019) also reported that BNST CRF-containing cells of CIA-exposed mice exhibited increased basal resting membrane potentials at acute (4 h) but not extended (4 days) withdrawal relative to alcohol-naive subjects. Additionally, a history of alcohol drinking enhanced the excitatory synaptic drive (that is, EPSC frequency × sEPSC amplitude) in BNST CRF-expressing neurons, during acute withdrawal in males but not female mice (Levine et al., 2021).

Opposingly, other studies did not detect alterations in BNST CRF levels or gene expression right after repeated alcohol exposure (Lowery-Gionta et al., 2012; Rinker et al., 2017) or during acute (Ketchesin et al., 2016) or protracted (Sommer et al., 2008; Sampedro-Piquero et al., 2022) periods of abstinence. In these studies, the absence of differences in BNST CRF gene or peptide expression between alcohol-exposed and alcohol-naive individuals might be due to some factors, including duration and type of alcohol exposure method, withdrawal period, animal species, strain, and sex used. Further studies directly testing these hypotheses would be very elucidating.

In the VTA, CRF immunoreactivity was increased immediately following one but not six cycles of alcohol drinking in the dark protocol in mice (Lowery-Gionta et al., 2012). Also, a four-day alcohol drinking protocol elevated CRF mRNA expression in mice VTA in response to the last intake session (Rinker et al., 2017). Contrarily, repeated alcohol drinking did not affect VTA CRF gene expression during acute (24 h) withdrawal (Ketchesin et al., 2016). Here, again, the time of exposure to alcohol drinking and the time elapsed from the final exposure to alcohol (i.e., withdrawal period) appear to be relevant for the observed outcome on the VTA CRF gene or peptide expression.

Considering the PFC and its subregions, most studies failed to detect the effects of a history of alcohol exposure on CRF transcript levels. Repeated alcohol consumption did not affect CRF gene expression in medial PFC (mPFC) during acute withdrawal (Ketchesin et al., 2016). Also, adolescence intermittent alcohol exposure did not impact PFC CRF mRNA levels during adulthood (Boutros et al., 2018). Chronic voluntary alcohol intake did not alter CRF gene expression in the prelimbic (PL) and infralimbic (IL) cortices of alcohol-preferring rats (Alhaddad et al., 2020). However, Conte et al. (2022) found that voluntary alcohol drinking elevated CRF immunoreactive cell counting in the rat anterior cingulate (ACG) cortex. Those studies' findings are complex to compare as they analyzed different PFC subregions, which might have significantly impacted the observed outcomes.

Studies have found considerably controversial data on the consequences of alcohol exposure on NAc CRF content. Binge-like repeated alcohol exposure via involuntary oral administration increased CRF immunoreactivity during one and ten days into withdrawal but not following a drinking session in rats (Brancato et al., 2021). In contrast, intermittent alcohol exposure during adolescence via ig administration reduced accumbal CRF mRNA expression, regardless of adult stress exposure, during adulthood (Boutros et al., 2018). Moreover, chronic alcohol intake did not change CRF gene expression in NAc of alcohol-preferring rats (Alhaddad et al., 2020), and accumbal CRF immunoreactivity was not altered following repeated alcohol drinking in mice (Lowery-Gionta et al., 2012).

As other brain regions have been less studied regarding alcohol-induced neuroadaptations of interest for this subtopic, we will compile literature findings according to the direction of observed alterations (up, down-regulation, or none). Intermittent alcohol exposure during adolescence increased CRF gene expression in the adult rat hippocampus (Sanchez-Marin et al., 2017). Chronic intermittent access to alcohol consumption increased CRF gene expression in the rostromedial tegmental nucleus (RMTg) of mice during withdrawal (Fu et al., 2024). More specifically, this study verified enhanced CRF gene expression in VTA-projecting RMTg neurons of alcohol withdrawal mice (Fu et al., 2024). Also, chronic alcohol drinking enhanced the number of cells immunoreactive to CRF in the dorsolateral septum of rats (Conte et al., 2022). However, CRF peptide or transcripts expression was not altered after alcohol exposure in the lateral septum (LS) (Lowery-Gionta et al., 2012), and nucleus incertus (Walker et al., 2017) of rodents. Finally, repeated exposure to high doses of alcohol vapor reduced CRF immunoreactivity in the median eminence of rats (Rivier et al., 1984).

3.2 CRFR1

Several studies have indicated that alcohol exposure or withdrawal may modulate CRFR1 protein or mRNA expression levels and CRFR1-positive neuron's activity/excitability in opposite directions—up or downregulating—depending on the analyzed brain region. Not rarely, different studies on the effects of alcohol within the same brain area may encounter diverse outcomes regarding drug-induced changes in CRFR1 signaling. This may be due to some factors, including acute versus repeated alcohol exposure, time into withdrawal, animal model, drug administration route and dose, and characteristics of experimental individuals (age, sex, strain, and species), as already discussed in the CRF section. This heterogeneity of findings unravels the complex nature of alcohol-induced neuroadaptations in the CRFergic system, which seems to be implied in the onset and development processes of alcohol use disorders.

The amygdala has been one of the most studied brain regions regarding neuroadaptations in CRFR1-mediated transmission in response to alcohol exposure or during alcohol withdrawal. However, the direction of findings significantly varies among studies. Both acute and repeated exposure to alcohol seem to modify CRFR1 expression in amygdala subregions. For example, exposure to a single CIA vapor session downregulated CRFR1 gene expression in mice CeA (Eisenhardt et al., 2015). Also, repeated alcohol intake reduced CRFR1 gene expression in the CeA and MeA, while no alteration was detected in the BLA of alcohol-preferring msP rats (Hansson et al., 2007). However, acute ip alcohol injection did not change CRFR1 gene expression in the rat amygdala—evaluated as the entire region (Lee & Rivier, 1997).

Studies have also suggested that variables such as the experimental subject's sex and even pattern of alcohol intake are relevant factors modulating CRFR1-related neuroadaptations in the amygdala. In this regard, Rodriguez et al. (2022) reported that chronic alcohol vapor exposure increased the proportion of CRFR1-positive cells in CeA only in female but not male rats. Also, Correia et al. (2015) found that moderate but not low or high voluntary alcohol intake levels increased CRFR1 mRNA expression in the amygdala of male mice.

Some studies reported that acute or prolonged alcohol withdrawal significantly affects—usually upregulating—CRFR1 signaling in the amygdala. This neuroadaptation is potentially one of the mechanisms underlying abstinence behavioral symptoms that contribute to relapse to alcohol drinking. Eisenhardt et al. (2015) reported upregulated CeA CRFR1 transcripts following protracted alcohol withdrawal, paralleling a post-dependent phenotype characterized by signs that include tremors, vocalizations, piloerection, and teeth chattering. Chronic alcohol vapor inhalation also increased CRFR1 mRNA levels in BLA and MeA, but not CeA, during protracted abstinence (Sommer et al., 2008). At 72 h into withdrawal, enhanced CeA CRFR1 protein and gene expression were detected in rats repeatedly injected with ip alcohol (Li et al., 2019). On the other hand, alcohol operant self-administration did not correlate with CFR1 gene expression in the rat amygdala during protracted withdrawal (Pickering et al., 2007b), and acute withdrawal from repeated alcohol drinking did not impact CRFR1 gene expression in the amygdala of mice (Ketchesin et al., 2016).

Some studies have indicated that alcohol exposure during early age may significantly increase CRFR1 levels later in life. Adolescent chronic intermittent exposure to alcohol via ig gavage enhanced CRFR1 transcripts and protein levels in the amygdala of adult rats (Sánchez-Marín et al., 2022a, 2022b). On the other hand, chronic intermittent alcohol exposure during adolescence via ip injections (3 g/kg) did not affect CRFR1 gene expression in the amygdala of adult rats (Sanchez-Marin et al., 2017). From those studies, one might infer that the route for alcohol administration (ig versus ip, for example) might also impact long-term consequences on CRFR1 gene expression.

Regarding prenatal alcohol exposure, impacts in amygdala CRFR1 protein or transcripts levels later in life may vary in a sex-dependent fashion. Decreased CRFR1 mRNA expression was observed in BLA of adult females but not male rats with a history of prenatal alcohol exposure (Raineki et al., 2016). Also, within the CeA, moderate prenatal alcohol exposure reduced the counting of CRFR1-positive cells in male rats but, opposingly, increased the number of CRFR1-labeled cells in females (Rouzer & Diaz, 2022).

Ex-vivo electrophysiology studies also have indicated that alcohol exposure leads to changes in amygdalar CRFR1-expressing neurons' electrophysiological properties. CIA vapor increased the excitability of CRFR1-positive neurons in the lateral amygdala (LA) (Agoglia et al., 2020). In turn, CeA CRFR1-expressing cells exhibited reduced phasic inhibition and loss of tonic inhibition (Herman et al., 2016). In CRFR1-expressing projection cells, CIA led to enhanced baseline activity and deficit in sensitivity to acute alcohol exposure (Herman et al., 2016).

Additionally, CeA CRFR1-containing neurons of mice previously exposed to alcohol drinking showed increased sensitivity to acute CRF relative to alcohol-naive animals on presynaptic GABA release (Agoglia et al., 2022).

Studies have also indicated sex differences in alcohol-induced neuroadaptation in electrophysiological measurements of amygdala CRFR1-containing cells. During acute withdrawal, increased sensitivity of GABA-A receptor-mediated sIPSCs (spontaneous inhibitory postsynaptic currents) to acute alcohol application was observed in CeA CRFR1-positive neurons of males but not female mice (Agoglia et al., 2022). Moreover, Agoglia et al. (2022) findings suggest that a history of alcohol intake reduced the frequency of sIPSCs in CeA CRFR1-expression neurons in response to the application of CRFR1 antagonist in females but not male mice.

Besides the amygdala, other brain regions present modifications in CRFR1 levels or activity of CRFR1-positive cells following alcohol exposure or during alcohol abstinence. In the hypothalamus, acute ip alcohol injection upregulates CRFR1 mRNA levels in the parvocellular PVN (Lee & Rivier, 1997). In the same way, alcohol operant self-administration was positively correlated with hypothalamic CFR1 gene expression during protracted abstinence (Pickering et al., 2007a). However, another study found that CIA exposure via oral intubation did not change CRFR1 mRNA expression in rat hypothalamus during protracted withdrawal (Marty et al., 2020).

Another component of the HPA axis, along with the previously mentioned hypothalamus, the pituitary, exhibits alcohol-related effects on CRFR1 transcript expression. For example, acute treatment with alcohol via oral gavage in a binge-like protocol reduced CRFR1 gene expression in the rat anterior pituitary (Zhou et al., 2000). In the same direction, alcohol prenatal exposure reduced CRFR1 mRNA levels in the pituitary of male but not female adult rats (Glavas et al., 2007). Repeated alcohol intake upregulated CRFR1 mRNA expression in single-housed rats' pituitary compared to water-drinking counterparts (Todkar et al., 2016). However, rats repeatedly treated with alcohol with oral gavage no longer exhibited changes in CRFR1 gene expression in the anterior pituitary relative to controls, while acutely treated mice did (Zhou et al., 2000).

For the BNST, some studies did not find alterations in CRFR1 gene expression during acute or protracted withdrawal following repeated alcohol exposure (Sommer et al., 2008; Ketchesin et al., 2016). However, as already discussed for the CRF, the species or strain used in the studies also significantly seem to impact outcomes on alcohol-induced modification in CRFR1. In this sense, CRFR1 gene expression in BNST of alcohol-drinking mSP rats was increased relative to their water-drinking counterparts, although the same alcohol-induced alteration was not observed in Wistar rats (Caputi et al., 2021).

In the VTA, repeated alcohol drinking did not modify CRFR1 transcript expression in mice during acute withdrawal (Ketchesin et al., 2016). Interestingly, although mRNA expression analysis did not find an effect of alcohol consumption on VTA CRFR1 transcripts, western blot analysis showed downregulation of CRFR1 protein in this region immediately after a final alcohol intake session (Rinker et al., 2017), indicating that caution has to be taken when comparing studies evaluating gene versus protein expression.

Considering the PFC and its subregions, although chronic intermittent access to alcohol intake did not impact CRFR1-positive cell counting or activation, in the PL and IL cortices of rats, a correlation between the activation of CRFR1 IL neurons and alcohol intake during the last drinking session was detected (Quadir et al., 2023). However, considering CRFR1 gene expression, studies have reported no alteration in PFC subregions, including cingulate (CG), IL, and PL cortices, upon repeated alcohol exposure or during acute withdrawal (Hansson et al., 2007; Ketchesin et al., 2016). During protracted withdrawal from CIA vapor exposure, CRFR1-positive pyramidal neurons exhibited decreased excitability relative to neurons from alcohol-naive mice (Patel et al., 2022).

Also, in the PFC or its subregions, adolescent and prenatal alcohol exposure appear to modify CRFR1 transcript levels when assessed in later stages of development. Adolescent intermittent alcohol exposure via ig administration enhanced PFC CRFR1 mRNA expression in adult rats (Boutros et al., 2018). In turn, prenatal alcohol exposure decreased CRFR1 mRNA expression in adult male rats' anterior CG and PL—but not in female or adolescent subjects (Raineki et al., 2016). Importantly, although prenatal alcohol exposure, per se, did not alter IL CRFR1 gene expression later in life, interaction with repeated stress may result in altered CRFR1 mRNA levels in a sex-dependent manner (Lam et al., 2018). The moment of analysis post-stress (immediately or after) also seems relevant for IL CRFR1 gene expression-related outcomes (Lam et al., 2018). Again, the studies also indicate that the alcohol exposure protocol and experimental subjects' features may considerably affect outcomes, which can make it difficult to compare and reconcile data across different studies.

Analyzing the NAc, voluntary alcohol intake reduced CRFR1 gene expression of alcohol-preferring rats (Hansson et al., 2007). Similarly, repeated exposure to binge-like alcohol intake downregulated accumbal CRFR1 mRNA levels in female, but not male, mice 24 h following the last drinking episode (Finn et al., 2018). However, adolescent intermittent ig alcohol administration did not alter accumbal CRFR1 mRNA levels during adulthood (Boutros et al., 2018).

In the hippocampus, chronic alcohol ig administration during adulthood reduced CRFR1 mRNA level (Xiao et al., 2018). Similarly, prenatal alcohol exposure reduced CRFR1 protein levels in the hippocampal formation of mice (Caldwell et al., 2015). Importantly, long-lasting modification in the hippocampal CRFR1 transcripts levels produced by alcohol prenatal exposure seems to vary in a sex-dependent manner (Raineki et al., 2018). In contrast, adolescence intermittent alcohol exposure did not impact CRFR1 mRNA levels in the hippocampus of adult rats (Sanchez-Marin et al., 2017).

Other regions also showed CRFR1 mRNA or protein level alterations in response to repeated alcohol consumption or acute alcohol withdrawal. Exposure to alcohol drinking, in a binge-like regimen, enhanced CRFR1 transcripts levels in the dorsal raphe nucleus (DRN) of adolescent alcohol-preferring rats (McClintick et al., 2015). Also, acute withdrawal from voluntary alcohol intake increased CRFR1 mRNA expression in the nucleus incertus of alcohol-preferring rats (Walker et al., 2017).

3.3 CRFR2

As well as with CRFR1, findings on alcohol-related changes in CRFR2 expression are controversial, possibly due to considerable heterogeneity of alcohol exposure protocols and experimental subjects.

Firstly, considering the amygdala region, chronic alcohol vapor inhalation downregulated CRFR2 mRNA levels in BLA but not in Cea nor MeA of adult rats during protracted withdrawal (Sommer et al., 2008). However, Pickering and colleagues (2007b) did not find a correlation between repeated alcohol self-administration and CRFR2 mRNA levels in rat amygdala in protracted abstinence (Pickering et al., 2007b). In the same way, neither acute alcohol exposure via vapor inhalation nor protracted withdrawal from CIA affected CRFR2 mRNA expression in the CeA of adult mice (Eisenhardt et al., 2015).

Repeated intermittent exposure to alcohol via ig gavage during adolescence, significantly increased amygdala CRFR2 protein levels in adult rats (Sánchez-Marín et al., 2022b), but failed to modify CRFR2 mRNA expression in another study (Sanchez-Marin et al., 2017).

Within the hypothalamus, acute ip alcohol injection did not induce changes in CRFR2 gene expression in rats' ventromedial hypothalamus nucleus (Lee & Rivier, 1997). During protracted alcohol withdrawal, a positive correlation between alcohol operant self-administration and CRFR2 mRNA levels was found in the rat hypothalamus (Pickering et al., 2007a). However, no changes in hypothalamic CRFR2 mRNA expression was detected during protracted abstinence from CIA exposure via oral intubation (Marty et al., 2020).

Other brain regions were also evaluated for alcohol or alcohol-withdrawal-induced changes in CRFR2 expression or binding, showing up- or down-regulations. Repeated alcohol drinking sessions reduced DRN CRFR2 mRNA levels in adolescent alcohol-preferring rats (McClintick et al., 2015). Opposedly, repeated systemic alcohol administration enhanced CRFR2 binding in adult mice's dorsomedial but not ventromedial dorsal raphe (Weitemier & Ryabinin, 2005). In the NAc, exposure to repeated alcohol drinking increased CRFR2 mRNA levels in males but reduced CRFR2 gene expression in female mice 24 h after the last binge session (Finn et al., 2018). And in the LS, repeated systemic alcohol administration slightly enhanced CRFR2 binding (Weitemier & Ryabinin, 2005).

However, no modification was reported for CRFR2 gene or protein levels following alcohol exposure or during drug withdrawal in other brain regions, such as the supraoptic nucleus (SON) (Lee & Rivier, 1997), VTA (Rinker et al., 2017), nucleus incertus (Walker et al., 2017), BNST (Sommer et al., 2008), and hippocampus (Sanchez-Marin et al., 2017).

3.4 Ucns and CRF-BP

Alcohol-related studies have notably focused more on CRF and their receptors than CRF-BP and Ucns. However, neuroadaptations in these molecules might be correlated with alcohol-induced behavioral changes and is worthy of further exploration.

Repeated systemic administration of alcohol reduced the number of Ucn1-fibers in the DRN and LS but did not change Ucn1-positive cell counting in the Edinger-Westphal nucleus (EWcp) of mice (Weitemier & Ryabinin, 2005). On the other hand, the EWcp Ucn cells exhibited higher neuronal activation levels in mice exposed to repeated alcohol/sucrose intake relative to water or sucrose solution (Ryabinin et al., 2003). Mice exposed to long-term intermittent alcohol access drinking protocol also presented enhanced Ucn mRNA expression in the EWcp relative to alcohol-naive subjects during acute withdrawal (Giardino et al., 2017).

Considering single alcohol exposure, Spangler et al. (2009) reported that acute ip alcohol administration enhanced neuronal activation in the perioculomotor Ucn-containing population of neurons of the EWcp. Nevertheless, acute alcohol ip administration downregulated EWcp Ucn1 peptide expression (Al-Omari et al., 2023).

Considering CRF-BP studies, repeated alcohol drinking reduced CRF-BP mRNA expression in IL and PL subregions of the mPFC of mice 24 h into withdrawal (Ketchesin et al., 2016). However, CRF-BP mRNA expression was unaffected in BNST, VTA, and amygdala (Ketchesin et al., 2016). In the same way, repeated alcohol exposure did not affect CRF-BP gene expression in the CeA (Läck et al., 2005) nor in the nucleus incertus (Walker et al., 2017).

Considering this collection of findings, alcohol exposure seems to more significantly impact Ucns rather than the CRF-BP expression. However, most studies on alcohol-induced changes in Ucn levels mainly focus on the EWcp, whereas the presented CRF-BP studies analyzed different regions.

4. Pharmacological manipulations of CRF/Ucn targets in preclinical models of alcohol use disorders

In the last decades, there was a growing interest in understanding the role of the CRF/Urocortins systems in regulating behaviors associated with alcohol seeking and consumption. In fact, numerous evidence indicates the influence of CRF and its receptors, particularly the CRFR1 receptor, on alcohol consumption in studies conducted in rodents, as extensively reviewed by Koob (2010), Koob and Zorrilla (2010), Quadros et al. (2016) and Phillips, Reed, & Pastor (2015). Table 1 was updated from Quadros and colleagues (2016) and summarizes the impacts of pharmacological manipulations of CRF/Ucn systems on alcohol-seeking and consumption behaviors assessed in preclinical studies.

The CRFR1 antagonism by antalarmin, a non-peptidergic antagonist, has shown efficacy in reducing alcohol intake in mice and rats across various experimental paradigms. Notably, it decreased free-choice alcohol consumption in rats with continuous intermittent access to alcohol (Cippitelli et al., 2012) and mice subjected to the Drinking in the Dark protocol (Rinker et al., 2017; Robinson et al., 2019). In primates, Lindell et al. (2017)

Table 1 Summarized effects of pharmacological manipulations of CRF/Urocortin targets on alcohol drinking preclinical studies.

Receptor	Mechanism	Drug	Drug administration	Results	References
CRFR1	Agonist	Stressin I	into LHb	Reduced escalated[a] alcohol drinking and total fluid consumption	Zuo et al. (2021)
CRFR1	Antagonist	Antalarmin	ip	Reduced escalated[a] drinking	Lodge and Lawrence (2003), Cippitelli et al. (2012), Hansson et al. (2006), Chu et al. (2007), Marinelli et al. (2007), Lowery-Gionta et al. (2012)
				Reduced alcohol seeking	Marinelli et al. (2007), Funk et al. (2007)
				No effect on escalated[a] drinking	Yang et al. (2008), Molander et al. (2012)
			im	No effect on continuous access to alcohol or escalated[a] drinking	Lindell et al. (2017)
			into CeA	Reduced escalated[a] drinking	Lowery-Gionta et al. (2012)
				Reduced stress-induced alcohol seeking	Gondré-Lewis et al. (2016)
			into mPFC	Reduced stress-induced alcohol seeking	Gondré-Lewis et al. (2016)

(continued)

Table 1 Summarized effects of pharmacological manipulations of CRF/Urocortin targets on alcohol drinking preclinical studies. (*cont'd*)

Receptor	Mechanism	Drug	Drug administration	Results	References
				Reduced alcohol and sucrose binge drinking	Robinson et al. (2019)
			into VTA	Reduced escalated[a] drinking	Rinker et al. (2017)
			into NAcc Core	Prevented the reduction of binge drinking induced by the activation of CeA-projecting neurons	Borrego et al. (2022)
CRFR1	Antagonist	CP 154,526	ip	Reduced stress-induced reinstatement of alcohol seeking	Lê et al. (2000)
				Reduced escalated[a] drinking	Agoglia et al. (2022)
				Reduced escalated[a] drinking	Correia et al. (2015), Hwa et al. (2016), Lowery et al. (2008), Lowery et al. (2010) Overstreet et al. (2007), Sparta et al. (2008, 2009)
			icv (*chronic minipump*)	No effect on escalated[a] drinking	Hwa et al. (2016)
			into VTA	Reduced escalated[a] drinking	Hwa et al. (2013), Sparta et al. (2013)

			into DRN	Reduced escalated[a] drinking	Hwa et al. (2013, 2016)
			into MRN	No effect on escalated[a] drinking	Hwa et al. (2016)
			into IL	Prevented stress-induced reinstatement of alcohol seeking	Flores-Ramirez et al. (2022)
			into PL	No effect on preventing stress-induced reinstatement of alcohol seeking	Flores-Ramirez et al. (2022)
CRFR1	Antagonist	CP 376,395	ip	Reduced escalated[a] drinking	Giardino e Ryabinin (2013), Hwa et al. (2016), Simms et al. (2014)
				Reduced escalated[a] drinking	Potretzke et al. (2020)
				Reduced stress-induced alcohol drinking	Newman et al. (2018), Nipper et al. (2024)
				No effect on escalated[a] drinking	Newman et al. (2018)
			icv (*chronic minipump*)	Reduced escalated[a] drinking only on Day 1; no effect in remaining days	Hwa et al. (2016)
			into BNST	Reduced escalated[a] drinking	Albrecht-Souza et al. (2017)

(continued)

Table 1 Summarized effects of pharmacological manipulations of CRF/Urocortin targets on alcohol drinking preclinical studies. (*cont'd*)

Receptor	Mechanism	Drug	Drug administration	Results	References
			into NI	Reduced Yohimbine-induced reinstatement of alcohol seeking	Walker et al. (2017)
CRFR1	Antagonist	CRA−1000	ip	Reduced escalated[a] drinking	Overstreet et al. (2004, 2007), Lowery-Gionta et al. (2012)
CRFR1	Antagonist	LWH−63	ip	Reduced escalated[a] drinking	Lowery-Gionta (2012)
			sc	Reduced escalated[a] drinking	Sabino et al. (2006)
				Modestly increased limited access alcohol and water drinking	Sabino et al. (2006)
CRFR1	Antagonist	MJL−1 −109 −2	ip	Reduced escalated[a] drinking	Funk et al. (2007)
				No effect on drinking	Sabino et al. (2006)
CRFR1	Antagonist	MPZP	sc	Reduced escalated[a] drinking	Gilpin et al. (2008), Richardson et al. (2008)
				No effect on escalated[a] drinking	Ji et al. (2008)
CRFR1	Antagonist	MTIP	ip	Reduced stress-induced reinstatement of alcohol seeking	Gehlert et al. (2007)

				Reduced escalated[a] drinking	Gehlert et al. (2007)
CRFR1	Antagonist	NBI–27914	ip	Reduced escalated[a] drinking	Lowery-Gionta et al. (2012), Giardino and Ryabinin (2013)
				No effect on escalated[a] drinking	Molander et al. (2012)
CRFR1	Antagonist	NBI–35965	into LHb	Reduced escalated[a] drinking	Zuo et al. (2021)
			into VTA	Reduced voluntary alcohol drinking only in the first 2 h; no effect in 4 h of test	Rinker et al. (2017)
			into mPFC	Reduced escalated[a] drinking	Robinson et al. (2019)
CRFR1	Antagonist	R121919	ip	No effect on escalated[a] drinking	Yang et al. (2008)
			sc	Increased escalated[a] drinking	Sabino et al. (2006)
				Prevented stress-induced suppression of drinking	Sabino et al. (2006)
				No effect on escalated[a] drinking	Sabino et al. (2013)
				Reduced escalated[a] drinking	Funk et al. (2007), Roberto et al. (2010), Roltsch et al. (2014)

(continued)

Table 1 Summarized effects of pharmacological manipulations of CRF/Urocortin targets on alcohol drinking preclinical studies. (*cont'd*)

Receptor	Mechanism	Drug	Drug administration	Results	References
			into CeA	Reduced stress-induced alcohol seeking	Weera et al. (2020)
			into vmPFC	No effect on stress-induced alcohol seeking	Schreiber et al. (2017)
CRFR1	Antagonist	SSR125543	into NAcc	Reduced escalated[a] drinking	Knapp et al. (2011)
			into AMY, DRN	No effect on escalated[a] drinking	Knapp et al. (2011)
CRF-R1/2	Agonist	CRF	icv	No effect on escalated[a] drinking	O'Callaghan et al. (2005)
				Reduced escalated[a] drinking	Bell et al. (1998), Thorsell et al. (2005)
				Reinstated CRF-induced ethanol seeking	Le et al. (2002)
			into DRN	No effect on drinking	Weitemier e Ryabinin (2006)
			into LS	Reduced escalated[a] drinking and water intake	Ryabinin et al. (2008)

			into MRN	Reinstated CRF-induced ethanol seeking	Le et al. (2002)
			into NAcc	Further augmented escalated[a] drinking	Knapp et al. (2011)
			into CeA, DRN, VTA, PVN	No effect on escalated[a] drinking	Knapp et al. (2011)
			into NAcc Core	Reduced alcohol binge drinking	Borrego et al. (2022)
			into LHb	Reduced escalated[a] drinking	Zuo et al. (2021)
CRF-R1/2	Agonist	Ucn 1	into DRN	No effect on drinking and reduced water intake	Weitemier and Ryabinin (2006)
			into LS	Reduced escalated[a] drinking	Ryabinin et al. (2008)
CRF-R1/2	Antagonist	D-Phe-CRF	icv	Reduced escalated[a] drinking	Valdez et al. (2002)
				Reduced reinstatement of alcohol seeking induced by the combination of stress and ethanol-cues	Liu and Weiss (2002)

(continued)

Table 1 Summarized effects of pharmacological manipulations of CRF/Urocortin targets on alcohol drinking preclinical studies. (*cont'd*)

Receptor	Mechanism	Drug	Drug administration	Results	References
				Reduced stress-induced reinstatement of alcohol seeking	Le et al. (2000), Liu and Weiss (2002)
				No effect on cue-induced reinstatement of alcohol seeking	Liu and Weiss (2002)
			into CeA	Reduced escalated[a] drinking	Finn et al. (2007), Funk et al. (2006)
			into MRN	Reduced yohimbine-induced reinstatement	Lê et al. (2013)
			into MRN	Reduced stress-induced reinstatement	Le et al. (2002)
CRF-R1/2	Antagonist	Alpha-helical CRF	icv	Increased drinking in low preference animals	O'Callaghan et al. (2005)
				Reduced escalated[a] drinking	Lowery et al. (2010)
				No effect on drinking in high preference animals	O'Callaghan et al. (2005)

CRF-R1/2	Antagonist	NBI 35965 + K−41498	into mPFC, VTA	No effect on escalated[a] drinking	Robinson et al. (2019), Rinker et al. (2017)
CRFR2	Agonist	Ucn 3	icv	Reduced escalated[a] drinking	Sharpe and Phillips (2009), Valdez et al. (2004), Lowery et al. (2010)
			into CeA	Reduced escalated[a] drinking	Funk and Koob (2007)
				Increased drinking in control animals	Funk and Koob (2007)
			into VTA	Reduced escalated[a] drinking	Rinker et al. (2017)
			into mPFC	Reduced escalated[a] drinking	Robinson et al. (2019)
			into LHb	Reduced escalated[a] drinking and water drinking	Zuo et al. (2021)
CRFR2	Antagonist	Astressi-n−2B	into VTA	Reduced escalated[a] drinking	Albrecht-Souza et al. (2015)
			into CeA	No effect on escalated[a] drinking	Albrecht-Souza et al. (2015)
			into LHb	Reduced escalated[a] drinking and total fluid consumption	Zuo et al. (2021)

(continued)

Table 1 Summarized effects of pharmacological manipulations of CRF/Urocortin targets on alcohol drinking preclinical studies. (*cont'd*)

Receptor	Mechanism	Drug	Drug administration	Results	References
			into NI	No effect on yohimbine-induced reinstatement of alcohol seeking	Walker et al. (2017)
			into BNST	Increased escalated[a] drinking in control and stressed mice	Albrecht-Souza et al. (2017)
CRFR2	Antagonist	Antisauvag-ine−30	into DRN	No effect on drinking	Weitemier and Ryabinin (2006)
CRFR2	Antagonist	K−41498	into mPFC, VTA	No effect on escalated[a] drinking	Rinker et al. (2017), Robinson et al. (2019)
CRF-BP	Antagonist	CRF(6–33)	into VTA	Reduced escalated[a] drinking	Albrecht-Souza et al. (2015)
			into CeA	No effect on escalated[a] drinking	Albrecht-Souza et al. (2015)

[a]***Escalated drinking*** refers to alcohol intake that is above baseline or control levels, as observed in protocols of alcohol deprivation, "binge" drinking (as in drinking in the dark protocol), intermittent/limited access conditions, withdrawal from alcohol vapor or liquid diet exposure, prolonged alcohol access conditions, etc. *ip, Intraperitoneal; sc, Subcutaneous; icv, Intracerebroventricular; im, intramuscular AMY, Amygdala; BLA, Basolateral amygdala; BNST, Bed nucleus of stria terminalis; CeA, Central nucleus of the amygdala;; DRN, Dorsal raphé nucleus; IL, Infralimbic cortex; LS, Lateral Septum; LHb, Lateral habenula; mPFC, Medial prefrontal cortex; MRN, Median raphé nucleus; NAcc, Nucleus accumbens; NI, Nucleus incertus; PL, prelimbic cortex; PVN, Paraventricular nucleus of the hypothalamus; vmPFC, Ventromedial prefrontal cortex; VTA, Ventral tegmental area. Updated from* Quadros et al. (2016).

Updated from Quadros, I. M., Macedo, G. C., Domingues, L. P., & Favoretto, C. A. (2016). An update on CRF mechanisms underlying alcohol use disorders and dependence. *Frontiers in Endocrinology*, 7, 134. https://doi.org/10.3389/fendo.2016.00134.

demonstrated that antalarmin did not affect alcohol consumption in groups with continuous or intermittent access to alcohol. However, when they assessed the behavior of primates with higher concentrations of CRF in the cerebrospinal fluid (CSF), antalarmin reduced alcohol solution consumption, regardless of the consumption protocol (Lindell et al., 2017).

Antalarmin administration before alcohol self-administration sessions diminished alcohol-seeking behaviors during withdrawal period only of mice exposed to chronic vapor, while control mice did not display alterations in alcohol self-administration when administered Antalarmin (Chu et al., 2007). Notably, these results suggest that CRFR1 function may be particularly important in the context of excessive drinking and particularly withdrawal-induced drinking.

The intraperitoneal administration of other CRFR1 antagonists was also effective in decreasing alcohol binge drinking in different self-administration protocols. For instance, in Drinking in the Dark protocols, it was observed that the antagonists CP 154,526, CP 376,395 e NBI 27914 reduced excessive alcohol consumption in C57BL/6J (Sparta et al., 2008; Lowery et al., 2010; Giardino & Ryabinin, 2013). Interestingly, CP 376,395 increased intake of sucrose solution, suggesting that the blockade of CRFR1 differentially affects self-administration of reinforcers depending on the motivational valence of a reinforcer (Giardino & Ryabinin, 2013). In addition, CP 154,526 reduced alcohol consumption in adrenalectomized and control mice, suggesting that HPA axis function is not required for binge-like drinking behaviors (Lowery et al., 2010).

In operant chambers, systemic administration of CP 154,526 prevented the reinstatement of alcohol seeking after deprivation periods, without affecting lever responding for a sucrose solution (Sparta et al., 2009). Studies using free-choice paradigms demonstrated that alcohol intake and preference were reduced in mice after CP 154,526 injections, with no alterations in water consumption (Correia et al., 2015; Agoglia et al., 2022). Conversely, CP 376,375 at doses of 10 and 20 mg/kg decreased voluntary alcohol consumption in male and female mice, while leaving preference for the substance unaffected (Potretzke et al., 2020). Funk et al. (2007) observed that intraperitoneal injections of antalarmin, MJL-1-109-2, or R121919 reduced excessive self-administration of alcohol during acute withdrawal from intermittent vapor exposure. Furthermore, CRFR1 antagonists such as CRA1000, CP 376,375, and CP 154,526 effectively prevented increased alcohol consumption after stress exposure when administered systemically (Overstreet et al., 2007; Lowery et al., 2008; Newman et al., 2018; Nipper et al., 2024).

CRFR1 antagonists have been shown to modulate important regions implicated in reward and stress-related behaviors. When administered into specific brain regions such as the ventral tegmental area (VTA) (Hwa et al., 2013; Rinker et al., 2017), medial prefrontal cortex (mPFC) (Robinson et al., 2019), nucleus accumbens (NAcc) (Knapp et al., 2011), and central amygdala (CeA) (Funk et al., 2006; Finn et al., 2007; Lowery-Gionta et al., 2012), CRFR1 antagonists like antalarmin, CP 154,526, NBI 35965, SSR125543, or nonspecific antagonists such as D-Phe-CRF (which block both CRFR1 and CRFR2), have consistently promoted decreases in voluntary alcohol consumption in both mice and rats. In the dorsal raphe nucleus (DRN), CRFR1 antagonism also reduced alcohol consumption, suggesting that CRFR1 activation may contribute to increased alcohol consumption by modulating serotonergic output to the prefrontal cortex (Hwa et al., 2013, 2016). Conversely, CRFR1 signaling in the medial raphe nucleus (MRN) appears to be less involved in alcohol consumption (Hwa et al., 2016); however, nonspecific antagonism by D-Phe-CRF has been shown to reduce stress-induced reinstatement of alcohol seeking (Le et al., 2002; Lê et al., 2013).

Systemic administration of selective CRFR1 antagonists such as CP 154,526, CP 376,395, MTIP, or intracerebroventricular injection of the nonspecific antagonist D-Phe-CRF has been shown to prevent the reinstatement of stress-induced alcohol seeking or voluntary alcohol consumption (Lê et al., 2000; Liu & Weiss, 2002; Gehlert et al., 2007; Newman et al., 2018; Nipper et al., 2024). Injections of antalarmin into the central amygdala (CeA) have been found to attenuate alcohol-seeking behaviors and impulsivity in maternally separated rats undergoing operant self-administration protocols (Gondré-Lewis et al., 2016). Similarly, Weera et al. (2020) demonstrated that injections of R121919 into the CeA reduced alcohol-seeking behavior in rats which presented a post-traumatic stress disorder (PTSD) phenotype following exposure to predator odor stress, while having no effect on the behavior of rats with no PTSD phenotype.

CRFR1 antagonism within the prefrontal cortex and its subregions yields diverse outcomes. Antalarmin effectively reduced alcohol-seeking behaviors in rats exposed to maternal separation (Gondré-Lewis et al., 2016), contrasting with R121919, which exhibited no impact on seeking behavior in rats exposed to predator odor stress (Schreiber et al., 2017). Flores-Ramirez et al. (2022) revealed that CP 154,526, by blocking CRFR1 receptors in the infralimbic cortex, prevented footshock stress-induced reinstatement of alcohol seeking, though such effects were absent in the prelimbic cortex. These discrepancies may be related to variations in stressors employed and the age of stress exposure in the animals.

Collectively, these findings indicate that CRFR1 signaling plays a crucial role in modulating stress-induced alcohol consumption and seeking behaviors, with outcomes depending on the stressor type, age at stress exposure, and specific brain regions implicated in these behaviors. However, an intriguing finding from the study by Albrechet-Souza and colleagues (2017) revealed that intra-BNST injections of CP 376,395 significantly reduced voluntary alcohol consumption only in mice unexposed to social defeat stress, but with a history of continuous access to alcohol.

A few studies have failed to demonstrate the attenuation of alcohol drinking and seeking, even after stress exposure or during withdrawal (Yang et al., 2008; Knapp et al., 2011; Molander et al., 2012; Sabino et al., 2013). Intriguing findings shed light on the complexity of CRFR1 receptor modulation in excessive alcohol consumption. Notably, a study by Zuo et al. (2021) showed that CRFR1 activation with agonists in the lateral habenula (LHb) mirrored the effects of CRFR1 antagonism, reducing escalated alcohol consumption. The inputs that converge to the LHb, along with its projections to various brain regions, suggest that modulating LHb CRFRs could lead to diverse behavioral outcomes via different neural pathways. Moreover, Borrego et al. (2022) demonstrated that CRFR1 antagonism prevented binge drinking reduction induced by chemogenetic stimulation of CeA-projecting neurons to the NAcc Core, indicating a regulatory role of CRFR1 signaling in excessive ethanol consumption driven by this projection.

Overall, while the role of CRFR1 in modulating excessive alcohol consumption and seeking behaviors is complex, studies employing CRFR1 antagonists suggest selective actions in mitigating these responses. Further research elucidating the mechanisms underlying CRFR1 modulation in different brain regions and pathways may pave the way for targeted interventions in alcohol use disorders.

The involvement of urocortin peptides and CRFR2 in the modulation of escalated alcohol drinking has been investigated by fewer studies. The selective agonism (activation) of CRFR2 by Ucn 3 has significant efficacy in reducing alcohol consumption in both mice and rats. This effect was observed following intracerebroventricular administration (Sharpe & Phillips, 2009; Valdez et al., 2004; Lowery et al., 2010) as well as in specific brain regions such as the VTA (Rinker et al., 2017) and mPFC (Robinson et al., 2019). Moreover, administration of Ucn 3 into the LHb has been shown to decrease both alcohol and water voluntary intake (Zuo et al., 2021). Funk and Koob (2007) demonstrated that intra-CeA administration of a CRFR2 agonist (Ucn 3) attenuated escalated alcohol consumption in dependent rats, during acute withdrawal from chronic alcohol vapor exposure. Similarly, Ryabinin et al. (2008) reported reduced alcohol intake following infusions of the nonspecific CRFR2 agonist Ucn 1 into the lateral septum. However, administration of the same agonist into the DRN did not impact alcohol drinking and resulted in decreased water intake (Weitemier & Ryabinin, 2006).

Partially supporting these findings, Albrechet-Souza et al. (2017) demonstrated that intra-BNST infusion of a CRFR2 antagonist increased alcohol consumption in both control and stressed mice. However, conflicting results have emerged from other studies, with some reporting no effect or even reduced alcohol consumption following administration of CRFR2 antagonists in different brain regions (Weitemier & Ryabinin, 2006; Albrechet-Souza et al., 2015; Walker et al., 2017; Rinker et al., 2017; Robinson et al., 2019; Zuo et al., 2021). Interestingly, two studies showed that the individual administration of NBI 35965, a CRFR1 antagonist, into the VTA or mPFC resulted in decreased alcohol intake in the drinking in the dark protocol in mice. However, when administered together with a CRFR2 antagonist (K-41498), there was no effect on alcohol consumption. This suggests that, in the VTA and the mPFC, the anti-drinking effects of CRFR1 antagonism may require intact CRFR2 receptor signaling (Rinker et al., 2017; Robinson et al., 2019).

CRFR1 and CRFR2 represent significant pharmacological targets with potential implications for treating AUDs. Emerging manipulations of other components of CRF/Ucn systems, such as the CRF-BP, have received attention as potential modulators of alcohol consumption, as recently suggested by Albrechet-Souza et al. (2015). Their study revealed the involvement of VTA CRF-BP in the initial stages of escalated alcohol drinking. Specifically, intra-VTA administration of CRF6–33, a CRF-BP inhibitor, reduced alcohol intake in mice during the Drinking in the Dark (DID) paradigm, but not during intermittent access to alcohol (Albrechet-Souza et al., 2015).

Collectively, these findings highlight the significance of CRF receptors in the context of escalated alcohol consumption and alcohol seeking behaviors. While there is an extensive body of evidence supporting the modulation of excessive alcohol intake by CRFR1, investigations into CRFR2 signaling suggest that this receptor may play distinct roles depending on the brain pathways involved. Nonetheless, it appears that the balance between CRFR1 and CRFR2 activation or blockade is crucial in determining the final outcome. Moreover, pharmacological studies shed light on the roles of both CRF receptors within brain reward pathways, as well as the modulation provided by CRF-BP in escalated alcohol drinking.

5. Therapeutic interventions and human studies with CRF/Ucn targets

Despite robust evidence supporting potential therapeutic effects of CRF/Ucn manipulations in preclinical studies of drug and alcohol use disorders, as well as other neuropsychiatric disorders, to this day, there are still no medications commercially available with direct effects on CRF/Ucn components. Human studies with molecules targeting CRFR1 signaling focused mostly in investigations of mood and anxiety disorders, with a few studies on PTSD, alcohol (and other drugs) use disorders, pain and others. Initially, the main challenge for human studies was the difficulty in finding potent and selective CRFR1 antagonist molecules that would cross the blood brain barrier and be safe and well tolerated (Heilig, 2023). We will discuss some of the clinical and human experimental studies which have investigated therapeutic and adverse effects of drugs targeting CRF type 1 receptors, including two clinical studies in patients with AUDs.

A pilot clinical trial provided promising results in the use of the CRFR1 antagonist, R121919 (NBI 30775), in a small number of depressed patients. Positive outcomes included a reduction in depression and anxiety scores, as well as improvement in sleep parameters (Zobel et al., 2000; Held et al., 2004). Remarkably, CRF-induced neuroendocrine responses of ACTH and cortisol were unaffected during the 30-day treatment with R121919, suggesting that HPA axis peripheral function was maintained despite chronic CRFR1 blockade (Zobel et al., 2000). However, these early CRFR1 antagonists, such as R121919 (NBI 30775) or antalarmin, for example, induced elevated liver enzymes, with risks of hepatic toxicity and were unfit candidates as medications (Zorrilla et al., 2013; Spierling & Zorrilla, 2017).

Subsequent clinical studies in depressed patients under more controlled conditions (double-blind; placebo-controlled; randomized allocation of participants) failed to detect efficacy of different CRFR1 antagonist molecules (Spierling & Zorrilla, 2017; Menke, 2024). Using second generation CRFR1 antagonists, with improved safety profile, Binnemam and coworkers tested CP-316,311, a selective CRFR1 antagonist, in a study involving 123 patients with major depression disorder. Over the course of 6-weeks of treatment, the CRFR1 antagonist was shown to be safe and well tolerated. However, changes in depression scores of patients treated with CP-316,311 were similar to placebo-treated subjects, indicating no clinical efficacy for the CRFR1 antagonist (Binneman et al., 2008). In a similar trend, Ono Pharmaceuticals tested ONO-2333Ms as a clinical compound and reported in July of 2008 that the program was discontinued because of "lack of efficacy" and data were not published. Other clinical trials for major depression employed CRFR1 antagonists such as SSR125543, verucerfont and pexacerfont, with no different outcomes (Zorrilla et al., 2013; Spierling & Zorrilla, 2017). So far, the development of CRFR1 antagonist molecules for treatment of depressive disorders has not yielded promising outcomes.

An experimental human study aimed to explore whether subchronic treatment with NBI-34041 (SB723620) would affect regulation of neuroendocrine HPA axis hormones in healthy male volunteers (Ising et al., 2007). There was no alteration in diurnal secretion of ACTH and cortisol during the 14-day treatment with the selective CRFR1 antagonist NBI-34041, as well as no changes in CRF-stimulated secretion of ACTH and cortisol. This suggests no impairment in regulation of HPA axis function due to treatment with the CRFR1 antagonist, in agreement with Zobel et al. (2000). Volunteers were further exposed to the Trier Social Stress Test, a psychosocial stress test that includes public speaking and a mental arithmetic task, known to produce emotional activation and stress responses. The group receiving the highest dose of NBI-34041 presented attenuated ACTH and cortisol responses to the psychosocial stress test (Ising et al., 2007), suggesting a protective effect of CRFR1 antagonist treatment on hormonal stress responses to social stress. This study suggested an interesting profile of action for CRFR1 antagonists in mitigating acute hormonal stress responses, at least in a human experimental setting.

A large multicenter, randomized, placebo-controlled study tested the effects of the selective CRFR1 antagonist, pexacerfont, for generalized anxiety disorder (Coric et al., 2010). After 8 weeks of treatment, no efficacy was detected for pexacerfont, which yielded similar changes in anxiety scores as the placebo-treated group, while escitalopram treatment (positive control) promoted reduction in anxiety scores (Coric et al., 2010). Once again, the CRFR1 antagonist did not change salivary cortisol levels in this study.

While this finding advocates for a good safety profile for this molecule, authors also raised the question of dosage being insufficient for therapeutic effects (Coric et al., 2010).

Using a human model of anxiety, Bailey et al. (2011b) reported promising anxiolytic effects of the CRFR1 antagonist, R317573. During 7 days, healthy participants received R317573 or placebo, and were later exposed to the 7.5% carbon dioxide-enriched inhalation test (Bailey et al., 2011a). A positive control condition included part of the placebo group receiving acute treatment with lorazepam (anxiolytic medication) prior to the CO2 inhalation test. Despite a limited sample size and high variability in some parameters, both groups treated with R317573 and lorazepam presented decreases in subjective anxiety-related symptoms, relative to placebo, after the CO2 inhalation test (Bailey et al., 2011b). Thus, the CRFR1 antagonist presented positive anti-anxiety effects in a human laboratory setting. Clinical studies with CRFR1 antagonists also occurred in the context of irritable bowel syndrome (IBS), where both central and peripheral CRF receptors may play a role in chronic, persistent gastrointestinal dysfunction that is often associated with stress (e.g., Sagami et al., 2004; Sweetser et al., 2009). Two doses of pexacerfont (BMS-562086) were used in a randomized, double-blind, placebo-controlled study in female IBS patients. After a two-week treatment, the CRFR1 antagonist showed no efficacy in improving gastrointestinal transit or bowel function, and no differences in anxiety or depression scores were observed between placebo and pexacerfont-treated groups (Sweetser et al., 2009).

IBS patients frequently experience anticipatory emotional and anxiety symptoms in expectation of abdominal pain. Hubbard et al. (2011) examined the effects of GW876008 (emicerfont) on state-trait anxiety, HPA-axis response as well as activation and connectivity of stress-related brain regions in female IBS patients and controls. Acute administration of emicerfont failed to alleviate anxiety-related scores in anticipation of visceral pain and failed to affect HPA axis hormones, despite significant signal reductions in brain regions associated with anxiety/stress modulation such as the hypothalamus, amygdala, hippocampus, insula, anterior cingulate, and orbitomedial prefrontal cortices in fMRI (Hubbard et al., 2011). Thus, although not correlated with changes in subjective emotional parameters, changes in brain region activity suggests a predominant central action of emicerfont.

Post-traumatic stress disorder (PTSD) is another stress-related disorder in which treatment with CRFR1 antagonist was assessed. A clinical trial was conducted to evaluate the efficacy of GSK561679 (verucerfont) using double-blind, placebo-controlled, randomized conditions in females with PTSD. Over the course of 6 weeks of treatment, no efficacy was detected for verucerfont on reducing PTSD scores, which were no different from placebo treatment (Dunlop et al., 2017). Interestingly, a subgroup of participants with a polymorphism on the CRHR1 gene who also experienced childhood abuse, presented beneficial effects of verucerfont reducing symptoms of PTSD. Although this finding requires further confirmation in larger samples, it indicates that particular subsets of patients may benefit from treatment with CRFR1 antagonists, despite an overall lack of efficacy (Dunlop et al., 2017).

Another experimental human trial tested the effects of two doses of GSK 561679, another selective CRFR1 antagonist, in healthy female volunteers submitted to protocols for fear- or anxiety-potentiated acoustic startle responses (Grillon et al., 2015). Contrary to the authors' hypothesis, the CRFR1 antagonist failed to promote anxiolytic effects on anxiety-potentiated startle. Nonetheless, the highest dose of GSK 561679 increased fear-potentiated startle, which is in accordance with rodent studies (Walker et al., 2009a, 2009b), and could be mediated by CRFR1 signaling between BNST and central amygdala.

Over the course of development of alcohol use disorders and alcohol dependence, a shift has been proposed to occur in the primary motivation for alcohol use, shifting from positive reward-seeking to a relief-seeking state. In more advanced stages of AUDs, alcohol use would be primarily motivated to alleviate aversive states, and stress-related brain systems would play a primary role in maintaining compulsive drug use, despite adverse consequences (e.g., Koob, 2021; Heilig, 2023). At this stage, stress and anxiety-related symptoms are a main component of negative affect that drives craving for alcohol and promotes drinking and relapse during early or protracted abstinence/withdrawal. Based on robust data stemming from preclinical studies, this would be the phase when interventions targeting CRF/Ucn, specially CRFR1 signaling, should be particularly helpful in reducing craving and relapse to alcohol use.

However, the first published clinical study with CRFR1 antagonist in the context of AUDs observed lack of efficacy of pexacerfont on alcohol craving, emotional responses, or anxiety in alcohol-dependent patients in early abstinence (Kwako et al., 2015). After 30-day treatment with pexacerfont, the double-blind, placebo-controlled study also reported no effects of pexacerfont on ACTH or cortisol levels, and no differential neural activation to alcohol-related or affective stimuli measured on fMRI (Kwako et al., 2015). The absence of anxiolytic effects of pexacerfont were consonant with the antagonist's lack of efficacy detected on generalized anxiety disorder, as previously mentioned (Coric et al., 2010). Based on CSF levels of pexacerfont in a separate cohort of healthy volunteers, receptor occupancy was estimated to approximately 90% of central CRFR1, confirming high levels of encephalic exposure (Kwako et al., 2015). Authors discussed the possibility that drug-receptor interaction dynamics could somehow contribute to the lack of therapeutic effects of pexacerfont, due to its rapid dissociation from CRFR1 (Kwako et al., 2015).

The same research group conducted a second double-blind, placebo-controlled study using verucerfont, a selective high-affinity CRFR1 antagonist with slow receptor dissociation kinetics, in alcohol-dependent females with high anxiety levels (Schwandt et al., 2016). Effects on the HPA axis were evaluated using the dexamethasone-CRF test, which showed that verucerfont blunted CRF-induced HPA axis activation. Amygdala activation in response to negative affect stimuli was also reduced in the verucerfont-treated subjects, although other brain regions presented increased activation to alcohol-cues in this group. (Schwandt et al., 2016). Conversely, scores of alcohol craving were unaltered by the CRFR1 antagonist treatment, similar to the results obtained with perucerfont (Kwako et al., 2015). Thus, verucerfont showed significant effects on HPA axis hormonal responses as well as brain activation due to aversive stimuli presentation, compared to pexacerfont, which could be due to differences in receptor binding kinetics. Nonetheless, effects on alcohol craving scores remained absent with both CRFR1 antagonists (Kwako et al., 2015; Schwandt et al., 2016).

A more recent pilot clinical study (2018) indicated an important effect of pexacerfont in craving and withdrawal symptoms for men with dependence to methamphetamine and/or heroin (Morabbi et al., 2018). Over the course of three weeks, the randomized, double-blind, placebo-controlled study reported significant reduction in craving scores and severity, as well as lower frequency of drug-temptation episodes and reduced depression symptoms induced by pexacerfont treatment. A favorable profile of action for pexacerfont as a potential therapeutic strategy for early withdrawal to methamphetamine/heroin dependence was found (Morabbi et al., 2018). However, further studies are required to expand sample sizes, patient profile (single-drug vs. polydrug users), treatment setting conditions (inpatient vs. outpatient settings), post-treatment follow-up, etc.

In summary, clinical trials involving CRFR1 antagonists in stress-related and alcohol and drug use disorders are still very limited, and have not yet provided strong evidence supporting clinical efficacy. As discussed in this section, potential medications tested are substances with high selectivity and affinity for CRFR1. The binding and kinetics for receptor interaction and dissociation profile of the different CRFR1 antagonists have more recently emerged as characteristics that could affect drug effects and clinical outcomes (e.g., Spierling & Zorrilla, 2017; Heilig, 2023). Furthermore, ligands tested so far have varied in their ability to affect HPA hormonal levels, raising the possibility that compensatory neural and hormonal mechanisms may come into play during chronic treatment with CRFR1 antagonists. In addition, the need to assess interventions directed at different targets of CRF/Ucn systems has been proposed, whether by testing non-selective substances that will modulate both CRFR1 and CRFR2, as well as drugs selectively targeting CRFR2 or even CRF-BP (Zorrilla et al., 2013; Shaham & de Wit, 2016; Spierling & Zorrilla, 2017). Considering the two studies on alcohol use disorders, it is important to note that they were both Phase II, inpatient experimental studies, which used as primary outcomes surrogate markers to predict for drug efficacy: craving scores for alcohol. Ideally, clinical efficacy should include drinking endpoints and also outpatient healthcare settings (Heilig, 2023).

6. Summary and perspectives

This chapter reviewed the role of CRF/Ucn signaling in alcohol use disorders, focusing mostly on abundant preclinical evidence indicating neuroadaptations in CRF/Ucn function and signaling induced by alcohol, as well as robust indications of potential therapeutic effects of CRF manipulations in reducing escalated alcohol consumption in rodent models. While most research has concentrated on CRF and its actions through CRFR1, there is increasing attention directed at Urocortin peptides, CRFR2, and even CRF-BP. Also, the scope of research on CRF/Ucn roles in stress-related brain regions, such as the extended amygdala and hypothalamus, has expanded to encompass areas more implicated in reward systems, amplifying the potential role of CRF/Ucn modulation in different stages of alcohol drinking, withdrawal and relapse.

Alcohol-induced changes in CRF/Ucn signaling seem to follow a dynamic regulation, which can vary according to the duration/extension of alcohol exposure (acute vs. chronic), the withdrawal period following alcohol exposure, brain region analyzed, as well as species/strain, age, sex, and protocols of alcohol exposure (voluntary intake vs. experimenter-administered injections vs. alcohol vapor chamber, etc.). Several studies indicate an upregulated expression of CRF and CRFR1 in amygdalar nuclei, especially the CeA, following acute and protracted withdrawal from chronic alcohol exposure, despite some conflicting findings (e.g., Pich et al., 1995; Hansson et al., 2007; Sommer et al., 2008; Roberto et al., 2010; Lowery-Gionta et al., 2012; Eisenhardt et al., 2015; Li et al., 2019; Caputi et al., 2021; Kreifeldt et al., 2022). Acute and protracted withdrawal from chronic alcohol exposure are timepoints associated with increased stress sensitivity, increased anxiety-like behavior and escalated alcohol self-administration that is likely motivated to alleviate withdrawal-related aversive states (e.g., Zorrilla et al., 2013; Koob, 2021; Heilig, 2023). These are timepoints in which drugs targeting CRFR1 seem to be particularly effective in reducing escalated alcohol intake and/or preventing reinstatement of alcohol-seeking in animal models (see Table 1). Administration of CRFR1-selective antagonists directly into regions of the extended amygdala, such as the CeA or BNST, is particularly effective in preventing escalated alcohol consumption or stress-induced reinstatement of alcohol-seeking (see Table 1). Interestingly, CRF-containing neuronal projections from the CeA to the nucleus accumbens core have been implicated in binge-like alcohol drinking in mice (Borrego et al., 2022). Pharmacological interventions suggest that both CRFR2 activation or blockade may regulate escalated alcohol intake in different brain regions (see Table 1). Moreover, pharmacological studies shed light on the roles of both CRF receptors within brain reward pathways, as well as the modulation provided by CRF-BP in escalated alcohol drinking in preclinical studies.

Human experimental and clinical studies have not lived up to the expectations raised by outcomes resulting from preclinical research, whether in drug- and alcohol-dependence trials or in trials for other stress-related disorders. Reasons for this alleged "failure" in clinical efficacy are unclear and can be of multiple perspectives, including limitations in clinical study design, number of participants and restriction to inpatient phase II AUD trials; species differences in CRF/Ucn components and function (rodents vs. humans), as well as in neuroadaptive and compensatory mechanisms; translational difficulty between acute actions of CRFR1 antagonists tested in AUD animal models, and chronic/subchronic treatment in human clinical setting; pharmacokinetic profile of antagonists tested in human studies so far; the use of selective, high-affinity CRFR1 antagonists, among other reasons (see (Spierling & Zorrilla, 2017; Heilig, 2023)). It is possible that CRF/Ucn pharmacological interventions may hold efficacy for specific subpopulations with selected gene vs. environment profiles. In addition, the need to assess other interventions directed at different targets of CRF/Ucn systems has been proposed, whether by testing non-selective substances that will modulate both CRFR1 and CRFR2, as well as drugs selectively targeting CRFR2 and/or CRF-BP (Zorrilla et al., 2013; Shaham & de Wit, 2016; Spierling & Zorrilla, 2017). While pharmaceutical companies have significantly reduced or discontinued investments on CRF/Ucn-related drug development (Heilig, 2023), there are still significant questions and lessons to be learned about the role of CRF/Ucn targets in the neurobiology of depression and anxiety disorders, alcohol and drug use disorders, and other stress-related disorders. It seems premature to abandon efforts on investigation of CRF/Ucn mechanisms in AUD and other stress-related disorders which still lack proper, effective treatment options.

References

Agoglia, A. E., Zhu, M. H., Quadir, S. G., Bluitt, M. N., Douglass, E., Hanback, T., ... Herman, M. A. (2022). Sex-specific plasticity in CRF regulation of inhibitory control in central amygdala CRF1 neurons after chronic voluntary alcohol drinking. *Addiction Biology, 27*(1), https://doi.org/10.1111/adb.13067.

Agoglia, A. E., Zhu, M., Ying, R., Sidhu, H., Natividad, L. A., Wolfe, S. A., ... Herman, M. A. (2020). Corticotropin-releasing factor receptor-1 neurons in the lateral amygdala display selective sensitivity to acute and chronic ethanol exposure. *ENeuro,* 7(2), https://doi.org/10.1523/ENEURO.0420-19.2020.

Albrechet-Souza, L., Hwa, L. S., Han, X., Zhang, E. Y., DeBold, J. F., & Miczek, K. A. (2015). Corticotropin releasing factor binding protein and CRF2 receptors in the ventral tegmental area: Modulation of ethanol binge drinking in C57BL/6J mice. *Alcoholism, Clinical and Experimental Research, 39*(9), 1609–1618. https://doi.org/10.1111/acer.12825.

Albrechet-Souza, L., Viola, T. W., Grassi-Oliveira, R., Miczek, K. A., & de Almeida, R. M. M. (2017). Corticotropin releasing factor in the bed nucleus of the stria terminalis in socially defeated and non-stressed mice with a history of chronic alcohol intake. *Frontiers in Pharmacology, 8*(OCT), https://doi.org/10.3389/fphar.2017.00762.

Alhaddad, H., Gordon, D. M., Bell, R. L., Jarvis, E. E., Kipp, Z. A., Hinds, T. D., & Sari, Y. (2020). Chronic ethanol consumption alters glucocorticoid receptor isoform expression in stress neurocircuits and mesocorticolimbic brain regions of alcohol-preferring rats. *Neuroscience, 437*. https://doi.org/10.1016/j.neuroscience.2020.04.033.

Allen, C. D., Grigoleit, J. S., Hong, J., Bae, S., Vaughan, J., & Lee, S. (2016). Exposure to alcohol during adolescence exerts long-term effects on stress response and the adult brain stress circuits. *Neuroscience, 339*. https://doi.org/10.1016/j.neuroscience.2016.09.031.

Allen, C. D., Rivier, C. L., & Lee, S. Y. (2011). Adolescent alcohol exposure alters the central brain circuits known to regulate the stress response. *Neuroscience, 182*. https://doi.org/10.1016/j.neuroscience.2011.03.003.

Al-Omari, A., Kecskés, M., Gaszner, B., Biró-Sütő, T., Fazekas, B., Berta, G., ... Kormos, V. (2023). Functionally active TRPA1 ion channel is downregulated in peptidergic neurons of the Edinger-Westphal nucleus upon acute alcohol exposure. *Frontiers in Cell and Developmental Biology, 10*. https://doi.org/10.3389/fcell.2022.1046559.

Bailey, J. E., Dawson, G. R., Dourish, C. T., & Nutt, D. J. (2011a). Validating the inhalation of 7.5% CO(2) in healthy volunteers as a human experimental medicine: A model of generalized anxiety disorder (GAD). *Journal of psychopharmacology (Oxford, England), 25*(9), 1192–1198. https://doi.org/10.1177/0269881111408455.

Bailey, J. E., Papadopoulos, A., Diaper, A., Phillips, S., Schmidt, M., van der Ark, P., ... Nutt, D. J. (2011b). Preliminary evidence of anxiolytic effects of the CRF(1) receptor antagonist R317573 in the 7.5% CO(2) proof-of-concept experimental model of human anxiety. *Journal of psychopharmacology (Oxford, England), 25*(9), 1199–1206. https://doi.org/10.1177/0269881111400650.

Behan, D. P., De Souza, E. B., Lowry, P. J., Potter, E., Sawchenko, P., & Vale, W. W. (1995a). Corticotropin releasing factor (CRF) binding protein: A novel regulator of CRF and related peptides. *Frontiers in Neuroendocrinology, 16*(4), 362–382. https://doi.org/10.1006/frne.1995.1013.

Behan, D. P., Heinrichs, S. C., Troncoso, J. C., Liu, X. J., Kawas, C. H., Ling, N., & De Souza, E. B. (1995b). Displacement of corticotropin releasing factor from its binding protein as a possible treatment for Alzheimer's disease. *Nature, 378*(6554), 284–287. https://doi.org/10.1038/378284a0.

Behan, D. P., Linton, E. A., & Lowry, P. J. (1989). Isolation of the human plasma corticotrophin-releasing factor-binding protein. *The Journal of Endocrinology, 122*(1), 23–31. https://doi.org/10.1677/joe.0.1220023.

Bell, S. M., Reynolds, J. G., Thiele, T. E., Gan, J., Figlewicz, D. P., & Woods, S. C. (1998). Effects of third intracerebroventricular injections of corticotropin-releasing factor (CRF) on ethanol drinking and food intake. *Psychopharmacology, 139*(1–2), https://doi.org/10.1007/s002130050697.

Binneman, B., Feltner, D., Kolluri, S., Shi, Y., Qiu, R., & Stiger, T. (2008). A 6-week randomized, placebo-controlled trial of CP-316,311 (a selective CRH1 antagonist) in the treatment of major depression. *The American Journal of Psychiatry, 165*(5), 617–620. https://doi.org/10.1176/appi.ajp.2008.07071199.

Blank, T., Nijholt, I., Grammatopoulos, D. K., Randeva, H. S., Hillhouse, E. W., & Spiess, J. (2003). Corticotropin-releasing factor receptors couple to multiple G-proteins to activate diverse intracellular signaling pathways in mouse hippocampus: Role in neuronal excitability and associative learning. *The Journal of Neuroscience: The Official Journal of the Society for Neuroscience, 23*(2), 700–707. https://doi.org/10.1523/JNEUROSCI.23-02-00700.2003.

Borrego, M. B., Grigsby, K. B., Townsley, K. G., Chan, A., Firsick, E. J., Tran, A., ... Ozburn, A. R. (2022). Central nucleus of the amygdala projections onto the nucleus accumbens core regulate binge-like alcohol drinking in a CRF-dependent manner. *Neuropharmacology, 203*. https://doi.org/10.1016/j.neuropharm.2021.108874.

Boutros, N., Der-Avakian, A., Kesby, J. P., Lee, S., Markou, A., & Semenova, S. (2018). Effects of adolescent alcohol exposure on stress-induced reward deficits, brain CRF, monoamines and glutamate in adult rats. *Psychopharmacology, 235*(3), https://doi.org/10.1007/s00213-017-4789-0.

Boutros, N., Der-Avakian, A., Semenova, S., Lee, S., & Markou, A. (2016). Risky choice and brain CRF after adolescent ethanol vapor exposure and social stress in adulthood. *Behavioural Brain Research, 311*. https://doi.org/10.1016/j.bbr.2016.05.038.

Boyson, C. O., Miguel, T. T., Quadros, I. M., Debold, J. F., & Miczek, K. A. (2011). Prevention of social stress-escalated cocaine self-administration by CRF-R1 antagonist in the rat VTA. *Psychopharmacology, 218*(1), 257–269. https://doi.org/10.1007/s00213-011-2266-8.

Brancato, A., Castelli, V., Lavanco, G., Tringali, G., Micale, V., Kuchar, M., ... Cannizzaro, C. (2021). Binge-like alcohol exposure in adolescence: Behavioural, neuroendocrine and molecular evidence of abnormal neuroplasticity ... and return. *Biomedicines, 9*(9), https://doi.org/10.3390/biomedicines9091161.

Caldwell, K. K., Goggin, S. L., Labrecque, M. T., & Allan, A. M. (2015). The impact of prenatal alcohol exposure on hippocampal-dependent outcome measures is influenced by prenatal and early-life rearing conditions. *Alcoholism: Clinical and Experimental Research, 39*(4), https://doi.org/10.1111/acer.12674.

Caldwell, K. K., Goggin, S. L., Tyler, C. R., & Allan, A. M. (2014). Prenatal alcohol exposure is associated with altered subcellular distribution of glucocorticoid and mineralocorticoid receptors in the adolescent mouse hippocampal formation. *Alcoholism: Clinical and Experimental Research, 38*(2), https://doi.org/10.1111/acer.12236.

Caputi, F. F., Stopponi, S., Rullo, L., Palmisano, M., Ubaldi, M., Candeletti, S., ... Romualdi, P. (2021). Dysregulation of nociceptin/orphanin fq and dynorphin systems in the extended amygdala of alcohol preferring marchigian sardinian (Msp) rats. *International Journal of Molecular Sciences, 22*(5), https://doi.org/10.3390/ijms22052448.

Castro, M. G., Rowe, J. M., Murray, C. A., Tomasec, P., Shering, A. F., Linton, E. A., ... Lowenstein, P. R. (1995). Generation and characterization of an antiserum reactive with a proteolytic processing site within rat procorticotrophin-releasing hormone. *Neuropeptides, 29*(4), 183–192. https://doi.org/10.1016/0143-4179(95)90060-8 PMID: 8584136.

Chalmers, D. T., Lovenberg, T. W., & De Souza, E. B. (1995). Localization of novel corticotropin-releasing factor receptor (CRF2) mRNA expression to specific subcortical nuclei in rat brain: Comparison with CRF1 receptor mRNA expression. *Journal of Neuroscience, 15*(10), 6340–6350. https://doi.org/10.1523/JNEUROSCI.15-10-06340.1995 PMID: 7472399.

Chen, A. M., Perrin, M. H., Digruccio, M. R., Vaughan, J. M., Brar, B. K., Arias, C. M., ... Vale, W. W. (2005). A soluble mouse brain splice variant of type 2alpha corticotropin-releasing factor (CRF) receptor binds ligands and modulates their activity. *Proceedings of the National Academy of Sciences of the United States of America, 102*(7), 2620–2625. https://doi.org/10.1073/pnas.0409583102.

Chu, K., Koob, G. F., Cole, M., Zorrilla, E. P., & Roberts, A. J. (2007). Dependence-induced increases in ethanol self-administration in mice are blocked by the CRF1 receptor antagonist antalarmin and by CRF1 receptor knockout. *Pharmacology, Biochemistry, and Behavior, 86*(4), https://doi.org/10.1016/j.pbb.2007.03.009.

Cippitelli, A., Damadzic, R., Singley, E., Thorsell, A., Ciccocioppo, R., Eskay, R. L., & Heilig, M. (2012). Pharmacological blockade of corticotropin-releasing hormone receptor 1 (CRH1R) reduces voluntary consumption of high alcohol concentrations in non-dependent Wistar rats. *Pharmacology, Biochemistry, and Behavior, 100*(3), https://doi.org/10.1016/j.pbb.2011.10.016.

Conte, R., Zangirolame, C. M. S., Gobbo, D. R., Pereira, L. D. S., Panfilio, C. E., Reginato, R. D., ... Céspedes, I. C. (2022). Effects of moderate alcohol consumption on behavior and neural systems of Wistar rats. *Anais da Academia Brasileira de Ciencias, 94*(3), https://doi.org/10.1590/0001-3765202220210673.

Coric, V., Feldman, H. H., Oren, D. A., Shekhar, A., Pultz, J., Dockens, R. C., ... Stock, E. G. (2010). Multicenter, randomized, double-blind, active comparator and placebo-controlled trial of a corticotropin-releasing factor receptor-1 antagonist in generalized anxiety disorder. *Depression and Anxiety, 27*(5), 417–425. https://doi.org/10.1002/da.20695.

Correia, D., Martynhak, B. J., Pereira, M., Siba, I. P., Ribeiro, A. F., Camarini, R., & Boerngen-Lacerda, R. (2015). Reduction of ethanol intake by corticotropin-releasing factor receptor-1 antagonist in "heavy-drinking" mice in a free-choice paradigm. *Psychopharmacology, 232*(15), https://doi.org/10.1007/s00213-015-3909-y.

de Guglielmo, G., Kallupi, M., Pomrenze, M. B., Crawford, E., Simpson, S., Schweitzer, P., ... George, O. (2019). Inactivation of a CRF-dependent amygdalofugal pathway reverses addiction-like behaviors in alcohol-dependent rats. *Nature Communications, 10*(1), https://doi.org/10.1038/s41467-019-09183-0.

Dunlop, B. W., Binder, E. B., Iosifescu, D., Mathew, S. J., Neylan, T. C., Pape, J. C., ... Mayberg, H. S. (2017). Corticotropin-releasing factor receptor 1 antagonism is ineffective for women with posttraumatic stress disorder. *Biological Psychiatry, 82*(12), 866–874. https://doi.org/10.1016/j.biopsych.2017.06.024.

Eisenhardt, M., Hansson, A. C., Spanagel, R., & Bilbao, A. (2015). Chronic intermittent ethanol exposure in mice leads to an up-regulation of CRH/CRHR 1 signaling. *Alcoholism: Clinical and Experimental Research, 39*(4), https://doi.org/10.1111/acer.12686.

Falco, A. M., Bergstrom, H. C., Bachus, S. E., & Smith, R. F. (2009). Persisting changes in basolateral amygdala mRNAs after chronic ethanol consumption. *Physiology and Behavior, 96*(1), https://doi.org/10.1016/j.physbeh.2008.09.019.

Finn, D. A., Hashimoto, J. G., Cozzoli, D. K., Helms, M. L., Nipper, M. A., Kaufman, M. N., ... Guizzetti, M. (2018). Binge ethanol drinking produces sexually divergent and distinct changes in nucleus accumbens signaling cascades and pathways in adult C57BL/6J mice. *Frontiers in Genetics, 9*. https://doi.org/10.3389/fgene.2018.00325.

Finn, D. A., Snelling, C., Fretwell, A. M., Tanchuck, M. A., Underwood, L., Cole, M., ... Roberts, A. J. (2007). Increased drinking during withdrawal from intermittent ethanol exposure is blocked by the CRF receptor antagonist D-Phe-CRF(12-41). *Alcoholism: Clinical and Experimental Research, 31*(6), https://doi.org/10.1111/j.1530-0277.2007.00379.x.

Flores-Ramirez, F. J., Matzeu, A., Sánchez-Marín, L., & Martin-Fardon, R. (2022). Blockade of corticotropin-releasing factor-1 receptors in the infralimbic cortex prevents stress-induced reinstatement of alcohol seeking in male Wistar rats: Evidence of interaction between CRF1 and orexin receptor signaling. *Neuropharmacology, 210*. https://doi.org/10.1016/j.neuropharm.2022.109046.

Fu, Y., Li, W., Mai, Y., Guan, J., Ding, R., Hou, J., ... Fu, R. (2024). Association between RMTg neuropeptide genes and negative effect during alcohol withdrawal in mice. *International Journal of Molecular Sciences, 25*(5), https://doi.org/10.3390/ijms25052933.

Funk, C. K., & Koob, G. F. (2007). A CRF2 agonist administered into the central nucleus of the amygdala decreases ethanol self-administration in ethanol-dependent rats. *Brain Research, 1155*(1), https://doi.org/10.1016/j.brainres.2007.04.009.

Funk, D., Li, Z., & Lê, A. D. (2006). Effects of environmental and pharmacological stressors on c-fos and corticotropin-releasing factor mRNA in rat brain: Relationship to the reinstatement of alcohol seeking. *Neuroscience, 138*(1), https://doi.org/10.1016/j.neuroscience.2005.10.062.

Funk, C. K., O'Dell, L. E., Crawford, E. F., & Koob, G. F. (2006). Corticotropin-releasing factor within the central nucleus of the amygdala mediates enhanced ethanol self-administration in withdrawn, ethanol-dependent rats. *Journal of Neuroscience, 26*(44), https://doi.org/10.1523/JNEUROSCI.3096-06.2006.

Funk, C. K., Zorrilla, E. P., Lee, M. J., Rice, K. C., & Koob, G. F. (2007). Corticotropin-releasing factor 1 antagonists selectively reduce ethanol self-administration in ethanol-dependent rats. *Biological Psychiatry, 61*(1), https://doi.org/10.1016/j.biopsych.2006.03.063.

Gehlert, D. R., Cippitelli, A., Thorsell, A., Lê, A. D., Hipskind, P. A., Hamdouchi, C., ... Heilig, M. (2007). 3-(4-Chloro-2-morpholin-4-yl-thiazol-5-yl)-8-(1-ethylpropyl)-2, 6-dimethyl-imidazo[1,2-b]pyridazine: A novel brain-penetrant, orally available corticotropin-releasing factor receptor 1 antagonist with efficacy in animal models of alcoholism. *Journal of Neuroscience, 27*(10), https://doi.org/10.1523/JNEUROSCI.4985-06.2007.

Giardino, W. J., Rodriguez, E. D., Smith, M. L., Ford, M. M., Galili, D., Mitchell, S. H., ... Ryabinin, A. E. (2017). Control of chronic excessive alcohol drinking by genetic manipulation of the Edinger-Westphal nucleus urocortin-1 neuropeptide system. *Translational Psychiatry,* 7(1), https://doi.org/10.1038/tp.2016.293.

Giardino, W. J., & Ryabinin, A. E. (2013). CRF1 receptor signaling regulates food and fluid intake in the drinking-in-the-dark model of binge alcohol consumption. *Alcoholism: Clinical and Experimental Research, 37*(7), https://doi.org/10.1111/acer.12076.

Gilpin, N. W., Richardson, H. N., & Koob, G. F. (2008). Effects of CRF1-receptor and opioid-receptor antagonists on dependence-induced increases in alcohol drinking by alcohol-preferring (P) rats. *Alcoholism: Clinical and Experimental Research, 32*(9), https://doi.org/10.1111/j.1530-0277.2008.00745.x.

Glavas, M. M., Ellis, L., Yu, W. K., & Weinberg, J. (2007). Effects of prenatal ethanol exposure on basal limbic-hypothalamic- pituitary-adrenal regulation: Role of corticosterone. *Alcoholism: Clinical and Experimental Research, 31*(9), https://doi.org/10.1111/j.1530-0277.2007.00460.x.

Gondré-Lewis, M. C., Warnock, K. T., Wang, H., June, H. L., Bell, K. A., Rabe, H., ... Aurelian, L. (2016). Early life stress is a risk factor for excessive alcohol drinking and impulsivity in adults and is mediated via a CRF/GABAA mechanism. *Stress (Amsterdam, Netherlands), 19*(2), https://doi.org/10.3109/10253890.2016.1160280.

Grillon, C., Hale, E., Lieberman, L., Davis, A., Pine, D. S., & Ernst, M. (2015). The CRH1 antagonist GSK561679 increases human fear but not anxiety as assessed by startle. *Neuropsychopharmacology: Official Publication of the American College of Neuropsychopharmacology, 40*(5), 1064–1071. https://doi.org/10.1038/npp.2014.316.

Hansson, A. C., Cippitelli, A., Sommer, W. H., Fedeli, A., Björk, K., Soverchia, L., ... Ciccocioppo, R. (2006). Variation at the rat Crhr1 locus and sensitivity to relapse into alcohol seeking induced by environmental stress. *Proceedings of the National Academy of Sciences of the United States of America, 103*(41), https://doi.org/10.1073/pnas.0604419103.

Hansson, A. C., Cippitelli, A., Sommer, W. H., Ciccocioppo, R., & Heilig, M. (2007). Region-specific down-regulation of Crhr1 gene expression in alcohol-preferring msP rats following ad lib access to alcohol. *Addiction Biology, 12*(1), https://doi.org/10.1111/j.1369-1600.2007.00050.x.

Hauger, R. L., Risbrough, V., Brauns, O., & Dautzenberg, F. M. (2006). Corticotropin releasing factor (CRF) receptor signaling in the central nervous system: New molecular targets. *CNS & Neurological Disorders Drug Targets, 5*(4), 453–479. https://doi.org/10.2174/187152706777950684.

Heilig, M. (2023). Stress-related neuropeptide systems as targets for treatment of alcohol addiction: A clinical perspective. *Journal of Internal Medicine, 293*(5), 559–573. https://doi.org/10.1111/joim.13636.

Held, K., Künzel, H., Ising, M., Schmid, D. A., Zobel, A., Murck, H., ... Steiger, A. (2004). Treatment with the CRH1-receptor-antagonist R121919 improves sleep-EEG in patients with depression. *Journal of Psychiatric Research, 38*(2), 129–136. https://doi.org/10.1016/s0022-3956(03)00076-1.

Henckens, M., Deussing, J., & Chen, A. (2016). Region-specific roles of the corticotropin-releasing factor–urocortin system in stress. *Nature Revews Neurosciece, 17*, 636–651. https://doi.org/10.1038/nrn.2016.94.

Herman, M. A., Contet, C., & Roberto, M. (2016). A functional switch in tonic GABA currents alters the output of central amygdala corticotropin releasing factor receptor-1 neurons following chronic ethanol exposure. *Journal of Neuroscience, 36*(42), https://doi.org/10.1523/JNEUROSCI.1267-16.2016.

Hubbard, C. S., Labus, J. S., Bueller, J., Stains, J., Suyenobu, B., Dukes, G. E., ... Mayer, E. A. (2011). Corticotropin-releasing factor receptor 1 antagonist alters regional activation and effective connectivity in an emotional-arousal circuit during expectation of abdominal pain. *The Journal of Neuroscience, 31*(35), 12491–12500. https://doi.org/10.1523/JNEUROSCI.1860-11.2011.

Hwa, L. S., Debold, J. F., & Miczek, K. A. (2013). Alcohol in excess: CRF_1 receptors in the rat and mouse VTA and DRN. *Psychopharmacology, 225*(2), 313–327. https://doi.org/10.1007/s00213-012-2820-z.

Hwa, L. S., Shimamoto, A., Kayyali, T., Norman, K. J., Valentino, R. J., DeBold, J. F., & Miczek, K. A. (2016). Dissociation of μ-opioid receptor and CRF-R1 antagonist effects on escalated ethanol consumption and mPFC serotonin in C57BL/6J mice. *Addiction Biology, 21*(1), https://doi.org/10.1111/adb.12189.

Inda, C., Armando, N. G., Dos Santos Claro, P. A., & Silberstein, S. (2017). Endocrinology and the brain: corticotropin-releasing hormone signaling. *Endocrine connections, 6*(6), R99–R120. https://doi.org/10.1530/EC-17-0111.

Ising, M., Zimmermann, U. S., Künzel, H. E., Uhr, M., Foster, A. C., Learned-Coughlin, S. M., ... Grigoriadis, D. E. (2007). High-affinity CRF1 receptor antagonist NBI-34041: Preclinical and clinical data suggest safety and efficacy in attenuating elevated stress response. *Neuropsychopharmacology: Official Publication of the American College of Neuropsychopharmacology, 32*(9), 1941–1949. https://doi.org/10.1038/sj.npp.1301328.

Ji, D., Gilpin, N. W., Richardson, H. N., Rivier, C. L., & Koob, G. F. (2008). Effects of naltrexone, duloxetine, and a corticotropin-releasing factor type 1 receptor antagonist on binge-like alcohol drinking in rats. *Behavioural Pharmacology, 19*(1), https://doi.org/10.1097/FBP.0b013e3282f3cf70.

Jiao, Y., Kim, S. C., Wang, Y., Wu, T., Jin, H., Lee, C. W., ... Zhao, R. (2021). Sauchinone blocks ethanol withdrawal-induced anxiety but spares locomotor sensitization: Involvement of nitric oxide in the bed nucleus of the stria terminalis. *Evidence-Based Complementary and Alternative Medicine, 2021*. https://doi.org/10.1155/2021/6670212.

Kavalakatt, S., Khadir, A., Madhu, D., Koistinen, H. A., Al-Mulla, F., Tuomilehto, J., ... Tiss, A. (2021). Urocortin 3 overexpression reduces ER stress and heat shock response in 3T3-L1 adipocytes. *Scientific Reports, 11*(1), 15666. https://doi.org/10.1038/s41598-021-95175-4.

Ketchesin, K. D., Stinnett, G. S., & Seasholtz, A. F. (2016). Binge drinking decreases corticotropin-releasing factor-binding protein expression in the medial prefrontal cortex of mice. *Alcoholism: Clinical and Experimental Research, 40*(8), https://doi.org/10.1111/acer.13119.

King, B. R., & Nicholson, R. C. (2007). Advances in understanding corticotrophin-releasing hormone gene expression. *Frontiers in Bioscience: A Journal and Virtual Library, 12*, 581–590. https://doi.org/10.2741/2084.

Knapp, D. J., Overstreet, D. H., Huang, M., Wills, T. A., Whitman, B. A., Angel, R. A., ... Breese, G. R. (2011). Effects of a stressor and corticotrophin releasing factor on ethanol deprivation-induced ethanol intake and anxiety-like behavior in alcohol-preferring P rats. *Psychopharmacology, 218*(1), https://doi.org/10.1007/s00213-011-2366-5.

Koob, G. F. (2010). The role of CRF and CRF-related peptides in the dark side of addiction. *Brain research, 1314*, 3–14. https://doi.org/10.1016/j.brainres.2009.11.008.

Koob, G. F. (2021). Drug addiction: Hyperkatifeia/negative reinforcement as a framework for medications development. *Pharmacological Reviews, 73*(1), 163–201. https://doi.org/10.1124/pharmrev.120.000083.

Koob, G. F., & Zorrilla, E. P. (2010). Neurobiological mechanisms of addiction: focus on corticotropin-releasing factor. *Current opinion in investigational drugs (London, England: 2000)*, *11*(1), 63–71.

Kreifeldt, M., Herman, M. A., Sidhu, H., Okhuarobo, A., Macedo, G. C., Shahryari, R., ... Contet, C. (2022). Central amygdala corticotropin-releasing factor neurons promote hyponeophagia but do not control alcohol drinking in mice. *Molecular Psychiatry, 27*(5), https://doi.org/10.1038/s41380-022-01496-9.

Kwako, L. E., Spagnolo, P. A., Schwandt, M. L., Thorsell, A., George, D. T., Momenan, R., et al. (2015). The corticotropin releasing hormone-1 (CRH1) receptor antagonist pexacerfont in alcohol dependence: A randomized controlled experimental medicine study. *Neuropsychopharmacology: Official Publication of the American College of Neuropsychopharmacology, 40*(5), 1053–1063. https://doi.org/10.1038/npp.2014.306.

Läck, A. K., Floyd, D. W., & McCool, B. A. (2005). Chronic ethanol ingestion modulates proanxiety factors expressed in rat central amygdala. *Alcohol (Fayetteville, N. Y.), 36*(2), https://doi.org/10.1016/j.alcohol.2005.07.004.

Lam, M. P., & Gianoulakis, C. (2011). Effects of acute ethanol on corticotropin-releasing hormone and β-endorphin systems at the level of the rat central amygdala. *Psychopharmacology, 218*(1), https://doi.org/10.1007/s00213-011-2337-x.

Lam, V. Y. Y., Raineki, C., Ellis, L., Yu, W., & Weinberg, J. (2018). Interactive effects of prenatal alcohol exposure and chronic stress in adulthood on anxiety-like behavior and central stress-related receptor mRNA expression: Sex- and time-dependent effects. *Psychoneuroendocrinology, 97*. https://doi.org/10.1016/j.psyneuen.2018.06.018.

László, F. A., Varga, C., Pávó, I., Gardi, J., Vecsernyés, M., Gálfi, M., ... Makara, G. B. (2001). Vasopressin pressor receptor-mediated activation of HPA axis by acute ethanol stress in rats. *American Journal of Physiology—Regulatory Integrative and Comparative Physiology, 280*(2), 49–2. https://doi.org/10.1152/ajpregu.2001.280.2.r458.

Le, A. D., Harding, S., Juzytsch, W., Fletcher, P. J., & Shaham, Y. (2002). The role of corticotropin-releasing factor in the median raphe nucleus in relapse to alcohol. *Journal of Neuroscience, 22*(18), https://doi.org/10.1523/jneurosci.22-18-07844.2002.

Lê, A. D., Funk, D., Coen, K., Li, Z., & Shaham, Y. (2013). Role of corticotropin-releasing factor in the median raphe nucleus in yohimbine-induced reinstatement of alcohol seeking in rats. *Addiction Biology, 18*(3), https://doi.org/10.1111/j.1369-1600.2011.00374.x.

Lê, A. D., Harding, S., Juzytsch, W., Watchus, J., Shalev, U., & Shaham, Y. (2000). The role of corticotrophin-releasing factor in stress-induced relapse to alcohol-seeking behavior in rats. *Psychopharmacology, 150*(3), https://doi.org/10.1007/s002130000411.

Lee, A. L., Ogle, W. O., & Sapolsky, R. M. (2002). Stress and depression: Possible links to neuron death in the hippocampus. *Bipolar Disorders, 4*(2), 117–128. https://doi.org/10.1034/j.1399-5618.2002.01144.x.

Lee, S., & Rivier, C. (1997). Alcohol increases the expression of type 1, but not type 2 α corticotropin-releasing factor (CRF) receptor messenger ribonucleic acid in the rat hypothalamus. *Molecular Brain Research, 52*(1), https://doi.org/10.1016/S0169-328X(97)00226-X.

Lee, S., Selvage, D., Hansen, K., & Rivier, C. (2004). Site of action of acute alcohol administration in stimulating the rat hypothalamic-pituitary-adrenal axis: Comparison between the effect of systemic and intracerebroventricular injection of this drug on pituitary and hypothalamic responses. *Endocrinology, 145*(10), https://doi.org/10.1210/en.2004-0110.

Levine, O. B., Skelly, M. J., Miller, J. D., Rivera-Irizarry, J. K., Rowson, S. A., DiBerto, J. F., ... Pleil, K. E. (2021). The paraventricular thalamus provides a polysynaptic brake on limbic CRF neurons to sex-dependently blunt binge alcohol drinking and avoidance behavior in mice. *Nature Communications, 12*(1), https://doi.org/10.1038/s41467-021-25368-y.

Li, L. B., Kim, Y. W., Wang, Y. H., Bai, L., Zhu, X. D., Zhao, Z. L., ... Zhao, R. J. (2019). Methanol extract of semen Ziziphi Spinosae attenuates ethanol withdrawal anxiety by improving neuropeptide signaling in the central amygdala. *BMC Complementary and Alternative Medicine, 19*(1), https://doi.org/10.1186/s12906-019-2546-0.

Lindell, S. G., Schwandt, M. L., Suomi, S. J., Rice, K. C., Heilig, M., & Barr, C. S. (2017). Intermittent access to ethanol induces escalated alcohol consumption in primates. *Journal of Addictive Behaviors, Therapy & Rehabilitation, 6*(1), 163. https://doi.org/10.4172/2324-9005.1000163.

Liu, X., & Weiss, F. (2002). Additive effect of stress and drug cues on reinstatement of ethanol seeking: Exacerbation by history of dependence and role of concurrent activation of corticotropin-releasing factor and opioid mechanisms. *Journal of Neuroscience, 22*(18), https://doi.org/10.1523/jneurosci.22-18-07856.2002.

Lodge, D. J., & Lawrence, A. J. (2003). The CRF1 receptor antagonist antalarmin reduces volitional ethanol consumption in isolation-reared fawn-hooded rats. *Neuroscience, 117*(2), https://doi.org/10.1016/S0306-4522(02)00793-5.

Logrip, M. L., Rivier, C., Lau, C., Im, S., Vaughan, J., & Lee, S. (2013). Adolescent alcohol exposure alters the rat adult hypothalamic-pituitary-adrenal axis responsiveness in a sex-specific manner. *Neuroscience, 235*. https://doi.org/10.1016/j.neuroscience.2012.12.069.

Lovenberg, T. W., Chalmers, D. T., Liu, C., & De Souza, E. B. (1995). CRF2 alpha and CRF2 beta receptor mRNAs are differentially distributed between the rat central nervous system and peripheral tissues. *Endocrinology, 136*(9), 4139–4142. https://doi.org/10.1210/endo.136.9.7544278.

Lowery, E. G., Spanos, M., Navarro, M., Lyons, A. M., Hodge, C. W., & Thiele, T. E. (2010). CRF-1 antagonist and CRF-2 agonist decrease binge-like ethanol drinking in c57bl/6j mice independent of the HPA axis. *Neuropsychopharmacology: Official Publication of the American College of Neuropsychopharmacology, 35*(6), https://doi.org/10.1038/npp.2009.209.

Lowery, E. G., Sparrow, A. M., Breese, G. R., Knapp, D. J., & Thiele, T. E. (2008). The CRF-1 receptor antagonist, CP-154,526, attenuates stress-induced increases in ethanol consumption by BALB/cJ mice. *Alcoholism: Clinical and Experimental Research, 32*(2), https://doi.org/10.1111/j.1530-0277.2007.00573.x.

Lowery-Gionta, E. G., Navarro, M., Li, C., Pleil, K. E., Rinker, J. A., Cox, B. R., ... Thiele, T. E. (2012). Corticotropin releasing factor signaling in the central amygdala is recruited during binge-like ethanol consumption in C57BL/6J mice. *Journal of Neuroscience, 32*(10), https://doi.org/10.1523/JNEUROSCI.6256-11.2012.

Marinelli, P. W., Funk, D., Juzytsch, W., Harding, S., Rice, K. C., Shaham, Y., & Lê, A. D. (2007). The CRF1 receptor antagonist antalarmin attenuates yohimbine-induced increases in operant alcohol self-administration and reinstatement of alcohol seeking in rats. *Psychopharmacology, 195*(3), https://doi.org/10.1007/s00213-007-0905-x.

Markovic, D., & Grammatopoulos, D. K. (2010). Characterization of structural determinants of type 1 corticotropin releasing hormone (CRH) receptor signalling properties. *Methods in Molecular Biology (Clifton, N. J.), 634*, 285–307. https://doi.org/10.1007/978-1-60761-652-8_21.

Marty, V. N., Mulpuri, Y., Munier, J. J., & Spigelman, I. (2020). Chronic alcohol disrupts hypothalamic responses to stress by modifying CRF and NMDA receptor function. *Neuropharmacology, 167*. https://doi.org/10.1016/j.neuropharm.2020.107991.

McCarthy, J. R., Heinrichs, S. C., & Grigoriadis, D. E. (1999). Recent advances with the CRF1 receptor: design of small molecule inhibitors, receptor subtypes and clinical indications. *Current Pharmaceutical Design, 5*(5), 289–315.

McClintick, J. N., McBride, W. J., Bell, R. L., Ding, Z. M., Liu, Y., Xuei, X., & Edenberg, H. J. (2015). Gene expression changes in serotonin, GABA-A receptors, neuropeptides and ion channels in the dorsal raphe nucleus of adolescent alcohol-preferring (P) rats following binge-like alcohol drinking. *Pharmacology, Biochemistry, and Behavior, 129*. https://doi.org/10.1016/j.pbb.2014.12.007.

Menke, A. (2024). The HPA axis as target for depression. *Current Neuropharmacology, 22*(5), 904–915. https://doi.org/10.2174/1570159X21666230811141557.

Miguel, T. T., & Nunes-de-Souza, R. L. (2011). Anxiogenic and antinociceptive effects induced by corticotropin-releasing factor (CRF) injections into the periaqueductal gray are modulated by CRF1 receptor in mice. *Hormones and behavior, 60*(3), 292–300. https://doi.org/10.1016/j.yhbeh.2011.06.004.

Molander, A., Vengeliene, V., Heilig, M., Wurst, W., Deussing, J. M., & Spanagel, R. (2012). Brain-specific inactivation of the Crhr1 gene inhibits post-dependent and stress-induced alcohol intake, but does not affect relapse-like drinking. *Neuropsychopharmacology: Official Publication of the American College of Neuropsychopharmacology, 37*(4), https://doi.org/10.1038/npp.2011.297.

Morabbi, M. J., Razaghi, E., Moazen-Zadeh, E., Safi-Aghdam, H., Zarrindast, M. R., Vousoghi, N., & Akhondzadeh, S. (2018). Pexacerfont as a CRF1 antagonist for the treatment of withdrawal symptoms in men with heroin/methamphetamine dependence: A randomized, double-blind, placebo-controlled clinical trial. *International Clinical Psychopharmacology, 33*(2), 111–119. https://doi.org/10.1097/YIC.0000000000000200.

Munier, J. J., Marty, V. N., & Spigelman, I. (2022). Sex differences in α-adrenergic receptor function contribute to impaired hypothalamic metaplasticity following chronic intermittent ethanol exposure. *Alcoholism: Clinical and Experimental Research, 46*(8), https://doi.org/10.1111/acer.14900.

Myers, D. A., Trinh, J. V., & Myers, T. R. (1998). Structure and function of the ovine type 1 corticotropin releasing factor receptor (CRF1) and a carboxyl-terminal variant. *Molecular and Cellular Endocrinology, 144*(1-2), 21–35. https://doi.org/10.1016/s0303-7207(98)00157-9 PMID: 9863624.

Neira, S., Lee, S., Hassanein, L. A., Sides, T., D'Ambrosio, S. L., Boyt, K. M., ... Kash, T. L. (2023). Impact and role of hypothalamic corticotropin releasing hormone neurons in withdrawal from chronic alcohol consumption in female and male mice. *Journal of Neuroscience, 43*(45), https://doi.org/10.1523/JNEUROSCI.1153-23.2023.

Newman, E. L., Albrechet-Souza, L., Andrew, P. M., Auld, J. G., Burk, K. C., Hwa, L. S., ... Miczek, K. A. (2018). Persistent escalation of alcohol consumption by mice exposed to brief episodes of social defeat stress: Suppression by CRF-R1 antagonism. *Psychopharmacology, 235*(6), https://doi.org/10.1007/s00213-018-4905-9.

Nipper, M. A., Helms, M. L., Finn, D. A., & Ryabinin, A. E. (2024). Stress-enhanced ethanol drinking does not increase sensitivity to the effects of a CRF-R1 antagonist on ethanol intake in male and female mice. *Alcohol (Fayetteville, N. Y.)*. https://doi.org/10.1016/j.alcohol.2024.01.001.

O'Callaghan, M. J., Croft, A. P., Jacquot, C., & Little, H. J. (2005). The hypothalamo-pituitary-adrenal axis and alcohol preference. *Brain Research Bulletin, 68*(3), https://doi.org/10.1016/j.brainresbull.2005.08.006.

Odeon, M. M., Yamauchi, L., Grosman, M., & Acosta, G. B. (2017). Long-term effects of repeated maternal separation and ethanol intake on HPA axis responsiveness in adult rats. *Brain Research, 1657*. https://doi.org/10.1016/j.brainres.2016.11.034.

Olive, M. F., Koenig, H. N., Nannini, M. A., & Hodge, C. W. (2002). Elevated extracellular CRF levels in the bed nucleus of the stria terminalis during ethanol withdrawal and reduction by subsequent ethanol intake. *Pharmacology, Biochemistry, and Behavior, 72*(1–2), https://doi.org/10.1016/S0091-3057(01)00748-1.

Overstreet, D. H., Knapp, D. J., & Breese, G. R. (2004). Modulation of multiple ethanol withdrawal-induced anxiety-like behavior by CRF and CRF1 receptors. *Pharmacology, Biochemistry, and Behavior, 77*(2), https://doi.org/10.1016/j.pbb.2003.11.010.

Overstreet, D. H., Knapp, D. J., & Breese, G. R. (2007). Drug challenges reveal differences in mediation of stress facilitation of voluntary alcohol drinking and withdrawal-induced anxiety in alcohol-preferring P rats. *Alcoholism: Clinical and Experimental Research, 31*(9), https://doi.org/10.1111/j.1530-0277.2007.00445.x.

Papadimitriou, A., & Priftis, K. N. (2009). Regulation of the hypothalamic-pituitary-adrenal axis. *Neuroimmunomodulation, 16*(5), 265–271. https://doi.org/10.1159/000216184.

Patel, R. R., Wolfe, S. A., Borgonetti, V., Gandhi, P. J., Rodriguez, L., Snyder, A. E., ... Roberto, M. (2022). Ethanol withdrawal-induced adaptations in prefrontal corticotropin releasing factor receptor 1-expressing neurons regulate anxiety and conditioned rewarding effects of ethanol. *Molecular Psychiatry, 27*(8), https://doi.org/10.1038/s41380-022-01642-3.

Pati, D., Marcinkiewcz, C. A., DiBerto, J. F., Cogan, E. S., McElligott, Z. A., & Kash, T. L. (2020). Chronic intermittent ethanol exposure dysregulates a GABAergic microcircuit in the bed nucleus of the stria terminalis. *Neuropharmacology, 168*. https://doi.org/10.1016/j.neuropharm.2019.107759.

Phillips, T. J., Reed, C., & Pastor, R. (2015). Preclinical evidence implicating corticotropin-releasing factor signaling in ethanol consumption and neuroadaptation. *Genes, brain. and behavior, 14*(1), 98–135. https://doi.org/10.1111/gbb.12189.

Pich, E. M., Lorang, M., Yeganeh, M., De Fonseca, F. R., Raber, J., Koob, G. F., & Weiss, F. (1995). Increase of extracellular corticotropin-releasing factor-like immunoreactivity levels in the amygdala of awake rats during restraint stress and ethanol withdrawal as measured by microdialysis. *Journal of Neuroscience, 15*(8), https://doi.org/10.1523/jneurosci.15-08-05439.1995.

Pickering, C., Avesson, L., Liljequist, S., Lindblom, J., & Schiöth, H. B. (2007a). The role of hypothalamic peptide gene expression in alcohol self-administration behavior. *Peptides, 28*(12), https://doi.org/10.1016/j.peptides.2007.09.011.

Pickering, C., Avesson, L., Lindblom, J., Liljequist, S., & Schiöth, H. B. (2007b). Identification of neurotransmitter receptor genes involved in alcohol self-administration in the rat prefrontal cortex, hippocampus and amygdala. *Progress in Neuro-Psychopharmacology and Biological Psychiatry, 31*(1), https://doi.org/10.1016/j.pnpbp.2006.06.010.

Potretzke, S., Robins, M. T., & Ryabinin, A. E. (2020). Differential sensitivity of alcohol drinking and partner preference to a CRFR1 antagonist in prairie voles and mice. *Hormones and Behavior, 120*. https://doi.org/10.1016/j.yhbeh.2020.104676.

Potter, E., Behan, D. P., Fischer, W. H., Linton, E. A., Lowry, P. J., & Vale, W. W. (1991). Cloning and characterization of the cDNAs for human and rat corticotropin releasing factor-binding proteins. *Nature, 349*(6308), 423–426. https://doi.org/10.1038/349423a0.

Potter, E., Behan, D. P., Linton, E. A., Lowry, P. J., Sawchenko, P. E., & Vale, W. W. (1992). The central distribution of a corticotropin-releasing factor (CRF)-binding protein predicts multiple sites and modes of interaction with CRF. *Proceedings of the National Academy of Sciences of the United States of America, 89*(9), 4192–4196. https://doi.org/10.1073/pnas.89.9.4192.

Quadir, S. G., Arleth, G. M., Cone, M. G., High, M. W., Ramage, M. C., Effinger, D. P., ... Herman, M. A. (2023). Chronic alcohol drinking drives sex-specific differences in affective behavior and medial prefrontal cortex activity in CRFR1: Cre: tdTomato transgenic rats. *Eneuro, 10*(7), https://doi.org/10.1523/ENEURO.0055-23.2023.

Quadros, I. M., Macedo, G. C., Domingues, L. P., & Favoretto, C. A. (2016). An update on CRF mechanisms underlying alcohol use disorders and dependence. *Frontiers in Endocrinology*, 7, 134. https://doi.org/10.3389/fendo.2016.00134.

Raineki, C., Chew, L., Mok, P., Ellis, L., & Weinberg, J. (2016). Short- and long-term effects of stress during adolescence on emotionality and HPA function of animals exposed to alcohol prenatally. *Psychoneuroendocrinology, 74*. https://doi.org/10.1016/j.psyneuen.2016.08.015.

Raineki, C., Ellis, L., & Weinberg, J. (2018). Impact of adolescent stress on the expression of stress-related receptors in the hippocampus of animals exposed to alcohol prenatally. *Hippocampus, 28*(3), https://doi.org/10.1002/hipo.22823.

Retson, T. A., Hoek, J. B., Sterling, R. C., & Van Bockstaele, E. J. (2015). Amygdalar neuronal plasticity and the interactions of alcohol, sex, and stress. *Brain Structure and Function, 220*(6), https://doi.org/10.1007/s00429-014-0851-4.

Richardson, H. N., Zhao, Y., Fekete, É. M., Funk, C. K., Wirsching, P., Janda, K. D., ... Koob, G. F. (2008). MPZP: A novel small molecule corticotropin-releasing factor type 1 receptor (CRF1) antagonist. *Pharmacology, Biochemistry, and Behavior, 88*(4), https://doi.org/10.1016/j.pbb.2007.10.008.

Rinker, J. A., Marshall, S. A., Mazzone, C. M., Lowery-Gionta, E. G., Gulati, V., Pleil, K. E., ... Thiele, T. E. (2017). Extended amygdala to ventral tegmental area corticotropin-releasing factor circuit controls binge ethanol intake. *Biological Psychiatry, 81*(11), https://doi.org/10.1016/j.biopsych.2016.02.029.

Rivier, C., Bruhn, T., & Vale, W. (1984). Effect of ethanol on the hypothalamic-pituitary-adrenal axis in the rat: role of corticotropin-releasing factor (CRF). *The Journal of Pharmacology and Experimental Therapeutics, 229*(1), 127–131. https://doi.org/10.1111/j.1530-0277.2001.tb02133.x.

Rivier, C., & Lee, S. (1996). Acute alcohol administration stimulates the activity of hypothalamic neurons that express corticotropin-releasing factor and vasopressin. *Brain Research, 726*, 1–2. https://doi.org/10.1016/0006-8993(96)00301-0.

Roberto, M., Cruz, M. T., Gilpin, N. W., Sabino, V., Schweitzer, P., Bajo, M., ... Parsons, L. H. (2010). Corticotropin releasing factor-induced amygdala gamma-aminobutyric acid release plays a key role in alcohol dependence. *Biological Psychiatry, 67*(9), https://doi.org/10.1016/j.biopsych.2009.11.007.

Robinson, S. L., Perez-Heydrich, C. A., & Thiele, T. E. (2019). Corticotropin releasing factor type 1 and 2 receptor signaling in the medial prefrontal cortex modulates Binge-like ethanol consumption in C57BL/6J mice. *Brain Sciences, 9*(7), https://doi.org/10.3390/brainsci9070171.

Rodriguez, L., Kirson, D., Wolfe, S. A., Patel, R. R., Varodayan, F. P., Snyder, A. E., ... Roberto, M. (2022). Alcohol dependence induces CRF sensitivity in female central amygdala GABA synapses. *International Journal of Molecular Sciences, 23*(14), https://doi.org/10.3390/ijms23147842.

Roltsch, E. A., Baynes, B. B., Mayeux, J. P., Whitaker, A. M., Baiamonte, B. A., & Gilpin, N. W. (2014). Predator odor stress alters corticotropin-releasing factor-1 receptor (CRF1R)-dependent behaviors in rats. *Neuropharmacology, 79*. https://doi.org/10.1016/j.neuropharm.2013.11.005.

Rouzer, S. K., & Diaz, M. R. (2022). Moderate prenatal alcohol exposure modifies sex-specific CRFR1 activity in the central amygdala and anxiety-like behavior in adolescent offspring. *Neuropsychopharmacology: Official Publication of the American College of Neuropsychopharmacology, 47*(12), https://doi.org/10.1038/s41386-022-01327-z.

Ruffaner-Hanson, C. D., Fernandez-Oropeza, A. K., Sun, M. S., Caldwell, K. K., Allan, A. M., Savage, D. D., ... Milligan, E. D. (2023). Prenatal alcohol exposure alters mRNA expression for stress peptides, glucocorticoid receptor function and immune factors in acutely stressed neonatal brain. *Frontiers in Neuroscience, 17*. https://doi.org/10.3389/fnins.2023.1203557.

Ryabinin, A. E., Galvan-Rosas, A., Bachtell, R. K., & Risinger, F. O. (2003). High alcohol/sucrose consumption during dark circadian phase in C57Bl/6J mice: Involvement of hippocampus, lateral septum and urocortin-positive cells of the Edinger-Westphal nucleus. *Psychopharmacology, 165*(3), https://doi.org/10.1007/s00213-002-1284-y.

Ryabinin, A. E., Yoneyama, N., Tanchuck, M. A., Mark, G. P., & Finn, D. A. (2008). Urocortin 1 microinjection into the mouse lateral septum regulates the acquisition and expression of alcohol consumption. *Neuroscience, 151*(3), https://doi.org/10.1016/j.neuroscience.2007.11.014.

Sabino, V., Cottone, P., Koob, G. F., Steardo, L., Lee, M. J., Rice, K. C., & Zorrilla, E. P. (2006). Dissociation between opioid and CRF1 antagonist sensitive drinking in Sardinian alcohol-preferring rats. *Psychopharmacology, 189*(2), https://doi.org/10.1007/s00213-006-0546-5.

Sabino, V., Kwak, J., Rice, K. C., & Cottone, P. (2013). Pharmacological characterization of the 20% alcohol intermittent access model in sardinian alcohol-preferring rats: A model of binge-like drinking. *Alcoholism: Clinical and Experimental Research, 37*(4), https://doi.org/10.1111/acer.12008.

Sagami, Y., Shimada, Y., Tayama, J., Nomura, T., Satake, M., Endo, Y., et al. (2004). Effect of a corticotropin releasing hormone receptor antagonist on colonic sensory and motor function in patients with irritable bowel syndrome. *Gut, 53*(7), 958–964. https://doi.org/10.1136/gut.2003.018911.

Sahuque, L. L., Kullberg, E. F., Mcgeehan, A. J., Kinder, J. R., Hicks, M. P., Blanton, M. G., ... Olive, M. F. (2006). Anxiogenic and aversive effects of corticotropin-releasing factor (CRF) in the bed nucleus of the stria terminalis in the rat: role of CRF receptor subtypes. *Psychopharmacology, 186*(1), 122–132. https://doi.org/10.1007/s00213-006-0362-y.

Sampedro-Piquero, P., Moreno-Fernández, R. D., Begega, A., López, M., & Santín, L. J. (2022). Long-term consequences of alcohol use in early adolescent mice: Focus on neuroadaptations in GR, CRF and BDNF. *Addiction Biology, 27*(2), https://doi.org/10.1111/adb.13158.

Sanchez-Marin, L., Pavon, F. J., Decara, J., Suarez, J., Gavito, A., Castilla-Ortega, E., ... Serrano, A. (2017). Effects of intermittent alcohol exposure on emotion and cognition: A potential role for the endogenous cannabinoid system and neuroinflammation. *Frontiers in Behavioral Neuroscience, 11*. https://doi.org/10.3389/fnbeh.2017.00015.

Sánchez-Marín, L., Flores-López, M., Gavito, A. L., Suárez, J., Pavón-Morón, F. J., de Fonseca, F. R., & Serrano, A. (2022a). Repeated restraint stress and binge alcohol during adolescence induce long-term effects on anxiety-like behavior and the expression of the endocannabinoid system in male rats. *Biomedicines, 10*(3), https://doi.org/10.3390/biomedicines10030593.

Sánchez-Marín, L., Flores-López, M., Pastor, A., Gavito, A. L., Suárez, J., de la Torre, R., ... Serrano, A. (2022b). Acute stress and alcohol exposure during adolescence result in an anxious phenotype in adulthood: Role of altered glutamate/endocannabinoid transmission mechanisms. *Progress in Neuro-Psychopharmacology and Biological Psychiatry, 113*. https://doi.org/10.1016/j.pnpbp.2021.110460.

Sapolsky, R. M. (2000). The possibility of neurotoxicity in the hippocampus in major depression: A primer on neuron death. *Biological Psychiatry, 48*(8), 755–765. https://doi.org/10.1016/s0006-3223(00)00971-9.

Schreiber, A. L., Lu, Y. L., Baynes, B. B., Richardson, H. N., & Gilpin, N. W. (2017). Corticotropin-releasing factor in ventromedial prefrontal cortex mediates avoidance of a traumatic stress-paired context. *Neuropharmacology, 113*. https://doi.org/10.1016/j.neuropharm.2016.05.008.

Schwandt, M. L., Cortes, C. R., Kwako, L. E., George, D. T., Momenan, R., Sinha, R., ... Heilig, M. (2016). The CRF1 antagonist verucerfont in anxious alcohol-dependent women: Translation of neuroendocrine, but not of anti-craving effects. *Neuropsychopharmacology: official publication of the American College of Neuropsychopharmacology, 41*(12), 2818–2829. https://doi.org/10.1038/npp.2016.61.

Seo, S. Y., Bang, S. K., Kang, S. Y., Cho, S. J., Choi, K. H., & Ryu, Y. H. (2021). Acupuncture alleviates anxiety and 22-khz ultrasonic vocalizations in rats subjected to repeated alcohol administration by modulating the brain-derived neurotrophic factor/corticotropin-releasing hormone signaling pathway. *International Journal of Molecular Sciences, 22*(8), https://doi.org/10.3390/ijms22084037.

Shaham, Y., & de Wit, H. (2016). Lost in translation: CRF1 receptor antagonists and addiction treatment. *Neuropsychopharmacology: Official Publication of the American College of Neuropsychopharmacology, 41*(12), 2795–2797. https://doi.org/10.1038/npp.2016.94.

Sharpe, A. L., & Phillips, T. J. (2009). Central urocortin 3 administration decreases limited-access ethanol intake in nondependent mice. *Behavioural Pharmacology, 20*(4), https://doi.org/10.1097/FBP.0b013e32832f01ba.

Silberman, Y., Fetterly, T. L., Awad, E. K., Milano, E. J., Usdin, T. B., & Winder, D. G. (2015). Ethanol produces corticotropin-releasing factor receptor-dependent enhancement of spontaneous glutamatergic transmission in the mouse central amygdala. *Alcoholism: Clinical and Experimental Research, 39*(11), https://doi.org/10.1111/acer.12881.

Simms, J. A., Nielsen, C. K., Li, R., & Bartlett, S. E. (2014). Intermittent access ethanol consumption dysregulates CRF function in the hypothalamus and is attenuated by the CRF-R1 antagonist, CP-376395. *Addiction Biology, 19*(4), https://doi.org/10.1111/adb.12024.

Simon, B., Buzás, A., Bokor, P., Csabafi, K., Ibos, K. E., Bodnár, É., ... Bagosi, Z. (2023). The effects of alcohol intoxication and withdrawal on hypothalamic neurohormones and extrahypothalamic neurotransmitters. *Biomedicines, 11*(5), https://doi.org/10.3390/biomedicines11051288.

Skelton, K. H., Owens, M. J., & Nemeroff, C. B. (2000). The neurobiology of urocortin. *Regulatory Peptides, 93*(1-3), 85–92. https://doi.org/10.1016/s0167-0115(00)00180-4.

Snyder, A. E., Salimando, G. J., Winder, D. G., & Silberman, Y. (2019). Chronic intermittent ethanol and acute stress similarly modulate BNST CRF neuron activity via noradrenergic signaling. *Alcoholism: Clinical and Experimental Research, 43*(8), https://doi.org/10.1111/acer.14118.

Sommer, W. H., Rimondini, R., Hansson, A. C., Hipskind, P. A., Gehlert, D. R., Barr, C. S., & Heilig, M. A. (2008). Upregulation of voluntary alcohol intake, behavioral sensitivity to stress, and amygdala Crhr1 expression following a history of dependence. *Biological Psychiatry, 63*(2), https://doi.org/10.1016/j.biopsych.2007.01.010.

Spangler, E., Cote, D. M., Anacker, A. M. J., Mark, G. P., & Ryabinin, A. E. (2009). Differential sensitivity of the perioculomotor urocortin-containing neurons to ethanol, psychostimulants and stress in mice and rats. *Neuroscience, 160*(1), https://doi.org/10.1016/j.neuroscience.2009.02.030.

Sparta, D. R., Ferraro, F. M., Fee, J. R., Knapp, D. J., Breese, G. R., & Thiele, T. E. (2009). The alcohol deprivation effect in C57BL/6J mice is observed using operant self-administration procedures and is modulated by CRF-1 receptor signaling. *Alcoholism: Clinical and Experimental Research, 33*(1), https://doi.org/10.1111/j.1530-0277.2008.00808.x.

Sparta, D. R., Hopf, F. W., Gibb, S. L., Cho, S. L., Stuber, G. D., Messing, R. O., ... Bonci, A. (2013). Binge ethanol-drinking potentiates corticotropin releasing factor R1 receptor activity in the ventral tegmental area. *Alcoholism: Clinical and Experimental Research, 37*(10), https://doi.org/10.1111/acer.12153.

Sparta, D. R., Sparrow, A. M., Lowery, E. G., Fee, J. R., Knapp, D. J., & Thiele, T. E. (2008). Blockade of the corticotropin releasing factor type 1 receptor attenuates elevated ethanol drinking associated with drinking in the dark procedures. *Alcoholism, Clinical and Experimental Research, 32*(2), 259–265. https://doi.org/10.1111/j.1530-0277.2007.00575.x.

Spierling, S. R., & Zorrilla, E. P. (2017). Don't stress about CRF: Assessing the translational failures of CRF_1antagonists. *Psychopharmacology, 234*(9-10), 1467–1481. https://doi.org/10.1007/s00213-017-4556-2.

Steckler, T., & Holsboer, F. (1999). Corticotropin-releasing hormone receptor subtypes and emotion. *Biological Psychiatry, 46*(11), 1480–1508. https://doi.org/10.1016/s0006-3223(99)00170-5.

Sweetser, S., Camilleri, M., Linker Nord, S. J., Burton, D. D., Castenada, L., Croop, R., et al. (2009). Do corticotropin releasing factor-1 receptors influence colonic transit and bowel function in women with irritable bowel syndrome? *American Journal of Physiology. Gastrointestinal and Liver Physiology, 296*(6), G1299–G1306. https://doi.org/10.1152/ajpgi.00011.2009.

Thorsell, A., Slawecki, C. J., & Ehlers, C. L. (2005). Effects of neuropeptide Y and corticotropin-releasing factor on ethanol intake in Wistar rats: Interaction with chronic ethanol exposure. *Behavioural Brain Research, 161*(1), https://doi.org/10.1016/j.bbr.2005.01.016.

Todkar, A., Granholm, L., Aljumah, M., Nilsson, K. W., Comasco, E., & Nylander, I. (2016). Hpa axis gene expression and DNA methylation profiles in rats exposed to early life stress, adult voluntary ethanol drinking and single housing. *Frontiers in Molecular Neuroscience, 8*. https://doi.org/10.3389/fnmol.2015.00090.

Ungless, M. A., Singh, V., Crowder, T. L., Yaka, R., Ron, D., & Bonci, A. (2003). Corticotropin-releasing factor requires CRF binding protein to potentiate NMDA receptors via CRF receptor 2 in dopamine neurons. *Neuron, 39*(3), 401–407. https://doi.org/10.1016/s0896-6273(03)00461-6.

Valdez, G. R., Roberts, A. J., Chan, K., Davis, H., Brennan, M., Zorrilla, E. P., & Koob, G. F. (2002). Increased ethanol self-administration and anxiety-like behavior during acute ethanol withdrawal and protracted abstinence: Regulation by corticotropin-releasing factor. *Alcoholism: Clinical and Experimental Research, 26*(10), https://doi.org/10.1111/j.1530-0277.2002.tb02448.x.

Valdez, G. R., Sabino, V., & Koob, G. F. (2004). Increased anxiety-like behavior and ethanol self-administration in dependent rats: Reversal via corticotropin-releasing factor-2 receptor activation. *Alcoholism: Clinical and Experimental Research, 28*(6), https://doi.org/10.1097/01.ALC.0000128222.29875.40.

Vale, W., Spiess, J., Rivier, C., & Rivier, J. (1981). Characterization of a 41-residue ovine hypothalamic peptide that stimulates secretion of corticotropin and beta-endorphin. *Science (New York, N. Y.), 213*(4514), 1394–1397. https://doi.org/10.1126/science.6267699.

Van Pett, K., Viau, V., Bittencourt, J. C., Chan, R. K., Li, H. Y., Arias, C., ... Sawchenko, P. E. (2000). Distribution of mRNAs encoding CRF receptors in brain and pituitary of rat and mouse. *The Journal of Comparative Neurology, 428*(2), 191–212. https://doi.org/10.1002/1096-9861(20001211)428:2<191::aid-cne1>3.0.co;2-u.

Vandael, D., & Gounko, N. V. (2019). Corticotropin releasing factor-binding protein (CRF-BP) as a potential new therapeutic target in Alzheimer's disease and stress disorders. *Translational Psychiatry, 9*(1), 272. https://doi.org/10.1038/s41398-019-0581-8.

Walker, L. C., Kastman, H. E., Koeleman, J. A., Smith, C. M., Perry, C. J., Krstew, E. V., ... Lawrence, A. J. (2017). Nucleus incertus corticotrophin-releasing factor 1 receptor signalling regulates alcohol seeking in rats. *Addiction Biology, 22*(6), https://doi.org/10.1111/adb.12426.

Walker, D. L., Miles, L. A., & Davis, M. (2009a). Selective participation of the bed nucleus of the stria terminalis and CRF in sustained anxiety-like versus phasic fear-like responses. *Progress in Neuro-Psychopharmacology & Biological Psychiatry, 33*(8), 1291–1308. https://doi.org/10.1016/j.pnpbp.2009.06.022.

Walker, D., Yang, Y., Ratti, E., Corsi, M., Trist, D., & Davis, M. (2009b). Differential effects of the CRF-R1 antagonist GSK876008 on fear-potentiated, light- and CRF-enhanced startle suggest preferential involvement in sustained vs phasic threat responses. *Neuropsychopharmacology: Official Publication of the American College of Neuropsychopharmacology, 34*(6), 1533–1542. https://doi.org/10.1038/npp.2008.210.

Weera, M. M., Schreiber, A. L., Avegno, E. M., & Gilpin, N. W. (2020). The role of central amygdala corticotropin-releasing factor in predator odor stress-induced avoidance behavior and escalated alcohol drinking in rats. *Neuropharmacology, 166.* https://doi.org/10.1016/j.neuropharm.2020.107979.

Weitemier, A. Z., & Ryabinin, A. E. (2005). Brain region-specific regulation of urocortin 1 innervation and corticotropin-releasing factor receptor type 2 binding by ethanol exposure. *Alcoholism: Clinical and Experimental Research, 29*(9), https://doi.org/10.1097/01.alc.0000179363.44542.05.

Weitemier, A. Z., & Ryabinin, A. E. (2006). Urocortin 1 in the dorsal raphe regulates food and fluid consumption, but not ethanol preference in C57BL/6J mice. *Neuroscience, 137*(4), https://doi.org/10.1016/j.neuroscience.2005.10.021.

Xiao, H. W., Ge, C., Feng, G. X., Li, Y., Luo, D., Dong, J. L., ... Fan, S. J. (2018). Gut microbiota modulates alcohol withdrawal-induced anxiety in mice. *Toxicology Letters, 287.* https://doi.org/10.1016/j.toxlet.2018.01.021.

Yang, X., Wang, S., Rice, K. C., Munro, C. A., & Wand, G. S. (2008). Restraint stress and ethanol consumption in two mouse strains. *Alcoholism: Clinical and Experimental Research, 32*(5), https://doi.org/10.1111/j.1530-0277.2008.00632.x.

Zhou, Y., Franck, J., Spangler, R., Maggos, C. E., Ho, A., & Kreek, M. J. (2000). Reduced hypothalamic POMC and anterior pituitary CRFR1 receptor mRNA levels after acute, but not chronic, daily "binge" intragastric alcohol administration. *Alcoholism: Clinical and Experimental Research, 24*(10), https://doi.org/10.1111/j.1530-0277.2000.tb04577.x.

Zobel, A. W., Nickel, T., Künzel, H. E., Ackl, N., Sonntag, A., Ising, M., & Holsboer, F. (2000). Effects of the high-affinity corticotropin-releasing hormone receptor 1 antagonist R121919 in major depression: The first 20 patients treated. *Journal of Psychiatric Research, 34*(3), 171–181. https://doi.org/10.1016/s0022-3956(00)00016-9.

Zorrilla, E. P., Heilig, M., de Wit, H., & Shaham, Y. (2013). Behavioral, biological, and chemical perspectives on targeting CRF(1) receptor antagonists to treat alcoholism. *Drug and Alcohol Dependence, 128*(3), 175–186. https://doi.org/10.1016/j.drugalcdep.2012.12.017.

Zuo, W., Zuo, Q., Wu, L., Mei, Q., Shah, M., Zheng, J., ... Ye, J. H. (2021). Roles of corticotropin-releasing factor signaling in the lateral habenula in anxiety-like and alcohol drinking behaviors in male rats. *Neurobiology of Stress, 15.* https://doi.org/10.1016/j.ynstr.2021.10039.

CHAPTER FIVE

Pharmacological activators of ALDH2: A new strategy for the treatment of alcohol use disorders

Sofía Adasme-Reyes[a], Juan Fuentes[a], Ignacio Gutiérrez-Vega[a], Eduardo Isla[a], Vicente Pérez[a], Carolina Ponce[a], María Elena Quilaqueo[a], Mario Herrera-Marschitz[b], María Elena Quintanilla[b], David Vásquez[a], and Mario Rivera-Meza[a,*]

[a]Department of Pharmacological and Toxicological Chemistry, Faculty of Chemical Sciences and Pharmacy, University of Chile, Santiago, Chile
[b]Program of Molecular and Clinical Pharmacology, Faculty of Medicine, University of Chile, Santiago, Chile
*Corresponding author. e-mail address: mario.rivera@ciq.uchile.cl

Contents

Abstract

In mammals, ethanol is metabolized to acetaldehyde mainly by the liver alcohol dehydrogenase (ADH), and acetaldehyde is subsequently oxidized to acetate by mitochondrial aldehyde dehydrogenase (ALDH2). The presence of an inactive variant of ALDH2 or the use of inhibitors of this enzyme leads to an accumulation of acetaldehyde after ethanol consumption, generating an aversive reaction that inhibits subsequent alcohol intake. However, experimental evidence shows that acetaldehyde has potent rewarding effects at the central level, suggesting that acetaldehyde would

International Review of Neurobiology, Volume 178
ISSN 0074-7742, https://doi.org/10.1016/bs.irn.2024.07.003

be responsible for the addictive effect of alcohol. Alda-1 is an organic molecule that acts as a pharmacological activator of ALDH2. Studies in animal models of alcohol use disorders (AUD; i.e. alcoholism) have shown that Alda-1 can inhibit the acquisition, the chronic intake, and the relapse of alcohol consumption. These effects are reversible without any effects on water consumption or other natural reinforcer such as saccharin. It has also been reported that Alda-1 can act as a protective agent from the toxic effects on various tissues and organs mediated by ethanol-derived acetaldehyde, including liver damage, cancer, and central nervous system (CNS) alterations. Using *in silico* tools such as molecular docking the identification of important molecular interactions between Alda-1 and ALDH2 has been demonstrated, identifying new molecules with higher pharmacological features. Thus, there is now preclinical evidence supporting the use of activators of ALDH2 as a pharmacological strategy for the treatment of AUD.

1. Introduction

Alcohol consumption is widespread in the world, dating back to immemorial times. The socio-sanitary consequences of harmful alcohol consumption are high, causing considerable social and economic losses, both for the individual and for the society in general. According to the World Health Organization, excessive alcohol consumption contributes annually to 3 million deaths globally, being responsible for 5.3% of total mortality (WHO, 2022). The current pharmacological treatment of alcoholism is based on the use of 2 main drugs: disulfiram and naltrexone. An inherent problem with these medications is their lack of effectiveness, significant adverse effects, and low compliance (Zindel & Kranzler, 2014). Therefore, it is necessary to develop new therapeutic alternatives to address the treatment of excessive alcohol consumption. This review addresses the main enzymes involved in ethanol metabolism and the properties of drugs targeting these enzymes to treat AUD, describing the development of a new type of drugs that acts through the activation of aldehyde dehydrogenase 2 (ALDH2), discussing their pharmacological effects on animal models of alcoholism and toxicological ethanol exposure.

2. Ethanol metabolism

Alcohol ingested orally is absorbed into the systemic circulation through the gastrointestinal mucosa, mainly in the large absorption surface of the small intestine (Umulis, Gürmen, Singh, & Fogler, 2005). The maximum blood concentration of alcohol is reached approximately 60 min after consumption, distributing quickly and uniformly throughout all the body's fluids. About 90% of the ingested alcohol is completely oxidized in the liver at a constant metabolization rate of approximately 100 mg/kg/h. In comparison, only 10% of the alcohol is excreted unmetabolized through the liver, urine, and in expired air (Lands, 1998).

The hepatic metabolization of alcohol is performed by three enzymatic pathways (Fig. 1), mainly by alcoholic dehydrogenase (ADH) and, to a lesser extent, by cytochrome P450 2E1 and catalase (Matsumoto & Fukui, 2002). All these enzymatic pathways produce ethanol oxidation to acetaldehyde, with hepatic ADH being the most active and the main responsible for its formation. The formed acetaldehyde is rapidly oxidized to acetate by mitochondrial aldehyde dehydrogenase ALDH2 (Klyosov, Rashkovetsky, Tahir, & Keung, 1996). Both enzymes ADH and ALDH2 require NAD^+ as a cofactor, generating NADH as a reaction product.

At the CNS, the blood-brain barrier (BBB) limits the diffusion of acetaldehyde into the brain due to the presence of ALDH in the cerebral microvasculature (Zimatkin, 1991). Furthermore, in variance with the liver, the expression of ADH in the CNS is low (Zimatkin, Pronko, Vasiliou, Gonzalez, & Deitrich, 2006). Therefore, the concentration of

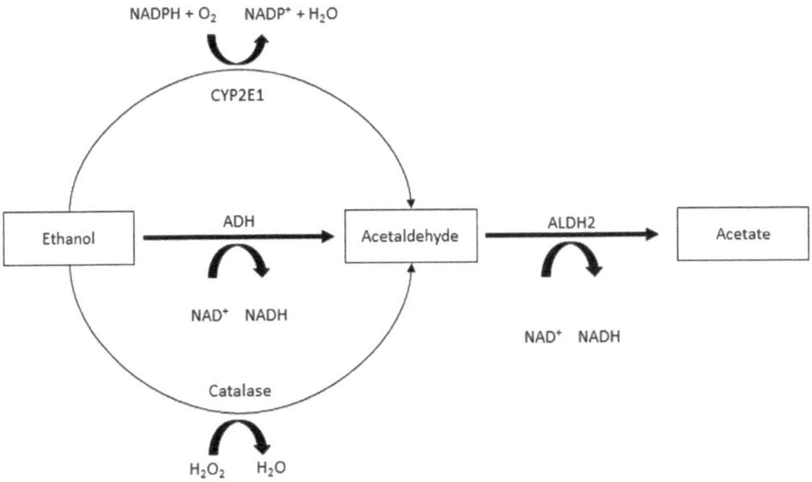

Fig. 1 Ethanol metabolism in the liver. Ethanol is metabolized in the liver through three enzymatic pathways. The main one occurs in the cytosol and is mediated by alcohol dehydrogenase (ADH). The other two pathways are proportionally smaller and occur in the endoplasmic reticulum (CYP2E1 microsomal system) and the peroxisomes (catalase). The acetaldehyde generated is oxidized to acetate in the mitochondria by the action of mitochondrial aldehyde dehydrogenase (ALDH2).

acetaldehyde in the CNS is mainly determined by local production from ethanol. Unlike hepatic metabolism, at the brain level, ethanol oxidation is carried out 60% by the enzyme catalase, with a minor contribution from the activity of CYP2E1.

3. ADH and ALDH polymorphisms

The main enzymes in the hepatic metabolism of alcohol, ADH and ALDH, have shown genetic polymorphisms, altering their activity and, therefore, the rate of transformation of ethanol and acetaldehyde, respectively. One of the main polymorphisms affects the mitochondrial ALDH enzyme (ALDH2), the most important for detoxifying acetaldehyde. A point mutation of the ALDH2 gene (ALDH2*2) generates an amino acid change at position 487 of the protein (glutamate to lysine). This amino acid change results in an enzyme with zero activity. Individuals carrying this mutation experience accumulation of acetaldehyde in the blood after alcohol consumption. Elevated levels of acetaldehyde in the bloodstream can lead to a range of discomforting symptoms, including heightened skin temperature, facial flushing, decreased blood pressure, nausea, headache, palpitations, and bronchoconstriction (Quertemont & Didone, 2006). The largest carriers are individuals of Asian ancestry, with 30% heterozygosity in Chinese, 40–50% in Japanese, and 30% in Koreans (Eng, Luczak, & Wall, 2007). It has been seen that the presence of this mutation is associated with a remarkable protection against alcohol consumption due to the aversive reaction generated by its consumption and increased blood acetaldehyde levels (Chen, Yang, Lai, & Yin, 2021; Li, Zhao, & Gelernter, 2012; Peng, Chen, Tsao, Wang, & Yin, 2007).

ADH, the primary liver enzyme responsible for transforming ethanol to acetaldehyde, corresponds to a family of proteins encoded by seven genes (ADH1-7) with different affinities for ethanol and other alcohols. It was found that ADH1B had the lowest Km for ethanol. Furthermore, three polymorphisms have been described for the ADH1B variant, resulting in enzymes with different rates of ethanol metabolization. One of these polymorphic variants is ADH1B*2, which unlike the normal ADH1B*1 allele, has a histidine instead of an arginine at position 47. This amino acid change results in an enzyme with the greatest catalytic activity, almost 90 times more active for transforming ethanol into acetaldehyde (Edenberg, 2007).

Genetic studies have shown that the ADH1B*2 variant is even more common than the ALDH2*2. A prevalence of 84–92% of heterozygotes for

ADH1B*2 has been seen in Chinese populations, 90% in Japanese populations, and 88–96% in Korean populations, versus 10% in white caucasians and 2% in Africans (Eng et al., 2007). This variant also confers protection against alcoholism. A meta-analysis showed that Chinese and Japanese ADH1B*1/*2 individuals had 1/5 the risk of becoming alcoholics compared to ADH1B*1/*1 individuals, and ADH1B*2/*2 individuals had 1/7 that risk (Luczak, Glatt, & Wall, 2006). The mechanism by which this mutation causes this protective effect is not known yet, since a higher acetaldehydemia after alcohol consumption is not found in ADH1B*2 individuals (Peng et al., 2007).

A study by Rivera-Meza et al. (2010) generated a model in rats, expressing (using adenoviral vectors) a murine variant analogous to ADH1B*2 in the liver. This study showed that these animals generated a significant increase in acetaldehyde within few minutes after receiving a dose of alcohol, but in arterial blood, not in venous blood, as determined in human studies. (Peng, Chen, Wang, Lai, & Yin, 2014). These animals expressing a variant analogous to ADH1B*2 reduced their voluntary alcohol consumption by more than 50%. These findings suggest that the protective effect of the ADH1B*2 variant would be due to a transient increase in peripheral acetaldehyde levels after alcohol consumption, triggering subsequent aversive reactions associated with this condition.

4. Inhibitors of ALDH2 in the treatment of alcoholism

The aldehyde dehydrogenase superfamily (ALDHs) represents a versatile group of $NAD(P)^+$-dependent enzymes, encompassing 19 distinct members, each playing indispensable roles in a broad spectrum of biological processes crucial for cellular survival and protection (Poturnajova, Kozovska, & Matuskova, 2021). These enzymes, distributed across various subcellular compartments such as the cytosol, mitochondria, endoplasmic reticulum, and nucleus, serve as pivotal catalysts in the intricate metabolism of endogenous and exogenous aldehydes (Jackson et al., 2011).

ALDHs are crucial responders to cellular challenges, particularly oxidative stress and lipid peroxidation, showcasing their adaptability and relevance for maintaining cellular homeostasis (Vasiliou, Pappa, & Estey, 2004). Additionally, the ALDH superfamily actively participates in vital enzymatic detoxification processes, converting aldehydes into essential molecules integral to cellular functions, such as retinoic acid, betaine and gamma-aminobutyric acid (Bilska-Wilkosz, Górny, & Iciek, 2019). Notably, ALDHs play a key role in acetaldehyde metabolism, particularly in the context of ethanol consumption. Acetaldehyde, primarily produced through the oxidation of ingested ethanol, stands as a well-known substrate for ALDHs, underscoring the pivotal role these enzymes play for mitigating the toxic effects of acetaldehyde generated by ethanol metabolism. Consequently, ALDH2-deficient individuals face lower risks of developing AUD (Chen et al., 2021; Li et al., 2012; Peng et al., 2007). Considering the protective role of decreased ALDH2 activity, pharmacological therapies based on inhibiting ALDH2 have been developed to treat AUD.

Disulfiram, known as Antabuse®, is an FDA-approved drug employed for treating chronic alcoholism since 1948. Disulfiram functions as a non-selective and irreversible ALDH1 and ALDH2 inhibitor through the carbamoylation of the catalytic Cys302 residue on ALDH2. This inhibition disrupts ethanol metabolism at the aldehyde stage, resulting in the accumulation of acetaldehyde in the bloodstream after alcohol intake, thereby inducing alcohol aversion (Omran, 2022). Administered orally, disulfiram is 80–90% absorbed due to its high lipophilicity; disulfiram is retained in adipose tissue and rapidly crosses the blood-brain barrier. Disulfiram is quickly metabolized by reduction to diethyldithiocarbamate; this process occurs spontaneously in solution and through enzymes by the action of thioesterases in the liver and kidney. The pharmacokinetics of disulfiram have high inter-subject variability; it may stem from differences in patient metabolism, lipid content, enterohepatic circulation, or protein-binding capacity in the plasma. While daily doses typically range from 250 to 500 mg, studies indicate significant variability in disulfiram responses. Lower doses do not lead to disulfiram–ethanol reactions in nearly half of the tested subjects. This variability between patients poses serious risks when prescribing disulfiram, as some individuals may experience toxic reactions even at the lowest dose (Lanz et al., 2023).

Although disulfiram has shown promise in the treatment of AUD, conflicting efficacy results in clinical practice combined with its adverse effects indicate that it may be best as an adjunct to other therapeutic measures (Haass-Koffler, Akhlaghi, Swift, & Leggio, 2017; Lehner, Gao, & Mackowiak, 2024). While the Food and Drug Administration has approved disulfiram for AUD treatment, its efficacy is limited due to compliance issues (Malcolm, Olive, & Lechner, 2008; Skinner, Lahmek, Pham, & Aubin, 2014). Moreover, disulfiram is not specific to ALDH2, inhibiting cytosolic ALDH1 (MacDonagh et al., 2017). Among the adverse effects of disulfiram it is possible to find headaches, fatigue, drowsiness, hepatotoxicity, and a metallic or garlic-like taste (Lanz et al., 2023). The most severe adverse effect is hepatotoxicity, caused by the formation of CS2 in hepatic cells due to the degradation of disulfiram.

The primary clinical variable typically assessed in disulfiram clinical studies is the duration of abstinence achieved by the patients. However, one of the main challenges arises from patient-generated bias, as a patient motivated to quit drinking maintains more extended abstinence regardless of the treatment. On the other hand, if the patient has already experienced the disulfiram-ethanol interaction, they may have 20% longer periods of abstinence (Banys, 1988). Numerous studies have been conducted to support the clinical benefit of disulfiram treatment in patients with AUD; however, most indicate that there are no significant increases in the duration of abstinence without parallel psychotherapy (Suh, Pettinati, Kampman, & O'Brien, 2006).

Cyanamide is a reversible inhibitor of the ALDH enzyme, which, like disulfiram, generates aversion to alcohol consumption due to the accumulation of acetaldehyde (Gao, Hao, Piao, & Gu, 2022). It is indicated for maintaining alcohol abstinence in some countries in Asia and Europe but has not been approved by the FDA. Cyanamide is administered orally, has an important hepatic first-pass effect, and is extensively metabolized in the liver, generating metabolites that are eliminated via the kidneys. Cyanamide is generally a well-tolerated drug, but it has side effects such as dermatitis, hepatitis, and neutropenia. Compared to disulfiram, its side effects are smaller and less frequent, but it has greater hepatotoxicity, so it should be used with caution in cases of hepatic and hematological disorders. In the treatment of alcohol addiction, cyanamide is a second-line drug relative to disulfiram owing to its greater capacity to induce liver injury compared with disulfiram (Niederhofer, Staffen, & Mair, 2003).

5. Acetaldehyde is a central rewarding metabolite of ethanol

Addictive drugs, including alcohol, have a common feature of activating the mesocorticolimbic system in the brain. This system is responsible for the response to natural rewards such as food and sexual activity. Studies conducted on rats by Di Chiara and Imperato in 1988 revealed that alcohol stimulates this system by activating dopaminergic neurons of the ventral tegmental area (VTA) that projects to the nucleus accumbens (NAc), increasing the release of dopamine in this region. The activation of this neuronal pathway is critical for the pleasurable effects induced by alcohol (Nestler, 2005).

The study of the neurobiological bases of AUD has considered the involvement of other ethanol-derived compounds as mediators of the pleasant effects of ethanol and, therefore, responsible for the development of dependence on consumption. Among these compounds, acetaldehyde has been proposed as an active ethanol-derived compound with central rewarding effects. Animal studies have shown that acetaldehyde can be self-administered directly to cerebral ventricles (Amit & Smith, 1985; Brown, Amit, & Rockman, 1979) or intravenously (Foddai, Dosia, Spiga, & Diana, 2004; Myers, Ng, & Singer, 1984), suggesting that acetaldehyde mediates the reinforcing properties generally attributed to ethanol. Further studies have reported the release of dopamine in the NAc by administering acetaldehyde directly into the VTA in murine models, potentially contributing to the pleasurable effects of ethanol consumption (Deehan, Engleman, Ding, McBride, & Rodd, 2013; McBride et al., 2002). This hypothesis is also supported by studies demonstrating that acetaldehyde has a greater rewarding effect than ethanol: acetaldehyde having an effect at concentrations of 10 μM, while ethanol at 10 mM, resulting in a difference of 3 orders of magnitude between these compounds (Israel et al., 2017; Rodd et al., 2005).

Consistent with this evidence, Tampier et al. (2013) reported a 60% decrease in relapse-like ethanol consumption in rats by reducing acetaldehyde production through silencing catalase enzyme in the VTA via lentiviral vector administration. Using a pharmacological strategy, Peana et al. (2015) reported decreased ethanol self-administration and relapse in rats by administering D-penicillamine, an acetaldehyde-sequestering agent. Moreover, the administration of a lentivector coding for ALDH2 into the VTA, aimed at enhancing the elimination of ethanol-derived acetaldehyde, resulted in an 85% reduction of ethanol intake in naïve UChB rats (Karahanian et al., 2015).

The mechanisms by which cerebral acetaldehyde would generate the pleasurable effects associated with ethanol consumption are still under study, establishing a relevant research area to clarify the mechanisms of alcohol consumption addiction. However, the experimental evidence suggests that reducing central ethanol-derived acetaldehyde levels represents a potential strategy to reduce the rewarding effects of ethanol.

6. Alda-1, a pharmacological activator of ALDH2

In 2008, Chen et al. (2008) described the generation of a small organic molecule called Alda-1 (*N*-(1,3-benzodioxol-5-ylmethyl)-2,6-dichlorobenzamide) (See Fig. 6A), which, at micromolar concentrations, acts as a pharmacological activator of ALDH2, doubling its catalytic activity. Further studies showed that a "chaperone-like effect" induced by the binding of Alda-1 to ALDH2 generates a conformational change in the enzyme, favoring its interaction with both the substrate (aldehyde) and the cofactor (NAD^+), increasing the catalytic activity of ALDH2 in vitro and in vivo (Belmont-Díaz, Yoval-Sánchez, Calleja-Castañeda, Pardo Vázquez, & Rodríguez-Zavala, 2016; Beretta et al., 2010; Perez-Miller et al., 2010). This property of Alda-1 has resulted in several preclinical studies on different pathologies in which ALDH2 dysfunction is suspected to play a role (Chen, Ferreira, Gross, & Mochly-Rosen, 2014; Chen, Joshi, & Mochly-Rosen, 2016). Animal studies have shown that Alda-1 administration exerts neuroprotection in models of ischemia/reperfusion, a pathological condition associated with increased oxidative stress (Chen et al., 2014). These neuroprotective effects of Alda-1 have been linked to the activation of ALDH2, resulting in a higher rate of elimination of cytotoxic aldehydes, including those derived from lipid peroxidation such as 4-hydroxynonenal (4-HNE) and malondialdehyde (MDA). A study by Chiu et al. (2015) showed the neuroprotective effect of Alda-1 in an animal model of Parkinson's disease by increasing the elimination of DOPAL, a neurotoxic aldehyde derived from the metabolism of dopamine, which is a substrate of ALDH2. Alda-1 is a lipophilic substance ($\log P = 4$); it can cross the blood-brain barrier, activating ALDH2 in the whole brain (Fu et al., 2014). The unique ability of Alda-1 to increase the activity of ALDH2 and its ability to reach the CNS easily makes this compound an interesting pharmacological alternative to increase acetaldehyde oxidation in the brain and, therefore, to reduce the initial motivational effects of ethanol.

7. Effects of Alda-1 in animal models of alcohol preference

Considering the importance of ALDH2 in ethanol metabolism and the involvement of acetaldehyde for its rewarding effects, several studies have focused on the beneficial effects that Alda-1 would have by reducing acetaldehyde levels and on how the activation of ALDH2 affects the consumption of ethanol in animal models of alcoholism.

Rivera-Meza et al. (2019) reported a study aimed at investigating the impact of Alda-1 on both the initiation and maintenance of alcohol intake in alcohol-preferring UChB rats. To evaluate the effects of ALDA-1 on alcohol acquisition, naïve UChB rats were given five daily doses of ALDA-1 (12.5, 25, or 50 mg/kg, i.p.) starting one day before the onset of ethanol exposure. In chronic intake studies, UChB rats with free access to 10% ethanol and water for 98 days were treated daily with ALDA-1 (12.5, 25, or 50 mg/kg, i.p.) for five days. The administration of ALDA-1 resulted in a dose-dependent (ranging from 72 to 90%) decrease in the acquisition of ethanol consumption in naïve rats (Fig. 2A). When Alda-1 was administered to UChB rats exposed to chronic ethanol consumption, a marked dose-dependent decrease in ethanol intake was observed (ranging from 61 to 82%) (Fig. 2B). In both cases, the administration of Alda-1 did not affect total fluid intake, and its effects on ethanol consumption were reversed upon discontinuation. The authors also showed that the administration of

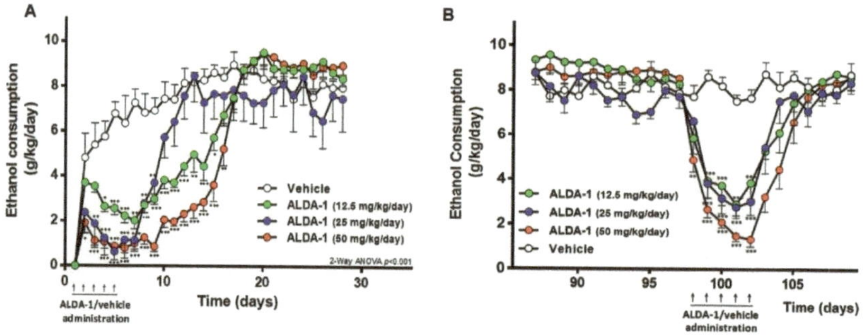

Fig. 2 Alda-1 administration reduces, dose-dependently, the acquisition and maintenance of voluntary ethanol intake in UChB-drinking rats. (A) Ethanol-naïve UChB rats were administered with Alda-1 (12.5, 25, and 50 mg/kg/day) or vehicle for five consecutive days. On day 1, animals were exposed to a free choice between water and 10% ethanol for approximately one month. (B) UChB rats previously exposed to voluntary ethanol consumption for approximately 100 days were administered with Alda-1 (12.5, 25, and 50 mg/kg/day) or vehicle for five consecutive days. After treatment discontinuation, animals continued voluntary drinking until day 110. *Figures are from Rivera-Meza, M., Vásquez, D., Quintanilla, M. E., Lagos, D., Rojas, B., Herrera-Marschitz, M., & Israel, Y. (2019). Activation of mitochondrial aldehyde dehydrogenase (ALDH2) by ALDA-1 reduces both the acquisition and maintenance of ethanol intake in rats: A dual mechanism? Neuropharmacology, 146, 175–183 with permission of the publisher.*

Alda-1 did not affect the consumption of a highly palatable substance such as saccharine, suggesting a selective action of Alda-1 on ethanol consumption. The analysis of the activity of ALDH2 in the brain and liver showed that it was increased by 3- and 2.3-fold, respectively, indicating that Alda-1 crosses the blood-brain barrier.

The treatment of alcohol use disorders is often limited due to relapse. As Rivera-Meza et al. (2019) demonstrated, the administration of Alda-1, by activation of ALDH2, markedly reduced the acquisition and maintenance of alcohol intake in rats. However, it was not clarified whether it also prevented relapse-like alcohol intake. In a recent study, Quilaqueo et al. (2023) investigated the effect of Alda-1 on post-deprivation and reaccess relapse-like ethanol intake in alcohol-preferring UChB rats. In these experiments, female UChB rats were exposed to voluntary ethanol intake for 100 days, followed by two weeks of ethanol withdrawal and one week of ethanol reaccess. Alda-1 or vehicle was administered daily for 14 days during the withdrawal/re-access period. Results showed that administration of Alda-1 during the deprivation and reaccess period significantly reduced ethanol consumption by 60% on the first day of reaccess and achieved almost complete abolition (96%) of ethanol intake after one week of reaccess. Notably, total fluid consumption was unaffected by the Alda-1 treatment, suggesting a specific effect on ethanol intake. The reduced ethanol intake seems to be associated with a decrease in the pleasurable effects of ethanol, as evidenced by an increased preference for higher ethanol concentrations (20%) in Alda-1-treated rats. In contrast, vehicle-treated rats did not change their pre-deprivation preference. This suggests that rats treated with Alda-1 tried to overcome the diminished pleasurable effects of ethanol caused by Alda-1 administration. Furthermore, the maximal inhibitory effect of Alda-1 on relapse-like ethanol consumption was not observed at the first session of ethanol reaccess; rather, the inhibitory effect was markedly increased as rats were exposed to further days of free ethanol availability and Alda-1 administration, in line with a decreased hedonic value of ethanol (Fig. 3).

7.1 Protective effects of Alda-1 to ethanol-induced organ and tissue damage

The harmful effects of ethanol consumption on organ and cell function are often linked to acetaldehyde, its main metabolite. Hepatic accumulation of acetaldehyde can cause liver damage and inflammation and potentially lead to conditions like alcoholic liver disease (Hyun, Han, Lee, Yoon, & Jung, 2021). Acetaldehyde itself is classified as a Group 1 carcinogen by the

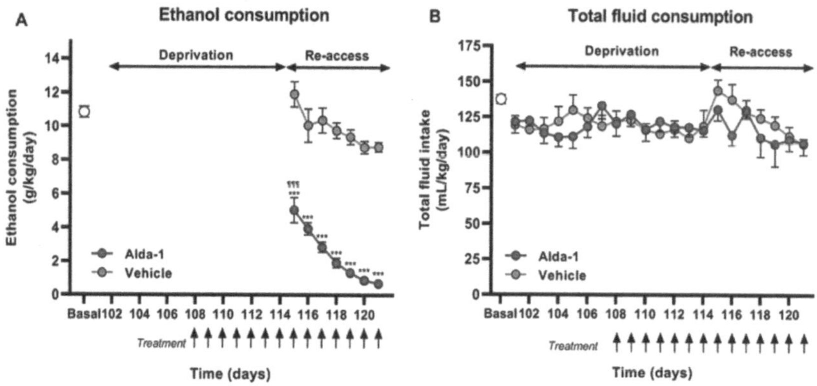

Fig. 3 Alda-1 administration blunted the relapse-like ethanol intake in UChB-drinking rats without affecting total fluid intake. UChB rats were exposed to chronic voluntary ethanol intake for approximately 100 days, followed by a 14-day deprivation period and subsequent reaccess to ethanol intake for seven days. (A) Effect of Alda-1 on relapse-like ethanol intake. The white circle represents the last 10-day average basal ethanol consumption of UChB rats before deprivation. On the 7th day of deprivation, animals were administered daily with Alda-1 25 mg/kg (i.g.) or vehicle for 14 consecutive days (black arrows). (B) Effect of Alda-1 on total fluid consumption. The white circle represents the last 10-day average fluid consumption (ethanol solution plus water) of UChB rats before deprivation. *The figure is from Quilaqueo, M. E., Adasme, S., Solís-Egaña, F., Quintanilla, M. E., Vásquez, D., Morales, P., … Rivera-Meza, M. (2023). The administration of Alda-1, an activator of ALDH2, inhibits relapse-like ethanol intake in female alcohol-preferring UChB rats. Life Sciences, 328, 121876 with permission of the publisher.*

International Agency for Research on Cancer (IARC), meaning it is considered carcinogenic to humans. Acetaldehyde can bind to proteins and DNA in the body, leading to DNA damage (Balbo et al., 2016; Brooks & Theruvathu, 2005; Tsuruta et al., 2020). This can increase the risk of mutations and the development of upper digestive tract cancers, including the mouth, throat, and esophagus (Seitz & Becker, 2007). Acetaldehyde can also induce oxidative stress, a condition implying an imbalance between the production of reactive oxygen species (ROS) and the body's ability for detoxification. This oxidative stress can lead to damage of cell membranes, proteins, and DNA (Comporti et al., 2010). Acetaldehyde also involves other health risks like cardiovascular diseases, neurological disorders, and respiratory problems (Kawano et al., 2004; Lee et al., 2005; Ren & Wold, 2008).

Acetaldehyde levels in the body are usually low since ALDH2 shows a high affinity for it, with a Km of approximately 0.2 μM (Eriksson, 1977). However, excessive alcohol consumption overwhelming the capacity of

ALDH2 to metabolize acetaldehyde or the presence of inactivating polymorphisms such as ALDH2*2 can lead to the accumulation of acetaldehyde in the blood, resulting in cellular damage and alterations of the normal functioning of organs, particularly the liver.

The capacity of Alda-1 to increase the activity of ALDH2 has been associated with protection against the harmful effects of acetaldehyde. For example, it has been shown that pharmacological activation of ALDH2 by Alda-1 reversed alcoholic steatosis and apoptosis by accelerating aldehyde clearance in mice chronically fed with a liquid diet containing ethanol (Zhong et al., 2015). Additionally, Alda-1 was found to reactivate certain transcription factors, increasing the activity of enzymes helping to break fatty acids (Zhong et al., 2015). Indeed, Alda-1 pre-treated mice receiving a single binge dose of ethanol (6 g/kg) showed lower levels of liver malondialdehyde-acetaldehyde adducts and reduced hepatic steatosis and liver injury (Zhong et al., 2014). On the other hand, when exposed to ethanol consumption, human ALDH2*2/*2 knock-in mice showed high levels of esophageal DNA damage, likely due to higher acetaldehyde concentrations. The administration of Alda-1 resulted in a marked decrease in the levels of N^2-ethylidene-dG DNA adducts in the esophagus, suggesting ALDH2 activation as a mechanism to prevent acetaldehyde-mediated esophageal carcinogenesis (Hirohashi et al., 2020).

Moreover, the administration of Alda-1 prevents the development of osteonecrosis of the femoral head (ONFH) in rats, a condition associated in humans with ethanol consumption (Lin et al., 2022). Although a clear mechanism for the protective effect of Alda-1 is lacking, the authors reported that Alda-1 administration was associated with new bone formation, reduced adipogenesis, and increased bone vascularization.

Excessive ethanol consumption is often associated with neurotoxicity and other CNS conditions, such as Alzheimer's disease (Brust, 2010; Rehm, Hasan, Black, Shield, & Schwarzinger, 2019). These deleterious central effects of ethanol have been related to the generation of cytotoxic aldehydes that are substrates of ALDH2, including acetaldehyde, MDA, and 4-HNE. Higher brain levels of these cytotoxic compounds have been associated with oxidative stress and neuroinflammation (Joshi et al., 2019; Sharma et al., 2022; Singh et al., 2013). In agreement, Kim, Hwang, Choi, Lim, and Kang (2018) reported in a rat model of binge exposure to ethanol, that the administration of Alda-1 significantly reduced neuroinflammation markers, increasing the number of viable neurons in the hippocampus compared to animals treated with saline. Ethanol-induced neuroinflammation has been linked to higher levels of Alzheimer's disease-associated toxic proteins like Aβ and phosphorylated tau in mice (Joshi et al., 2019). Alda-1 treatment significantly improved mitochondrial function by reducing Aβ42, phosphorylated tau accumulation in the brain, and blunted synaptic loss after chronic ethanol exposure (Joshi et al., 2019).

7.2 Possible mechanisms associated with the protective effects of Alda-1 against ethanol preference

As indicated previously, Alda-1 administration reduced the acquisition, maintenance, and relapse of ethanol intake in UChB-drinking rats (Quilaqueo et al., 2023; Rivera-Meza et al., 2019). As indicated by Quilaqueo et al. (2023), the decrease in ethanol consumption caused by the administration of Alda-1 was not associated with a reduction in neuroinflammatory markers or cytotoxic aldehydes such as 4-HNE and MDA. Therefore, these findings suggest that the protective effect of Alda-1 against ethanol consumption may be due to its ability to lower acetaldehyde levels in the brain.

Regarding the rewarding effects of acetaldehyde, it has been reported that ethanol-derived acetaldehyde can react with endogenous monoamines to form alkaloids known as tetrahydroisoquinolines (THIQs) (Cohen, 1976; Davis & Walsh, 1970; Deehan et al., 2013). One of the most studied THIQs is salsolinol. It has been described that this compound can be formed spontaneously by a non-enzymatic Pictet-Spengler condensation reaction of acetaldehyde and dopamine, resulting in a racemic mixture of R-salsolinol and S-salsolinol. In addition, this compound can be produced enzymatically with the participation of the enzyme R-salsolinol synthase through a stereospecific reaction that produces R-salsolinol (Villageliú, Borts, & Lyte, 2018).

Salsolinol has been associated with an increase in dopamine release from VTA to NAc and with the promotion of voluntary ethanol consumption in rats (Deehan et al., 2013; Quintanilla et al., 2014; Quintanilla et al., 2016). In this regard, it has been reported that R-salsolinol is responsible for inducing conditioned place preference, increasing locomotor activity, and increasing voluntary ethanol consumption in rats (Quintanilla et al., 2016). Regarding the mechanism of action of salsolinol, experimental evidence suggests that the rewarding effects of salsolinol are mediated by μ-opioid receptors. The studies by Matsuzawa, Suzuki, and Misawa (2000) and Hipólito, Sánchez-Catalán, Zornoza, Polache, and Granero (2010) demonstrated that the administration of the μ-opioid antagonist β-funaltrexamine attenuated the place preference conditioning and locomotor activity increase induced by salsolinol, respectively. Furthermore, Quintanilla et al. (2014) found that the sensitization to ethanol intake, mediated by salsolinol, was fully blocked by administering the opioid antagonist naltrexone in the VTA. This suggests that the VTA is a primary site of action of salsolinol, being the activation μ-opioid receptors the mechanism of action. The proposed mechanism suggests that salsolinol binds to μ-opioid receptors on GABAergic neurons in the VTA. This binding induces hyperpolarization, leading to disinhibition of dopaminergic neurons (Xie, Krnjević, & Ye, 2013). This disinhibition results in the activation of the dopaminergic system, resembling the action of opioids (Johnson & North, 1992). In support of this mechanism, in vitro studies demonstrated that salsolinol activates μ-opioid receptors, and this effect was blocked by co-incubation with the μ-opioid antagonist naltrexone (Berríos-Cárcamo et al., 2017).

Therefore, it is proposed that Alda-1 activates ALDH2, resulting in lower concentrations of ethanol-derived acetaldehyde in the brain. This reduction in acetaldehyde levels results in decreased production of salsolinol, which would be associated with reduced pleasurable effects associated with alcohol consumption (Fig. 4).

8. Future directions for ALDH2 activators

Only a few small molecules have been identified as metabolic enzyme activators, with Alda-1 being a noteworthy example of a novel class of compounds expressly designed to activate aldehyde dehydrogenase.

Alda-1, as a representative member of the Alda family, has exhibited remarkable effectiveness in curtailing ethanol consumption in animal models while preserving the intake of other palatable substances, such as saccharine (Rivera-Meza et al., 2019). However, in vivo studies have revealed some pharmacokinetic constraints of Alda-1, including a low oral bioavailability of 4.62% and a higher plasma clearance rate of 4.01 ± 0.72 L/h/kg (Taneja et al., 2015). These challenges, which include a high elimination rate and low bioavailability, create significant obstacles to the clinical use of Alda-1. Therefore, developing analogs with improved pharmacokinetic features is imperative while maintaining their pharmacological effectiveness.

Considering the remarkable efficacy of Alda-1 in activating the ALDH2 enzyme, a thorough understanding of the complex mechanisms underlying

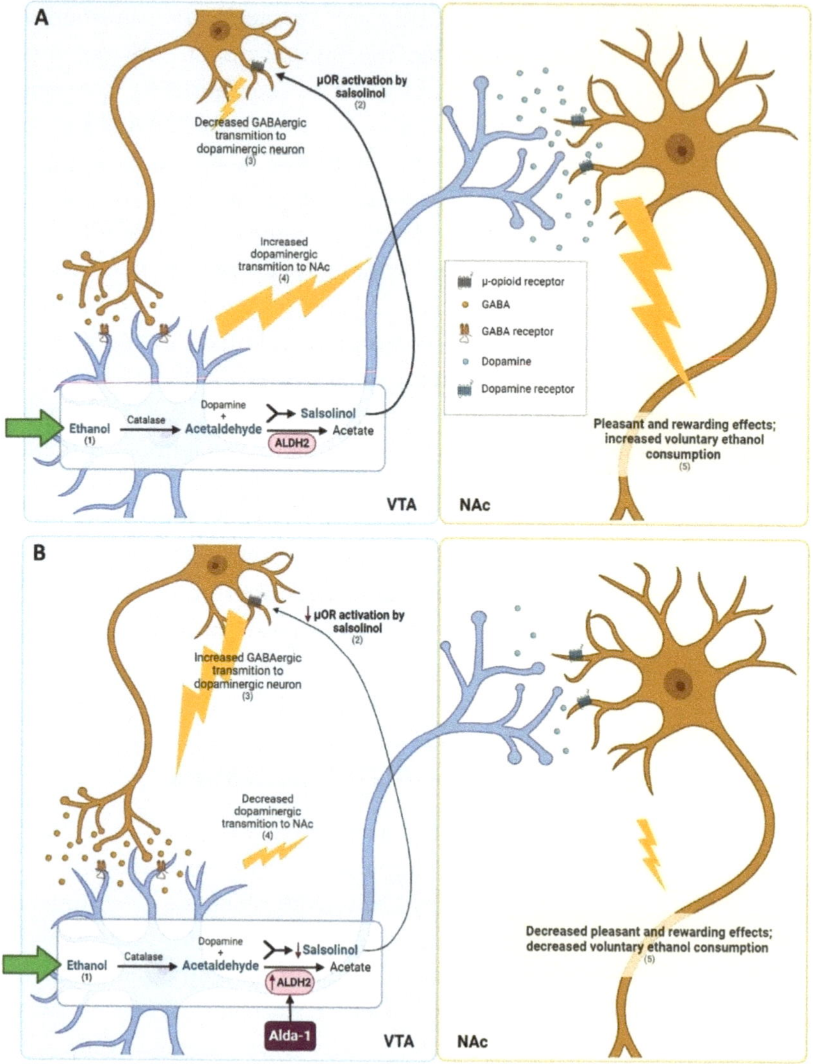

Fig. 4 Proposed mechanism by which Alda-1 administration reduces ethanol consumption. (A) In the brain, catalase oxidizes ethanol (1) to acetaldehyde, which reacts with dopamine to produce salsolinol; this compound could activate μ-opioid receptors (μOR) in GABAergic interneurons of VTA (2), leading to a decrease in GABAergic transmission to dopaminergic neurons in VTA (3). The reduced GABAergic tone disinhibits VTA dopamine neurons (4), increasing dopamine release in the NAc to generate the pleasant and rewarding effects associated with ethanol consumption (5). (B) In the presence of Alda-1, the increased activity of ALDH2 would reduce acetaldehyde levels and decrease salsolinol formation. A reduced activation by salsolinol of μOR on GABA neurons would result in increased release of GABA and, therefore, the inhibition

the interaction between Alda-1 and ALDH2 is critical to designing optimal compounds. Insights from studies by Perez-Miller et al. (2010) reveal the arrangement of π–π interactions between the benzodioxol group of Alda-1 and pivotal amino acids of the active site of ALDH2 (Phe 296, Asp 457, and Phe 459) (Fig. 5). Utilization of molecular docking techniques has facilitated exploring interactions between drugs with current clinical use and the ALDH2 enzyme.

In investigations about pre-existing medications and the ALDH2 enzyme, Hosoi et al. (2014) reported significant findings about flurbiprofen, a nonsteroidal anti-inflammatory drug (NSAID) derived from propionic acid. They found that flurbiprofen can increase the catalytic velocity of ALDH2 without causing changes in enzyme expression levels or stability. These findings are remarkable since they closely resemble the observed effects of Alda-1 on ALDH2. Expanding this line of research, Torres (2022) reported that Alda-1 and flurbiprofen increased ALDH2 activity in PC12 cells with a similar potency (3-fold at 40 μM). Interestingly, in the same in vitro experiments, it was reported that ibuprofen, another NSAID structurally related to flurbiprofen, did not increase the activity of ALDH2 in PC12 cells. In animal studies developed in UCHB-drinking rats, Torres (2022) found that oral administration of flurbiprofen or Alda-1 was effective for reducing the acquisition of alcohol consumption in UCHB-drinking rats. Moreover, it was found that the protective effects against ethanol consumption of flurbiprofen were dose-dependent, with the brain ALDH2 activity negatively correlated with ethanol intake in UChB rats.

A recent study by Tian et al. (2023) reported that a docking-based screening identified tadalafil, a phosphodiesterase type 5 (PDE5) inhibitor approved clinically for treating erectile dysfunction, benign prostatic hyperplasia, and pulmonary arterial hypertension, as a potential activator of ALDH2. The authors found that tadalafil showed a 29% higher capacity than Alda-1 to increase the ALDH2 activity in an in vitro assay, confirming the in silico prediction that tadalafil interacts with ALDH2 similarly to Alda-1. Interestingly, Blednov, Benavidez, Black, and Harris (2014) reported a study comparing the effect of 9 PDE inhibitors with different subtype selectivity

of VTA DA neurons and reduced release of dopamine in the NAc. Upon ethanol consumption, reduced dopaminergic neurotransmission in the NAc will result in lower rewarding effects and reduced interest in following by consuming ethanol. Keys. Blue neuron: dopaminergic neuron; Orange neuron: GABAergic neuron. The green arrow indicates the start point of the scheme. *The scheme was created in Biorender.com.*

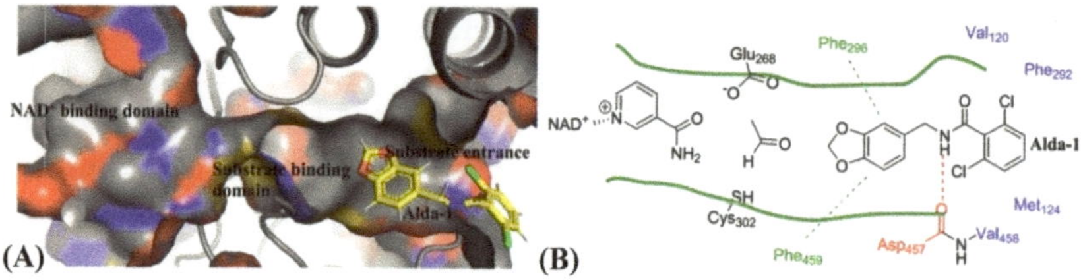

Fig. 5 (A) Binding pose of Alda-1 in the substrate entrance of ALDH2, with reference to its PDB ID 3INL. (B) Significant amino acid residues in the catalytic domain and the Alda-1 binding cavity of ALDH2. *The figure is from Cheng et al. (2020), with permission from the publisher.*

on voluntary alcohol intake by C57BL/6J male mice. The authors found that only selective PDE4 inhibitors reduced ethanol consumption, but the PDE4 inhibitor mesopram had the most potent effects. The capacity of mesopram, or other PDE4 inhibitors, to activate ALDH2 is unknown.

The search for compounds with superior bioavailability to Alda-1 has prompted a focused research approach toward developing Alda-1 analogs. The strategy involves introducing structural modifications that preserve the essential interactions with the ALDH2 enzyme. Alda-1 analog studies by Hu et al. (2019) have delved into substitutions to amplify water solubility and efficacy relative to Alda-1. Notably, substituting the benzodioxol group with pyridine rings has augmented water solubility. Additionally, introducing hydrophobic moieties, such as alkyl chains at positions 2 and 6 of the benzamide ring, has yielded heightened ALDH2 activity compared to Alda-1.

Moreover, Cheng et al. (2020) reported that changes in the halogen substitution of the benzamide significantly resulted in important changes in the capacity of the molecules to activate ALDH2. The nature of the halogen atom was identified as a decisive factor in the activation of ALDH2, as the introduction of iodine caused a loss of the activating capacity of the molecule. Additionally, substituting for fluorine resulted in a marginal reduction in activity compared to chlorine, while incorporating bromine led to a substantial augmentation of ALDH2 activity relative to chlorine. Concomitantly, substitutions of the benzodioxol group were correlated with diminished activity, as evidenced by the substitution with thiophene, yielding a discernible loss of π–π interactions with the amino acids of the active site of ALDH2. Conversely, substituting the benzamide ring with alkyl chains exhibited a pronounced escalation in ALDH2 activity, highlighting the critical role of the hydrogen bond interaction between the amide nitrogen proton and the carbonyl oxygen atom of the

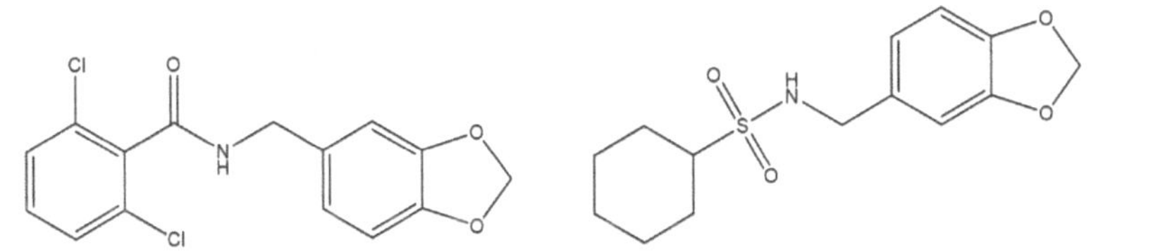

N-(benzo[d][1,3]dioxol-5-ylmethyl)-2,6-dichlorobenzamide
(Alda-1)

N-(benzo[d][1,3]dioxol-5-ylmethyl)cyclohexanesulfonamide
(FE-2)

Fig. 6 Chemical structures of Alda-1 (A) and FE-2 (B).

main chain of Asp457. Notably, removing or altering this proton's orientation resulted in a substantial diminution of enzymatic activity.

Our laboratory investigated this approach by introducing strategic structural modifications to the Alda-1 molecule. Specifically, Pérez (2024) reported studies describing the substitution of the amide group with a sulfonamide and replacing the 2,6-dichlorobenzene moiety with a cyclohexane structural element to generate the compound FE-2 [*N*-(benzo[*d*][1,3]dioxol-5-ylmethyl)cyclohexanesulfonamide] (Fig. 6B).

Using a cell-based assay of ALDH2 activity, the author showed that FE-2 showed a 50% higher capacity than Alda-1 to activate ALDH2, measuring both compounds at a concentration of 20 μM. Interestingly, the structural modifications introduced in FE-2 did not change the affinity to the enzyme since both compounds, FE-2 and Alda-1, showed the same EC_{50} (25 μM) for the activation of ALDH2. Further studies are needed to determine if this new analog of Alda-1 shows better pharmacokinetics properties than Alda-1 and their comparative efficacy in reducing alcohol consumption in preclinical models.

9. Conclusions

The studies included in this review demonstrate that the ALDH2 enzyme represents an important pharmacological target for the treatment of AUD. In this regard, ALDH2 inhibitors have been utilized for years as a therapeutic approach for treating AUD. Furthermore, individuals carrying the inactive ALDH2*2 are markedly protected from developing AUD. The rationale behind this is based on the aversive effects that result from acetaldehyde accumulation at the peripheral level after alcohol consumption.

However, acetaldehyde seems to be a "*two-sided coin*" since an important body of experimental evidence shows that acetaldehyde has rewarding effects at a central level. Therefore, ethanol would be a prodrug, and acetaldehyde its active metabolite. Furthermore, acetaldehyde is capable of giving rise to other molecules, such as salsolinol, which in turn shows even more powerful addictive effects and could constitute the final effector of the pleasurable effects of alcohol.

Based on these central effects of acetaldehyde (or salsolinol), the pharmacological activation of ALDH2 appears as a new therapeutic strategy for the treatment of AUD. This strategy would be based on reducing the bioavailability of these active metabolites of alcohol and the corresponding decrease in the hedonic value of drinking.

Alda-1 was the first pharmacological activator of ALDH2 reported in the literature. Its use in animal models of alcoholism has shown that the administration of Alda-1 results in a marked reduction in the acquisition, maintenance, and relapse of alcohol consumption in animals. These effects caused by Alda-1 are not only limited to a decrease in ethanol consumption but also extend to those toxic effects of alcohol intake that are mediated by acetaldehyde or other derived aldehydes (4-HNE, MDA), such as liver damage, esophageal cancer, and neurological disorders.

However, Alda-1 has some limitations regarding its pharmacokinetic properties, making its optimization necessary to constitute a more viable therapeutic alternative. In this sense, using *in silico* techniques such as molecular docking has allowed the identification of new molecules or others already in clinical use that act as activators of ALDH2, some molecules presenting even higher potency than Alda-1.

Some opinions have arisen that the clinical use of ALDH2 activators, such as Alda-1, would have ethical limitations since it could allow genetically protected individuals against alcoholism, such as ALDH2*2 gene carriers, to drink without presenting the aversive effects of acetaldehyde accumulation (Lehner et al., 2024). We believe that this situation would not be possible since the acceleration of ALDH2 activity systemically would decrease not only acetaldehyde levels at the peripheral level but also at the central level, thus reducing the motivation for alcohol consumption in these individuals.

Overall, pharmacological activators of ALDH2 emerge as a promising strategy for treating AUDs. However, more studies are needed to optimize the pharmacodynamic and pharmacokinetic properties of these new drugs and advance toward clinical projections of efficacy and safety.

Acknowledgments

This work was supported by the following grants: FONDECYT #1201577 (MRM); #1231443 (MHM) and Agencia Nacional de Investigación y Desarrollo ANID ACT210012.

References

Amit, Z., & Smith, B. R. (1985). A multi-dimensional examination of the positive reinforcing properties of acetaldehyde. *Alcohol (Fayetteville, N. Y.)*, *2*, 367–370.

Balbo, S., Cervera Juanes, R., Khariwala, S., Baker, E. J., Daunais, J. B., & Grant, K. A. (2016). Increased levels of the acetaldehyde-derived DNA adduct N 2-ethyldeoxyguanosine in oral mucosa DNA from Rhesus monkeys exposed to alcohol. *Mutagenesis*, *31*, 553–558.

Banys, P. (1988). The clinical use of disulfiram (Antabuse): A review. *Psychoactive Drugs*, *20*, 243–261.

Belmont-Díaz, J. A., Yoval-Sánchez, B., Calleja-Castañeda, L. F., Pardo Vázquez, J. P., & Rodríguez-Zavala, J. S. (2016). Alda-1 modulates the kinetic properties of mitochondrial aldehyde dehydrogenase (ALDH2). *The FEBS Journal*, *283*, 3637–3650.

Beretta, M., Gorren, A. C. F., Wenzl, M. V., Weis, R., Russwurm, M., Koesling, D., ... Mayer, B. (2010). Characterization of the East Asian variant of aldehyde dehydrogenase-2: Bioactivation of nitroglycerin and effects of Alda-1. *The Journal of Biological Chemistry*, *285*, 943–952.

Berríos-Cárcamo, P., Quintanilla, M. E., Herrera-Marschitz, M., Vasiliou, V., Zapata-Torres, G., & Rivera-Meza, M. (2017). Racemic salsolinol and its enantiomers act as agonists of the μ-opioid receptor by activating the Gi protein-adenylate cyclase pathway. *Frontiers in Behavioral Neuroscience*, *10*, 253.

Bilska-Wilkosz, A., Górny, M., & Iciek, M. (2019). Inactivation of aldehyde dehydrogenase by disulfiram in the presence and absence of lipoic acid or dihydrolipoic acid: An in vitro study. *Biomolecules*, *9*, 375.

Blednov, Y. A., Benavidez, J. M., Black, M., & Harris, R. A. (2014). Inhibition of phosphodiesterase 4 reduces ethanol intake and preference in C57BL/6J mice. *Frontiers in Neuroscience*, *8*, 129.

Brooks, P. J., & Theruvathu, J. A. (2005). DNA adducts from acetaldehyde: Implications for alcohol-related carcinogenesis. *Alcohol (Fayetteville, N. Y.)*, *35*, 187–193.

Brown, Z. W., Amit, Z., & Rockman, G. E. (1979). Intraventricular self-administration of acetaldehyde, but not ethanol, in naive laboratory rats. *Psychopharmacology*, *64*, 271–276.

Brust, J. C. M. (2010). Ethanol and cognition: Indirect effects, neurotoxicity and neuroprotection: A review. *International Journal of Environmental Research and Public Health*, 7, 1540–1557.

Chen, C. H., Budas, G. R., Churchill, E. N., Disatnik, M. E., Hurley, T. D., & Mochly-Rosen, D. (2008). Activation of aldehyde dehydrogenase-2 reduces ischemic damage to the heart. *Science (New York, N. Y.)*, *321*, 1493–1495.

Chen, C. H., Ferreira, J. C., Gross, E. R., & Mochly-Rosen, D. (2014). Targeting aldehyde dehydrogenase 2: New therapeutic opportunities. *Physiological Reviews*, *94*, 1–34.

Chen, C. H., Joshi, A. U., & Mochly-Rosen, D. (2016). The role of mitochondrial aldehyde dehydrogenase 2 (ALDH2) in neuropathology and neurodegeneration. *Acta Neurologica Taiwanica*, *25*, 111–123.

Chen, Y. C., Yang, L. F., Lai, C. L., & Yin, S. J. (2021). Acetaldehyde enhances alcohol sensitivity and protects against alcoholism: Evidence from alcohol metabolism in subjects with variant ALDH2*2 gene allele. *Biomolecules*, *11*, 1183.

Cheng, M. C., Lo, W. C., Chang, Y. W., Lee, S. S., & Chang, C. C. (2020). Design, synthesis and the structure-activity relationship of agonists targeting on the ALDH2 catalytic tunnel. *Bioorganic Chemistry*, *104*, 104166.

Chiu, C. C., Yeh, T. H., Lai, S. C., Wu-Chou, Y. H., Chen, C. H., Mochly-Rosen, D., ... Lu, C. S. (2015). Neuroprotective effects of aldehyde dehydrogenase 2 activation in rotenone-induced cellular and animal models of parkinsonism. *Experimental Neurology*, *263*, 244–253.

Cohen, G. (1976). Alkaloid products in the metabolism of alcohol and biogenic amines. *Biochemical Pharmacology*, *25*, 1123–1128.

Comporti, M., Signorini, C., Leoncini, S., Gardi, C., Ciccoli, L., Giardini, A., ... Arezzini, B. (2010). Ethanol-induced oxidative stress: Basic knowledge. *Genes & Nutrition, 5*, 101–109.

Davis, V. E., & Walsh, M. J. (1970). Alcohol, amines, and alkaloids: A possible biochemical basis for alcohol addiction. *Science (New York, N. Y.), 167*, 1005–1007.

Deehan, G. A., Engleman, E. A., Ding, Z. M., McBride, W. J., & Rodd, Z. A. (2013). Microinjections of acetaldehyde or salsolinol into the posterior ventral tegmental area increase dopamine release in the nucleus accumbens shell. *Alcoholism, Clinical and Experimental Research, 37*, 722–729.

Edenberg, H. J. (2007). The genetics of alcohol metabolism: Role of alcohol dehydrogenase and aldehyde dehydrogenase variants. *Alcohol Research & Health: The Journal of the National Institute on Alcohol Abuse and Alcoholism, 30*, 5–13.

Eng, M. Y., Luczak, S. E., & Wall, T. L. (2007). ALDH2, ADH1B, and ADH1C genotypes in Asians: A literature review. *Alcohol Research & Health: The Journal of the National Institute on Alcohol Abuse and Alcoholism, 30*, 22–27.

Eriksson, C. J. (1977). Acetaldehyde metabolism in vivo during ethanol oxidation. *Advances in Experimental Medicine and Biology, 85A*, 319–341.

Foddai, M., Dosia, G., Spiga, S., & Diana, M. (2004). Acetaldehyde increases dopaminergic neuronal activity in the VTA. *Neuropsychopharmacology: Official Publication of the American College of Neuropsychopharmacology, 29*, 530–536.

Fu, S. H., Zhang, H. F., Yang, Z. B., Li, T. B., Liu, B., Lou, Z., ... Peng, J. (2014). Alda-1 reduces cerebral ischemia/reperfusion injury in rat through clearance of reactive aldehydes. *Naunyn-Schmiedeberg's Archives of Pharmacology, 387*, 87–94.

Gao, J., Hao, Y., Piao, X., & Gu, X. (2022). Aldehyde dehydrogenase 2 as a therapeutic target in oxidative stress-related diseases: Post-translational modifications deserve more attention. *International Journal of Molecular Sciences, 23*, 2682.

Haass-Koffler, C. L., Akhlaghi, F., Swift, R. M., & Leggio, L. (2017). Altering ethanol pharmacokinetics to treat alcohol use disorder: Can you teach an old dog new tricks? *Journal of Psychopharmacology (Oxford, England), 31*, 812–818.

Hipólito, L., Sánchez-Catalán, M. J., Zornoza, T., Polache, A., & Granero, L. (2010). Locomotor stimulant effects of acute and repeated intrategmental injections of salsolinol in rats: Role of mu-opioid receptors. *Psychopharmacology, 209*, 1–11.

Hirohashi, K., Ohashi, S., Amanuma, Y., Nakai, Y., Ida, T., Baba, K., ... Muto, M. (2020). Protective effects of Alda-1, an ALDH2 activator, on alcohol-derived DNA damage in the esophagus of human ALDH2*2 (Glu504Lys) knock-in mice. *Carcinogenesis, 41*, 194–202.

Hosoi, T., Yamaguchi, R., Noji, K., Matsuo, S., Baba, S., Toyoda, K., ... Ozawa, K. (2014). Flurbiprofen ameliorated obesity by attenuating leptin resistance induced by endoplasmic reticulum stress. *EMBO Molecular Medicine, 6*, 335–346.

Hu, J., Tian, W., Zhou, R., Zhang, Y., Lv, J., Zhu, J., ... Zheng, C. (2019). Design, synthesis, and biological evaluation of new ALDH2 activators. *Journal of Saudi Chemical Society, 23*, 255–262.

Hyun, J., Han, J., Lee, C., Yoon, M., & Jung, Y. (2021). Pathophysiological Aspects of Alcohol Metabolism in the Liver. *International Journal of Molecular Sciences, 22*, 5717.

Israel, Y., Karahanian, E., Ezquer, F., Morales, P., Ezquer, M., Rivera-Meza, M., ... Quintanilla, M. E. (2017). Acquisition, maintenance and relapse-like alcohol drinking: Lessons from the UChB rat Line. *Frontiers in Behavioral Neuroscience, 11*, 57.

Jackson, B., Brocker, C., Thompson, D. C., Black, W., Vasiliou, K., Nebert, D. W., & Vasiliou, V. (2011). Update on the aldehyde dehydrogenase gene (ALDH) superfamily. *Human Genomics, 5*, 283–303.

Johnson, S. W., & North, R. A. (1992). Opioids excite dopamine neurons by hyperpolarization of local interneurons. *The Journal of Neuroscience, 12*, 483–488.

Joshi, A. U., Van Wassenhove, L. D., Logas, K. L., Minhas, P. S., Andreasson, K. I., Weinberg, K. I., ... Mochly-Rosen, D. (2019). Aldehyde dehydrogenase 2 activity and aldehydic load contribute to neuroinflammation and Alzheimer's disease related pathology. *Acta Neuropathologica Communications, 7*, 190.

Karahanian, E., Rivera-Meza, M., Tampier, L., Quintanilla, M. E., Herrera-Marschitz, M., & Israel, Y. (2015). Long-term inhibition of ethanol intake by the administration of an aldehyde dehydrogenase-2 (ALDH2)-coding lentiviral vector into the ventral tegmental area of rats. *Addiction Biology, 20*, 336–344.

Kawano, T., Matsuse, H., Kondo, Y., Machida, I., Saeki, S., Tomari, S., ... Kohno, S. (2004). Acetaldehyde induces histamine release from human airway mast cells to cause bronchoconstriction. *International Archives of Allergy and Immunology, 134*, 233–239.

Kim, C., Hwang, S., Choi, H. J., Lim, T. H., & Kang, J. S. (2018). The effect of aldehyde dehydrogenase activator, Alda-1®, on the ethanol-induced brain damage in a rat of binge ethanol intoxication. *bioRxiv, 353854*.

Klyosov, A. A., Rashkovetsky, L. G., Tahir, M. K., & Keung, W. M. (1996). Possible role of liver cytosolic and mitochondrial aldehyde dehydrogenases in acetaldehyde metabolism. *Biochemistry, 35*, 4445–4456.

Lands, W. E. (1998). A review of alcohol clearance in humans. *Alcohol (Fayetteville, N. Y.), 15*, 147–160.

Lanz, J., Biniaz-Harris, N., Kuvaldina, M., Jain, S., Lewis, K., & Fallon, B. A. (2023). Disulfiram: Mechanisms, applications, and challenges. *Antibiotics, 12*, 524.

Lee, R. D., An, S. M., Kim, S. S., Rhee, G. S., Kwack, S. J., Seok, J. H., ... Park, K. L. (2005). Neurotoxic effects of alcohol and acetaldehyde during embryonic development. *Journal of Toxicology and Environmental Health. Part A, 68*, 2147–2162.

Lehner, T., Gao, B., & Mackowiak, B. (2024). Alcohol metabolism in alcohol use disorder: A potential therapeutic target. *Alcohol and Alcoholism (Oxford, Oxfordshire), 59*.

Li, D., Zhao, H., & Gelernter, J. (2012). Strong protective effect of the aldehyde dehydrogenase gene (ALDH2) 504lys (*2) allele against alcoholism and alcohol-induced medical diseases in Asians. *Human Genetics, 131*, 725–737.

Lin, X., Zhu, D., Wang, K., Luo, P., Rui, G., Gao, Y., ... Yu, H. (2022). Activation of aldehyde dehydrogenase 2 protects ethanol-induced osteonecrosis of the femoral head in rat model. *Cell Proliferation, 55*, e13252.

Luczak, S. E., Glatt, S. J., & Wall, T. L. (2006). Meta-analyses of ALDH2 and ADH1B with alcohol dependence in Asians. *Psychological Bulletin, 132*, 607–621.

MacDonagh, L., Gallagher, M. F., Ffrench, B., Gasch, C., Breen, E., Gray, S. G., ... Barr, M. P. (2017). Targeting the cancer stem cell marker, aldehyde dehydrogenase 1, to circumvent cisplatin resistance in NSCLC. *Oncotarget, 8*, 72544–72563.

Malcolm, R., Olive, M. F., & Lechner, W. (2008). The safety of disulfiram for the treatment of alcohol and cocaine dependence in randomized clinical trials: Guidance for clinical practice. *Expert Opinion on Drug Safety, 7*, 459–472.

Matsumoto, H., & Fukui, Y. (2002). Pharmacokinetics of ethanol: A review of the methodology. *Addiction Biology, 7*, 5–14.

Matsuzawa, S., Suzuki, T., & Misawa, M. (2000). Involvement of mu-opioid receptor in the salsolinol-associated place preference in rats exposed to conditioned fear stress. *Alcoholism, Clinical and Experimental Research, 24*, 366–372.

McBride, W. J., Li, T. K., Deitrich, R. A., Zimatkin, S., Smith, B. R., & Rodd-Henricks, Z. A. (2002). Involvement of acetaldehyde in alcohol addiction. *Alcoholism, Clinical and Experimental Research, 26*, 114–119.

Myers, W., Ng, K., & Singer, G. (1984). Ethanol preference in rats with a prior history of acetaldehyde self-administration. *Experientia, 40*, 1008–1010.

Nestler, E. J. (2005). Is there a common molecular pathway for addiction? *Nature Neuroscience, 8*, 1445–1449.

Niederhofer, H., Staffen, W., & Mair, A. (2003). Comparison of cyanamide and placebo in the treatment of alcohol dependence of adolescents. *Alcohol and Alcoholism (Oxford, Oxfordshire), 38*, 50–53.

Omran, Z. (2022). Novel disulfiram derivatives as ALDH1a1-selective inhibitors. *Molecules (Basel, Switzerland), 27*, 480.

Peana, A. T., Porcheddu, V., Bennardini, F., Carta, A., Rosas, M., & Acquas, E. (2015). Role of ethanol-derived acetaldehyde in operant oral self-administration of ethanol in rats. *Psychopharmacology, 232*, 4269–4276.

Peng, G. S., Chen, Y. C., Tsao, T. P., Wang, M. F., & Yin, S. J. (2007). Pharmacokinetic and pharmacodynamic basis for partial protection against alcoholism in Asians, heterozygous for the variant ALDH2*2 gene allele. *Pharmacogenetics and Genomics, 17*, 845–855.

Peng, G. S., Chen, Y. C., Wang, M. F., Lai, C. L., & Yin, S. J. (2014). ALDH2*2 but not ADH1B*2 is a causative variant gene allele for Asian alcohol flushing after a low-dose challenge: Correlation of the pharmacokinetic and pharmacodynamic findings. *Pharmacogenetics and Genomics, 24*, 607–617.

Pérez V. (2024). Design, synthesis, in vitro and in silico evaluation of new activators of ALDH2. Undergrad thesis, University of Chile.

Perez-Miller, S., Younus, H., Vanam, R., Chen, C. H., Mochly-Rosen, D., & Hurley, T. D. (2010). Alda-1 is an agonist and chemical chaperone for the common human aldehyde dehydrogenase 2 variant. *Nature Structural & Molecular Biology, 17*, 159–164.

Poturnajova, M., Kozovska, Z., & Matuskova, M. (2021). Aldehyde dehydrogenase 1A1 and 1A3 isoforms-mechanism of activation and regulation in cancer. *Cellular Signalling, 87*, 110120.

Quertemont, E., & Didone, V. (2006). Role of acetaldehyde in mediating the pharmacological and behavioral effects of alcohol. *Alcohol Research & Health: The Journal of the National Institute on Alcohol Abuse and Alcoholism, 29*, 258–265.

Quilaqueo, M. E., Adasme, S., Solís-Egaña, F., Quintanilla, M. E., Vásquez, D., Morales, P., ... Rivera-Meza, M. (2023). The administration of Alda-1, an activator of ALDH2, inhibits relapse-like ethanol intake in female alcohol-preferring UChB rats. *Life Sciences, 328*, 121876.

Quintanilla, M. E., Rivera-Meza, M., Berrios-Cárcamo, P. A., Bustamante, D., Buscaglia, M., Morales, P., ... Israel, Y. (2014). Salsolinol, free of isosalsolinol, exerts ethanol-like motivational/sensitization effects leading to increases in ethanol intake. *Alcohol (Fayetteville, N. Y.), 48*, 551–559.

Quintanilla, M. E., Rivera-Meza, M., Berríos-Cárcamo, P., Cassels, B. K., Herrera-Marschitz, M., & Israel, Y. (2016). R)-Salsolinol, a product of ethanol metabolism, stereospecifically induces behavioral sensitization and leads to excessive alcohol intake. *Addiction Biology, 21*, 1063–1071.

Rehm, J., Hasan, O. S. M., Black, S. E., Shield, K. D., & Schwarzinger, M. (2019). Alcohol use and dementia: A systematic scoping review. *Alzheimer's Research & Therapy, 11*, 1.

Ren, J., & Wold, L. E. (2008). Mechanisms of alcoholic heart disease. *Therapeutic Advances in Cardiovascular Disease, 2*, 497–506.

Rivera-Meza, M., Quintanilla, M. E., Tampier, L., Mura, C. V., Sapag, A., & Israel, Y. (2010). Mechanism of protection against alcoholism by an alcohol dehydrogenase polymorphism: Development of an animal model. *The FASEB Journal, 24*, 266–274.

Rivera-Meza, M., Vásquez, D., Quintanilla, M. E., Lagos, D., Rojas, B., Herrera-Marschitz, M., & Israel, Y. (2019). Activation of mitochondrial aldehyde dehydrogenase (ALDH2) by ALDA-1 reduces both the acquisition and maintenance of ethanol intake in rats: A dual mechanism? *Neuropharmacology, 146*, 175–183.

Rodd, Z. A., Bell, R. L., Zhang, Y., Murphy, J. M., Goldstein, A., Zaffaroni, A., ... McBride, W. J. (2005). Regional heterogeneity for the intracranial self-administration of ethanol and acetaldehyde within the ventral tegmental area of alcohol-preferring (P) rats: Involvement of dopamine and serotonin. *Neuropsychopharmacology: Official Publication of the American College of Neuropsychopharmacology, 30*, 330–338.

Seitz, H. K., & Becker, P. (2007). Alcohol metabolism and cancer risk. *Alcohol Research & Health: The Journal of the National Institute on Alcohol Abuse and Alcoholism, 30*, 38–41.

Sharma, S., Sharma, P., Bailey, T., Bhattarai, S., Subedi, U., Miller, C., ... Miriyala, S. (2022). Electrophilic aldehyde 4-hydroxy-2-nonenal mediated signaling and mitochondrial dysfunction. *Biomolecules, 12*, 1555.

Singh, S., Brocker, C., Koppaka, V., Chen, V., Jackson, B. C., Matsumoto, A., ... Vasiliou, V. (2022). Aldehyde dehydrogenases in cellular responses to oxidative/electrophilic stress. *Free Radical Biology & Medicine, 56*, 89–101.

Skinner, M. D., Lahmek, P., Pham, H., & Aubin, H. J. (2014). Disulfiram efficacy in the treatment of alcohol dependence: A meta-analysis. *PLoS One, 9*, e87366.

Suh, J. J., Pettinati, H. M., Kampman, K. M., & O'Brien, C. P. (2006). The status of disulfiram: A half of a century later. *Journal of Clinical Psychopharmacology, 26*, 290–302.

Tampier, L., Quintanilla, M. E., Karahanian, E., Rivera-Meza, M., Herrera-Marschitz, M., & Israel, Y. (2013). The alcohol deprivation effect: Marked inhibition by anticatalase gene administration into the ventral tegmental area in rats. *Alcoholism, Clinical and Experimental Research, 37*, 1278–1285.

Taneja, I., Raju, K. S. R., Mittal, M., Dev, K., Khan, M. F., Mauryaad, R., & Wahajuddin, M. (2015). Bioavailability, plasma protein binding and metabolic stability studies of a ALDH2 activator, alda-1, using a validated LC-ESI-MS/MS method in rat plasma. *RSC Advances, 5*, 54395–54402.

Tian, W., Sun, X., Gao, X., Wang, W., Chen, Z., Chen, X., & Zheng, C. (2023). Discovery of novel small-molecule aldehyde dehydrogenase 2 activators based on drug repurposing. *Journal of Saudi Chemical Society, 27*, 101727.

Torres, J. M. (2022). Effect of flurbiprofen and Alda-1 in the expression and activity of ALDH2 in cultures of PC12 cells. Undergrad thesis, University of Chile. https://repositorio.uchile.cl/handle/2250/189513.

Tsuruta, H., Sonohara, Y., Tohashi, K., Aoki Shioi, N., Iwai, S., & Kuraoka, I. (2020). Effects of acetaldehyde-induced DNA lesions on DNA metabolism. *Genes and Environment, 42*, 2.

Umulis, D. M., Gürmen, N. M., Singh, P., & Fogler, H. S. (2005). A physiologically based model for ethanol and acetaldehyde metabolism in human beings. *Alcohol (Fayetteville, N. Y.), 35*, 3–12.

Vasiliou, V., Pappa, A., & Estey, T. (2004). Role of human aldehyde dehydrogenases in endobiotic and xenobiotic metabolism. *Drug Metabolism Reviews, 36*, 279–299.

Villageliú, D. N., Borts, D. J., & Lyte, M. (2018). Production of the neurotoxin salsolinol by a gut-associated bacterium and its modulation by alcohol. *Frontiers in Microbiology, 9*, 3092.

WHO (World Health Organization). (2022). *World health statistics 2022: Monitoring health for the SDGs, sustainable development goals*. Geneva: World Health Organization 2022. Licence: CC BY-NC-SA 3.0 IGO.

Xie, G., Krnjević, K., & Ye, J. H. (2013). Salsolinol modulation of dopamine neurons. *Frontiers in Behavioral Neuroscience, 7*, 52.

Zhong, Z., Ramshesh, V. K., Rehman, H., Liu, Q., Theruvath, T. P., Krishnasamy, Y., & Lemasters, J. J. (2014). Acute ethanol causes hepatic mitochondrial depolarization in mice: Role of ethanol metabolism. *PLoS One, 9*, e91308.

Zhong, W., Zhang, W., Li, Q., Xie, G., Sun, Q., Sun, X., ... Zhou, Z. (2015). Pharmacological activation of aldehyde dehydrogenase 2 by Alda-1 reverses alcohol-induced hepatic steatosis and cell death in mice. *Journal of Hepatology, 62*, 1375–1381.

Zimatkin, S. M. (1991). Histochemical study of aldehyde dehydrogenase in the rat CNS. *Journal of Neurochemistry, 56*, 1–11.

Zimatkin, S. M., Pronko, S. P., Vasiliou, V., Gonzalez, F. J., & Deitrich, R. A. (2006). Enzymatic mechanisms of ethanol oxidation in the brain. *Alcoholism, Clinical and Experimental Research, 30*, 1500–1505.

Zindel, L. R., & Kranzler, H. R. (2014). Pharmacotherapy of alcohol use disorders: Seventy-five years of progress. *Journal of Studies on Alcohol and Drugs, 75*, 79–88.

CHAPTER SIX

Mesenchymal stem cells as a promising therapy for alcohol use disorder

Javiera Gallardo[a], Pablo Berríos-Cárcamo[a], and Fernando Ezquer[a,b,*]
[a]Center for Regenerative Medicine, Faculty of Medicine, Clínica Alemana-Universidad del Desarrollo, Santiago, Chile
[b]Research Center for the Development of Novel Therapeutics Alternatives for Alcohol Use Disorders, Santiago, Chile
*Corresponding author. e-mail address: eezquer@udd.cl

Contents

Abstract

Alcohol Use Disorder (AUD) is a highly prevalent medical condition characterized by impaired control over alcohol consumption, despite negative consequences on the individual's daily life and health. There is increasing evidence suggesting that chronic alcohol intake, like other addictive drugs, induces neuroinflammation and oxidative stress, disrupting glutamate homeostasis in the main brain areas related to drug addiction. This review explores the potential application of mesenchymal stem cells (MSCs)-based therapy for the treatment of AUD. MSCs secrete a broad array of anti-inflammatory and antioxidant molecules, thus, the administration of MSCs, or their

International Review of Neurobiology, Volume 178
ISSN 0074-7742, https://doi.org/10.1016/bs.irn.2024.07.002

secretome, could reduce neuroinflammation and oxidative stress in the brain. These effects correlate with an increase in the expression of the main glutamate transporter, GLT1, which, through the normalization of the extracellular glutamate levels, could mediate the inhibitory effect of MSCs' secretome on chronic alcohol consumption, thus highlighting GLT1 as a central target to reduce chronic alcohol consumption.

1. Introduction

Over the past few years, the understanding of Alcohol Use Disorder (AUD) has advanced beyond behavioral observations to recognize its significant impact on neurobiology. Among all substance use disorders, AUD is the most prevalent with an estimated of 100.4 million cases worldwide in 2016 (Degenhardt et al., 2018). The progression of AUD, like other substance use disorders, begins with initial stages characterized by a rewarding effect, which increase motivation for alcohol consumption (Koob & Volkow, 2016). As consumption continues, AUD transitions to stages marked by dysphoric effects and negative reinforcement including withdrawal-associated symptoms, inducing an impaired control over alcohol consumption (Volkow, Michaelides, & Baler, 2019). With limited treatment options approved by the FDA (Kranzler & Soyka, 2018; Zindel & Kranzler, 2014), AUD poses a serious global health concern. Researchers are increasingly recognizing that chronic alcohol consumption induces neuroinflammation, oxidative stress, and a disrupted glutamate balance in crucial brain areas associated with reward processing and addiction including the nucleus accumbens and the prefrontal cortex (Berríos-Cárcamo, 2020; Crews & Vetreno, 2016; Das & Vasudevan, 2007; Ezquer, 2018, 2019a,b; Gano, Deak, & Pautassi, 2023; Machado da Silva, De Souza Ferreira, & Giustina, 2024; Ramalingam & Kim, 2012; Schneider, 2017). This review sheds light on a promising frontier in AUD treatment: mesenchymal stem cells (MSCs)-based therapy. MSCs are adult stem cells that exhibit remarkable potential for self-renewal and multipotent differentiation (Dominici et al., 2006; Salehinejad, Moshrefi, & Eslaminejad, 2020; Wei et al., 2013). MSCs secrete a wide array of biologically relevant molecules with anti-inflammatory and antioxidant properties (Gao et al., 2016; González-González, García-Sánchez, Dotta, Rodríguez-Rey, & Pérez-Campo, 2020; Han et al., 2022; Stavely & Nurgali, 2020), offering a multitarget approach to combat the neuropathological effects of chronic alcohol consumption (Lindvall, Barker, Brüstle, Isacson, & Svendsen, 2012). The administration of living MSCs or their secreted components (exosomes or secretome) in animal models of chronic alcohol consumption leads to a strong reduction in ethanol-induced neuroinflammation and oxidative stress, resulting in a significant reduction of voluntary alcohol intake and the prevention of alcohol relapse after alcohol deprivation and re-access (Ezquer et al., 2018; Ezquer et al., 2019a; 2019b; Mellado et al., 2023; Quintanilla et al., 2019), offering potential relief from the detrimental effects of AUD. This chapter explores the intricate landscape of AUD treatment and the potential of MSC-based therapy in mitigating neuroinflammation, oxidative stress, and alterations in glutamate homeostasis resulting from chronic alcohol consumption.

2. Alcohol use disorder

Alcohol use disorder (AUD) falls within the spectrum of substance use disorders, which encompass a variety of conditions associated with problematic use of addictive substance. These disorders arise when the consumption of psychoactive substances, including legal substances like alcohol and nicotine, illicit substances or prescription medications, leads to significant difficulties and challenges in an individual's daily life (Degenhardt et al., 2018; Whiteford, Ferrari, & Degenhardt, 2016). Substance use disorder can vary in severity and manifest in different ways, from harmful use to physical and psychological dependence (Machado da Silva, De Souza Ferreira, & Giustina, 2024; McLellan, 2017).

AUD is a medical condition characterized by an impaired ability to control or stop alcohol consumption, despite facing adverse social, occupational, or health consequences (McLellan, 2017). It involves a problematic pattern of alcohol use resulting in clinically significant impairment or distress. A systematic analysis, using data from the Global Burden of Disease Study from 1990 to 2016, encompassing 195 countries, indicates significant geographical disparities in both the extent and the proportional impact of alcohol and drug use on the overall burden of disease, which varies substantially between countries and is strongly associated with social development (Degenhardt et al., 2018). Unfortunately, these numbers are on the rise. In 1990, alcohol use was the 13th most relevant risk factor for deaths worldwide; in 2019, it has risen to the 8th most relevant risk factor, responsible for over 2.4 million deaths from various alcohol-attributable causes (Institute for Health Metrics and Evaluation, 2020). Furthermore, 49.3% of cirrhosis and other liver diseases are attributable to excessive alcohol use. These alarming statistics underscore the urgent need for comprehensive strategies and global efforts to address the growing impact of alcohol and drug-related health issues worldwide.

Despite the well-documented high prevalence and harmful effects of AUD, effective therapy for this condition remains elusive. Currently, only a limited number of FDA-approved medications targeting specific aspects of AUD treatment, such as reducing cravings, blocking the rewarding effects of alcohol, and preventing alcohol relapse are available. These medications include disulfiram, naltrexone, and acamprosate (Kranzler & Soyka, 2018). (i) Disulfiram was the first FDA-approved drug, and it has been utilized since the 1950s for the treatment of AUD. The mechanism of action of disulfiram is the non-selective inhibition of the enzyme aldehyde dehydrogenase, resulting in an increase in plasmatic concentration of acetaldehyde, which produces unpleasant sensations whenever the patient drinks alcohol, including tachycardia, shortness of breath, tachypnea, sensation of heat, anxiety, panic, headache, nausea, and vomiting (Batra, Müller, Mann, & Heinz, 2016; Stokłosa et al., 2023). (ii) Naltrexone has been FDA-approved for treating AUD since 1984 and for opioid addiction since 1994. Its proposed mechanism of action involves reducing the pleasurable effects of alcohol (Volpicelli, Clay, Watson, & Volpicelli, 1994; Wardle, Bershad, & De Wit, 2016). As a μ-opioid receptor antagonist, naltrexone achieves this effect by blocking the activity of μ-opioid receptors in the brain (Bello, 2022; Ko, Butelman, Traynor, & Woods, 1998). Studies have shown that naltrexone inhibits the alcohol-induced release of dopamine and endorphins, both of which play a central role in the brain's reward system (Lee et al., 2005; Ray, Chin, & Miotto, 2010; Ray, Green, Roche, Magill, & Bujarski, 2019; Zalewska-Kaszubska, Gorska, Dyr, & Czarnecka, 2006). (iii) Acamprosate has been available for treating alcohol dependence syndrome since 1989 in Europe (Kumar et al., 2020), and the FDA approved its use for promoting abstinence in alcohol-dependent individuals in 2004, along with long-acting injectable formulation of naltrexone in 2006 (Littleton, 2007). Literature indicates that acamprosate modulates glutamatergic transmission by inhibiting *N*-methyl-D-aspartate (NMDA) receptors and metabotropic glutamate-5 receptors (mGluR5). These mechanisms of action, when used in combination with psychosocial therapy, could enable the restoration of normal glutamatergic neuronal activity, allowing for a physiological response to glutamate activation without triggering pathological overactivation (Littleton, 2007).

While some studies have shown promise for certain medications, the inconsistency in findings underscores the complexity of treating AUD. For instance, disulfiram treatment did not show superiority over the placebo group in randomized clinical trials (Jørgensen, Pedersen, & Tønnesen, 2011;

Skinner, Lahmek, Pham, & Aubin, 2014). While for naltrexone and acamprosate, there are studies with completely opposite results. In a double-blind placebo-controlled clinical trial comparing the efficacy of both drugs, the Kiefer group supported the efficacy of both naltrexone and acamprosate in preventing alcohol use relapse (Kiefer et al., 2003). While on the other hand, Morley et al. indicated that neither naltrexone nor acamprosate treatment significantly improved outcomes compared to the placebo group (Morley et al., 2006). Moreover, acamprosate has demonstrated modest efficacy in encouraging abstinence when used in conjunction with psychosocial interventions after successful withdrawal from alcohol (Plosker, 2015), while naltrexone exhibits a slight advantage in reducing heavy drinking and craving (Maisel, Blodgett, Wilbourne, Humphreys, & Finney, 2013). However, achieving the necessary period of abstinence before starting treatment with these medications is often challenging for patients struggling with problematic alcohol consumption (Maisel et al., 2013). Given the widespread nature of AUD and the limited effectiveness of existing treatments, there is a pressing need to explore and develop innovative therapeutic interventions.

3. Glutamatergic pathways as major inducers of ethanol consumption

The progression of AUD, as well as disorders arising from the consumption of other drug of abuse, evolves from initial stages characterized by the drug-induced reinforcing effects to neuroadaptations that induce conditioning to drug-predicting cues and reduce the ability of self-regulate drug consumption (Machado da Silva, De Souza Ferreira, & Giustina, 2024; Volkow et al., 2019). The initial mechanisms of addiction are similar among several abused drugs such as cocaine, amphetamines, opioids, nicotine, and alcohol. Despite their distinct modes of action, upon consumption, all these substances elevate dopamine levels in the nucleus accumbens (NAc) (Imperato & Di Chiara, 1986), a brain hub that interprets the value of the stimuli to guide behavior (Floresco, 2015). However, while alcohol's reinforcing effects are controlled by dopaminergic projections to the NAc, the development of drug-induced conditioning involves the pathological activation of glutamatergic pathways that activate the NAc, resulting in the failure of the prefrontal cortex (PFC) to regulate drug-seeking behavior (Kalivas, 2009). Glutamate synaptic and extracellular levels are usually tightly regulated by proteins that control glutamate release and uptake, defined as glutamate homeostasis (Kalivas, 2009). Released glutamate in the NAc promotes the action potential in the post-synaptic neuron mostly through the activation of ionotropic receptors: *N*-methyl-D-aspartate (NMDA) and α-amine-3-hydroxy-5-methyl-4-isoxazolepropionic acid (AMPA) receptors (Niciu, Kelmendi, & Sanacora, 2012). Glutamatergic signaling is terminated by glutamate uptake, mainly by its transport through excitatory amino acid transporters (EAAT) in astrocytes, which, together with neurons, are part of the tripartite synapsis (Huang & Bergles, 2004). EAAT1, also called glutamate aspartate transporter (GLAST) in rodents is expressed in astrocytes in the cortex and cerebellum (Niciu et al., 2012). EAAT2, called glutamate transporter 1 (GLT1) in rodents, is a symporter of glutamate and sodium expressed throughout the brain mostly in astrocytes and less abundantly in axon terminals (Danbolt, Furness, & Zhou, 2016). Analysis of variants lacking GLT1 shows that this transporter accounts for ~90% of glutamate synaptic uptake in the brain (Danbolt, 2001; Tanaka et al., 1997). Mice globally lacking GLT1, or specifically lacking astrocytic GLT1, suffer lethal seizures before reaching four weeks of age (Danbolt et al., 2016). Consequently, direct inhibition of glutamate transporters in adult animals increases the extracellular glutamate levels (Melendez, Vuthiganon, & Kalivas, 2005). Interestingly, reduced uptake of glutamate has been observed in the NAc of rats that consumed ethanol chronically (Schreiber & Freund, 2000) and reduced levels of GLT1 have also been determined in this brain region of rats and mice after chronic ethanol consumption and relapse compared to control animals (Alhaddad et al., 2014; Gass, Sinclair, Cleva, Widholm, & Olive, 2011; Goodwani, Rao, Bell, & Sari, 2015; Griffin, Haun, Hazelbaker, Ramachandra, & Becker, 2014). In animal models of ethanol dependence, significantly increased levels of basal extracellular glutamate have been observed in the NAc (Ding et al., 2013; Griffin, Ramachandra, Knackstedt, & Becker, 2015; Lallemand, Ward, De Witte, & Verbanck, 2011; Pati, Kelly, Stennett, Frazier, & Knackstedt, 2016). These data suggest that reducing glutamate synaptic levels could be beneficial to curb glutamate neurotransmission and diminish ethanol cravings and relapse.

4. Modulation of GLT1 expression as therapy for the alcohol use disorder

The therapeutic potential of molecules capable of increasing GLT1 levels has been widely recognized (Alhaddad et al., 2014, 2022;

Ezquer et al., 2022; Rao, Saternos, Goodwani, & Sari, 2015; Rothstein et al., 2005; Sari, Smith, Ali, & Rebec, 2009). Rothstein et al. conducted a screening of 1040 FDA-approved compounds and found that mice treated with β-lactam antibiotics for five days exhibited increased GLT1 gene expression, leading to higher GLT1 protein levels (Rothstein et al., 2005). In the same study, the efficacy of the β-lactam ceftriaxone was assessed in vitro. They observed that the co-culture of primary astrocytes and neurons treated with 100 μM ceftriaxone for seven days significantly elevated GLT1 levels (Rothstein et al., 2005). Other groups have obtained similar results, showing that the co-culture of astrocytes and neurons treated for five days with 100 μM ceftriaxone (Beller, Gurkoff, Berman, & Lyeth, 2011) or primary astrocyte cultures treated with 100 μM ceftriaxone for 48 h (Zhang, Zhang, & Qu, 2015) increased GLT1 protein levels. Moreover, administration of ceftriaxone at 100 mg/kg/day for two and five days resulted in a five-fold reduction in ethanol intake in rat model of ethanol dependence (Rao et al., 2015). This effect was associated with significantly enhanced expression of total GLT1 and its isoforms, GLT1a and GLT1b, in the NAc and PFC after five days of ceftriaxone treatment (Rao et al., 2015). Interestingly, this mechanism is not limited to alcohol addiction. Ceftriaxone also restored GLT1 levels in the NAc core of rats trained to self-administer cocaine and prevented cue- and cocaine-induced reinstatement of cocaine seeking (Knackstedt, Melendez, & Kalivas, 2010; Sari et al., 2009). In addition, ceftriaxone treatment prevents reinstatement of methamphetamine and nicotine seeking in conditioned place preference paradigms (Abulseoud, Miller, Wu, Choi, & Holschneider, 2012; Alajaji, Bowers, Knackstedt, & Damaj, 2013). However, despite these promising preclinical results, there are several limitations that prevent the clinical use of ceftriaxone to reduce ethanol consumption or the consumption of other addictive drugs: (i) The antibiotic effect of ceftriaxone would be a major side effect of its therapeutical use, resulting in gastrointestinal symptoms such as diarrhea (Cudkowicz et al., 2014). In addition, its antibiotic effect could also promote a rise in the development of β-lactam resistant bacterial strains and would confound results since the alteration of the gut microbiota can modify ethanol consumption by itself (Ezquer et al., 2021; Segovia-Rodríguez et al., 2022) and (ii) Ceftriaxone has a poor penetration of the blood-brain-barrier (Nau et al., 1993), requiring a large dose to increase GLT1 levels in the brain (Smaga, Fierro, Mesa, Filip, & Knackstedt, 2020), which could translate into toxic doses required for its use in humans. On the other hand, a new β-lactam with no antibiotic activity, called MC-100093, has been shown to also increase GLT1 mRNA and protein levels and to reduce ethanol and cocaine self-administration (Alhaddad et al., 2022). However, MC-100093 treatment had a similar potency to ceftriaxone treatment in rats to reduce ethanol consumption, requiring high doses to achieve therapeutic effects (Alhaddad et al., 2022; Knackstedt et al., 2021). That is a problem since using new drugs at high concentrations can lead to the emergence of new adverse effects. Nevertheless, the discussed effectiveness of drugs that increase GLT1 expression indicates that this mechanism could be a fruitful target to design new therapies to achieve clinical success in reducing ethanol consumption.

5. Contribution of oxidative stress and neuroinflammation to the mechanism of ethanol consumption

Recent evidence suggest that brain oxidative stress and neuroinflammation are crucial contributors to the alteration of glutamate homeostasis observed after chronic ethanol consumption (Berríos-Cárcamo et al., 2020). While oxidative stress directly inhibits glutamate transport activity by the cross-linking of relevant GLT1 cysteine residues (Trotti, Danbolt, & Volterra, 1998), neuroinflammation increases the levels of oxidative stress (see below). Ethanol increases brain oxidative stress and neuroinflammation via the activation of pattern recognition receptors (i.e. Toll-like receptors, TLRs), after promoting the release of high mobility group box 1 (HMGB1), a nuclear protein that is also a TLR2 and TLR4 agonist (Crews & Vetreno, 2016). In addition, ethanol disrupts the intestinal barrier, allowing bacterial toxins like lipopolysaccharides (LPS, another TLR4 agonist) to enter the bloodstream (Ferrier et al., 2006). Ethanol-induced LPS stimulates immune cells to produce proinflammatory cytokines (Mandrekar & Szabo, 2009; Szabo, 2015), which can cross the blood-brain-barrier exacerbating neuroinflammation (Pan & Kastin, 2002). Consequently, studies in animal models of ethanol consumption have shown a significant rise in brain oxidative stress and neuroinflammation following chronic intake of ethanol (Ezquer et al., 2019a; 2019b; Quintanilla et al., 2018; Quintanilla et al., 2019), after a period of abstinence (Fernández-Rodríguez et al., 2022, 2023; Freeman et al., 2012), or after a forced ethanol exposure (Alfonso-Loeches, Pascual-Lucas, Blanco, Sanchez-Vera, & Guerri, 2010;

Costa, 2015; Crews, Walter, Coleman, & Vetreno, 2017; Ehrlich, Pirchl, & Humpel, 2012; Kane, 2014; Marshall, Geil, & Nixon, 2016; Montesinos, Pascual, Rodríguez-Arias, Miñarro, & Guerri, 2016; Niedzwiedz-Massey, 2023; Obernier, White, Swartzwelder, & Crews, 2002; Qin & Crews, 2012a,b; Vetreno & Crews, 2012; Whitman, Knapp, Werner, Crews, & Breese, 2013). This inflammatory response involves the activation of microglia and astrocytes (Colombo & Farina, 2016; Kraft & Harry, 2011), which produce cytokines and express TLRs to modulate an inflammatory response (Fischer & Maier, 2015; Kraft & Harry, 2011). Under oxidative stress and inflammation, the transcription factor NF-κB is activated in microglia and astrocytes (Hsieh, Wang, Wu, Chu, & Yang, 2010; Park et al., 2015), which promotes the expression of proinflammatory cytokines and oxidant enzymes. The release of oxidative stress and proinflammatory cytokines further activate microglia and astrocytes creating a vicious self-potentiating cycle (Blaser, Dostert, Mak, & Brenner, 2016). This sustained inflammation can persist long after the initial trigger has affected the brain, as evidenced by increased levels of proinflammatory cytokines in the brain of mice up to 10 months after an acute LPS administration (Qin et al., 2007). LPS treatment also promotes glutamate uptake inhibition and increases ethanol consumption in mice even two months after LPS injection (Blednov et al., 2011; Zhao et al., 2004). TLR4 activation is crucial for ethanol-induced oxidative stress and neuroinflammation since the activation of astrocytes and microglia, the activation of NF-κB, and the production of proinflammatory cytokines following forced chronic ethanol consumption are blunted in TLR-KO mice (Alfonso-Loeches et al., 2010).

6. Anti-inflammatory and antioxidant therapies to reduce ethanol consumption

The interaction between alcohol-induced neuroinflammation and oxidative stress provides a pathway for therapeutic exploration in treating AUD. Indeed, the use of anti-inflammatory and antioxidant compounds prevents ethanol-induced neuroinflammation, oxidative stress, and reduces ethanol consumption. For instance, minocycline, an inhibitor of microglial activation, reduces voluntary ethanol consumption in rats and mice (Agrawal, Hewetson, George, Syapin, & Bergeson, 2011; Gajbhiye, Tripathi, Petare, Potey, & Shankar, 2018). *N*-acetylcysteine (NAC), a small molecule with anti-inflammatory and antioxidant properties, decreases relapse consumption in rats after prolonged ethanol intake, while concurrently reducing oxidative stress levels and neuroinflammation markers (Cano-Cebrián et al., 2021; Fernández-Rodríguez et al., 2023; Fredriksson, Jayaram-Lindström, Kalivas, Melas, & Steensland, 2023; Garcia-Keller et al., 2020). However, the administration of two doses of NAC in a period of abstinence failed to restore GLT1 protein levels in the NAc of ethanol-consuming rats (Fernández-Rodríguez et al., 2023). In addition, studies in humans have shown that NAC treatment can be effective in reducing self-administration of drugs of abuse, such as cocaine and tobacco smoking (Grant et al., 2014; Prado et al., 2015; Schulte et al., 2018), and gambling (Grant et al., 2014). Interestingly, studies using animal models of cocaine self-administration show that NAC administration reduces cocaine reinstatement and normalizes cocaine-induced reduction of GLT1 levels in the NAc (Reissner et al., 2015). However, while preclinical addiction models have shown promising results, a clinical trial aimed at studying the effect of NAC treatment to reduce alcohol consumption in AUD patients yielded modest results, with no significant effects observed in reducing drinking or heavy drinking days (Morley et al., 2023). Another anti-inflammatory drug, the phosphodiestarase 4 inhibitor, ibudilast, has shown promise to reduce ethanol consumption in preclinical models. Indeed, systemic administration of different doses of ibudilast reduced ethanol consumption and relapse in rats (Bell et al., 2015). Ibudilast is a well-tolerated drug that has been approved by the FDA for the treatment of asthma (Rolan, Hutchinson, & Johnson, 2009), facilitating its examination as a therapeutic agent in humans. Clinical trials have shown that ibudilast treatment reduces the number of heavy drinking days in patients suffering AUD (Grodin et al., 2021). In addition, anti-inflammatory treatment with acetylsalicylic acid (ASA) reduces neuroinflammation and chronic ethanol intake and relapse in alcohol-consuming rats (Israel et al., 2021). Furthermore, the combination of NAC and ASA significantly inhibited chronic alcohol intake and relapse-like ethanol intake in alcohol-consuming rats, with greater efficacy than either drug separately (Quintanilla et al., 2020). This treatment also reduced hippocampal oxidative stress and neuroinflammation and inhibited both chronic and relapse ethanol intake (Quintanilla et al., 2020). Furthermore, NAC and ASA combined treatment increase GLT1 levels in the NAc (Ezquer et al., 2022). These results suggest that antioxidant and anti-inflammatory treatments with multiple targets can be beneficial for the treatment of AUD. In summary, these findings underscore the intricate interplay between neuroinflammation, oxidative stress, and chronic alcohol consumption, highlighting potential therapeutic avenues for mitigating AUD that could be exploited using potent antioxidant and anti-inflammatory therapies.

7. Mesenchymal stem cells as potential antioxidant and anti-inflammatory therapy for addressing AUD

Substantial investigations have focused on novel treatments involving the use of mesenchymal stem cells and their secreted products for addressing AUD due to their potent antioxidant (Kadekar, Rangole, Kale, & Limaye, 2016; Lee et al., 2016; Valle-Prieto & Conget, 2010) and anti-inflammatory (Gebler, Zabel, & Seliger, 2012; Prockop & Oh, 2012; Yagi et al., 2010) capabilities through the action of several mechanisms (see below) (Gao et al., 2016; González-González et al., 2020; Stavely & Nurgali, 2020).

Stem cells are a unique cell type with the ability to self-renew through cell division and to differentiate into multiple cell lineages. Among the different types of adult stem cells, mesenchymal stem cells, also known as mesenchymal stromal cells (MSCs) (Caplan, 2017), exhibit remarkable potential for multidirectional differentiation into various cell lineages of mesenchymal origin, such as adipocytes, chondrocytes and osteoblasts, in response to specific growth signals in vitro (Dominici et al., 2006). The ability of MSCs to be expanded extensively in vitro while retaining their multipotent potential makes them a highly attractive source for cell-based therapy research, owing to their pleiotropic mechanisms of action, including cell replacement, trophic support via paracrine signaling, and modulation of inflammation (Lindvall et al., 2012). Importantly, MSCs show low immunogenicity allowing autologous and allogeneic transplantations without immunosuppression (Ryan, Barry, Murphy, & Mahon, 2005), and have exhibited no major toxicity or tumorigenicity in rodents or human patients post-transplantation (Ra et al., 2011), underscoring their remarkable safety profile and suitability for a broad array of clinical applications. Given these precedents, MSCs are considered a promising cell source for the treatment of inflammatory diseases (Ullah, Subbarao, & Rho, 2015).

MSCs can be easily isolated from various sources such as bone marrow (Gronthos, Graves, Ohta, & Simmons, 1994; Mamidi et al., 2012; Otsuru, Hofmann, Olson, Dominici, & Horwitz, 2013; Stewart et al., 1999), umbilical cord (Kita, Gauglitz, Phan, Herndon, & Jeschke, 2010; Moretti, 2009; Zhang et al., 2011) and adipose tissue (Baglioni et al., 2009; Gronthos et al., 2001; Pendleton et al., 2013; Zhang et al., 2006), among others sources (Ullah et al., 2015). Regardless of their tissular origin, the International Society for Cellular Therapy (ISCT) has defined a general characterization criteria and proposed that MSCs must (i) be plastic adherent (e.g. culture flask), (ii) express the putative MSCs surface markers

CD73, CD90 and CD105 (>95%), (iii) not express the classic markers of other cell lineages CD14, CD34, CD45, CD19 and HLA-DR (<2%), and (iv) be able to undergo multi-lineage differentiation in vitro (Dominici et al., 2006) (Fig. 1).

Originally, MSCs were isolated from bone marrow and have been widely used in pre-clinical and clinical application. However, the process of harvesting MSCs from bone marrow (BM-MSCs) is highly invasive and painful, with low efficiency and increased senescence with age (Koobatian, Liang, Swartz, & Andreadis, 2015). In contrast, adipose tissue-derived MSCs (AT-MSCs) have emerged as a more attractive cell type due to their ease of isolation from cosmetic liposuction procedures, high abundance, rapid proliferation, and enhanced immunosuppressive properties (Elman, Li, Wang, Gimble, & Parekkadan, 2014; Frese, Dijkman, & Hoerstrup, 2016). Numerous studies have highlighted the comparable characteristics shared between BM-MSCs and AT-MSCs, encompassing morphological features and cell surface antigen expression (Danišovič et al., 2009; De Ugarte et al., 2003; Puissant et al., 2005; Vishnubalaji, Al-Nbaheen, Kadalmani, Aldahmash, & Ramesh, 2012). However, distinct biological differences have emerged concerning progenitor cell abundance, proliferation rates, and differentiation abilities. For instance, while approximately 40 mL of bone marrow aspirate contains roughly 2.4×10^4 MSCs in a skeletally mature adult individual, the processing of 300 mL of lipoaspirate (waste material) typically yields between 1×10^7 and 6×10^8 AT-MSCs (Aust et al., 2004; Locke, Windsor, & Dunbar, 2009), representing a remarkable 400-fold increase compared to BM-MSCs. Additionally, AT-MSCs exhibit easier culturing and faster growth rates compared to BM-MSCs (Heidari et al., 2013). Another noteworthy comparison is that, during early expansion, BM-MSCs frequently manifest signs of senescence, contrasting with the longer culture lifespan of AT-MSCs before senescence sets in (Dmitrieva et al., 2012; Kern, Eichler, Stoeve, Klüter, & Bieback, 2006). These distinguishing features underscore the value of AT-MSCs as a significant source of MSCs.

MSCs administration can reduce neuroinflammation and oxidative stress, thereby aiding in the recovery of neurological functions (Farfán et al., 2020; Galindo et al., 2011; Laroni, De Rosbo, & Uccelli, 2015). MSCs constitutively express high levels of the antioxidant glutathione (GSH) and antioxidant enzymes such as superoxide dismutase 1 and 2 (SOD1 and SOD2), catalase and glutathione peroxidase (Kim et al., 2008; Stavely & Nurgali, 2020; Valle-Prieto & Conget, 2010). Furthermore, MSCs demonstrated a potent anti-inflammatory effect through the secretion of

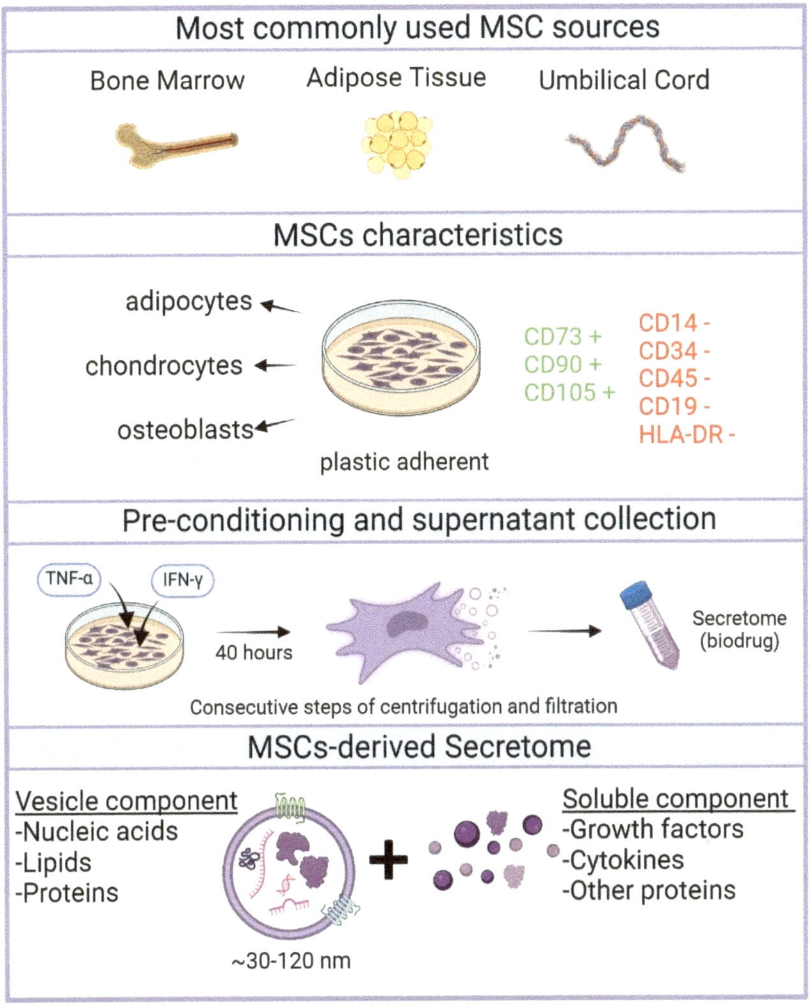

Fig. 1 Schematic representation of the most common MSCs source, characterization, preconditioning, and purification process to obtain the MSC-derived secretome. MSCscan be obtained from various tissues. However, the most common sources forregenerative medicine purposes are bone marrow, adipose tissue, and umbilicalcord blood. Regardless of the tissue source, MSCs are characterized by theirability to adhere to plastic surfaces (e.g., culture plates) and by theexpression of the putative MSC markers CD73, CD90, and CD105, and the absenceof expression of markers characteristics of other cell lineages such as CD14,CD34, CD45, CD19 and HLA-DR. Additionally, they possess the capacity todifferentiate into adipocytes, osteoblasts, and chondrocytes. To enhance theirtherapeutic properties, MSCs can be preconditioned by in vitro incubation with pro-inflammatory cytokines such as TNF-α and IFN-γ. Thisexposure primes MSCs to produce a more robust anti-inflammatory and antioxidantresponse, thereby improving their effectiveness in modulating immune reactionsand tissue repair processes. Following

(Continued)

anti-inflammatory cytokines reducing microglial and astrocyte activation (Da Silva Meirelles, Fontes, Covas, & Caplan, 2009; Yan et al., 2013; Zhong et al., 2020). MSCs also have an important role in the regulation of macrophages, lymphocytes and natural killer (NK) cells, whose action is critical for inflammation response and tissue regeneration. Among the molecules secreted by MSCs with immunomodulatory capacity are: prostaglandin E2 (PGE2), indoleamine 2,3-dioxygenase (IDO), transforming growth factor–β1 (TGF-β1) and interleukin-10 (IL-10) (Da Silva Meirelles et al., 2009; Gao et al., 2016; González-González et al., 2020; Zhang et al., 2013). Interestingly, it has been reported that the therapeutic potential of MSCs can be enhanced by altering its composition through the in vitro exposure of the cells to specific preconditioning stimuli. These stimuli have been categorized into seven groups: cytokines; immune receptor agonists; modification of culture conditions; hypoxia; autophagy; genetic modifications; and other agents that cannot be categorized (Sarsenova et al., 2022). Of particular interest is the preconditioning of MSCs with the pro-inflammatory cytokines interferon gamma (IFN-γ) and tumor necrosis factor alpha (TNF-α) (Fig. 1), since it has been reported that in vitro preconditioning of MSCs with both pro-inflammatory cytokines, induces the secretion of high levels of anti-inflammatory and antioxidant molecules, generating a potentiated biodrug that can be used as a therapeutic tool against inflammatory diseases (Ferreira et al., 2018; Oses et al., 2017; Schäfer, Spohn, & Baer, 2016), such as AUD (Ezquer et al., 2018; Ezquer et al., 2019a; Ezquer, et al., 2019b; Quintanilla et al., 2019). Thus, because of their potent effects on antioxidant and immune responses, MSCs could offer a multitarget approach to combat the neuropathological effects of chronic alcohol consumption.

8. Preclinical evidence of the use of mesenchymal stem cells to reduce ethanol consumption

In a pioneer proof-of-concept study, Israel and colleagues demonstrated that a single intracerebroventricular administration of rat MSCs (rMSCs) significantly reduced chronic alcohol intake and relapse-like

Fig. 1—Cont'd preconditioning, MSCs are cultured for 40 h under controlled conditions to allow them to release a variety ofbioactive molecules, including proteins, nucleic acids, and lipids in a solubleform or inside extracellular vesicles, collectively known as the MSC secretome. The secretome is then collected from the cultured MSCsfor further analysis or therapeutic applications. *Created with BioRender.com.*

drinking by up to 50% in an animal model that had voluntarily consumed high levels of alcohol for three months (Israel et al., 2016). Furthermore, the decrease in alcohol consumption was observed up to 40 days after a single intracerebroventricular administration, suggesting a long-term effect of MSCs (Israel et al., 2016). Interestingly, intracerebroventricular administration of human MSCs (hMSCs), which had been in vitro preconditioned by incubation with IFN-γ and TNF-α to increase their anti-inflammatory and antioxidative potential, also reduced chronic alcohol intake and relapse-like drinking by up to 85% in high-drinking rats that had voluntarily consumed alcohol for 17 weeks (Ezquer et al., 2019a). Also, it was shown that preconditioned hMSCs decrease neuroinflammation and oxidative stress induced by chronic alcohol consumption. This was evidenced by the reduction of ethanol-induced astrocyte and microglia activation and the restoration of the normal hippocampal GSSG/GSH ratio, respectively (Ezquer et al., 2019a) (Fig. 2). In a follow-up investigation, aiming for a translational approach in MSCs delivery that could avoid the need for intracerebroventricular administration, MSCs were cultured under 3D-conditions to form MSCs spheroids (Ezquer et al., 2018). This culture method significantly enhanced the anti-inflammatory capacity of MSCs and, when disaggregated, reduced their cell volume by 90% compared to MSCs cultured in 2D, facilitating the entry of the cells into the brain after intravenous administration (Ezquer et al., 2018). In this research, it was observed that a single intravenous administration of MSCs spheroids in rats voluntarily consuming alcohol for 12 weeks reduced their alcohol intake by 93% in 48 h, an effect that remained for at least three weeks, and induced an 80% reduction in relapse-like ethanol intake after 14 days of ethanol deprivation and re-access. In addition, MSCs spheroid administration fully reversed neuroinflammation and oxidative stress in the hippocampus (Ezquer et al., 2018). Interestingly, GLT1 levels were increased in both the NAc and PFC following MSCs spheroid administration compared to vehicle-treated rats, which could explain the therapeutic effect of MSCs spheroids in reducing alcohol intake by promoting glutamate uptake (Ezquer et al., 2018) (Fig. 2). MSCs have also been utilized in the treatment of other addictive drugs, such as cocaine. Placenta-derived MSCs attenuated cocaine-seeking behavior through their capacity to migrate to specific mesolimbic regions, target the parenchyma in the dentate gyrus of the hippocampus, and promote neurogenesis in a rat model of cocaine addiction (Pe'er-Nissan et al., 2022). The decrease in alcohol or cocaine consumption observed with MSC-based therapy in

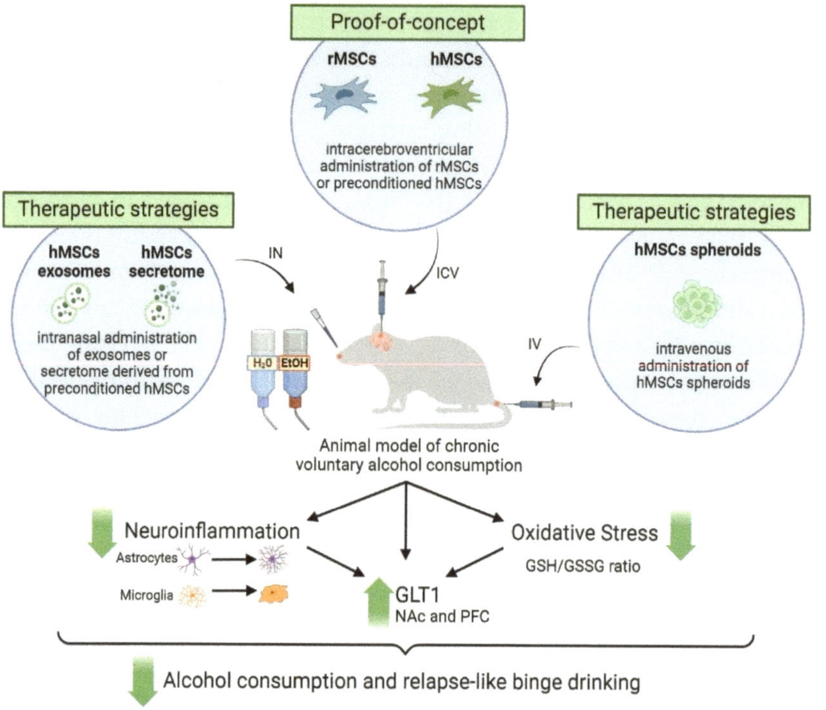

Fig. 2 Preclinical studies employing adipose tissue-derived MSCs or their secreted products have demonstrated a reduction in both ethanol intake and relapse-like drinking behavior in an animal model of chronic voluntary ethanol consumption. (Top) Intracerebroventricular (ICV) administration of rat (rMSCs) or preconditioned human MSCs (hMSCs) reduce ethanol intake and prevent ethanol relapse after deprivation and re-access. (Left) Intranasal (IN) administration of the complete secretome derived from preconditioned hMSCs or the exosome fraction of the secretome derived from preconditioned hMSCs decrease alcohol consumption. (Right) Intravenous (IV) administration of hMSC spheroids reduces chronic ethanol consumption. (Bottom) A decrease in ethanol-induced neuroinflammation and oxidative stress is observed concurrently with the reduction in ethanol consumption after the administration of rMSCs, hMSCs or their secreted products. In the case of the hMSC secretome, it was demonstrated that the effect of reducing ethanol consumption is mediated through the induction of the expression of the main glutamate transporter, GLT1, in the brain (nucleus accumbens, NAc; prefrontal cortex, PFC), shedding light on the potential mechanism by which MSCs treatments could exert their therapeutic effect. *Created with BioRender.com.*

these pre-clinical model is presumed to be linked to the potent anti-inflammatory and anti-oxidative properties of MSCs. Integrating MSCs, often referred to as the "guardians of inflammation" (Prockop & Oh, 2012), into the therapeutic landscape for AUD may offer a potential avenue for addressing not only the behavioral and cognitive consequences of alcohol addiction but also the underlying neurobiological processes that are triggering this behavior.

9. Secretome derived from mesenchymal stem cells as an improved antioxidant and anti-inflammatory biodrug

While MSCs offer promising benefits, their use in cellular therapies is constrained by various limitations. Notably, factors like tissue source, donor age, and inherent differences among donors can significantly influence the expansion and differentiation of these cells, adding complexity to their therapeutic efficacy. For example, MSCs isolated from aged donors (>50 years) exhibit senescent features, leading to reduced viability, proliferation and differentiation compared to cells obtained from younger donors (<40 years) (Choudhery, Badowski, Muise, Pierce, & Harris, 2014; Phinney et al., 1999). Additionally, the administration of living cells has the potential risk of emboli formation and the differentiation of the cells into unwanted structures (Tatsumi et al., 2013). Considering these challenges, exploring new alternative approaches becomes imperative. Given that MSCs primarily contribute to tissue repair via their paracrine factors and by stimulating host cells, rather than through direct cell replacement (Caplan & Correa, 2011; Caplan & Dennis, 2006; Da Silva Meirelles et al., 2009), studying the products secreted by MSCs, known as their secretome, is an auspicious alternative for more targeted regenerative therapies to minimize adverse secondary effects. Therefore, transitioning to the use of MSC-derived acellular products instead of living MSCs in the future represents a shift that could streamline management and enhance safety.

The MSCs-secretome consists of an heterogeneous pool of proteins, nucleic acids, lipids and extracellular vesicles (EVs) that are released from cells into the extracellular space (Maguire, 2013; Vizoso, Eiro, Cid, Schneider, & Perez-Fernandez, 2017). The soluble component of the secretome comprises a diverse array of cytokines, chemokines, growth factors, and immunomodulatory compounds. These components serve essential functions, including immunomodulation, anti-apoptosis, angiogenesis, support for the growth and differentiation of local stem and progenitor cells, anti-fibrosis, and chemoattraction (Da Silva Meirelles et al., 2009). The vesicle component of the MSCs-secretome can be categorized into apoptotic bodies, microparticles, and exosomes, each distinguished by their size, density, surface markers, and origin. Among these, exosomes have emerged as pivotal elements in the therapeutic mechanisms of MSCs. Notably, MSCs-derived exosomes have garnered considerable attention due to their ability to interact with target cells and modulate cell signaling (Kourembanas, 2015), and have been found to play a significant role in MSC-mediated immune (Wang et al., 2020) and antioxidant regulation (Xia, Dai, Jin, & Chen, 2021).

MSCs-derived exosomes are small, spherical vesicles ranging from 30 to 120 nm in size, enclosed by a lipid bilayer. They originate from late endocytic compartments, specifically multivesicular bodies, through inward invagination of endosomal membranes, and are subsequently released into the extracellular space (Koniusz et al., 2016). During their formation, transmembrane proteins become incorporated into the invaginating membrane, enabling trafficking, adhesion and endocrine effects of exosomes (Vizoso et al., 2017). In addition, cytosolic components are encapsulated within the vesicles including genetic materials (DNA, RNA fragments and microRNAs(miRNAs)), enzymes, proteins and growth factors (Maguire, 2013; Mathivanan, Ji, & Simpson, 2010). Using the complete MSCs-secretome presents significant advantages over the administration of living MSCs, as it is simpler, quicker, safer, and more cost-effective to produce. Additionally, the MSCs-secretome can be lyophilized, resulting in an off-the-shelf soluble powder containing EVs, molecules and proteins, ensuring long-term stability and straightforward reconstitution for its clinical applications (Bari et al., 2018). Additionally, MSCs products such as exosomes and secretome may present translational advantages owing to their potential for intranasal administration, offering a direct delivery route to the central nervous system while bypassing the blood-brain barrier, minimizing systemic absorption, and limiting potential peripheral side effects (Vyas, Shahiwala, Marathe, & Misra, 2005).

10. Therapeutic potential of MSCs secretome in the treatment of alcohol use disorder

MSCs-derived acellular products have been successfully used to reduce ethanol consumption in animal models (see below). Interestingly, fluorescently-tagged exosomes derived from preconditioned hMSCs have been shown to reach the rat brain efficiently six hours after their intranasal administration, with a maximum signal level 12 h after their administration (Ezquer et al., 2019b). The intranasal administration of exosomes also reversed alcohol-induced neuroinflammation and oxidative stress, replicating the effect observed with the intracerebroventricular administration of MSCs. However, to sustain the reduction of chronic ethanol intake, exosomes need to be administered once a week (Ezquer et al., 2019b) (Fig. 2). The reduction in oxidative stress levels observed in this study aligns with previous studies indicating that exosomes derived from MSCs can reduce oxidative stress induced by different stressors (Arslan et al., 2013). Additionally, MSCs exosomes induced a marked increase in GLT1 gene expression in the NAc, suggesting that the increase in GLT1 resulting from exosome administration could restore glutamate homeostasis in the brain, allowing for a reduction in alcohol consumption (Ezquer et al., 2019b). Furthermore, the intranasal administration of the complete secretome derived from preconditioned MSCs (comprised of exosomes and soluble components) to rats that chronically consumed alcohol resulted in an 85% reduction in ethanol intake after 48 h. Secretome administration also completely inhibited relapse-like drinking after ethanol deprivation and re-access and fully abolished ethanol-induced oxidative stress and neuroinflammation (Quintanilla et al., 2019) (Fig. 2). In this work, to demonstrate that the therapeutic effect observed after MSCs or MSCs-derived secretome administration is attributable to the increase in GLT1 levels, the increase in GLT1 levels was prevented by the intracerebral administration of a GLT1 antisense oligonucleotide, given 24 h before the intranasal administration of the secretome. In these experiments, the administration of the anti-GLT1 oligonucleotide fully abolished the inhibitory effect of the secretome on ethanol intake, confirming the central role of GLT1 in the anti-addictive effects of the secretome (Quintanilla et al., 2019). These experiments demonstrate that administering products secreted by MSCs, in addition to the intracerebroventricular delivery of MSCs themselves, are capable of significantly reducing alcohol consumption in an animal model bred for high ethanol intake. These serve as proof-of-concept of the use of potent biodrugs that reduce neuroinflammation and oxidative stress, thereby increasing GLT1 levels and consequently reducing alcohol consumption and relapse-like binge drinking (Fig. 2). These biodrugs hold promise as potential therapeutic agents against AUD.

Recently, another research group independently showed that the administration of extracellular vesicles derived from MSCs also reduced neuroinflammation induced by binge-like ethanol treatment over a two-week period in mice, in addition to reversing ethanol-induced synaptic alterations and alleviating cognitive dysfunction (Mellado et al., 2023).

Interestingly, the intranasal administration of MSCs-derived secretome has also resulted in the reduction of the voluntary consumption of nicotine (Quintanilla et al., 2019) or morphine (Quintanilla et al., 2022), accompanied by a reduction of inflammation and an increase of GLT1 expression in the NAc, showing that this therapeutic mechanism may be of relevance for the treatment of multiple drugs of abuse. Overall, these studies support that the products secreted by MSCs can replicate the effects achieved by the administration of living MSCs, reducing ethanol intake, neuroinflammation, and oxidative stress observed in rodents that voluntarily consumed

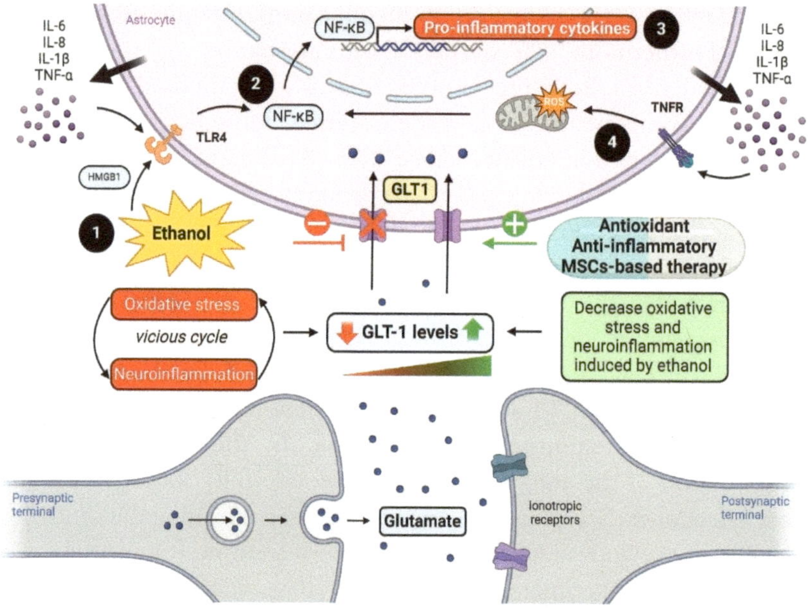

Fig. 3 Ethanol induces oxidative stress and neuroinflammation, and reduces GLT1 levels. Both phenomena are restored after MSC-based therapy. Under normal conditions, glutamate is rapidly removed from the synaptic cleft through the glutamate transporter GLT1 present in astrocytes forming the tripartite synapse. Ethanol exposure promotes an increase in oxidative stress and neuroinflammation, via TLR4 activation in astrocytes and microglia (1), initiating a signaling cascade that leads to the nuclear translocation of NF-κB (2) and the secretion of proinflammatory cytokines (3). Astrocytes and microglia activation also increase the levels of ROS in the cell, which in turn induces further inflammation (4). High levels of oxidative stress induce GLT1 oxidation and the reduction in GLT1 levels, leading to the accumulation of glutamate in the synaptic space, generating the overactivation of glutamatergic signaling. In contrast, antioxidant, anti-inflammatory and MSC-based treatments decrease ethanol-induced oxidative stress and neuroinflammation and restore glutamate homeostasis by increasing GLT1 levels. *Created with BioRender.com.*

alcohol chronically. The reduction of oxidative stress and neuroinflammation may be one of the mechanisms by which MSC-based therapy is able to decrease alcohol consumption and relapse-like binge in these animal models. Additionally, the increase in GLT1 expression, which facilitates the restoration of glutamate homeostasis in the NAc, represents another mechanism by which MSC-based therapy contributes to the reduction in chronic ethanol intake and the prevention of ethanol relapse after deprivation and re-access (Fig. 3).

11. Conclusion

It has been increasingly recognized that chronic alcohol intake induces neuroinflammation and brain oxidative stress in specific brain regions, a phenomenon that self-perpetuates in a long acting a vicious cycle. This results in altered brain glutamate homeostasis through the reduction of glutamate transporter levels, inducing an increase in drug-seeking behaviors. Preclinical evidence has demonstrated that preconditioned MSCs or MSCs-derived secretome (i) reduces ethanol consumption, (ii) prevents ethanol relapse after ethanol deprivation and reaccess, (iii) reduces neuroinflammation and oxidative stress induced by chronic alcohol consumption, and (iv) increases GLT1 levels, restoring normal brain glutamate homeostasis (Fig. 3).

Further studies are needed to demonstrate which specific molecules produced by MSCs are directly responsible for the observed therapeutic effects and to evaluate the safety and effectiveness of this procedure in a clinical setting. However, due to the limited treatment options currently available for the treatment of AUD, the non-invasive administration of the secretome derived from MSCs appears as a promising tool for the treatment of this devastating disease.

Acknowledgment

This work was supported by FONDECYT 1240162 and ACT210012 grants to Fernando Ezquer.

References

Abulseoud, O. A., Miller, J. D., Wu, J., Choi, D. S., & Holschneider, D. P. (2012). Ceftriaxone upregulates the glutamate transporter in medial prefrontal cortex and blocks reinstatement of methamphetamine seeking in a condition place preference paradigm. *Brain Research, 1456*, 14–21. https://doi.org/10.1016/j.brainres.2012.03.045.

Agrawal, R. G., Hewetson, A., George, C. M., Syapin, P. J., & Bergeson, S. E. (2011). Minocycline reduces ethanol drinking. *Brain, Behavior, and Immunity, 25*, S165–S169. https://doi.org/10.1016/j.bbi.2011.03.002.Minocycline.

Alajaji, M., Bowers, M. S., Knackstedt, L., & Damaj, M. I. (2013). Effects of the beta-lactam antibiotic ceftriaxone on nicotine withdrawal and nicotine-induced reinstatement of preference in mice. *Psychopharmacology, 228*(3), 419–426. https://doi.org/10.1007/s00213-013-3047-3.

Alfonso-Loeches, S., Pascual-Lucas, M., Blanco, A. M., Sanchez-Vera, I., & Guerri, C. (2010). Pivotal role of TLR4 receptors in alcohol-induced neuroinflammation and brain damage. *Journal of Neuroscience, 30*(24), 8285–8295. https://doi.org/10.1523/JNEUROSCI.0976-10.2010.

Alhaddad, H., et al. (2014). Effects of MS-153 on chronic ethanol consumption and GLT1 modulation of glutamate levels in male alcohol-preferring rats. *Frontiers in Behavioral Neuroscience, 8*(366)), 1–13. https://doi.org/10.3389/fnbeh.2014.00366.

Alhaddad, H., et al. (2022). Effects of a novel beta lactam compound, MC-100093, on the expression of glutamate transporters/receptors and ethanol drinking behavior of alcohol-preferring rats. *Journal of Pharmacology and Experimental Therapeutics, 383*(3), 208–216. https://doi.org/10.1124/jpet.122.001147.

Arslan, F., et al. (2013). Mesenchymal stem cell-derived exosomes increase ATP levels, decrease oxidative stress and activate PI3K/Akt pathway to enhance myocardial viability and prevent adverse remodeling after myocardial ischemia/reperfusion injury. *Stem Cell Research, 10*(3), 301–312. https://doi.org/10.1016/j.scr.2013.01.002.

Aust, L., et al. (2004). Yield of human adipose-derived adult stem cells from liposuction aspirates. *Cytotherapy, 6*(1), 7–14. https://doi.org/10.1080/14653240310004539.

Baglioni, S., et al. (2009). Characterization of human adult stem-cell populations isolated from visceral and subcutaneous adipose tissue. *The FASEB Journal, 23*(10), 3494–3505. https://doi.org/10.1096/fj.08-126946.

Bari, E., et al. (2018). Pilot production of mesenchymal stem/stromal freeze-dried secretome for cell-free regenerative nanomedicine: A validated GMP-compliant process. *Cells,* 7(11), 1–22. https://doi.org/10.3390/cells7110190.

Batra, A., Müller, C. A., Mann, K., & Heinz, A. (2016). Alcohol dependence and harmful use of alcohol. *Deutsches Arzteblatt International, 113*(17), 301–310. https://doi.org/10.3238/arztebl.2016.0301.

Bell, R. L., et al. (2015). Ibudilast reduces alcohol drinking in multiple animal models of alcohol dependence. *Addiction Biology, 20*(1), 38–42. https://doi.org/10.1111/adb.12106.

Beller, J. A., Gurkoff, G. G., Berman, R. F., & Lyeth, B. G. (2011). Pharmacological enhancement of glutamate transport reduces excitotoxicity in vitro. *Restorative Neurology and Neuroscience, 29*(5), 331–346. https://doi.org/10.3233/RNN-2011-603.

Bello, M. (2022). Molecular basis of inhibitory mechanism of naltrexone and its metabolites through structural and energetic analyses. *Molecules (Basel, Switzerland), 27*(15), https://doi.org/10.3390/molecules27154919.

Berríos-Cárcamo, P., et al. (2020). Oxidative stress and neuroinflammation as a pivot in drug abuse. A focus on the therapeutic potential of antioxidant and anti-inflammatory agents and biomolecules. *Antioxidants, 9*(9), 1–26. https://doi.org/10.3390/antiox9090830.

Blaser, H., Dostert, C., Mak, T. W., & Brenner, D. (2016). TNF and ROS crosstalk in inflammation. *Trends in Cell Biology, 26*(4), 249–261. https://doi.org/10.1016/j.tcb.2015.12.002.

Blednov, Y. A., Benavidez, J. M., Geil, C., Perra, S., Morikawa, H., & Harris, R. A. (2011). Activation of inflammatory signaling by lipopolysaccharide produces a prolonged increase of voluntary alcohol intake in mice. *Brain, Behavior, and Immunity, 25*, S92–S105. https://doi.org/10.1016/j.bbi.2011.01.008.

Cano-Cebrián, M. J., Fernández-Rodríguez, S., Hipólito, L., Granero, L., Polache, A., & Zornoza, T. (2021). Efficacy of N-acetylcysteine in the prevention of alcohol relapse-like drinking: Study in long-term ethanol-experienced male rats. *Journal of Neuroscience Research, 99*(2), 638–648. https://doi.org/10.1002/jnr.24736.

Caplan, A. I. (2017). Mesenchymal stem cells: Time to change the name!. *Stem Cells Translational Medicine, 6*(6), 1445–1451. https://doi.org/10.1002/sctm.17-0051.

Caplan, A. I., & Correa, D. (2011). The MSC: An injury drugstore. *Cell Stem Cell, 9*(1), 11–15. https://doi.org/10.1016/j.stem.2011.06.008.

Caplan, A. I., & Dennis, J. E. (2006). Mesenchymal stem cells as trophic mediators. *Journal of Cellular Biochemistry, 98*(5), 1076–1084. https://doi.org/10.1002/jcb.20886.

Choudhery, M. S., Badowski, M., Muise, A., Pierce, J., & Harris, D. T. (2014). Donor age negatively impacts adipose tissue-derived mesenchymal stem cell expansion and differentiation. *Journal of Translational Medicine, 12*(1), 1–14. https://doi.org/10.1186/1479-5876-12-8.

Colombo, E., & Farina, C. (2016). Astrocytes: Key regulators of neuroinflammation. *Trends in Immunology, 37*(9), 608–620. https://doi.org/10.1016/j.it.2016.06.006.

Costa, P. A., et al. (2015). Brain DNA damage and behavioral changes after repeated intermittent acute ethanol withdrawal by young rats. *Psychopharmacology, 232*(19), 3623–3636. https://doi.org/10.1007/s00213-015-4015-x.

Crews, F. T., & Vetreno, R. P. (2016). Mechanisms of neuroimmune gene induction in alcoholism. *Psychopharmacology, 233*(9), 1543–1557. https://doi.org/10.1007/s00213-015-3906-1.

Crews, F. T., Walter, T. J., Coleman, L. G., Jr, & Vetreno, R. P. (2017). Toll-like receptor signaling and stages of addiction. *Psychopharmacology, 234*(9–10), 1483–1498. https://doi.org/10.1007/s00213-017-4560-6.

Cudkowicz, M. E., et al. (2014). Safety and efficacy of ceftriaxone for amyotrophic lateral sclerosis: A multi-stage, randomised, double-blind, placebo-controlled trial. *The Lancet Neurology, 13*(11), 1083–1091. https://doi.org/10.1016/S1474-4422(14)70222-4.

Da Silva Meirelles, L., Fontes, A. M., Covas, D. T., & Caplan, A. I. (2009). Mechanisms involved in the therapeutic properties of mesenchymal stem cells. *Cytokine and Growth Factor Reviews, 20*(5–6), 419–427. https://doi.org/10.1016/j.cytogfr.2009.10.002.

Danbolt, N. C. (2001). Glutamate uptake. *Progress in Neurobiology, 65*(1), 1–105. https://doi.org/10.1016/S0301-0082(00)00067-8.

Danbolt, N. C., Furness, D. N., & Zhou, Y. (2016). Neuronal vs glial glutamate uptake: Resolving the conundrum. *Neurochemistry International, 98*, 29–45. https://doi.org/10.1016/j.neuint.2016.05.009.

Danišovič, L., et al. (2009). Comparison of in vitro chondrogenic potential of human mesenchymal stem cells derived from bone marrow and adipose tissue. *General Physiology and Biophysics, 28*(1), 56–62. https://doi.org/10.4149/gpb_2009_01_56.

Das, S. K., & Vasudevan, D. M. (2007). Alcohol-induced oxidative stress. *Life Sciences, 81*(3), 177–187. https://doi.org/10.1016/j.lfs.2007.05.005.

De Ugarte, D. A., et al. (2003). Comparison of multi-lineage cells from human adipose tissue and bone marrow. *Cells, Tissues, Organs, 174*(3), 101–109. https://doi.org/10.1159/000071150.

Degenhardt, L., et al. (2018). The global burden of disease attributable to alcohol and drug use in 195 countries and territories, 1990–2016: A systematic analysis for the Global Burden of Disease Study 2016. *The Lancet Psychiatry, 5*(12), 987–1012. https://doi.org/10.1016/S2215-0366(18)30337-7.

Ding, Z. M., Rodd, Z. A., Engleman, E. A., Bailey, J. A., Lahiri, D. K., & McBride, W. J. (2013). Alcohol drinking and deprivation alter basal extracellular glutamate concentrations and clearance in the mesolimbic system of alcohol-preferring (P) rats. *Addiction Biology, 18*(2), 297–306. https://doi.org/10.1111/adb.12018.

Dmitrieva, R. I., Minullina, I. R., Bilibina, A. A., Tarasova, O. V., Anisimov, S. V., & Zaritskey, A. Y. (2012). Bone marrow- and subcutaneous adipose tissue-derived mesenchymal stem cells: Differences and similarities. *Cell Cycle (Georgetown, Tex.), 11*(2), 377–383. https://doi.org/10.4161/cc.11.2.18858.

Dominici, M., et al. (2006). Minimal criteria for defining multipotent mesenchymal stromal cells. The International Society for Cellular Therapy position statement. *Cytotherapy, 8*(4), 315–317. https://doi.org/10.1080/14653240600855905.

Ehrlich, D., Pirchl, M., & Humpel, C. (2012). Effects of long-term moderate ethanol and cholesterol on cognition, cholinergic neurons, inflammation, and vascular impairment in rats. *Neuroscience, 205*, 154–166. https://doi.org/10.1016/j.neuroscience.2011.12.054.

Elman, J. S., Li, M., Wang, F., Gimble, J. M., & Parekkadan, B. (2014). A comparison of adipose and bone marrow-derived mesenchymal stromal cell secreted factors in the treatment of systemic inflammation. *Journal of Inflammation (United Kingdom), 11*(1), 4–11. https://doi.org/10.1186/1476-9255-11-1.

Ezquer, F., et al. (2018). Intravenous administration of anti-inflammatory mesenchymal stem cell spheroids reduces chronic alcohol intake and abolishes binge-drinking. *Scientific Reports, 8*(1), 1–15. https://doi.org/10.1038/s41598-018-22750-7.

Ezquer, F., et al. (2019a). Activated mesenchymal stem cell administration inhibits chronic alcohol drinking and suppresses relapse-like drinking in high-alcohol drinker rats. *Addiction Biology, 24*(1), 17–27. https://doi.org/10.1111/adb.12572.

Ezquer, F., et al. (2019b). Intranasal delivery of mesenchymal stem cell-derived exosomes reduces oxidative stress and markedly inhibits ethanol consumption and post-deprivation relapse drinking. *Addiction Biology, 24*(5), 994–1007. https://doi.org/10.1111/adb.12675.

Ezquer, F., et al. (2021). Innate gut microbiota predisposes to high alcohol consumption. *Addiction Biology, 26*(4), 1–14. https://doi.org/10.1111/adb.13018.

Ezquer, F., et al. (2022). A dual treatment blocks alcohol binge-drinking relapse: Microbiota as a new player. *Drug and Alcohol Dependence, 236*(109466), https://doi.org/10.1016/j.drugalcdep.2022.109466.

Farfán, N., et al. (2020). Intranasal administration of mesenchymal stem cell secretome reduces hippocampal oxidative stress, neuroinflammation and cell death, improving the behavioral outcome following perinatal asphyxia. *International Journal of Molecular Sciences, 21*(20), 1–27. https://doi.org/10.3390/ijms21207800.

Fernández-Rodríguez, S., et al. (2022). Different brain oxidative and neuroinflammation status in rats during prolonged abstinence depending on their ethanol relapse-like drinking behavior: Effects of ethanol reintroduction. *Drug and Alcohol Dependence, 232*(109284), https://doi.org/10.1016/j.drugalcdep.2022.109284.

Fernández-Rodríguez, S., et al. (2023). N-Acetylcysteine normalizes brain oxidative stress and neuroinflammation observed after protracted ethanol abstinence: A preclinical study in long-term ethanol-experienced male rats. *Psychopharmacology, 240*(4), 725–738. https://doi.org/10.1007/s00213-023-06311-z.

Ferreira, J. R., Teixeira, G. Q., Santos, S. G., Barbosa, M. A., Almeida-Porada, G., & Gonçalves, R. M. (2018). Mesenchymal stromal cell secretome: Influencing therapeutic potential by cellular pre-conditioning. *Frontiers in Immunology, 9*(2837), 1–17. https://doi.org/10.3389/fimmu.2018.02837.

Ferrier, L., et al. (2006). Impairment of the intestinal barrier by ethanol involves enteric microflora and mast cell activation in rodents. *American Journal of Pathology, 168*(4), 1148–1154. https://doi.org/10.2353/ajpath.2006.050617.

Fischer, R., & Maier, O. (2015). Interrelation of oxidative stress and inflammation in neurodegenerative disease: Role of TNF. *Oxidative Medicine and Cellular Longevity*610813. https://doi.org/10.1155/2015/610813.

Floresco, S. B. (2015). The nucleus accumbens: An interface between cognition, emotion, and action. *Annual Review of Psychology, 66*, 25–32. https://doi.org/10.1146/annurev-psych-010213-115159.

Fredriksson, I., Jayaram-Lindström, N., Kalivas, P. W., Melas, P. A., & Steensland, P. (2023). N-acetylcysteine improves impulse control and attenuates relapse-like alcohol intake in long-term drinking rats. *Behavioural Brain Research, 436*(114089), https://doi.org/10.1016/j.bbr.2022.114089.

Freeman, K., et al. (2012). Temporal changes in innate immune signals in a rat model of alcohol withdrawal in emotional and cardiorespiratory homeostatic nuclei. *Journal of Neuroinflammation, 9*, 1–11. https://doi.org/10.1186/1742-2094-9-97.

Frese, L., Dijkman, P. E., & Hoerstrup, S. P. (2016). Adipose tissue-derived stem cells in regenerative medicine. *Transfusion Medicine and Hemotherapy, 43*(4), 268–274. https://doi.org/10.1159/000448180.

Gajbhiye, S. V., Tripathi, R. K., Petare, A., Potey, A. V., & Shankar, A. (2018). Minocycline in alcohol withdrawal induced anxiety and alcohol relapse in rats. *Current Clinical Pharmacology, 13*(1), 65–72. https://doi.org/10.2174/1574884713666180228110310.

Galindo, L. T., et al. (2011). Mesenchymal stem cell therapy modulates the inflammatory response in experimental traumatic brain injury. *Neurology Research International, 2011*(564089), https://doi.org/10.1155/2011/564089.

Gano, A., Deak, T., & Pautassi, R. M. (2023). A review on the reciprocal interactions between neuroinflammatory processes and substance use and misuse, with a focus on alcohol misuse. *American Journal of Drug and Alcohol Abuse, 49*(3), 269–282. https://doi.org/10.1080/00952990.2023.2201944.

Gao, F., et al. (2016). Mesenchymal stem cells and immunomodulation: Current status and future prospects. *Cell Death and Disease, 7*(1), https://doi.org/10.1038/cddis.2015.327.

Garcia-Keller, C., Smiley, C., Monforton, C., Melton, S., Kalivas, P. W., & Gass, J. (2020). 'N-Acetylcysteine treatment during acute stress prevents stress-induced augmentation of addictive drug use and relapse. *Addiction Biology, 25*(5), 1–12. https://doi.org/10.1111/adb.12798.

Gass, J. T., Sinclair, C. M., Cleva, R. M., Widholm, J. J., & Olive, M. F. (2011). Alcohol-seeking behavior is associated with increased glutamate transmission in basolateral amygdala and nucleus accumbens as measured by glutamate-oxidase-coated biosensors. *Addiction Biology, 16*(2), 215–228. https://doi.org/10.1111/j.1369-1600.2010.00262.x.

Gebler, A., Zabel, O., & Seliger, B. (2012). The immunomodulatory capacity of mesenchymal stem cells. *Trends in Molecular Medicine, 18*(2), 128–134. https://doi.org/10.1016/j.molmed.2011.10.004.

González-González, A., García-Sánchez, D., Dotta, M., Rodríguez-Rey, J. C., & Pérez-Campo, F. M. (2020). Mesenchymal stem cells secretome: The cornerstone of cell-free regenerative medicine. *World Journal of Stem Cells, 12*(12), 1439–1690. https://doi.org/10.4252/wjsc.v12.i12.1529.

Goodwani, S., Rao, P. S., Bell, R. L., & Sari, Y. (2015). Amoxicillin and amoxicillin/clavulanate reduce ethanol intake and increase GLT-1 expression as well as AKT phosphorylation in mesocorticolimbic regions. *Brain Research, 1622*, 397–408. https://doi.org/10.1016/j.brainres.2015.07.008.

Grant, J. E., et al. (2014). A randomized, placebo-controlled trial of N-acetylcysteine plus imaginal desensitization for nicotine-dependent pathological gamblers. *The Journal of Clinical Psychiatry, 75*(1), 39–45. https://doi.org/10.4088/JCP.13m08411.

Griffin, W. C., 3rd, Haun, H. L., Hazelbaker, C. L., Ramachandra, V. S., & Becker, H. C. (2014). Increased extracellular glutamate in the nucleus accumbens promotes excessive ethanol drinking in ethanol dependent mice. *Neuropsychopharmacology: Official Publication of the American College of Neuropsychopharmacology, 39*(3), 707–717. https://doi.org/10.1038/npp.2013.256.

Griffin, W. C., Ramachandra, V. S., Knackstedt, L. A., & Becker, H. C. (2015). Repeated cycles of chronic intermittent ethanol exposure increases basal glutamate in the nucleus accumbens of mice without affecting glutamate transport. *Frontiers in Pharmacology, 6*(27)), 1–7. https://doi.org/10.3389/fphar.2015.00027.
Grodin, E. N., et al. (2021). Ibudilast, a neuroimmune modulator, reduces heavy drinking and alcohol cue-elicited neural activation: A randomized trial. *Translational Psychiatry, 11*(1), 1–8. https://doi.org/10.1038/s41398-021-01478-5.
Gronthos, S., Franklin, D. M., Leddy, H. A., Robey, P. G., Storms, R. W., & Gimble, J. M. (2001). Surface protein characterization of human adipose tissue-derived stromal cells. *Journal of Cellular Physiology, 189*(1), 54–63. https://doi.org/10.1002/jcp.1138.
Gronthos, S., Graves, S. E., Ohta, S., & Simmons, P. J. (1994). The STRO-1+ fraction of adult human bone marrow contains the osteogenic precursors. *Blood, 84*(12), 4164–4173. https://doi.org/10.1182/blood.v84.12.4164.bloodjournal84124164.
Han, Y., et al. (2022). The secretion profile of mesenchymal stem cells and potential applications in treating human diseases. *Signal Transduction and Targeted Therapy,* 7(1), https://doi.org/10.1038/s41392-022-00932-0.
Heidari, B., et al. (2013). Comparison of proliferative and multilineage differentiation potential of sheep mesenchymal stem cells derived from bone marrow, liver, and adipose tissue. *Avicenna Journal of Medical Biotechnology, 5*(2), 104–117.
Hsieh, H. L., Wang, H. H., Wu, W. B., Chu, P. J., & Yang, C. M. (2010). Transforming growth factor-β1 induces matrix metalloproteinase-9 and cell migration in astrocytes: Roles of ROS-dependent ERK- and JNK-NF-κB pathways. *Journal of Neuroinflammation,* 7, 1–17. https://doi.org/10.1186/1742-2094-7-88.
Huang, Y. H., & Bergles, D. E. (2004). Glutamate transporters bring competition to the synapse. *Current Opinion in Neurobiology, 14*(3), 346–352. https://doi.org/10.1016/j.conb.2004.05.007.
Imperato, A., & Di Chiara, G. (1986). Preferential stimulation of dopamine release in the nucleus accumbens of freely moving rats by ethanol. *The Journal of Pharmacology and Experimental Therapeutics, 239*(1), 219–228.
Institute for Health Metrics and Evaluation. (2020). Alcohol use. https://www.healthdata.org/research-analysis/health-risks-issues/alcohol-use#:~:text=1.34%20billion%20people%20consumed%20harmful,amounts%20of%20alcohol%20were%20male.
Israel, Y., et al. (2021). Aspirin and N-acetylcysteine co-administration markedly inhibit chronic ethanol intake and block relapse binge drinking: Role of neuroinflammation-oxidative stress self-perpetuation. *Addiction Biology, 26*(1), 1–13. https://doi.org/10.1111/adb.12853.
Israel, Y., Ezquer, F., Quintanilla, M. E., Morales, P., Ezquer, M., & Herrera-Marschitz, M. (2016). Intracerebral stem cell administration inhibits relapse-like alcohol drinking in rats. *Alcohol and Alcoholism, 52*(1), 1–4. https://doi.org/10.1093/alcalc/agw068.
Jørgensen, C. H., Pedersen, B., & Tønnesen, H. (2011). The efficacy of disulfiram for the treatment of alcohol use disorder. *Alcoholism: Clinical and Experimental Research, 35*(10), 1749–1758. https://doi.org/10.1111/j.1530-0277.2011.01523.x.
Kadekar, D., Rangole, S., Kale, V., & Limaye, L. (2016). Conditioned medium from placental mesenchymal stem cells reduces oxidative stress during the cryopreservation of ex vivo expanded umbilical cord blood cells. *PLoS One, 11*(10), 1–21. https://doi.org/10.1371/journal.pone.0165466.
Kalivas, P. W. (2009). The glutamate homeostasis hypothesis of addiction. *Nature Reviews. Neuroscience, 10*(8), 561–572. https://doi.org/10.1038/nrn2515.
Kane, C. J. M., et al. (2014). Effects of ethanol on immune response in the brain: Region specific changes in adolescent versus adult mice. *Alcoholism, Clinical and Experimental Research, 38*(2), 384–391. https://doi.org/10.1111/acer.12244.Effects.

Kern, S., Eichler, H., Stoeve, J., Klüter, H., & Bieback, K. (2006). Comparative analysis of mesenchymal stem cells from bone marrow, umbilical cord blood, or adipose tissue. *Stem Cells, 24*(5), 1294–1301. https://doi.org/10.1634/stemcells.2005-0342.

Kiefer, F., et al. (2003). Comparing and combining naltrexone and acamprosate in relapse prevention of alcoholism. A double-blind, placebo-controlled study. *Archives of General Psychiatry, 60*, 92–99. https://doi.org/10.1001/archpsyc.60.1.92.

Kim, W. S., et al. (2008). Evidence supporting antioxidant action of adipose-derived stem cells: Protection of human dermal fibroblasts from oxidative stress. *Journal of Dermatological Science, 49*(2), 133–142. https://doi.org/10.1016/j.jdermsci.2007.08.004.

Kita, K., Gauglitz, G. G., Phan, T. T., Herndon, D. N., & Jeschke, M. G. (2010). Isolation and characterization of mesenchymal stem cells from the sub-amniotic human umbilical cord lining membrane. *Stem Cells and Development, 19*(4), 491–501. https://doi.org/10.1089/scd.2009.0192.

Knackstedt, L. A., et al. (2021). MC-100093, a novel β-lactam glutamate transporter-1 enhancer devoid of antimicrobial properties, attenuates cocaine relapse in rats. *Journal of Pharmacology and Experimental Therapeutics, 378*(2), 51–59. https://doi.org/10.1124/jpet.121.000532.

Knackstedt, L. A., Melendez, R. I., & Kalivas, P. W. (2010). Ceftriaxone restores glutamate homeostasis and prevents relapse to cocaine-seeking. *Biological Psychiatry, 67*(1), 81–84. https://doi.org/10.1016/j.biopsych.2009.07.018.Ceftriaxone.

Ko, M. C., Butelman, E. R., Traynor, J. R., & Woods, J. H. (1998). Differentiation of kappa opioid agonist-induced antinociception by naltrexone apparent pA2 analysis in rhesus monkeys. *The Journal of Pharmacology and Experimental Therapeutics, 285*(2), 518–526.

Koniusz, S., Andrzejewska, A., Muraca, M., Srivastava, A. K., Janowski, M., & Lukomska, B. (2016). Extracellular vesicles in physiology, pathology, and therapy of the immune and central nervous system, with focus on extracellular vesicles derived from mesenchymal stem cells as therapeutic tools. *Frontiers in Cellular Neuroscience, 10*(109), 1–20. https://doi.org/10.3389/fncel.2016.00109.

Koob, G. F., & Volkow, N. D. (2016). Neurobiology of addiction: A neurocircuitry analysis. *The Lancet Psychiatry, 3*(8), 760–773. https://doi.org/10.1016/S2215-0366(16)00104-8.

Koobatian, M. T., Liang, M. S., Swartz, D. D., & Andreadis, S. T. (2015). Differential effects of culture senescence and mechanical stimulation on the proliferation and leiomyogenic differentiation of MSC from different sources: Implications for engineering vascular grafts. *Tissue Engineering - Part A, 21*(7–8), 1364–1375. https://doi.org/10.1089/ten.tea.2014.0535.

Kourembanas, S. (2015). Exosomes: Vehicles of intercellular signaling, biomarkers, and vectors of cell therapy. *Annual Review of Physiology,* 77, 13–27. https://doi.org/10.1146/annurev-physiol-021014-071641.

Kraft, A. D., & Harry, G. J. (2011). Features of microglia and neuroinflammation relevant to environmental exposure and neurotoxicity. *International Journal of Environmental Research and Public Health, 8*(7), 2980–3018. https://doi.org/10.3390/ijerph8072980.

Kranzler, H. R., & Soyka, M. (2018). Diagnosis and pharmacotherapy of alcohol use disorder a review. *JAMA - Journal of the American Medical Association, 320*(8), 815–824. https://doi.org/10.1001/jama.2018.11406.

Kumar, A., Sharma, A., Bansal, P. D., Bahetra, M., Gill, H. K., & Kumar, R. (2020). A comparative study on the safety and efficacy of naltrexone versus baclofen versus acamprosate in the management of alcohol dependence. *Indian Journal of Psychiatry,* 650–658. https://doi.org/10.4103/psychiatry.IndianJPsychiatry.

Lallemand, F., Ward, R. J., De Witte, P., & Verbanck, P. (2011). Binge drinking +/- chronic nicotine administration alters extracellular glutamate and arginine levels in the nucleus accumbens of adult male and female wistar rats. *Alcohol and Alcoholism, 46*(4), 373–382. https://doi.org/10.1093/alcalc/agr031.

Laroni, A., De Rosbo, N. K., & Uccelli, A. (2015). Mesenchymal stem cells for the treatment of neurological diseases: Immunoregulation beyond neuroprotection. *Immunology Letters, 168*(2), 183–190. https://doi.org/10.1016/j.imlet.2015.08.007.

Lee, Y. K., et al. (2005). Effects of naltrexone on the ethanol-induced changes in the rat central dopaminergic system. *Alcohol and Alcoholism, 40*(4), 297–301. https://doi.org/10.1093/alcalc/agh163.

Lee, H. J., et al. (2016). Long-term effects of bone marrow-derived mesenchymal stem cells in dextran sulfate sodium-induced murine chronic colitis. *Gut and Liver, 10*(3), 412–419. https://doi.org/10.5009/gnl15229.

Lindvall, O., Barker, R. A., Brüstle, O., Isacson, O., & Svendsen, C. N. (2012). Clinical translation of stem cells in neurodegenerative disorders. *Cell Stem Cell, 10*(2), 151–155. https://doi.org/10.1016/j.stem.2012.01.009.

Littleton, J. M. (2007). Acamprosate in alcohol dependence: Implications of a unique mechanism of action. *Journal of Addiction Medicine, 1*(3), 115–125. https://doi.org/10.1097/ADM.0b013e318156c26f.

Locke, M., Windsor, J., & Dunbar, P. R. (2009). Human adipose-derived stem cells: Isolation, characterization and applications in surgery. *ANZ Journal of Surgery, 79*(4), 235–244. https://doi.org/10.1111/j.1445-2197.2009.04852.x.

Machado da Silva, M. C., De Souza Ferreira, L. P., & Giustina, A. D. (2024). Could immunotherapy be a hope for addiction treatment? *Clinics (Sao Paulo, Brazil), 79*, 100347. https://doi.org/10.1016/j.clinsp.2024.100347.

Maguire, G. (2013). Stem cell therapy without the cells. *Communicative and Integrative Biology, 6*(6), https://doi.org/10.4161/cib.26631.

Maisel, N. C., Blodgett, J. C., Wilbourne, P. L., Humphreys, K., & Finney, J. W. (2013). Meta-analysis of naltrexone and acamprosate for treating alcohol use disorders: When are these medications most helpful? Natalya'. *Addiction (Abingdon, England), 108*(2), 275–293. https://doi.org/10.1111/j.1360-0443.2012.04054.x.Meta-analysis.

Mamidi, M. K., et al. (2012). Comparative cellular and molecular analyses of pooled bone marrow multipotent mesenchymal stromal cells during continuous passaging and after successive cryopreservation. *Journal of Cellular Biochemistry, 113*(10), 3153–3164. https://doi.org/10.1002/jcb.24193.

Mandrekar, P., & Szabo, G. (2009). Signalling pathways in alcohol-induced liver inflammation. *Journal of Hepatology, 50*(6), 1258–1266. https://doi.org/10.1016/j.jhep.2009.03.007.

Marshall, S. A., Geil, C. R., & Nixon, K. (2016). Prior binge ethanol exposure potentiates the microglial response in a model of alcohol-induced neurodegeneration. *Brain Sciences, 6*(2), https://doi.org/10.3390/brainsci6020016.

Mathivanan, S., Ji, H., & Simpson, R. J. (2010). Exosomes: Extracellular organelles important in intercellular communication. *Journal of Proteomics, 73*(10), 1907–1920. https://doi.org/10.1016/j.jprot.2010.06.006.

McLellan, A. T. (2017). Substance misuse and substance use disorders: Why do they matter in healthcare?'. *Transactions of the American Clinical and Climatological Association, 128*, 112–130.

Melendez, R. I., Vuthiganon, J., & Kalivas, P. W. (2005). Regulation of extracellular glutamate in the prefrontal cortex: Focus on the cystine glutamate exchanger and group I metabotropic glutamate receptors. *Journal of Pharmacology and Experimental Therapeutics, 314*(1), 139–147. https://doi.org/10.1124/jpet.104.081521.

Mellado, S., et al. (2023). Therapeutic role of mesenchymal stem cell-derived extracellular vesicles in neuroinflammation and cognitive dysfunctions induced by binge-like ethanol treatment in adolescent mice. *CNS Neuroscience and Therapeutics, 29*(12), 4018–4031. https://doi.org/10.1111/cns.14326.

Montesinos, J., Pascual, M., Rodríguez-Arias, M., Miñarro, J., & Guerri, C. (2016). Involvement of TLR4 in the long-term epigenetic changes, rewarding and anxiety effects induced by intermittent ethanol treatment in adolescence. *Brain, Behavior, and Immunity, 53*, 159–171. https://doi.org/10.1016/j.bbi.2015.12.006.

Moretti, P. (2009). Mesenchymal stromal cells derived from human umbilical cord tissues: Primitive cells with potential for clinical and tissue engineering applications. *Advances in Biochemical Engineering/Biotechnology, 123*, 29–54. https://doi.org/10.1007/10_2009_15.

Morley, K. C., et al. (2006). Naltrexone versus acamprosate in the treatment of alcohol dependence: A multi-centre, randomized, double-blind, placebo-controlled trial. *Addiction (Abingdon, England), 101*(10), 1451–1462. https://doi.org/10.1111/j.1360-0443.2006.01555.x.

Morley, K. C., et al. (2023). N acetylcysteine in the treatment of alcohol use disorder: A randomized, double-blind, placebo-controlled trial. *Alcohol and Alcoholism, 58*(5), 553–560. https://doi.org/10.1093/alcalc/agad044.

Nau, R., et al. (1993). Passage of cefotaxime and ceftriaxone into cerebrospinal fluid of patients with uninflamed meninges. *Antimicrobial Agents and Chemotherapy, 37*(7), 1518–1524. https://doi.org/10.1128/AAC.37.7.1518.

Niciu, M. J., Kelmendi, B., & Sanacora, G. (2012). Overview of glutamatergic neurotransmission in the nervous system. *Pharmacology, Biochemistry, and Behavior, 100*(4), 656–664. https://doi.org/10.1016/j.pbb.2011.08.008.Overview.

Niedzwiedz-Massey, V. M., et al. (2023). Effects of chronic and binge ethanol administration on mouse cerebellar and hippocampal neuroinflammation. *The American Journal of Drug and Alcohol Abuse, 49*(3), 345–358. https://doi.org/10.1080/00952990.2022.2128361.

Obernier, J. A., White, A. M., Swartzwelder, H. S., & Crews, F. T. (2002). Cognitive deficits and CNS damage after a 4-day binge ethanol exposure in rats. *Pharmacology, Biochemistry, and Behavior, 72*(3), 521–532. https://doi.org/10.1016/S0091-3057(02)00715-3.

Oses, C., et al. (2017). Preconditioning of adipose tissue-derived mesenchymal stem cells with deferoxamine increases the production of pro-angiogenic, neuroprotective and anti-inflammatory factors: Potential application in the treatment of diabetic neuropathy. *PLoS One, 12*(5), 1–22. https://doi.org/10.1371/journal.pone.0178011.

Otsuru, S., Hofmann, T. J., Olson, T. S., Dominici, M., & Horwitz, E. M. (2013). Improved isolation and expansion of bone marrow mesenchymal stromal cells using a novel marrow filter device. *Cytotherapy, 15*(2), 146–153. https://doi.org/10.1016/j.jcyt.2012.10.012.

Pan, W., & Kastin, A. J. (2002). TNFα transport across the blood-brain barrier is abolished in receptor knockout mice. *Experimental Neurology, 174*(2), 193–200. https://doi.org/10.1006/exnr.2002.7871.

Park, J., et al. (2015). Mitochondrial ROS govern the LPS-induced pro-inflammatory response in microglia cells by regulating MAPK and NF-κB pathways. *Neuroscience Letters, 584*, 191–196. https://doi.org/10.1016/j.neulet.2014.10.016.

Pati, D., Kelly, K., Stennett, B., Frazier, C. J., & Knackstedt, L. A. (2016). Alcohol consumption increases basal extracellular glutamate in the nucleus accumbens core of Sprague–Dawley rats without increasing spontaneous glutamate release. *European Journal of Neuroscience, 44*(2), 1896–1905. https://doi.org/10.1111/ejn.13284.

Pe'er-Nissan, H., et al. (2022). Placenta-derived mesenchymal-like adherent stromal cells as an effective cell therapy for cocaine addiction in a rat model. *Pharmaceutics, 14*(7), https://doi.org/10.3390/pharmaceutics14071311.

Pendleton, C., Li, Q., Chesler, D. A., Yuan, K., Guerrero-Cazares, H., & Quinones-Hinojosa, A. (2013). Mesenchymal stem cells derived from adipose tissue vs bone marrow: In vitro comparison of their tropism towards gliomas. *PLoS One, 8*(3), https://doi.org/10.1371/journal.pone.0058198.

Phinney, D. G., Kopen, G., Righter, W., Webster, S., Tremain, N., & Prockop, D. J. (1999). Donor variation in the growth properties and osteogenic potential of human marrow stromal cells. *Journal of Cellular Biochemistry, 75*(3), 424–436.

Plosker, G. L. (2015). Acamprosate: A review of its use in alcohol dependence. *Drugs, 75*(11), 1255–1268. https://doi.org/10.1007/s40265-015-0423-9.

Prado, E., et al. (2015). 'N-acetylcysteine for therapy-resistant tobacco use disorder: A pilot study. *Redox Report, 20*(5), 215–222. https://doi.org/10.1179/1351000215Y.0000000004.

Prockop, D. J., & Oh, J. Y. (2012). Mesenchymal stem/stromal cells (MSCs): Role as guardians of inflammation. *Molecular Therapy, 20*(1), 14–20. https://doi.org/10.1038/mt.2011.211.

Puissant, B., et al. (2005). Immunomodulatory effect of human adipose tissue-derived adult stem cells: Comparison with bone marrow mesenchymal stem cells. *British Journal of Haematology, 129*(1), 118–129. https://doi.org/10.1111/j.1365-2141.2005.05409.x.

Qin, L., et al. (2007). Systemic LPS causes chronic neuroinflammation and progressive neurodegeneration. *Glia, 55*, 453–462. https://doi.org/10.1002/glia.20467.

Qin, L., & Crews, F. T. (2012a). Chronic ethanol increases systemic TLR3 agonist-induced neuroinflammation and neurodegeneration. *Journal of Neuroinflammation, 9*(1), 1–18. https://doi.org/10.1186/1742-2094-9-130.

Qin, L., & Crews, F. T. (2012b). NADPH oxidase and reactive oxygen species contribute to alcohol-induced microglial activation and neurodegeneration. *Journal of Neuroinflammation, 9*, 1–19. https://doi.org/10.1186/1742-2094-9-5.

Quintanilla, M. E., et al. (2019). Intranasal mesenchymal stem cell secretome administration markedly inhibits alcohol and nicotine self-administration and blocks relapse-intake: Mechanism and translational options. *Stem Cell Research and Therapy, 10*(1), 1–16. https://doi.org/10.1186/s13287-019-1304-z.

Quintanilla, M. E., et al. (2020). N-Acetylcysteine and Acetylsalicylic acid inhibit alcohol consumption by different mechanisms: Combined protection. *Frontiers in Behavioral Neuroscience, 14*(122), 1–15. https://doi.org/10.3389/fnbeh.2020.00122.

Quintanilla, M. E., et al. (2022). Effect of human mesenchymal stem cell secretome administration on morphine self-administration and relapse in two animal models of opioid dependence. *Translational Psychiatry, 12*(1), 1–14. https://doi.org/10.1038/s41398-022-02225-0.

Quintanilla, M. E., Morales, P., Ezquer, F., Ezquer, M., Herrera-Marschitz, M., & Israel, Y. (2018). Commonality of ethanol and nicotine reinforcement and relapse in wistar-derived UChB rats: Inhibition by N-acetylcysteine. *Alcoholism: Clinical and Experimental Research, 42*(10), 1988–1999. https://doi.org/10.1111/acer.13842.

Ra, J. C., et al. (2011). Safety of intravenous infusion of human adipose tissue-derived mesenchymal stem cells in animals and humans. *Stem Cells and Development, 20*(8), 1297–1308. https://doi.org/10.1089/scd.2010.0466.

Ramalingam, M., & Kim, S. J. (2012). Reactive oxygen/nitrogen species and their functional correlations in neurodegenerative diseases. *Journal of Neural Transmission, 119*(8), 891–910. https://doi.org/10.1007/s00702-011-0758-7.

Rao, P. S., Saternos, H., Goodwani, S., & Sari, Y. (2015). Effects of ceftriaxone on GLT1 isoforms, xCT and associated signaling pathways in P rats exposed to ethanol. *Psychopharmacology, 132*(13), 2333–2342. https://doi.org/10.1007/s00213-015-3868-3.Effects.

Ray, L. A., Chin, P. F., & Miotto, K. (2010). Naltrexone for the treatment of alcoholism: Clinical findings, mechanisms of action, and pharmacogenetics. *CNS & Neurological Disorders - Drug Targets, 9*(1), 13–22. https://doi.org/10.2174/187152710790966704.

Ray, L. A., Green, R., Roche, D. J. O., Magill, M., & Bujarski, S. (2019). Naltrexone effects on subjective responses to alcohol in the human laboratory: A systematic review and meta-analysis. *Addiction Biology, 24*(6), 1138–1152. https://doi.org/10.1111/adb.12747.

Reissner, K. J., Gipson, C. D., Tran, P. K., Knackstedt, L. A., Scofield, M. D., & Kalivas, P. W. (2015). Glutamate transporter GLT-1 mediates N-acetylcysteine inhibition of cocaine reinstatement. *Addiction Biology, 20*(2), 316–323. https://doi.org/10.1111/adb.12127.

Rolan, P., Hutchinson, M., & Johnson, K. (2009). Ibudilast: A review of its pharmacology, efficacy and safety in respiratory and neurological disease. *Expert Opinion on Pharmacotherapy, 10*(17), 2897–2904. https://doi.org/10.1517/14656560903426189.

Rothstein, J. D., et al. (2005). β-Lactam antibiotics offer neuroprotection by increasing glutamate transporter expression. *Nature, 433*(7021), 73–77. https://doi.org/10.1038/nature03180.

Ryan, J. M., Barry, F. P., Murphy, J. M., & Mahon, B. P. (2005). Mesenchymal stem cells avoid allogeneic rejection. *Journal of Inflammation, 2*, 1–11. https://doi.org/10.1186/1476-9255-2-8.

Salehinejad, P., Moshrefi, M., & Eslaminejad, T. (2020). An overview on mesenchymal stem cells derived from extraembryonic tissues: Supplement sources and isolation methods. *Stem Cells and Cloning: Advances and Applications, 13*, 57–65. https://doi.org/10.2147/SCCAA.S248519.

Sari, Y., Smith, K. D., Ali, P. K., & Rebec, G. V. (2009). Upregulation of Glt1 attenuates cue-induced reinstatement of cocaine-seeking behavior in rats. *Journal of Neuroscience, 29*(29), 9239–9243. https://doi.org/10.1523/JNEUROSCI.1746-09.2009.

Sarsenova, M., Kim, Y., Raziyeva, K., Kazybay, B., Ogay, V., & Saparov, A. (2022). Recent advances to enhance the immunomodulatory potential of mesenchymal stem cells. *Frontiers in Immunology, 13*(1010399), 1–22. https://doi.org/10.3389/fimmu.2022.1010399.

Schäfer, R., Spohn, G., & Baer, P. C. (2016). Mesenchymal stem/stromal cells in regenerative medicine: Can preconditioning strategies improve therapeutic efficacy? *Transfusion Medicine and Hemotherapy, 43*(4), 256–267. https://doi.org/10.1159/000447458.

Schneider, R., et al. (2017). N-acetylcysteine prevents alcohol related neuroinflammation in rats. *Neurochemical Research, 42*(8), 2135–2141. https://doi.org/10.1007/s11064-017-2218-8.

Schreiber, R., & Freund, W. D. (2000). Glutamate transport is downregulated in the cerebral cortex of alcohol-preferring rats. *Medical Science Monitor, 6*(4), 649–652.

Schulte, M. H. J., et al. (2018). Reprint of the effect of N-acetylcysteine and working memory training on cocaine use, craving and inhibition in regular cocaine users: Correspondence of lab assessments and Ecological Momentary Assessment. *Addictive Behaviors, 83*(July 2017), 79–86. https://doi.org/10.1016/j.addbeh.2018.03.023.

Segovia-Rodríguez, L., et al. (2022). Gut microbiota and voluntary alcohol consumption. *Translational Psychiatry, 12*(1), https://doi.org/10.1038/s41398-022-01920-2.

Skinner, M. D., Lahmek, P., Pham, H., & Aubin, H. J. (2014). Disulfiram efficacy in the treatment of alcohol dependence: A meta-analysis. *PLoS One, 9*(2), https://doi.org/10.1371/journal.pone.0087366.

Smaga, I., Fierro, D., Mesa, J., Filip, M., & Knackstedt, L. A. (2020). Molecular changes evoked by the beta-lactam antibiotic ceftriaxone across rodent models of substance use disorder and neurological disease. *Neuroscience and Biobehavioral Reviews, 115*, 116–130. https://doi.org/10.1007/978-981-16-1037-0_41.

Stavely, R., & Nurgali, K. (2020). The emerging antioxidant paradigm of mesenchymal stem cell therapy. *Stem Cells Translational Medicine, 9*(9), 985–1006. https://doi.org/10.1002/sctm.19-0446.

Stewart, K., et al. (1999). Further characterization of cells expressing STRO-1 in cultures of adult human bone marrow stromal cells. *Journal of Bone and Mineral Research, 14*(8), 1345–1356. https://doi.org/10.1359/jbmr.1999.14.8.1345.

Stokłosa, I., Więckiewicz, G., Stokłosa, M., Piegza, M., Pudlo, R., & Gorczyca, P. (2023). Medications for the treatment of alcohol dependence—Current state of knowledge and future perspectives from a public health perspective. *International Journal of Environmental Research and Public Health, 20*(3), https://doi.org/10.3390/ijerph20031870.

Szabo, G. (2015). Gut-liver axis in alcoholic liver disease. *Gastroenterology, 148*(1), 30–36. https://doi.org/10.1053/j.gastro.2014.10.042.
Tanaka, K., et al. (1997). Epilepsy and exacerbation of brain injury in mice lacking the glutamate transporter GLT-1. *Science (New York, N. Y.), 276*, 1699–1702. https://doi.org/10.1126/science.276.5319.1699.
Tatsumi, K., et al. (2013). Tissue factor triggers procoagulation in transplanted mesenchymal stem cells leading to thromboembolism. *Biochemical and Biophysical Research Communications, 431*(2), 203–209. https://doi.org/10.1016/j.bbrc.2012.12.134.
Trotti, D., Danbolt, N. C., & Volterra, A. (1998). Glutamate transporters are oxidant-vulnerable: A molecular link between oxidative and excitotoxic neurodegeneration? *Trends in Pharmacological Sciences, 19*(8), 328–334. https://doi.org/10.1016/S0165-6147(98)01230-9.
Ullah, I., Subbarao, R. B., & Rho, G. J. (2015). Human mesenchymal stem cells - Current trends and future prospective. *Bioscience Reports, 35*(2), e00191. https://doi.org/10.1042/BSR20150025.
Valle-Prieto, A., & Conget, P. (2010). Human mesenchymal stem cells efficiently manage oxidative stress. *Stem Cells and Development, 19*(12), 1885–1893. https://doi.org/10.1016/s0141-0229(97)83511-9.
Vetreno, R. P., & Crews, F. T. (2012). Adolescent binge drinking increases expression of the danger signal receptor agonist HMGB1 and toll-like receptors in the adult prefrontal cortex. *Neuroscience, 226*, 475–488. https://doi.org/10.1016/j.neuroscience.2012.08.046.
Vishnubalaji, R., Al-Nbaheen, M., Kadalmani, B., Aldahmash, A., & Ramesh, T. (2012). Comparative investigation of the differentiation capability of bone-marrow- and adipose-derived mesenchymal stem cells by qualitative and quantitative analysis. *Cell and Tissue Research, 347*(2), 419–427. https://doi.org/10.1007/s00441-011-1306-3.
Vizoso, F. J., Eiro, N., Cid, S., Schneider, J., & Perez-Fernandez, R. (2017). Mesenchymal stem cell secretome: Toward cell-free therapeutic strategies in regenerative medicine. *International Journal of Molecular Sciences, 18*(9), https://doi.org/10.3390/ijms18091852.
Volkow, N. D., Michaelides, M., & Baler, R. (2019). The neuroscience of drug reward and addiction. *Physiological Reviews, 99*(4), 2115–2140. https://doi.org/10.1152/physrev.00014.2018.
Volpicelli, J. R., Clay, K. L., Watson, N. T., & Volpicelli, L. A. (1994). Naltrexone and the treatment of alcohol dependence. *Alcohol Health & Research World, 18*(4), 272–278.
Vyas, T. K., Shahiwala, A., Marathe, S., & Misra, A. (2005). Intranasal drug delivery for brain targeting. *Current Drug Delivery, 2*(2), 165–175. https://doi.org/10.2174/1567201053586047.
Wang, Z. G., He, Z. Y., Liang, S., Yang, Q., Cheng, P., & Chen, A. M. (2020). Comprehensive proteomic analysis of exosomes derived from human bone marrow, adipose tissue, and umbilical cord mesenchymal stem cells. *Stem Cell Research and Therapy, 11*(1), 1–11. https://doi.org/10.1186/s13287-020-02032-8.
Wardle, M. C., Bershad, A. K., & De Wit, H. (2016). Naltrexone alters the processing of social and emotional stimuli in healthy adults. *Social Neuroscience, 11*(6), 579–591. https://doi.org/10.1080/17470919.2015.1136355.Naltrexone.
Wei, X., Yang, X., Han, Z. P., Qu, F. F., Shao, L., & Shi, Y. F. (2013). Mesenchymal stem cells: A new trend for cell therapy. *Acta Pharmacologica Sinica, 34*(6), 747–754. https://doi.org/10.1038/aps.2013.50.
Whiteford, H., Ferrari, A., & Degenhardt, L. (2016). Global burden of disease studies: Implications for mental and substance use disorders. *Health Affairs, 35*(6), 1114–1120. https://doi.org/10.1377/hlthaff.2016.0082.
Whitman, B. A., Knapp, D. J., Werner, D. F., Crews, F. T., & Breese, G. R. (2013). The cytokine mRNA increase induced by withdrawal from chronic ethanol in the sterile environment of brain is mediated by CRF and HMGB1 release. *Alcoholism: Clinical and Experimental Research, 37*(12), 2086–2097. https://doi.org/10.1111/acer.12189.

Xia, C., Dai, Z., Jin, Y., & Chen, P. (2021). Emerging antioxidant paradigm of mesenchymal stem cell-derived exosome therapy. *Frontiers in Endocrinology, 12*(November), 1–13. https://doi.org/10.3389/fendo.2021.727272.

Yagi, H., et al. (2010). Reactive bone marrow stromal cells attenuate systemic inflammation via sTNFR1. *Molecular Therapy, 18*(10), 1857–1864. https://doi.org/10.1038/mt.2010.155.

Yan, K., et al. (2013). Bone marrow-derived mesenchymal stem cells maintain the resting phenotype of microglia and inhibit microglial activation. *PLoS One, 8*(12), 6–13. https://doi.org/10.1371/journal.pone.0084116.

Zalewska-Kaszubska, J., Gorska, D., Dyr, W., & Czarnecka, E. (2006). Effect of acute administration of ethanol on beta-endorphin plasma level in ethanol preferring and non-preferring rats chronically treated with naltrexone. *Pharmacology, Biochemistry, and Behavior, 85*(1), 155–159. https://doi.org/10.1016/j.pbb.2006.07.028.

Zhang, X., et al. (2006). Runx2 overexpression enhances osteoblastic differentiation and mineralization in adipose - Derived stem cells in vitro and in vivo. *Calcified Tissue International, 79*(3), 169–178. https://doi.org/10.1007/s00223-006-0083-6.

Zhang, X., et al. (2011). Isolation and characterization of mesenchymal stem cells from human umbilical cord blood: Reevaluation of critical factors for successful isolation and high ability to proliferate and differentiate to chondrocytes as compared to mesenchymal stem cells from bone marrow and adipose tissue. *Journal of Cellular Biochemistry, 112*(4), 1206–1218. https://doi.org/10.1002/jcb.23042.

Zhang, R., et al. (2013). Anti-inflammatory and immunomodulatory mechanisms of mesenchymal stem cell transplantation in experimental traumatic brain injury. *Journal of Neuroinflammation, 10*, 106. https://doi.org/10.1186/1742-2094-10-106.

Zhang, Y., Zhang, X., & Qu, S. (2015). Ceftriaxone protects astrocytes from MPP+ via suppression of NF-κB/JNK/c-Jun signaling. *Molecular Neurobiology, 52*(1), 78–92. https://doi.org/10.1007/s12035-014-8845-z.

Zhao, W., et al. (2004). Activated microglia initiate motor neuron injury by a nitric oxide and glutamate-mediated mechanism. *Journal of Neuropathology and Experimental Neurology, 63*(9), 964–977. https://doi.org/10.1093/jnen/63.9.964.

Zhong, Z., et al. (2020). Bone marrow mesenchymal stem cells upregulate PI3K/AKT pathway and down-regulate NF-κB pathway by secreting glial cell-derived neurotrophic factors to regulate microglial polarization and alleviate deafferentation pain in rats. *Neurobiology of Disease, 143*(104945), https://doi.org/10.1016/j.nbd.2020.104945.

Zindel, L. R., & Kranzler, H. R. (2014). Pharmacotherapy of alcohol use disorders: Seventy-five years of progress. *Journal of Studies on Alcohol and Drugs, Supplement, 13736*(s17), 79–88. https://doi.org/10.15288/jsads.2014.s17.79.

CHAPTER SEVEN

Sodium oxybate: A comprehensive review of efficacy and safety in the treatment of alcohol withdrawal syndrome and alcohol dependence

Julien Guiraud[a,b,*] **and Wim van den Brink**[a]
[a]Department of Psychiatry, Amsterdam Neuroscience, Amsterdam UMC, Academic Medical Center, University of Amsterdam, Amsterdam, The Netherlands
[b]Vergio, Clichy, France
*Corresponding author. e-mail address: jguiraud@vergioconsulting.com

Contents

International Review of Neurobiology, Volume 178
ISSN 0074-7742, https://doi.org/10.1016/bs.irn.2024.07.005

Abstract

Alcohol dependence (AD) significantly impacts public health, affecting 3.4% of people aged 18–64 and contributing to around 12% of overall mortality. Individuals with AD have a markedly reduced life expectancy, dying up to 28 years earlier than the general population. Current treatments for AD show limited efficacy, with many patients not responding to these interventions, highlighting the need for new therapeutic options with novel mechanisms of action.

Sodium oxybate (SMO), the sodium salt of GHB, is one such candidate, pharmacologically similar to alcohol; it acts on several neurotransmitters including GABA, potentially mitigating withdrawal symptoms and craving for alcohol.

SMO has been clinically used in Italy and Austria since the 1990s, approved for treating alcohol withdrawal syndrome (AWS) and for maintaining abstinence in AD patients. Several randomized clinical trials (RCTs) and meta-analyses showed evidence of SMO to be effective and safe in these indications. For AWS, SMO was more effective than placebo and as effective as benzodiazepines in reducing withdrawal symptoms. For maintaining abstinence, SMO significantly improved continuous abstinence duration and abstinence rate compared to placebo. Comprehensive clinical data indicate that SMO is well-tolerated, with main adverse effects being mild, such as dizziness and vertigo, and serious adverse events being rare. The effectiveness and safety of SMO, coupled with its approval in two EU countries affirm its potential as a treatment option for AD, particularly in severe cases.

Further RCTs, especially with stratification by severity of dependence, are suggested to refine our understanding of its efficacy across different patient subgroups.

1. Introduction

1.1 Alcohol dependence and alcohol use disorder

Alcohol dependence (AD) is generally defined as a chronic (relapsing) condition with genetic and environmental factors affecting its development, expression, and persistence (European Medicines Agency, 2010a). According to the ICD-11, AD is characterized by impaired control over drinking, a perceived urge to use alcohol (craving), continued use of alcohol despite adverse consequences, distortions in thinking, most notably denial, a higher priority given to alcohol use than to other activities and obligations, and finally alcohol withdrawal syndrome (AWS) and increased tolerance (World Health Organization, 2019). In contrast with ICD-11, AD is no longer mentioned in the fifth edition of the Diagnostic and Statistical Manual of Mental Disorders (DSM-5). DSM–5 integrates two DSM–IV disorders, alcohol abuse and AD, into a single disorder called alcohol use disorder (AUD) with mild, moderate, and severe sub-classifications based on the number of AUD-criteria that are met: mild AUD with 2–3 of the 11 AUD-criteria; moderate AUD with 4–5 of the 11 AUD-criteria; and severe AUD with ≥6 of the 11 AUD-criteria (National Institute on Alcohol Abuse and Alcoholism, 2021). Studies have shown that DSM-5 mild AUD and moderate-severe AUD are largely comparable to DSM-IV alcohol abuse and alcohol dependence, respectively (and particularly so in clinical samples) (Goldstein et al., 2015). The European Medicines Agency (EMA) currently considers that alcohol abuse is not an appropriate target and that the developments of new medications should concentrate on the treatment of AD (European Medicines Agency, 2010a). Similarly, the US Food Drug Administration (FDA) recommends to recruit moderate to severe AUD patients for clinical trials (FDA, 2015).

Current research indicates several pathways involved in the development and persistence of AD, including the opioid and dopamine systems, which appear to cause alcohol craving and relapse due to positive reinforcing effects of alcohol consumption, especially in earlier stages of the disease. A second pathway involves several components of the gamma-aminobutyric acid (GABA)/glutamate system, which is involved in the development of alcohol craving and relapse due to a hyperglutamatergic state. A further pathway seems to be a hypodopaminergic state, especially after chronic alcohol intake, which is associated with a state of dysphoria that promotes resumption of alcohol intake (European Medicines Agency, 2010a).

Aetiology of AD or AUD have not yet been fully established and there are most certainly environmental and genetic contributions. To date, the most commonly recognized causes of AD or AUD are: the way the person was raised, his character as well as other factors such as (childhood) traumatic experiences, chronic stress and genetics (Lale, Sklar, Wooldridge, & Sarkin, 2014).

The average time of progression from initial problem drinking to AD has been reported to be in the 6 to 8-year range. In general, AD patients constitute a very heterogeneous population with respect to genetic factors, personality disorders, co-morbidities, severity, cognitive impairment, age of onset, gender, motivation concerning readiness to change, social support for drinking or abstinence, craving and other factors. Especially psychiatric co-morbidity like depression, anxiety, posttraumatic stress disorder (PTSD) and attention deficit/hyperactivity disorder(ADHD) is common in people with alcohol problems and is an essential factor influencing the course of disease and treatment (European Medicines Agency, 2010a; Groenman, Janssen, & Oosterlaan, 2017; Langeland, Draijer, & van den Brink, 2004).

AD affects 4.1 and 3.7% of the general population aged 15+ years in the region of the Americas and Europe, respectively (World Health Organization, 2018). AD is associated with high alcohol consumption, and thus leads to altered metabolic pathways as a consequence of ethanol oxidation in the liver hepatocytes resulting in an increase in fatty acids leading to steatosis (Aithal & Grove, 2015). Ethanol's metabolite acetaldehyde can cause deoxyribonucleic acid (DNA) damage and block DNA synthesis and repair, whilst both ethanol and acetaldehyde can disrupt DNA methylation. Ethanol can also induce inflammation and oxidative stress leading to lipid peroxidation and further DNA damage (Rumgay, Murphy, Ferrari, & Soerjomataram, 2021). Therefore, AD is associated with acute and chronic health consequences, most notably various cancers, liver cirrhosis, pancreatitis, cardiovascular diseases, intentional and unintentional injuries and alcohol withdrawal syndrome (AWS) (Rehm et al., 2010; World Health Organization, 2018). AWS is a clinically significant cluster of symptoms, behaviors and/or physiological features, varying in severity and duration, that occurs upon abrupt cessation or reduction of alcohol use in individuals who have developed AD or have used alcohol for a prolonged period and/or in large amounts (World Health Organization, 2019). AD is also highly disabling (World Health Organization, 2018) and the disabling effects can manifest themselves in absenteeism from work, failure to fulfill one's social roles, and other interpersonal problems and functional problems in daily life (Ormel, 1994; Samokhvalov, Irving, Mohapatra, & Rehm, 2010).

As a consequence, about 12% of the overall mortality in the European Union (EU) is due to AD (Rehm, Shield, Rehm, Gmel, & Frick, 2012) and subjects with alcohol use disorders have an average life expectancy of 47–65 years and die 12–28 years earlier than people in the general population (Schwarzinger, Thiébaut, Baillot, Mallet, & Rehm, 2017; Westman et al., 2015). This reduced but broad range of life expectancy is largely explained by the age of onset of AD and the volume of alcohol consumed.

There is strong evidence that alcohol-related harm is mainly determined by the volume of alcohol consumed (Rehm et al., 2010, 2012). Therefore, the volume of alcohol consumption has been categorized in different Risk Drinking Levels (RDL) by the World Health Organization (WHO) (World Health Organization, 2000). The WHO RDLs were defined based on 16 cohort studies which showed increased mortality/morbidity risk at each of the sex-specific levels of alcohol consumption (Table 1) (World Health Organization, 2000).

The prevalence of AUD with a Very High (VH) RDL in the United States of America (USA) is estimated at 2.4% (Shmulewitz et al., 2021), i.e., 6.2 million adults ages 18 and over. In Europe, the estimated prevalence of VH RDL was 0.74–0.85% based on data from 13 EU countries (Rehm, Guiraud, Poulnais, & Shield, 2018). Subjects with VH RDL are responsible for 53.6% of all liver cirrhosis, 43.8% of all pancreatitis and 41.1% of oral cavity and pharyngeal cancers and have a life expectancy estimated at 47–61 years, which represents a reduction by 21–35 years compared to the general population. Therefore, AD with a VH RDL has been considered a life-threatening and seriously debilitating disease (European Medicines Agency, 2020a; Rehm et al., 2018).

1.2 Available treatments for AD and medical need

The goals of AD treatment include the achievement of sustained abstinence, reduction in frequency and severity of relapse, and improvement in health and psychosocial functioning. AD patients targeting abstinence from alcohol are detoxified and then follow a maintenance of abstinence phase directed at the prevention of relapse in alcohol use.

Table 1 WHO risk drinking levels.

Drinking level category	Average daily consumption of ethanol (g/day)	
	Male	Female
Low risk	>0 to ≤40	>0 to ≤20
Moderate risk	>40 to ≤60	>20 to ≤40
High risk	>60 to ≤100	>40 to ≤60
Very high risk	>100	>60

From World Health Organization, 2000

During the detoxification phase, withdrawal symptoms can appear and begin within 4–12 h after cessation or reduction of alcohol use, peak in intensity during the second day of abstinence, and generally resolve within several days. Patients with severe withdrawal symptoms are usually treated for 5–7 days in an in-patient or out-patient setting often with benzodiazepines to reduce central nervous system irritability, fluids to prevent dehydration, thiamine (vitamin B1) to prevent Wernicke/Korsakov and if necessary further medications such as anticonvulsants and antipsychotic agents to prevent or treat withdrawal seizures and delirium (Bahji, Bach, Danilewitz, Crockford, El-Guebaly, Devoe, Saitz et al., 2022; European Medicines Agency, 2010a). Clomethiazole is also used in the treatment of AWS in some countries, such as Austria (Nimmerrichter, Walter, Gutierrez-Lobos, & Lesch, 2002).

Benzodiazepines use in the treatment of AWS is well-established (Mayo-Smith, 1997). Efficacy of benzodiazepines in the management of alcohol withdrawal, in the prevention of complications such as seizures and delirium, and in ameliorating the severity of alcohol withdrawal has been shown (Bahji, Bach, Danilewitz, Crockford, El-Guebaly, Devoe, Saitz et al., 2022). However, the quality of the evidence was considered to be generally low to moderate. Studies were generally underpowered and did not report all relevant outcomes (Amato, Minozzi, Vecchi, & Davoli, 2010; Bahji, Bach, Danilewitz, Crockford, El-Guebaly, Devoe, Saitz et al., 2022). Moreover, the use of benzodiazepines is associated with multiple side effects, such as risk of excess sedation, memory deficits, and respiratory depression in particular in patients with liver impairment, as is often the case in AD patients (Skala et al., 2014). Finally, the prevalence of benzodiazepine dependence among patients treated for AUD has been reported at 21.6% and benzodiazepines are not effective in the long-term treatment of AD (Guiraud, Spanagel, & van den Brink, 2024; van den Brink et al., 2018). Therefore, there is a need for alternative non-benzodiazepine GABAergic compounds effective in the treatment of AWS and also in the maintenance of abstinence in AD (van den Brink et al., 2018).

In Europe, only about 8–10% of patients with AD are treated, indicating a large treatment gap (Rehm et al., 2012). Treatment of AD is generally delivered under a continuing care model and includes either detoxification followed by relapse prevention or harm reduction aiming at reducing the level of alcohol consumption (van den Brink et al., 2018). For the maintenance of abstinence phase and together with the traditionally offered variety of psychosocial interventions designed to optimize the chances of achieving long-term abstinence from alcohol, several pharmacological agents have been used as an adjunct to psychosocial interventions to optimize long-term abstinence (Mann, Lehert, & Morgan, 2004).

Approved medicines in the maintenance of alcohol abstinence in the United States of America and in Europe include acamprosate, naltrexone and disulfiram. Nalmefene and baclofen have been registered for the reduction of alcohol consumption in the EU and in France, respectively. However, these medicines only show modest efficacy with mixed-results and with many patients not responding to these treatments (European Medicines Agency, 2010a; Litten et al., 2012; van den Brink et al., 2018). In addition, disulfiram and naltrexone are contraindicated in patients with hepatic impairment and acamprosate in patients with renal impairment (van den Brink et al., 2018). Beyond the efficacy of existing interventions, the risk for relapse after initiated abstinence may also depend on multiple variables, such as general personal characteristics (like demographic background, personality factors, degree of dependence, family history), degree of stress and perceived self-efficacy and level of confidence respectively with regard to cue-situations (European Medicines Agency, 2010a).

Given the burden caused by AD and the modest (and inconsistent) efficacy of existing interventions, the WHO considers the development of new medicines for the treatment of alcohol use disorders to be a priority for Europe and the World (Kaplan et al., 2013). Therefore, additional treatment options are needed, preferably with a different mechanism of action than currently available medications (Guiraud, 2023).

One of the most successful approaches for the treatment of patients with substance use disorders is substitution therapy which has shown to be most effective for the treatment of opioid and nicotine dependence. Substitution therapies for AD may at least in part fulfill the medical need in the treatment of AD. Substitution drugs for AD should (i) bind to alcohol-sensitive sites at ionotropic receptor or ion channels, (ii) should substitute for ethanol in a dug discrimination test; i.e. should produce ethanol-like effects, (iii) should reduce or even prevent the emergence of AWS, (iv) should maintain abstinence from alcohol or reduce alcohol consumption, and (v) should have a positive benefit-risk ratio in the treatment of AD (Guiraud et al., 2024). Sodium oxybate (SMO) has been reported to be probably the closest to a substitution medication for AD and may, thus, be a valuable therapeutic option for the treatment of AD (Guiraud et al., 2024).

Therefore, we here review and make a critical analysis of clinical data related to the efficacy and safety of SMO in the treatment of AD. The chapter is structured according to the recommendations of the International Council for Harmonization of Technical Requirements for Pharmaceuticals for Human Use for preparing clinical overviews of medicinal products (ICH, 2016). A literature search was carried out in the database of the "National Centre for Biotechnology Information PubMed" (http://www.ncbi.nlm.nih.gov/pubmed/) up to March 2024. The search criteria included the following search string: "SMO" or "sodium oxybate" or "GHB" AND "pharmacokinetics" AND "pharmacodynamics" AND "mechanism" AND "pharmacology" AND "interactions" AND "alcohol dependence" AND "alcohol withdrawal" AND "efficacy OR safety" AND "adverse". In addition, this search was complemented with an open internet search to identify any relevant data related to SMO or γ-hydroxybutyric acid (GHB). The authors of this chapter also had access to clinical study reports of phase 2 and 3 trials (GATE 1, GATE 2, SMO032, GHB/GP 29–96) related to SMO in the treatment of AD. Moreover, reference has been made to the summary of product characteristics (SmPC) of the medicinal product using SMO as an active substance and registered in Austria for the treatment of AWS and for the maintenance of abstinence (i.e., Alcover® syrup). Since SMO is also registered as an oral solution in Europe for the treatment of narcolepsy with cataplexy, under the tradename Xyrem®, reference has also been made to the SmPC of Xyrem®.

2. History of sodium oxybate

2.1 Sodium oxybate and GHB

SMO was first synthetised by Dr Henri Laborit with the aim of obtaining a precursor of GABA which can cross the blood-brain barrier (Laborit, 1964). It was later discovered that SMO is the sodium salt of GHB (Fig. 1), a short-chain fatty acid which is naturally present in human and animal tissues as a neurotransmitter (Kamal et al., 2016).

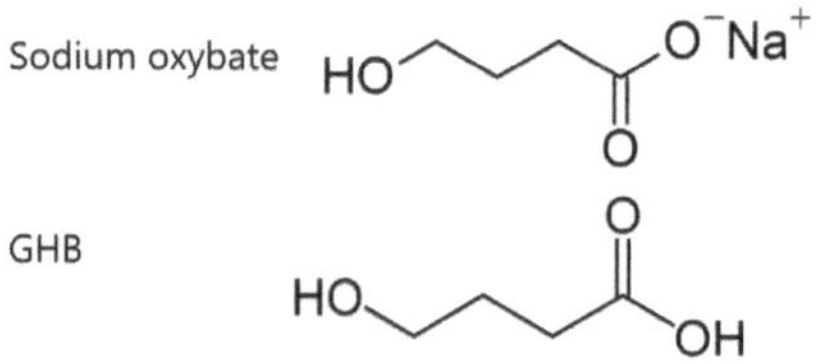

Fig. 1 Chemical structures of SMO and GHB.

At low doses, SMO seems to have alcohol-mimetic euphoric effects and it has been approved as an oral solution or a syrup (Alcover® syrup) at low doses (i.e., 50 mg/kg divided into three administrations per day) for the treatment of AWS and for the maintenance of abstinence in AD patients in Italy and Austria since 1991 and 1999, respectively (Table 2, Fig. 2). At moderate doses (i.e., 2.25–4.5 g), SMO exhibits a sedative effect and it is approved for the treatment of narcolepsy with cataplexy in the USA and in the EU since 2002 and 2005, respectively (Table 2). It has also been approved at 60 mg/kg i.v. as an anesthetic adjuvant in surgery and obstetrics in some European countries (Table 2) for several decades (Guiraud, 2023; Kamal et al., 2016; van den Brink et al., 2018).

GHB was also authorized as a food supplement and was used to induce sleep or increase muscle mass in the 1980's and 1990's but due to safety concerns, including cases of abuse and misuse, GHB has been banned and currently, only pharmaceutical SMO is authorized for medical use (Giorgetti, Busardò, & Giorgetti, 2022; Kamal et al., 2016; Wang, Swick, Carter, Thorpy, & Benowitz, 2009).

SMO is transparent, odorless, and slightly salty when dissolved in water and may become undetectable if not formulated. Therefore, an oral dry granules formulation with abuse/misuse deterrent properties (Hopveus®) was developed. The granules are flavored (apple), partly insoluble, floating,

Table 2 Sodium oxybate approved indications and posology.

Indications	Posology and route of administration	Countries
Treatment of AWS Maintenance of abstinence in AD	50 mg/kg/day p.o. t.i.d.	Italy and Austria
Treatment of narcolepsy with cataplexy	4.5 to 9 g/night p.o. twice a night	USA, Europe, Canada, Australia
Anesthetic adjuvant surgery/ obstetrics Sedation in neurotraumatology	60 mg/kg i.v. 20-50 mg /kg i.v. per hour	France, Germany, Latvia

p.o., per os (orally); *i.v.*, intravenaouly; *t.i.d.*, three times per day.
Data from Carter, Pardi, Gorsline, and Griffiths (2009); Guiraud (2023); UCB (2022).

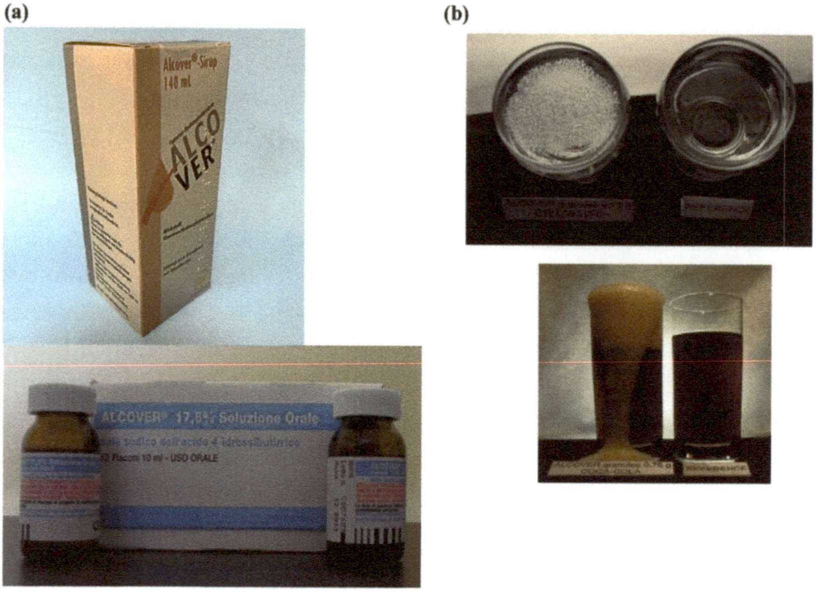

Fig. 2 (A) Sodium oxybate (Alcover® syrup in 140 mL bottle or 10 mL vials) approved in Austria and Italy for the treatment of alcohol use disorder; (B) Hopveus® oral dry granules in sachet formulation with abuse/misuse deterrent properties.

effervescent and filled in sachets (Guiraud, 2023). They are noticeable when put in a drink preventing the risk of criminal misuse (Fig. 2B). This formulation has not yet been registered at the time of this review.

2.2 Exogeneous and endogenous GHB synthesis

Endogenous GHB can be synthetized from different precursors: from GABA, through transamination and reduction in the neuronal compartment by GABA transaminase and succinic semialdehyde reductase, from γ-butyrolactone (GBL), as detected in the rat brain and from 1,4-butanediol (1,4-BD), which exists among lipid-diols in the human brain (Kamal et al., 2016). Exogeneous GHB shares the same precursors (Fig. 3), with SMO being synthetized from GBL (Kamal et al., 2016). Both GBL and 1,4-BD have cosmetic and industrial uses, making them difficult to categorize as controlled substances. Therefore, the organic solvents GBL and 1,4-BD are utilized for recreational use as alternatives for GHB. These precursors are quickly metabolized in-vivo to GHB with an absorption half-life of approximately 1 min. The pharmacological and behavioral effects of GBL and 1,4-BD derive from their swift transformation into GHB (Busardò & Jones, 2015).

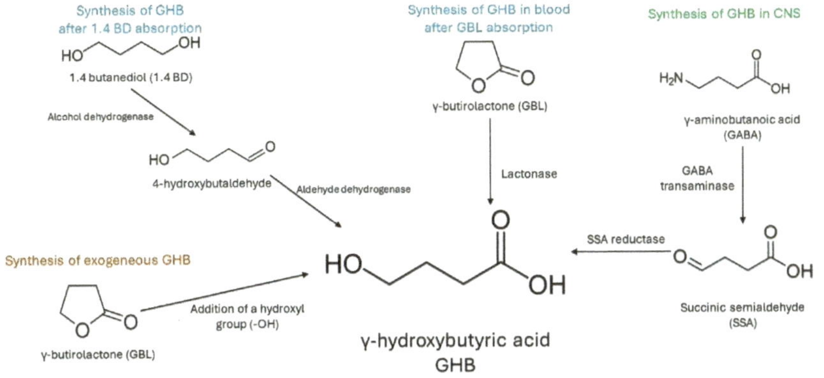

Fig. 3 Synthesis of endogenous and exogenous GHB based on data from Kamal et al. (2016) and Busardò and Jones (2015).

3. Sodium oxybate clinical pharmacology

3.1 Pharmacokinetics

3.1.1 Absorption

Orally administered SMO is quickly absorbed in the upper gastro-intestinal tract, though it has a relatively low bioavailability of 25–40%. This bioavailability significantly depends on whether SMO is taken with or without food (Busardò, Kyriakou, Napoletano, Marinelli, & Zaami, 2015). When administered twice daily at 25 mg/kg for a week to AD patients, SMO achieves peak plasma concentrations within 20–45 min. This regimen does not lead to SMO accumulation or alter its pharmacokinetics (PK) over time. SMO showed non-linear elimination kinetics, and increasing the dose to 50 mg/kg significantly raised the area under the curve (AUC) without much change in maximum concentration (Cmax), indicating capacity-limited absorption (Ferrara et al., 1992).

Studies in healthy males confirm that SMO's absorption and elimination are capacity-limited (Palatini et al., 1993). Additionally, a study on two oral doses of GHB (25 and 35 mg/kg) demonstrated non-linear dose-exposure relationships, with the higher dose resulting in about 40% greater plasma exposure than expected (Liechti et al., 2016). Further, in a study with even higher doses (40–72 mg/kg), mean peak GHB concentrations in plasma increased disproportionately with the dose administered, highlighting a similar non-linear relationship and capacity-limited pharmacokinetics (Abanades et al., 2006). Therefore, dose increases should remain under the guidance of physicians with strict adherence to the dose increments specified in the product information (i.e., 0.75 g/dose for the treatment of narcolepsy) (UCB, 2022).

3.1.2 Distribution

Over the therapeutic concentration range from 12.5 to 50 mg/kg administered to healthy volunteers, SMO was not shown to significantly bind to plasma proteins (Palatini et al., 1993). Due to the negligible binding of the drug to plasma proteins, the distribution of SMO in the body fluids and tissues depends on the distribution of total body water. The volume of distribution (Vd) of SMO in healthy subjects ranged from 0.19 to 0.33 l/kg. In patients suffering from narcolepsy or liver failure the Vd was not different (0.23–0.32 l/kg) from healthy controls. The speed and extent of distribution of SMO from the bloodstream to tissue compartments depends on water content and ratio of blood flow to tissue mass (Busardò et al., 2015). Patient's hydration status and body composition may thus have an effect on SMO efficacy and safety.

3.1.3 Metabolism

SMO undergoes extensive first-pass hepatic metabolism by a series of enzymatic pathways. A major metabolic route is ß-oxidation by GHB dehydrogenase resulting in succinic semialdehyde, which by further oxidation is converted into succinic acid by succinic semialdehyde/aldehyde dehydrogenase in the central nervous system (CNS), entering the intermediary metabolism and being transformed via the Krebs cycle into end-products carbon dioxide and water (Busardò et al., 2015). The non-linear kinetics of SMO in humans is consistent with capacity-limited SMO metabolism by these dehydrogenases (Liechti et al., 2016).

The biosynthesis and metabolism of SMO is also closely linked to the GABA metabolism and SMO can be considered both a metabolite and a precursor of GABA. The amino acid glutamate passes the blood-brain barrier and is decarboxylated to produce GABA. Metabolism of GABA is accomplished by an aminotransferase enzyme to give succinic semialdehyde, which can either become oxidized to give succinic acid or alternatively reduced to form GHB (Fig. 3) (Busardò et al., 2015). Therefore, no involvement of cytochrome P450 (CYP) isoenzymes has been reported for SMO metabolism, which minimizes the risk of drug interactions with active substances metabolized by CYP isoenzymes.

The terminal plasma elimination half-life (t1/2) for SMO in humans ranges from 30 to 50 min (Busardò et al., 2015). Given this short t1/2, SMO is administered three times a day for the treatment of AD (Guiraud, 2023), and twice a night for the treatment of narcolepsy with cataplexy (UCB, 2022). There is no evidence that the metabolic enzymes that clear SMO from the blood are boosted after chronic exposure to substrate (Busardò et al., 2015).

3.1.4 Excretion

SMO is rapidly eliminated, with an elimination half-life (t½) of 30–50 min. It is almost completely converted into carbon dioxide and water through the Krebs cycle. Only about 1% of administered SMO is eliminated as unchanged drug in the urine (Ferrara et al., 1992). Fecal excretion is negligible (UCB, 2022).

GHB and/or its metabolites seem to pass into breast milk in humans (Barker, Puchowicz, Letterio, Higgins, & Sharkey, 2017; Busardò et al., 2016). Therefore, SMO should not be used during breastfeeding and its use is not recommended during pregnancy (UCB, 2022).

3.1.5 Pharmacokinetics in special populations

PK of SMO has been studied in various populations, including AD patients without decompensated liver disease and healthy volunteers.

For AD patients hospitalized for treatment of AWS and AD and who were administered 25 mg/kg SMO every 12 h for 7 days, there was no plasma accumulation of SMO nor any time-dependent changes in its PK. The drug's clearance showed capacity limitations with increased doses, and its metabolism did not involve the hepatic P450 isozyme system, hence no auto-induction or auto-inhibition of metabolism was observed (Ferrara et al., 1992). The PK profile in sober alcoholics was comparable to that of healthy controls, characterized by rapid absorption, early Tmax, and short half-life (Busardò et al., 2015).

Patients with liver cirrhosis exhibited a slower elimination rate when administered a single 25 mg/kg dose of SMO; both the half-life and AUC were doubled compared to controls, though the half-life did not exceed 2 h, making drug accumulation unlikely under a thrice-daily dosing regimen (Ferrara et al., 1996). It is thus recommended to start with a halved dose of SMO in patients with hepatic impairment and to monitor closely when adjusting doses (UCB, 2022).

Since SMO is metabolized in the liver and minimal amounts are excreted via urine, no dose adjustment is necessary for patients with renal impairment, though they should be cautious about their sodium intake (Keating, 2014).

In patients over the age of 65, the PK of SMO did not differ significantly from those in younger patients. Similarly, the major pharmacokinetic parameters of SMO in pediatric subjects align with those observed in adults (UCB, 2022).

An open-label study in healthy volunteers receiving a single 4.5-g dose of SMO found no significant differences in PK between sexes (Borgen, Okerholm, Morrison, & Lai, 2003). This was supported by similar concentration-time profiles observed in healthy men and women in another study (Busardò et al., 2015).

3.1.6 Pharmacokinetic drug interactions

SMO is primarily metabolized in the liver but exhibits no PK interactions with a variety of medications, as shown in several studies.

In vitro studies showed that SMO does not significantly inhibit cytochrome P450 enzymes such as CYP1A2, CYP2C9, CYP2C19, CYP2D6, CYP2E1, or CYP3A, even at concentrations higher than therapeutic levels, indicating minimal risk of metabolic drug interactions (Guerzoni, Pellesi, Pini, & Caputo, 2018). In this respect, no PK interactions were identified between SMO and the following drugs: duloxetine, tramadol, modafinil, omeprazole, ibuprofen, lorazepam, diclofenac, zolpidem, and protriptyline (European Medicines Agency, 2006; UCB, 2022). Due to a different hepatic pathway, SMO does not interfere with anti-retroviral medications or interferons, which are used in treating HIV, HBV, and HCV (Guerzoni et al., 2018).

The combined effects of SMO and ethanol were investigated in a double-blind, placebo-controlled, crossover study where 22 healthy subjects were given 50 mg/kg SMO (Xyrem®) and 0.6 g/kg ethanol in 2 doses. No significant pharmacokinetic interaction was detected (Thai, Dyer, Benowitz, & Haller, 2006). This finding is relevant for situations where SMO treatment might be considered for AD patients who are not yet abstinent. However, while there may be no pharmacokinetic interaction, SMO and alcohol could still potentially exhibit additive pharmacodynamic effects, which are detailed further in Section 3.2.3 below.

When divalproex sodium was administered concurrently with SMO, there was a significant increase in SMO's AUC and renal clearance, suggesting a potential interaction affecting GHB dehydrogenase and monocarboxylate transporter pathways (Eller et al., 2013). Furthermore, a single, supposed PK interaction has been reported in a case of chronic cluster headache, where the concomitant intake of SMO and topiramate resulted in a neurological coma (Guerzoni et al., 2018).

Last, two single-center, randomized, open-label studies in healthy volunteers receiving single oral 4.5 g doses of SMO evaluated effects of food on its oral bioavailability. Food significantly altered the bioavailability of SMO by decreasing mean Cmax, increasing median time-to-peak concentration, and decreasing the AUC. Food did not affect elimination and urinary excretion of unchanged drug (Borgen et al., 2003). Therefore, SMO should be taken at least half an hour before meals or at least 2–3 h after meals (Borgen et al., 2003; D&A Pharma, 2018; UCB, 2022).

3.2 Pharmacodynamics

3.2.1 SMO mechanism of action

The best-established target for GHB is the GABA-B receptor, at which GHB displays low affinity and is a partial agonist. GABA-B receptor activation accounts for some of the reported in vivo effects of SMO, including sedation (Bay, Eghorn, Klein, & Wellendorph, 2014). GHB also acts through high-affinity post-synaptic binding sites popularly referred to as GHB receptors and which may be extra-synaptic GABA-A receptors that contain α4/δ-subunits (Absalom et al., 2012; Bay et al., 2014; Connelly, Errington, & Crunelli, 2013). Activation of post-synaptic GABA-A receptors is responsible for inhibition of GABAergic neurons with an indirect effect on dopamine release. Importantly, these post-synaptic GABA-A receptor subtypes are potentially primary targets for ethanol (Förstera, Castro, Moraga-Cid, & Aguayo, 2016; Olsen & Liang, 2017; Spanagel, 2009). Thus, SMO has a dual action on GHB/GABA-A and GABA-B receptors. At low doses, SMO seems to have alcohol-mimetic euphoric effects due to the predominant effect of SMO on GHB/GABA-A receptors. This also explains why SMO showed efficacy and is approved in the maintenance of abstinence in AD patients at a posology of 50 mg/kg/day (2.5–4.5 g/day) divided into three administrations over the day. At medium to high SMO doses, a sedative effect and a reduced efficacy in the maintenance of abstinence in AD patients (as compared to low doses) have been observed which may be due to a predominant effect of SMO on the GABA-B receptors (Guiraud et al., 2021; Guiraud, 2023; Skala et al., 2014).

Data also suggest that low doses of GHB activate NMDA receptors resulting in an increase of hippocampal glutamate transmission. Conversely, GHB could indirectly reduce glutamate transmission via a bidirectional effect on the pre-synaptic glutamatergic GABA-B receptors, where the heteroreceptors are more sensitive to GHB than the autoreceptors (Kamal et al., 2016). Through an indirect effect on nucleus accumbens neurons with the activation of post- and pre-synaptic GABA-B receptors, GHB is also supposed to decrease the glutamate-mediated synaptic responses and to increase the release of dopamine in the reward mediating nucleus accumbens area (Kamal et al., 2016).

In addition to the action on the GABAergic and glutamatergic systems, data suggest that exogeneous GHB or SMO also induces an increase in serotonin turnover and a reduction in cholinergic neurotransmission in stress conditions contributing to its anxiolytic effect (Kamal et al., 2016). Last, the GHB-induced dopamine release in the striatum is possibly accompanied by the release of endogenous opioids. This indicates a certain opioidergic analgesic and sedative effect of GHB which can be antagonized by naloxone (Kamal et al., 2016). Finally, SMO has shown some evidence of efficacy in the treatment of opioid withdrawal syndrome in heroin dependent patients (Gallimberti et al., 1993).

Thus, SMO appears to be pharmacologically related to alcohol as data suggest that low SMO doses inhibit GABAergic neurons, increases dopamine release, reduces glutamate transmission, decreases serotonin, and is possibly accompanied by the release of endogenous opioids (Guiraud et al., 2024). These mechanisms suggest a potential for abuse of SMO. However, as detailed in Section 6.5.9 below, instances of SMO abuse have been rare both in clinical trials for AD and in clinical practice in Italy and Austria. Effective prevention strategies involve controlled prescription and medication dispensing, alongside vigilant monitoring of any increases in dosage.

3.2.2 Pharmacodynamic effects

In one pilot pharmacology phase I, double-blind, randomized, crossover and controlled study conducted in 8 volunteers receiving increasing single oral SMO doses (40, 50, 60, and 72 mg/kg), SMO produced dose-related changes in subjective effects measured by questionnaires and visual analogue scales. SMO showed a mixed stimulant-sedative pattern, with initially increased scores in subjective feeling of euphoria, high, and liking followed by mild-moderate symptoms of sedation with impairment of performance and balance. SMO-mediated physiological and subjective effects were dose dependent and related to GHB plasma concentrations (Abanades et al., 2006). Similar results were also observed after lower oral doses of SMO (25 and 35 mg/kg) administered to 32 healthy male subjects in a randomized, placebo-controlled, cross-over study, where SMO produced mixed stimulant-sedative effects, with a dose-dependent increase in sedation and dizziness and without altering heart rate or blood pressure (Liechti et al., 2016).

The onset of SMO's effects on performance and behavior has been reported to occur 15–30 min after ingestion and lasts for up to 1–6 h depending on the dose, owing to the fast metabolism (Busardò et al., 2015).

3.2.3 Pharmacodynamic drug interactions

SMO has been studied for interactions with various medications and with alcohol.

No pharmacodynamic interactions were observed between SMO (2.25 g) and tramadol (100 mg), methadone, protriptyline (10 mg) and duloxetine (60 mg) (D&A Pharma, 2018; Guerzoni et al., 2018). However, the concomitant administration of SMO and divalproex sodium resulted in significant cognitive deficits (e.g., in working memory, reaction times) compared to SMO alone, suggesting that divalproex sodium affects SMO-induced cognitive impairments (Eller et al., 2013). Similarly, co-administration of lorazepam and SMO (2.25 g) increased SMO-induced sleepiness (Guerzoni et al., 2018).

The interaction between SMO and alcohol has been investigated in 3 studies. In a double-blind, placebo-controlled, crossover study in 22 healthy subjects, the co-administration of a dose of 50 mg/kg of SMO (e.g., 3.5 g/70 kg) and 0.6 g/kg ethanol (42 g/70 kg, e.g., 3.2 cans of beer of 330 mL) resulted in a decrease by 2% in oxygen saturation compared with the administration of GHB alone and ethanol alone. In addition, more episodes of vomiting and hypotension were observed with the co-administration of SMO and ethanol than with SMO alone and ethanol alone (Thai et al., 2006). Another study evaluating the effects of oral SMO 1.0 and 2.0 g, alone or in combination with 0.5 g/kg of ethyl alcohol (35 g/70 kg, e.g., 2.7 cans of beer of 330 mL) in healthy volunteers reported that the combination of SMO and alcohol impaired reactive skills more than SMO 2.0 g alone, but no potentiation of effect was seen (Mattila, Palva, Seppälä, & Ostrovskaya, 1978). Similarly, in a randomized, double-blind, double-dummy, crossover trial assessing the pharmacodynamic interaction between oral SMO 2.25 g and alcohol 49 g/70 kg (e.g., 3.8 cans of beer of 330 mL) in healthy volunteers, results showed that there was no reciprocal reinforcement between SMO and alcohol. There was a significant interaction between SMO and alcohol at 15 min (i.e., increase in alertness and stimulation and decrease in sedation), which antagonized the subjective sedative effects of alcohol. Also, an isolated mild decrease in digit vigilance accuracy occurred at 165 min post-dose after the combination. No deleterious clinically significant interaction between alcohol and SMO on cognitive function was observed. The low level of interaction between alcohol and (adequate, but relatively low doses of) SMO may account for the good safety profile of SMO even in the case of alcohol relapses in alcohol-dependent patients (Pross, Patat, Vivet, Bidaut, & Fauchoux, 2015).

Despite these reassuring data and given SMO's mechanism of action, there is a need for cautious use of SMO with other CNS-active substances, especially at doses higher than those tested in the above studies, due to the potential for enhanced sedative effects and respiratory depression (D&A Pharma, 2018).

4. Efficacy of SMO in the treatment of AWS

The main objectives of the clinical management of AWS are to decrease the severity of symptoms, prevent more severe clinical manifestations such as seizures and delirium tremens, and facilitate entry of the patient into a treatment program to achieve and maintain long-term abstinence from alcohol (Leggio, Kenna, & Swift, 2008). Objective measurements for monitoring the course and severity of uncomplicated AWS can be achieved with the revised version of the Clinical Institute Withdrawal Assessment for Alcohol (CIWA-Ar) which is considered to be the gold standard measurement of uncomplicated AWS (Amato et al., 2010; Leone, Vigna-Taglianti, Avanzi, Brambilla, & Faggiano, 2010; Liu & Wang, 2017; Minozzi, Amato, Vecchi, & Davoli, 2010; Pribék et al., 2021). Some findings suggest that the CIWA-Ar scores of 0–8, 9–15 and 16 or more indicate mild, moderate, and severe withdrawal syndrome, respectively (Mayo-Smith, 1997; Saitz & O'Malley, 1997).

RCTs in the treatment of moderate to severe AWS include a treatment period of usually five days to three weeks (Amato et al., 2010; Leone et al., 2010). For ethical reasons (potentially life-threatening condition), placebo-controlled studies are no longer used to assess efficacy of investigational medicines in the treatment of AWS and, since effective treatments are available, almost all recent trials are benzodiazepine-controlled. As clomethiazole is approved and effective in the treatment of AWS (Mann, Batra, Fauth-Bühler, & Hoch, 2017), trials for AWS can also be clomethiazole-controlled. To have assay sensitivity, equivalence or non-inferiority active-controlled trials should include patient populations with moderate or severe AWS at baseline as these patients do not improve without effective treatment (Adinoff, 1994; Krupitsky et al., 2007; Lyon, Khan, Gessert, Larson, & Renier, 2011).

The efficacy of SMO in the treatment of acute AWS has been evaluated in seven published studies, including two phase 3 RCTs, with a total of 670 adult patients. A total of five RCTs compared SMO to placebo (Gallimberti et al., 1989), clomethiazole (Nimmerrichter et al., 2002), oxazepam (Caputo et al., 2014) and/or diazepam (Addolorato, Balducci et al., 1999; Nava et al., 2007). The remaining studies were non-controlled and conducted in hospital settings (Korninger, Roller, & Lesch, 2003; Moncini, Masini, Gambassi, & Mannaioni, 2000). RCTs included patients with moderate to severe AWS at baseline without (history of) complications (e.g., without seizures or delirium tremens) (Table 3).

4.1 Pilot RCTs testing the efficacy of SMO in uncomplicated AWS

The effect of SMO on AWS was initially investigated in a randomized, double-blind, placebo-controlled study conducted in Italy in 23 alcohol-dependent patients who met the DSM III-R criteria of AWS. Patients were randomized in two groups: SMO 50 mg/kg/day (oral syrup preparation) ($n = 11$) or placebo ($n = 12$). The primary endpoint was the AWS score, a scale similar to the CIWA-Ar that measures the six main withdrawal symptoms, each scored on a 4-point scale (0–3): tremors, sweating, nausea, depression, anxiety, and restlessness. The score of each symptom was measured at each time point and was added for each patient to determine the AWS score, the maximum being 18 points. The double-blind treatment duration was 7 h. After completion of this treatment phase, patients treated with placebo were assigned to a treatment schedule with pharmacotherapy, as indicated by their clinical state. Patients in the SMO group received further doses of the drug every 8 h up to the 3rd day. Subsequently, the total daily dose was reduced by 30% per day until the 7th day when SMO was discontinued. The mean (SD) AWS score at baseline was 12.6 (6.1) in the SMO group and 11.8 (5.7) in the placebo group, indicating a patient population with moderate AWS severity. In the SMO group, nearly all withdrawal symptoms disappeared within 2–7 h after treatment initiation. By contrast, withdrawal scores of control patients did

Table 3 Study characteristics and populations in studies for the treatment of AWS.

Criteria	GATE 1 Caputo et al. (2014)	GHB/GP 29-96 Nimmerichter et al. (2002)	Gallimberti et al. (1989)	Addolorato et al. (1999)	Nava et al. (2007)	Korninger et al. (2003)	Moncini et al. (2000)
Study design	Phase 3 Double-blind RCT	Phase 3 Double-blind RCT	Pilot Double-blind RCT	Pilot Single-blind RCT	Pilot Open RCT	Observational	Observational
Sample size (*N*)	126	98	23	60	42	299	22
Male/Female (%)	84/16	90/10	70/30	85/15	NR	NR	NR
Mean age (years)	43	41	44	43	NR	51	43
Disease	Moderate to severe uncomplicated AWS	Moderate to severe uncomplicated AWS	Moderate to severe uncomplicated AWS	Moderate to severe uncomplicated AWS	Severe uncomplicated AWS	Mild to severe AWS	Complicated & uncomplicated AWS
Severity at baseline	CIWA-Ar of 18	CIWA-Ar of 16	AWS score of 12-13	CIWA-Ar of 14	CIWA-Ar of 28	NR	AWS score of 16
Concomitant illness	None	None	None	None	None	Liver/hypertension /gastritis/heart	NR
SMO daily dose	70 mg/kg	50 mg/kg 100 mg/kg	50 mg/kg	50 mg/kg	50 mg/kg	50 mg/kg	50–150 mg/kg

Control group (daily dose)	OXA (210 mg)	CLO (1000 mg)	PBO	DZP (0.5-0.75 mg/kg)	DZP (0.5 mg/kg)	NA	NA
Treatment duration (days)	10	6	7	10	21	7	6
Use of rescue drug	No	No	No	No	No	NR	NR
Endpoint measured	CIWA-Ar	CIWA-Ar	AWS score	CIWA-Ar	CIWA-Ar	Improvement of AWS	AWS score
Timing of assessment	Days 10/20	Days 2–8	1–7 h, days 2–7	Days 1–18	Days 7/14/21	NR	Days 1–10

PBO, placebo; *OXA*, oxazepam; *CLO*, clomethiazole; *DZP*, diazepam; *NA*, not applicable; *NR*, not reported; severity at baseline expressed in mean score.

not decrease, and even significantly increased after 7 h. The mean difference in AWS score was significantly ($p < 0.05$) in favor of SMO at each time point (1, 2, 3, 5 and 7 h) of the double-blind treatment phase. The mean AWS score in SMO group remained below 2 until the end of the study. The authors reported that no complications occurred. These results showed that SMO at 50 mg/kg/day was significantly superior to placebo in reducing the withdrawal score (Gallimberti et al., 1989).

Given these positive findings, Addolorato et al. investigated in a randomized, diazepam-controlled, single-blind study the efficacy and safety of SMO (Alcover® syrup) for the treatment of AWS. Participants were 60 adult alcohol dependent patients with mean alcohol consumption 238 g/day at baseline, with a baseline CIWA-Ar score ≥10 and needing pharmacological treatment for AWS. Patients were randomized in two groups: 30 patients received oral diazepam (0.5–0.75 mg/kg body weight for 6 days, tapering the dose 25% daily until day 10) and 30 patients were administered oral SMO (50 mg/kg body weight for 10 days). The 10-day treatment period was followed by an 8-day untreated follow-up period. The CIWA-Ar was used to evaluate the AWS physical symptoms at days 1, 2, 3, 4, 5, 10, and 18. The State Anxiety Inventory test for current anxiety assessment and the Zung self-rating Depression Scale for current depression assessment were also performed. Investigators who measured the outcomes were blind to treatment assignment. Analyses were conducted on treatment completers (i.e., drop-outs excluded from the analysis). There was no clinically relevant difference in baseline characteristics between treatment groups. Mean (SD) baseline CIWA-Ar score was 14.4 (4.4) in the diazepam group and 13.7 (2.6) in the SMO arm. Drop-outs were higher in the diazepam group than in the SMO group: 8 versus 4 patients. Both treatments were effective in reducing AWS. Mean score (SD) at Day 10 was 3.2 (2.5) in diazepam group and 2.6 (3.5) in SMO group resulting in mean difference (95% CI) in CIWA-Ar reduction numerically in favor of SMO: −0.1 (−2.3; 1.1). No rebound effects were observed during the follow-up period. No significant difference was found between the groups in CIWA-Ar score at the different times of outcome measurement. Considering the CIWA-Ar subscore and score on the Zung scale, a significant larger reduction of anxiety on day 4 ($p < 0.02$), agitation on day 5 ($p < 0.02$) and time of recovery of depression on day 5 ($p < 0.02$) was observed in the SMO group compared to the diazepam group. The authors concluded that SMO was as effective in the management of AWS as a benzodiazepine and it seemed to be quicker in reducing anxiety, agitation, and depression (Addolorato, Balducci et al., 1999).

Similarly, an open, randomized, diazepam-controlled study investigated the efficacy and safety of SMO (Alcover® syrup) for the suppression of severe AWS (CIWA-AR ≥ 20) and hypercortisolism. A total of 42 alcohol-dependant inpatients were randomized to diazepam (0.5 mg/kg bodyweight, four times a day (q.i.d.), $n = 21$) or SMO (50 mg/kg bodyweight, divided q.i.d., $n = 21$). Treatments were orally administered for three weeks. Starting from the first week of treatment, and only if the CIWA-Ar score was <5 medications were gradually tapered according to clinical response. Analyses were conducted on the intent-to-treat population. The authors reported no clinically relevant difference in baseline characteristics between treatment groups. All randomized subjects completed the study. Baseline mean CIWA-Ar score was 28 in the ITT population. At each time point, SMO was statistically more effective than diazepam in reducing both the withdrawal syndrome (CIWA-Ar total score and CIWA-Ar mean subscores, i.e., tremor, paroxysmal sweats, anxiety and agitation) and hypercortisolism (reducing cortisol levels at different times of observation). End of treatment mean CIWA-Ar score was 1.2 in the SMO group and 4.7 in the diazepam group. In the SMO group, a significant relationship between a reduction of plasma cortisol levels and a decrease of alcohol withdrawal symptoms was identified ($r = 0.95$, $p = 0.02$). In the diazepam group, a similar but non-significant correlation was observed ($r = 0.82$, $p = 0.09$). No medical complications were observed. The authors concluded that these results confirmed the presence of a correlation between high levels of plasma cortisol and alcohol withdrawal symptoms and they showed a slight superiority of SMO over diazepam in the suppression of both alcohol withdrawal and hypercortisolism (Nava et al., 2007).

4.2 Phase 3 studies testing SMO efficacy in uncomplicated AWS

To confirm these positive findings of SMO efficacy in the pilot studies on the treatment of uncomplicated AWS, two phase 3 studies were conducted.

4.2.1 The GHB/GP 29-96 Phase 3 study

The GHB/GP 29–96 phase 3 study was a randomized, double-blind, double-dummy, clomethiazole-controlled study that tested the equivalence of SMO (50 or 100 mg/kg/day) and clomethiazole in the treatment of uncomplicated AWS in adult AD patients in an Austrian center. Some data of this phase 3 study were reported by Nimmerichter et al. (2002). Here, we also report additional data from the GHB/GP 29–96 clinical study report. The trial was conducted in accordance with the ethical principles of the

Declaration of Helsinki and the Good Clinical Practices (GCP). Methods were pre-specified in the GHB/GP 29–96 study protocol and its amendment. Patients with a CIWA-Ar score ≥9 were randomized using a block design to receive either oral clomethiazole 1000 mg daily ($n = 33$), or oral SMO 50 mg/kg/day ($n = 33$), or oral SMO 100 mg/kg/day ($n = 32$). Each eligible patient was assigned to the lowest free (i.e., not yet assigned) number on the randomization list. The medication was administered for 5 days, halved on day 6, and on days 7 and 8 only placebo was given. SMO syrup and placebo syrup were identical in appearance and taste and the same applied for clomethiazole capsules and placebo capsules. The primary efficacy endpoint was the CIWA-Ar score at Day 6. The equivalence margin was ± 1.5 points of CIWA-Ar. Analyses were conducted on the ITT population. Missing data were imputed with Last Observation Carried Forward (LOCF) approach. The procedures and measurements were documented in the Case Report Forms. Data audits were performed by the Sponsor (GL Pharma) and a complete verification of all efficacy and safety data was carried out. A total of 98 and 65 patients were included in the ITT and in the per protocol population, respectively. The clomethiazole group had a higher drop-out rate than the two SMO groups: six, four, and one patients discontinued the study in the clomethiazole, the SMO 100 mg/kg, and the SMO 50 mg/kg groups, respectively. The clomethiazole group showed a significantly higher drop-out rate ($p = 0.046$) than the SMO 50 mg/kg group. Two patients discontinued the study due to adverse events (epileptic seizures): one in the clomethiazole group and one in the SMO 100 mg/kg group. Four and five patients did not complete the study due to "no decrease or increase in withdrawal syndrome" and other reasons, respectively. A total of 11 patients did not complete the study and were excluded from the per protocol population and an additional 22 patients were excluded from the per protocol population due to protocol deviations leaving 66 (98−33) patients in the per protocol group. There was no clinically relevant difference in baseline characteristics between treatment groups. Mean (SD) CIWA-Ar scores at screening were 16.1 (3.2), 16.2 (5.0) and 16.2 (4.9) in the SMO 50 mg/kg, the SMO 100 mg/kg, and the clomethiazole groups, respectively, indicating a patient population with a moderate AWS severity. All groups achieved a clinically relevant improvement of their withdrawal symptoms on CIWA-Ar (Fig. 4).

There was no statistically significant difference between the three treatment groups at the different measured time points (Fig. 5A). Mean (90%CI) CIWA-Ar scores at day 6 were close to the minimum score of zero (no symptom of

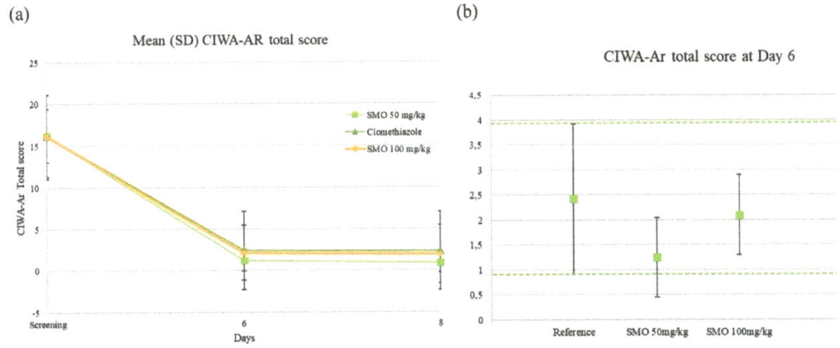

Fig. 4 GHB/GP 29–96 efficacy results in the ITT population: (A) Mean CIWA-Ar score over the study period; (B) equivalence test in CIWA-Ar score at Day 6.

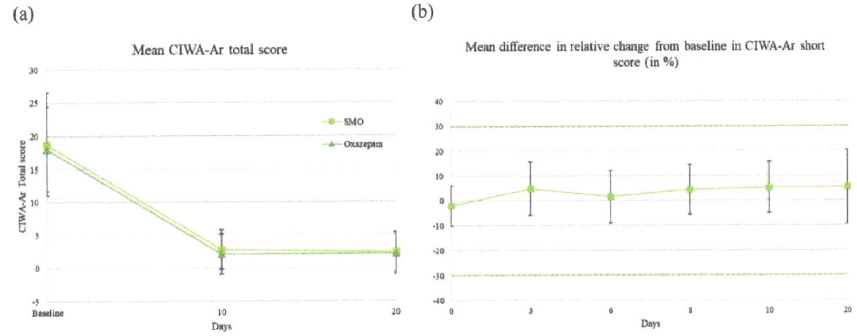

Fig. 5 GATE 1 efficacy results in ITT population: (A) Mean CIWA-Ar score over the study period; (B) Mean difference in CIWA-Ar score.

withdrawal at all) in all groups in the ITT population: 1.2 (0.4, 2.0), 2.1 (1.3, 2.9) and 2.4 (0.9, 3.9) for the SMO 50 mg/kg, the SMO 100 mg/kg and the clomethiazole groups, respectively. Mean (90%CI) CIWA-Ar scores at Day 6 were slightly lower in the per protocol population: 0.9 (0.5, 1.3), 1.6 (1.2, 2.0) and 1.1 (−0.4, 2.6) in the SMO 50 mg/kg, the SMO 100 mg/kg and the clomethiazole groups, respectively. The equivalences of SMO 100 mg/kg and clomethiazole in CIWA-Ar at Day 6 were shown in both the ITT (Fig. 4) and the per protocol populations. In the per protocol population, SMO 50 mg/kg was equivalent to clomethiazole in CIWA-Ar at Day 6 whereas in the ITT, SMO 50 mg/kg was non inferior to clomethiazole (Fig. 4B). The SMO 50 mg/kg group was not equivalent to the clomethiazole group with an estimated mean CIWA-Ar at Day 6 that was numerically lower than the clomethiazole group with a 90%CI which included the lower limit of the equivalence margin. Overall, the authors of this study concluded that SMO was as effective as clomethiazole in terms of CIWA-Ar score reduction (Nimmerrichter et al., 2002).

4.2.2 The GATE 1 Phase 3 study

Further to this first positive phase 3, an international, randomized, double-blind, double-dummy, oxazepam-controlled GATE 1 phase 3 equivalence study evaluated the efficacy of SMO (Alcover® syrup) in the treatment of AWS in four EU countries: Austria, Germany, Italy, and Poland. The aim of this equivalence study was to confirm that SMO was as effective as oxazepam, to treat and manage uncomplicated AWS. The trial was conducted in accordance with the ethical principles of the Declaration of Helsinki and the GCP. Methods were pre-specified in the GATE 1 study protocol. A total of 126 alcohol-dependent patients affected by AWS (CIWA-Ar score > 10) were randomized in two groups: 61 patients received oral SMO (1.75 g t.i.d. during days 1–5 for patients with body weight under 75 kg and 2.1 g t.i.d. during days 1–5 for the others; dose reduction by 50% during days 6 and 7 and by 75% during days 8 to 10), and 65 patients received oral oxazepam (210 mg during days 1–5, 30 mg t.i.d. on days 6–7, and then 15 mg t.i.d. until the end of the treatment phase). Randomization was stratified by site and the random numbers were computer-generated using a pseudo random uniform distribution with a block size of 4 patients. Each eligible subject received the lowest unassigned randomization number available in the site. Sponsor, investigators, and patients were blind to treatment assignment during the full study period. Active and dummy medications were identical in appearance and taste. The oxazepam daily dosage administered in GATE 1 was above the dosages applied in prior studies with oxazepam (Amato et al., 2010), preventing the risk of under dosage. The 10-day treatment period was followed by a 10-day untreated follow-up period. The primary efficacy variable was the relative change from baseline (rCfB) to Day 10 in CIWA-Ar-short score, the secondary endpoints included the CIWA-Ar-short score and the CIWA-AR total score at Day 10 and at Day 20. The CIWA-Ar short score was used in this study as one reported disadvantage to the CIWA-Ar total score is that it may be difficult to administer and interpret the score in severely agitated patients as 7 of the 10 items require reliance on the patients' response (Perry, 2014; Wetterling et al., 1997). Therefore, the CIWA-Ar-short score was developed including the three core AWS symptoms which can be evaluated by physicians without patient self-assessment. The CIWA-Ar-short score is the sum of paroxysmal sweats, tremor and anxiety scores (Nimmerrichter et al., 2002). The equivalence margin in rCfB to Day 10 in CIWA-Ar-short score was [−30%; +30%] and was defined as the smallest effect that would be important to detect based on past performance of benzodiazepines and a clinically relevant threshold for a successful detoxification. The 95% CI of the mean difference in the rCfB to Day 10 in CIWA-Ar-short score between SMO and oxazepam needed to be within the predefined equivalence margin [−30%; +30%]. At planning stage, it was assumed that there would be a rCfB in CIWA-Ar-short score of −50% with a related SD of 50–70% leading to a planned sample size of 104 patients per group. In case the equivalence was shown with the above assumption, the related standardized effect size (Cohen's d) would have been <0.20, also supporting a small difference. Analyses were conducted on the modified ITT population (N= 126), which included all recruited subjects to whom at least one dose of the medications had been administered. Missing data were imputed with LOCF (Caputo et al., 2014; European Medicines Agency, 2020a). After 7 years of enrollment, only 126 patients were randomized, therefore the study was stopped prematurely. The power was not importantly impacted because the anticipated standard deviation was more than twice the actual standard deviation (European Medicines Agency, 2020a). The GATE 1 study was positive in the planned analysis of the primary endpoint. Results showed a rCfB to Day 10 in CIWA-Ar-short score of −79.70% (SD 33.06) for the SMO arm and of −84.89% (SD 25.80) for the oxazepam arm resulting in a treatment difference of 5.2 points with a 95% CI ranging from −5.2 to 15.6. These results indicated that the 95% CI on the primary endpoint is within the predefined equivalence margin [−30%; +30%], indicating that equivalence was observed (Fig. 5). This treatment difference corresponds to a Cohen's d of 0.17, indicating a very small effect size. Given the mean baseline CIWA-Ar short score of approximately 11, the upper limit of the 95% confidence interval for the mean difference represents 1.68 points of CIWA-Ar-short score. Considering the CIWA-Ar severity scale for each symptom, such a difference is not clinically relevant. Specifically, a score of 0 indicates no symptoms, 1 indicates a barely perceptible symptom, and scores of 2 and 3 are not defined by specific severity but are more severe than lowers score. A score of 4 indicates a moderate symptom, and 7 indicates a severe symptom (Nimmerrichter et al., 2002). Overall, these results indicated a small difference in the primary endpoint between the two treatment groups. The equivalence was also observed at all other assessed time points (Fig. 5). Both results in the per protocol analysis and the estimand analyses showed that the 95% CI of the mean difference in rCfB in CIWA-AR short score falls within the predefined equivalence margin [−30%; +30%] and the estimate was in the same magnitude as the estimate in the ITT population (European Medicines Agency, 2020a).

A significant decrease of the mean CIWA-Ar total score from the baseline to the end of treatment was found both in the SMO group (adjusted mean (SE) change from baseline −15.62 (0.38), $p < 0.0001$) and in the oxazepam group (adjusted mean (SE) change from baseline −16.27 (0.38), $p < 0.0001$), with no significant differences between the two treatment groups: adjusted mean difference (95% CI) 0.65 (−0.37; 1.66), $p = 0.210$. The mean CIWA-Ar score remained stable at the end of the follow-up period in both the SMO group and in the oxazepam group with no significant differences between the two treatments: adjusted mean difference (95% CI) 0.21 (−0.84; 1.26), $p = 0.692$ (Fig. 5). The authors of the study concluded that SMO was as effective as oxazepam in the treatment of uncomplicated AWS (Caputo et al., 2014).

4.3 Observational studies investigating SMO effectiveness in AWS

The effectiveness of SMO in the management of AWS was assessed in a hospital setting in patients who were admitted to hospital for reasons primarily unrelated to their AD. A total of 299 patients were treated with SMO when AWS symptoms occurred. Fifty percent of included patients were admitted to hospital for internal diseases (hypertension, gastritis, coronary heart disease, liver disease, COPD), 22% for neurological/psychiatric diseases (depression, polyneuropathy), 20% for trauma, and 4% for surgery. Oral SMO was usually given at a daily dose of 50 mg/kg in 3 divided doses, the clinical course of the patients was followed for 7 days or until discharge from hospital. At the start of SMO treatment, tremor was present in 81% of patients, sweating in 76% and unrest in 92%. Symptoms occurred 1–72 h after admission. The effect of SMO to ameliorate or suppress the symptoms of the AWS was judged to be excellent in 57%, good in 34%, fair in 18%, insufficient in 3% of patients. The authors concluded that SMO was an attractive alternative to tranquilizers in the management of the AWS in a general hospital setting (Korninger et al., 2003).

Moncini and colleagues investigated the effectiveness of oral SMO (50–150 mg/kg/day Alcover® syrup for 6 days) in the treatment of 22 AD patients with AWS with or without seizures/delirium tremens. The effect was measured with the alcohol withdrawal scale which is similar to CIWA-Ar but with the minimum and the maximum being 0 and 38 points, respectively. Mean (SE) alcohol withdrawal score at baseline was 16.3 (2.6) with symptoms of hallucinations in 8 patients and seizures in 3 patients. During the treatment period with SMO, there was a rapid decrease in the mean alcohol withdrawal score in all patients, with a significant effect after the first administration of the drug. Nearly all withdrawal signs and symptoms disappeared after 4 days (mean (SE) score, 0.96 (0.5)), and total remission of the alcohol withdrawal syndrome was observed after 10 days. The authors reported that in (indirect) comparison with benzodiazepines, treatment with SMO reduced both the mortality rate and the duration of stay in intensive therapy with a length of stay in intensive care decreasing by almost 50% (Moncini et al., 2000).

4.4 Meta-analyses of SMO for the treatment of AWS

In a meta-analysis performed in 2010 by the Cochrane Collaboration, SMO was found to be significantly superior to placebo and to clomethiazole in reducing AWS. No significant difference was found between SMO and diazepam in the CIWA-Ar total score. When looking at CIWA-Ar sub-scores, SMO showed a significantly lower sub-score than diazepam for tremor, paroxysmal sweet, anxiety and agitation (Leone et al., 2010). In a network meta-analysis comparing efficacy and safety of pharmacotherapies for alcohol withdrawal, SMO was much more effective than placebo in reducing the CIWA-Ar score ($d = -1.80$; 95%CI, −3.35 to −0.26) (Bahji, Bach, Danilewitz, Crockford, El-Guebaly, Devoe, Saitz et al., 2022).

4.5 Discussion on SMO efficacy in the treatment of AWS

The pharmacological profile of SMO with its agonist action on GABA-A receptors and its anxiolytic property is the rationale for testing it in the treatment of AWS (Guiraud et al., 2024; Nimmerrichter et al., 2002). The efficacy of SMO in the treatment of AWS is supported by data from 7 trials, including two phases 3 trials, and by two meta-analyses with one meta-analysis conducted by the Cochrane Collaboration. Studies included a total of 670 patients.

The Phase 3 studies were positive in the planned analysis of their primary endpoint and SMO efficacy was found to be equivalent to clomethiazole and oxazepam in the treatment of uncomplicated AWS. These phase 3 findings are corroborated by the results of pilot trials where SMO 50–70 mg/kg/day was more effective than placebo in a double-blind RCT and at least as effective as diazepam in reducing alcohol withdrawal symptoms in two RCTs. These active-controlled studies included patient populations which do not improve without effective treatment. Treatment durations were in accordance with standard clinical practice. Efficacy was assessed with the CIWA-Ar, which allows for monitoring the course and severity of uncomplicated AWS (Amato et al., 2010; Leone et al., 2010; Liu & Wang, 2017; Minozzi et al., 2010; Pribék et al., 2021). No rebound effects during the untreated follow-up periods were observed. Efficacy of SMO in alcohol withdrawal RCTs was further supported by results from observational studies which showed effectiveness of SMO in the treatment of complicated AWS and in patients with comorbidities and by (network) meta-analyses where SMO was superior to placebo in the treatment of uncomplicated AWS.

The most important limitation is that - except for a very small observational study - SMO efficacy in the treatment of complicated AWS has not been investigated. In addition, GATE 1 recruitment was prematurely stopped due to a very long recruitment period. However, the power was not importantly impacted and GATE 1 results were consistent with those from other RCTs testing SMO efficacy in AWS. We, therefore, conclude that SMO can be an alternative to benzodiazepines for the treatment of uncomplicated AWS.

5. Efficacy of SMO in maintenance of abstinence

For pivotal (phase 3) trials directed at maintenance of abstinence in AD patients, EMA recommends to conduct placebo-controlled RCTs with a treatment duration of 3–6 months followed by a medication-free period, for a total study duration of 12 to 15 months (European Medicines Agency, 2010a, 2010b). The recommended primary endpoint is the abstinence rate which is the proportion of patients with continuous abstinence (no relapse to any drinking) over the treatment period. The Cumulative Abstinence Duration (CAD) was the primary endpoint recommended in European guidelines prior to the implementation of the EMA guideline in 2010 (Plinius Maior Society, 1994). It has thus been widely used as primary endpoint in acamprosate and disulfiram studies and was defined as primary endpoint in the Cochrane meta-analysis of acamprosate in the treatment of AD (Rösner et al., 2010). It is calculated as the number of days with no alcohol use for each patient during the treatment period. At treatment group level, CAD measures the differences in abstinence rate as well as the differences in abstinence duration in relapsing patients. It can therefore be conceptualized as a composite endpoint with the current recommended primary endpoint (abstinence rate) as one of its components (Guiraud et al., 2022). CAD is now an important secondary endpoint (European Medicines Agency, 2010a). The Percentage of Days Abstinent (PDA) has also been used in certain studies and is very closely related to CAD as it measures the ratio between the CAD and the planned treatment period duration for each patient.

Development of medications for the maintenance of abstinence is challenging. Dropouts are high (mean 46% of enrolled patients in acamprosate studies – Nice, 2011) and the demonstration of efficacy of treatments approved for this indication is based on a mix of positive and negative studies (European Medicines Agency, 2010a; Litten et al., 2012; Witkiewitz, Litten, & Leggio, 2019). One of the main reasons for these mixed results is the unpredictable variability of the placebo response in RCTs for AD and its negative correlation with the treatment effect size (Litten et al., 2013). It is recognized that studies often fail when the placebo response is high and unpredictable (European Medicines Agency, 2007). There has been recently a convergence of evidence that the placebo response in double-blind RCTs is lower and pharmacological treatment effect sizes are larger in AD patients with a High or Very High RDL (>60 g alcohol/day for men and >40 g alcohol/day for women; at baseline - Rehm et al., 2018) and with less than 14 consecutive days of abstinence before randomization ('high severity population') than in the complement population with Low or Medium RDL at baseline or more than 14 consecutive abstinent days before randomization ("mild severity population") (Gual, He, Torup, van den Brink, & Mann, 2013; Gueorguieva et al., 2011, 2012; Mann et al., 2016; Scherrer et al., 2021; van den Brink et al., 2013, 2018).

This context should be considered when the efficacy of SMO in the maintenance of alcohol abstinence was evaluated in 12 studies including a total of 1548 alcohol-dependent patients, of whom 1146 received SMO. Seven RCTs compared efficacy and safety of SMO with placebo (Di Bello, Gambassi, Mugnai, Masini, & Mannaioni, 1995; Gallimberti, Ferri, Ferrara, Fadda, & Gessa, 1992; Guiraud et al., 2021, 2022), naltrexone (Caputo et al., 2003, 2007; Nava, Premi, Manzato, & Lucchini, 2006) or disulfiram (Nava et al., 2006) (Table 4). In addition, five open and noncontrolled studies investigated the efficacy of SMO on maintenance of alcohol abstinence treatment (Addolorato et al., 1996, 1998; Caputo et al., 2009, 2011; Maremmani, Lamanna, & Tagliamonte, 2001).

Table 4 Study characteristics and populations in RCTs testing the effect of SMO on maintenance of abstinence.

Criteria	GATE 2 Guiraud et al. (2022)	SMO032 Guiraud et al. (2021)	Gallimberti et al. (1992)	Di Bello et al. (1995)	Nava et al. (2006)	Caputo et al. (2007)	Caputo et al. (2003)
Study design	Phase 3 Double-blind RCT	Phase 2b Double-blind RCT	Pilot Double-blind RCT	Pilot Double-blind RCT	Pilot Open RCT	Pilot Open RCT	Pilot Open RCT
Sample size (*N*)	314	509	82	17	86	55	35
Male/female (%)	80/20	75/25	66/34	NR	85/15	78/22	77/23
Mean age (years)	44	48	37	NR	41	48	49
Disease	Alcohol dependence	Alcohol dependence	Alcohol dependence	Alcohol dependence	Alcohol dependence	Alcohol dDependence	Alcohol dependence
Severity at baseline	Mild-severity	31% high-severity 69% mild-severity	High-severity	Mild-severity	Mild-severity	High-severity	High-severity
Concomitant illness	None	None	None	None	None	None	None
SMO daily dose	~50 mg/kg	0.75, 1.25, 1.75, 2.25 g t.i.d.	50 mg/kg	50 mg/kg	50 mg/kg	50 mg/kg	50 mg/kg

Control group (daily dose)	PBO	PBO	PBO	PBO	NTX (50 mg) DSF (200 mg)	NTX (50 mg) NTX (50 mg) + SMO	NTX (50 mg)
Treatment duration (months)	6	3	3	6	12	3	3
Endpoint measured	CAD, abstinence rate	PDA, abstinence rate	PDA, abstinence rate	Abstinence rate	Abstinence rate	Abstinence rate	Abstinence rate

PBO, placebo; *NTX*, naltrexone; *DSF*, disulfiram; *PDA*, percentage of days abstinent; *CAD*, cumulative abstinent days.

Treatment duration and reported endpoints were in agreement with recommended endpoints and their definition in European guidelines (European Medicines Agency, 2010a). All RCTs included detoxified adult AD patients with a similar proportion of male subjects across studies. However, there was heterogeneity across trials in the population severity at baseline and in the duration of the treatment period (Table 4), two factors reported in the literature to be predictors of the placebo response and possible effect modifiers (Scherrer et al., 2021; van den Brink et al., 2018). Observational studies included AD patients with comorbidities (Caputo et al., 2009), AD patients with poly-addiction (Caputo et al., 2011), treatment-resistant AD subjects (Maremmani et al., 2001) and AD patients still drinking at treatment initiation (Addolorato et al., 1998).

5.1 Pilot RCTs exploring SMO efficacy in maintenance of abstinence

The first randomized, double-blind, placebo-controlled study investigating the efficacy and safety of SMO (Alcover® syrup) for the maintenance of abstinence in AD outpatients was a study in Italy (Gallimberti et al., 1992). A total of 82 high severity AD patients (see above definition) were randomized in two groups: SMO (50 mg/kg/day, divided into three daily doses) or placebo (three doses). The treatment duration was 3 months. Alcohol consumption was assessed by self-report. Efficacy was measured with the PDA during the treatment period, the number of drinks/day during the treatment period, and the abstinence rate. Analyses were conducted on the ITT population for the abstinence rate (dropout considered as relapse) and on treatment completers for the other endpoints. There was no clinically relevant difference in baseline characteristics between treatment groups. 71 patients completed the treatment period, 36 in the SMO and 35 in the placebo group. Results showed a significantly higher PDA in the SMO group: mean (SE) 25.9% (3.1) in the SMO group compared with 8.4% (1.6) in the placebo group, $p < 0.001$. The number of drinks per day was significantly lower in the SMO group: mean (SE) of 4.7 (0.4) in SMO group compared with 9.3 (0.7) in the placebo group, $p < 0.01$. The abstinence rate was 26.8% in the SMO group compared with 4.9% in the placebo group, resulting in a risk ratio [95% CI] of 5.50 [1.30; 23.2], $p = 0.02$. These results showed efficacy of SMO in the maintenance of abstinence in AD patients (Gallimberti et al., 1992; van den Brink et al., 2018).

Another randomized, double-blind, placebo-controlled study evaluated the effect of SMO (Alcover® syrup) on the craving for alcohol and the relapse to drinking in detoxified AD outpatients (Di Bello et al., 1995). A total of 17 mild severity AD patients were randomly assigned to two groups: SMO 50 mg/kg/day ($n = 9$) or placebo ($n = 8$). Subjects were treated for 6 months, and analyses were conducted on the ITT population. Patients receiving SMO more often drank in a controlled way and had fewer relapses to heavy drinking at 6 months than patients receiving placebo, but the observed differences were not statistically significant. The alcohol craving score at end of treatment was significantly lower in the SMO group ($p < 0.05$). The abstinence rate was numerically in favor of SMO but with a small effect size and not statistically significant (Di Bello et al., 1995).

An open randomized study compared the efficacy of SMO (Alcover® syrup) with naltrexone (NTX) in maintaining abstinence from alcohol after 3 months of treatment (Caputo et al., 2003). A total of 35 detoxified high severity AD outpatients were randomly enrolled in one of two groups: the SMO group consisted of 18 patients treated with oral doses of SMO (50 mg/kg/day of body weight fractioned in three doses) for 3 months; the NTX group consisted of 17 patients treated with oral doses of NTX (50 mg/day) for 3 months. Analyses were conducted on the ITT population. Dropouts were considered as failures in the abstinence rate analysis. There was no clinically relevant difference in baseline characteristics between treatment groups. Eight patients (23%) did not complete the 3-month treatment phase. At the end of the study, a statistically significant difference in favor of SMO was found in the abstinence rate: excess abstinence rate +31.4% ($p = 0.02$), indicating that SMO is more effective than NTX in maintaining abstinence from alcohol in a 3-month treatment period (Caputo et al., 2003).

Similarly, another open randomized study compared the efficacy of SMO (Alcover syrup® – 50 mg/kg/day), NTX, and their combination in maintaining abstinence (Caputo et al., 2007). A group of 55 detoxified high severity AD patients was randomly assigned to one of three groups and treated for 3 months with SMO ($n = 20$), SMO plus NTX ($n = 18$), or NTX ($n = 17$). Analyses were conducted on the ITT population. Dropouts were considered as failures in the abstinence rate analysis. There was no clinically relevant difference in baseline characteristics between treatment groups. Nine patients (16%) did not complete the 3-month treatment phase. At the end of treatments, abstinence was maintained by 13 patients (72.2%) in the SMO plus NTX group, 8 patients (40%) in the SMO group, and one patient (5.9%) in NTX group. SMO was statistically significantly superior to NTX in the abstinence rate ($p = 0.04$) with a large effect size (excess rate: +34.1%; relative risk = 6.8). The SMO/NTX combination was also more effective than either drug given alone; this suggests that the two drugs combine their different actions synergistically without suppressing the favorable effects of each other (Caputo et al., 2007).

Another open randomized study was conducted to compare different AD treatments on ethanol intake, craving, and on biochemical measures of alcohol consumption (Nava et al., 2006). A total of 86 detoxified mild severity AD patients were randomly assigned to SMO (Alcover® syrup – 50 mg/kg of body weight fractioned in three doses., $n = 28$), NTX (50 mg/day, $n = 27$) or disulfiram (DSF – 200 mg/day, $n = 31$) treatment for 12 months. Analyses were conducted on the ITT population. Dropouts were considered as failures in the abstinence rate analysis. There was no clinically relevant difference in baseline characteristics between treatment groups. 22 patients (26%) did not complete the 12-month treatment phase. SMO showed a numerically higher abstinence rate compared to the other treatments: +16.0% vs NTX, +34.1% vs DSF ($p = 0.08$). Patients in the SMO group had significantly larger decreases in alcohol craving and in laboratory markers of alcohol abuse than did patients in the NTX and DSF groups (Nava et al., 2006).

5.2 Phase 2 and 3 trials in maintenance of abstinence

5.2.1 The SMO032 Phase 2b study

A multinational, double-blind, placebo-controlled Phase 2b study (SMO032) investigated the efficacy and the safety of new abuse and misuse deterrent oral formulation of SMO in the maintenance of abstinence in detoxified AD outpatients (Guiraud et al., 2021). A total of 509 AD patients were randomized in a ratio of 1:1:1:1:1 to 12 weeks of placebo or one of four SMO doses (0.75, 1.25, 1.75 or 2.25 g t.i.d.) followed by a one-week medication-free period. The primary endpoint was the PDA at end of treatment. Analyses were conducted on the ITT population. Missing data were considered as relapse for the PDA and the abstinence rate.

There were no clinically relevant differences in baseline characteristics between treatment groups. Most enrolled patients (69%) had mild severity AD at baseline. A total of 189 of the 511 randomized patients (37.0%) did not complete the 12-week treatment phase. An unexpectedly high placebo response (median 92% PDA) was observed. This probably compromised the demonstration of efficacy in PDA, but several secondary endpoints still showed statistically significant improvements in one or more SMO-dose groups: (1) during the treatment period significant group differences were observed in the number of heavy drinking days, the end-of-treatment craving scores, and the %CDT at end-of-treatment, and (2) during the one-week follow-up significant group differences were observed in abstinence rate, time-to-relapse, the percentage of subjects with no heavy drinking days and the number of heavy drinking days. An inverted U-shape dose-response relation influenced by the level of alcohol consumption at baseline and body weight was identified: the more alcohol consumed at baseline, the higher the SMO dose in mg/kg had to be administered to have a beneficial effect. Importantly, a significant interaction ($p = 0.001$) was detected between treatment condition and AD severity on PDA, indicating that the treatment effect was significantly dependent on population severity. A post hoc analysis was carried out comparing the effects of SMO in the subgroup of mild severity ($n = 339$) and in the subgroup of high severity patients ($n = 154$). These subgroup analyses were conducted in accordance with the European Medicines Agency guideline on exploratory subgroup analyses in confirmatory trials (European Medicines Agency, 2019a). In the high severity population, pooled SMO doses showed statistically significant higher PDA during the 12-week treatment period compared to placebo: treatment difference +15.0%, $p = 0.022$, Cohen's $d = 0.42$. In this severe subpopulation, SMO treatment was also associated with significantly better key secondary outcomes compared to placebo: PDA last four weeks (treatment difference + 24.3%, $p = 0.003$, Cohen's $d = 0.55$), abstinence rate (risk difference + 18.1%, $p = 0.04$, risk ratio = 2.22). In contrast, no improvements were observed in the mild severity group (Guiraud et al., 2021).

5.2.2 The GATE 2 Phase 3 study

Guiraud et al. (2022) reported the results of a Phase 3 (GATE 2) double-blind, randomized, placebo-controlled trial aiming at confirming the efficacy and safety of SMO (Alcover®-syrup) in the maintenance of abstinence in alcohol dependent patients. The trial included patients from 11 sites in Austria, Italy, Germany, and Poland. It was conducted in accordance with the ethical principles of the Declaration of Helsinki, the GCP, and the European guideline for the development of alcohol dependence treatment (Plinius Maior Society, 1994). Before initiating the study procedures, the study protocol and the attached documentation were approved by ethics committees/institutional review boards at all sites. Written informed consent was obtained from all patients prior to their inclusion in the study. Inclusion criteria were: age 21–75 years, a clinical diagnosis of DSM-IV and ICD-10 alcohol dependence, availability of a responsible relative or caregiver, and a successful detoxification, including a 10-day treatment period and a subsequent 10-day untreated abstinent period. Randomization was stratified by site and the random numbers were computer-generated using a pseudo random uniform distribution with a block size of 4 patients. Sponsor, investigators, and patients were blind to treatment assignment during the full study period. Blinding was not broken for any patient during the trial. SMO (175 mg/mL) and placebo oral solutions were identical in appearance and taste. The primary efficacy outcome was the CAD during the 6-month treatment phase. Key secondary outcome measures included the CAD during the 12-month study period. Analyses were conducted on the ITT population. Missing data for the primary endpoint were considered as relapse to any drinking. In addition, a sensitivity analysis using multiple imputation was conducted.

A total of 314 patients were randomized to 6 months of placebo or SMO (3.3–3.9 g/day, corresponding to ≈50 mg/kg/day) followed by a 6-month medication-free period. Of the 314 alcohol-dependent patients randomized, 154 received SMO and 160 received placebo. There was no clinically relevant difference in baseline characteristics between treatment groups. A total of 182 patients (58%) did not complete the 6-month treatment phase with a higher dropout rate in the placebo group than in the SMO group (64% vs 52%). Based on the pre-specified primary analysis (fixed effect factorial ANOVA with site-by-treatment interaction), SMO showed efficacy for the primary endpoint, CAD at 6 months treatment, with a mean difference of +43.1 days, 95% CI [17.6–68.5; $p = 0.001$] between the two treatment groups (Fig. 6).

The statistically significant beneficial effect of SMO in CAD was explained by a numerically higher abstinence rate and a longer abstinence duration in relapsing patients. The sensitivity analysis with multiple imputation was statistically significant and supported results of the primary analysis. Since significant heterogeneity of effect across sites and unequal sample sizes across centers ($n = 3$–66) were identified, a site-level random effect meta-analysis was performed with results supporting the pre-specified analysis: overall mean difference between treatment conditions in CAD was +32.4 days ($p = 0.014$). Interestingly, it was found that the estimated treatment effect in CAD at 6 months at site level was negatively correlated ($r = -0.63$; $p = 0.04$) with the placebo response in the sites. The SMO effect was sustained during the post-treatment period with a mean difference of +82.7 days of CAD at 12 months, 95% CI [36.47–128.83; $p < 0.001$] (Fig. 6). The

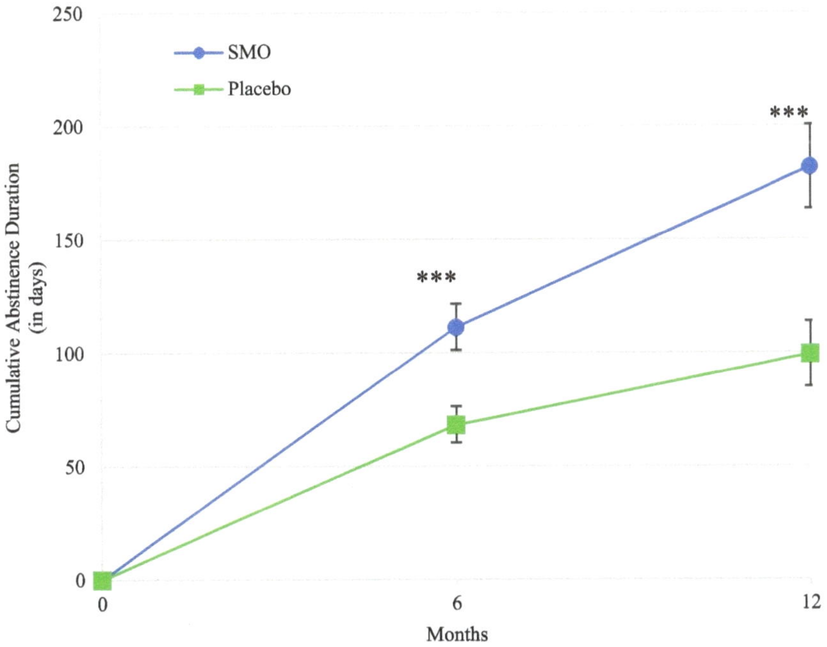

Fig. 6 GATE 2 efficacy results in ITT population: Adjusted mean CAD over the study period. Bars indicate standard error; ***$p \leq 0.001$. *Data from Guiraud, J., Addolorato, G., Antonelli, M., Aubin, H.-J., Bejczy, A., Benyamina, A., … van den Brink, W. (2022). Sodium oxybate for the maintenance of abstinence in alcohol dependent patients: An international, multicenter, randomized, double-blind, placebo-controlled trial. Journal of Psychopharmacology.*

authors reported that results of this large randomized controlled trial in alcohol-dependent patients demonstrated a significant and clinically relevant sustained effect of SMO on CAD (Guiraud et al., 2022).

5.3 Meta-analyses

In the meta-analysis performed in 2010 by The Cochrane Collaboration, SMO was superior to placebo (RR = 5.35; 95%CI: 1.28–22.4; $p = 0.02$) and to naltrexone (RR = 2.59; 95%CI: 1.35–4.98; $p = 0.004$) in abstinence rate after a 3 months treatment period (Leone et al., 2010). Similarly, a meta-analysis including the 7 above mentioned RCTs showed a significant effect of SMO (50 mg/kg) compared to placebo (RR = 1.35; 95%CI: 1.05–1.74) and compared to NTX (RR = 1.79; 95%CI: 1.20–2.68) in the abstinence rate (van den Brink et al., 2018).

In addition, in a network meta-analysis including 64 trials (43 interventions), SMO was significantly more effective than placebo in maintaining abstinence (odd ratio = 2.31) and was better ranked than acamprosate, naltrexone and disulfiram in achieving abstinence (Cheng et al., 2020). Last, SMO showed a significant efficacy in the maintenance with the largest effect size among the interventions included (risk ratio = 1.90) in another recent network meta-analysis including 156 trials (27,334 participants) and 37 unique treatment comparisons (Bahji, Bach, Danilewitz, Crockford, Devoe, El-Guebaly, Saitz et al., 2022). Overall, SMO demonstrated an efficacy in the maintenance of abstinence but heterogeneity of SMO treatment effect was observed within and across trials with effect sizes negatively correlated with placebo response (Guiraud, 2023).

5.4 Subgroup analyses and network meta-regression

In the Phase 2b SMO32 study, a significant interaction ($p = 0.001$) was detected between treatment conditions and AD severity on PDA, indicating that the treatment effect was significantly dependent on population severity. In the high severity population and compared to placebo, pooled SMO doses showed statistically significant higher PDA during the 12-week treatment period and significantly higher abstinence rate, probably due to a low placebo response. In contrast, no improvements were observed in the mild severity group where the placebo response was high (Guiraud et al., 2021; van den Brink et al., 2018). Furthermore, planned treatment duration was found to be a predictor of the placebo response in trials for the maintenance of abstinence in AD patients (Scherrer et al., 2021).

Given these findings, a network meta-regression analysis was conducted to investigate the effect of these two potential moderators of SMO efficacy in the maintenance of abstinence: population severity and treatment duration. Network meta-regression allows for testing the effect of potential moderators on intervention efficacy at study level. It was decided to investigate the effect of the above factors on SMO efficacy in abstinence rate and PDA. Randomized controlled trials for AD with at least one SMO group conducted in high-severity and mild-severity populations were assigned to a high-severity and mild-severity group of studies, respectively. Eight studies (1082 patients) were retained: four in the high-severity group and four in the mild-severity group. The high-severity group was associated with larger SMO effect sizes than the mild-severity group: abstinence rate *RR* 3.16, $p = 0.004$; PDA +26.9%, $p < 0.001$. For PDA, longer treatment duration was associated with larger SMO effect size: +11.3% per extra month, $p < 0.001$. In the high-severity group, SMO showed clinical benefit compared to placebo: abstinence rate RR 2.91, $p = 0.03$; PDA +16.9%,

$p < 0.001$. In the mild-severity group, SMO showed benefit compared to placebo only in PDA for longer treatment duration: +23.9%, $p < 0.001$. Thus, in the retained studies with alcohol-dependent patients, high-severity population and longer treatment duration were associated with larger SMO effect sizes (Guiraud et al., 2023).

5.5 Observational studies

SMO effectiveness was also explored in a series of uncontrolled observational studies.

In a first open, uncontrolled multicenter study, Addolorato et al. (1996) evaluated the efficacy and safety of SMO in inducing and maintaining abstinence. A total of 179 AD outpatients were recruited from 19 sites in Italy and were treated with SMO (Alcover® syrup - 50 mg/kg/day) for 6 months followed by a 12-month medication-free period. Patients were not detoxified and were still drinking at treatment initiation. A total of 70 patients (39%) did not complete the study. SMO led to complete abstinence during drug administration in 78.0% of patients who completed the treatment phase. Of these abstainers, 43 (51%) and 30 (36%) subjects remained abstinent at 6 months and at 12 months, respectively, of the medication-free follow-up phase (Addolorato et al., 1996).

Caputo et al. (2011) investigated the effect of SMO on alcohol abstinence and the risk of craving for and abuse of SMO in AD subjects with and without psychiatric co-morbidity in an open, uncontrolled study. A total of 48 high severity AD patients were enrolled and classified into two groups: group A (20 AD patients without psychiatric co-morbidity) and group B (28 AD patients with psychiatric co-morbidity). All patients were treated with SMO (Alcover® syrup - 50 mg/kg/day) for 12 weeks. A significant reduction of alcohol intake in both groups was observed ($p < 0.0001$). Alcohol abstinence as well as alcohol drinking during the 12 weeks of treatment did not differ between the two groups at the end of treatment ($p = 0.9$) (Caputo et al., 2011).

Another open, uncontrolled study investigated the effect of SMO on alcohol abstinence and the risk of craving for and abuse of SMO in AD patients with poly-addiction (Caputo et al., 2009). A total of 47 high severity AD patients was enrolled and divided into four groups: group A (pure alcohol dependent patients), group B (alcohol dependent patients with a sustained full remission from cocaine dependence), group C (alcohol dependent patients with a sustained full remission from heroin dependence) and group D (alcohol dependent patients in a methadone maintenance treatment (MMT) program).

All patients were treated with SMO (Alcover® syrup - 50 mg/kg/day) for three months. At the end of treatment, continuous abstinence from alcohol was maintained by 25 patients (53.2%): 9 patients (64.3%) in group A, 5 patients (50.0%) in group B, 8 patients (61.5%) in group C and 3 patients (30.0%) in group D. There were no significant differences between the groups (Caputo et al., 2009).

Addolorato et al. (1998) investigated the effect of greater fractioning of SMO (Alcover® syrup - 50 mg/kg/day) dose in the maintenance of abstinence in AD patients not responding to the usual three SMO administrations per day in an open study. A total of 154 high severity AD patients was treated with SMO (50 mg/kg/day divided in three administrations) for 8 weeks (phase 1). Patients who were not abstinent at the end of phase 1 were administered the same daily SMO dose but divided in six administrations for a subsequent 8-week period. Of the 154 patients, 115 completed phase 1; 78 (67.8%) of these began and maintained abstinence (group A) while 37 subjects (32.2%) continued to drink alcohol (group B) showing a craving significantly higher than group A at the end of phase 1 ($p < 0.001$); in these patients the greater fractioning of the study medication in phase 2 caused a significant further reduction in craving ($p < 0.005$) and 26 patients (70.2%) began and maintained abstinence. Moreover, no significant differences in final craving score between group A and B was observed. Within the limits of an open study, these data showed that non-responder subjects to the conventional fractioning of SMO seemed to benefit from a greater fractioning of the drug (Addolorato et al., 1998).

Maremmani et al. (2001) explored the effect of SMO (Alcover® syrup) in treatment-resistant chronic AD patients in an open, uncontrolled study. Thirty-five AD patients who met criteria for treatment resistance were administered SMO with doses ranging between 25 and 100 mg/kg/day depending on patients' clinical response for one year. About 26% of these patients successfully completed the protocol and were considered responders: 11.4% showed complete abstinence (full responder patients) and 14.3% strongly reduced their alcohol intake (partial responder patients). About one-third (34.3%) of the patients was still in treatment after one year and 40% of the patients were non-responders. The retention rate under treatment of the studied sample was statistically higher than that found during the last treatment of the same subjects (Maremmani et al., 2001).

5.6 Discussion on SMO efficacy in the maintenance of abstinence

The efficacy of SMO in the maintenance of abstinence in AD patients is supported by 7 RCTs, including one phase 3 trial, 5 observational studies, and 4 (network) meta-analyses and one network meta-regression analysis.

The phase 3 GATE 2 study was international, double-blind, randomized, placebo-controlled with a 6-month treatment period followed by a 6-month medication-free period. The study was conducted in accordance with GCP and methods were prespecified in a protocol. The primary endpoint was the CAD during the 6-month treatment period. This design complied with the EMA recommendations for confirmatory trials in the maintenance of abstinence in AD patients (European Medicines Agency, 2010a), with the exception of the primary endpoint which differed from the abstinence rate (Guiraud, 2023). However, CAD can be conceptualized as a composite endpoint that measures both the abstinence rate and the abstinence duration in relapsing patients (Guiraud et al., 2022). According to ICH E9, composite endpoints can be used as a primary outcome in confirmatory trials as long as the method of combining the multiple measurement is specified in the protocol, and an interpretation of the resulting scale is provided in terms of the size of a clinically relevant benefit (ICH, 1998). The CAD computation method was specified in the GATE 2 protocol and was compliant with the recommendation from the European guideline for the development of treatment of AD at that time (Plinius Maior Society, 1994). In GATE 2, SMO efficacy in CAD was explained by a higher continuous abstinence rate as well as a longer abstinence duration in relapsing patients in the SMO group. Therefore, CAD captures the beneficial effect of SMO in the continuous abstinence rate. However, it is acknowledged that, although precise and clinically relevant, CAD is more difficult to interpret than abstinence rate. The study was positive in the prespecified primary analysis of the primary endpoint in the ITT population: SMO (≈50 mg/kg/day) showed a statistically highly significant ($p = 0.001$) and clinically relevant higher CAD after 6 months treatment: adjusted mean difference +43.1 days. There was a high dropout rate but this is consistent with those commonly observed in AD trials and those from RCTs that were used to establish efficacy of approved compounds in the treatment of AD (Guiraud et al., 2022). In addition, the robustness of the efficacy result in the GATE 2 study was supported by various sensitivity analyses, with different imputation methods or using different statistical models, that all showed a significant effect of SMO with similar point-estimates. The risk of bias for the results of the primary outcome of this study was considered to be low (Guiraud et al., 2023). Furthermore, SMO efficacy in CAD was sustained during the 6-month untreated follow-up period (Guiraud et al., 2022). An assessment using the GRADE criteria supports a high evidence quality of SMO efficacy in CAD in this study (Guiraud, 2023). Therefore, these data support that SMO has demonstrated statistically significant and clinically relevant efficacy in CAD in GATE 2 with a high evidence quality. These results are supported by results of smaller trials and (network) meta-analyses which consistently showed evidence of SMO efficacy in abstinence rate with clinically relevant effect sizes.

However, heterogeneity of SMO treatment effect was detected both within and across trials. In the phase 2b SMO032 study, the efficacy demonstration in the primary endpoint, PDA, was probably compromised by an unexpectedly high placebo response (mean 73%). This high placebo effect contrasted with a lower placebo response seen in earlier studies, such as the GATE 2 study, which had a mean placebo response of 38%. A subgroup analysis within the SMO032 trial revealed significant SMO efficacy in both PDA and abstinence rate in high-severity AD patients, likely due to lower placebo responses. Similarly, in the GATE 2 study, a significant and substantial heterogeneity of treatment effect on CAD across sites was identified and the estimated SMO treatment effect in CAD was larger in sites with a lower placebo response. Heterogeneity of medication effect with efficacy negatively correlated with placebo response is common in this therapeutic area (European Medicines Agency, 2010a; Litten et al., 2013). To explore heterogeneity of SMO treatment effect, a network meta-regression analysis of 8 studies tested the effect of population severity and treatment duration, i.e., two potential moderators of medication effect and/or predictors of placebo response in AD, on SMO efficacy in the maintenance of abstinence. Results showed significantly larger SMO effect sizes in high-severity populations and with longer treatment durations (Guiraud et al., 2023). However, the presence of (residual) bias and/or confounding cannot be excluded in meta-regression analysis and, therefore, data were provided supporting the fulfillment of the EMA and Cochrane criteria for the credibility and plausibility of exploratory subgroup findings/meta-regressions (Guiraud et al., 2023). Based on these findings, there is a probable causal relationship between population severity and treatment duration and the SMO treatment effect (Guiraud, 2023). These findings underline the importance of considering patient severity and treatment duration in SMO's clinical use, potentially aiding in more tailored and effective treatment strategies for AD.

Exploratory analyses showed a dose-response relationship: the more alcohol consumed at baseline, the higher the SMO dose in mg/kg was needed for a positive effect. In heavy drinking AD patients, an SMO dose of 60 mg/kg/day had the best response. This dose-response is consistent with the pharmacology of SMO and the dose-response observed in preclinical studies.

Overall, SMO showed efficacy in the maintenance of abstinence and in longer abstinence duration in the phase 3 GATE 2 study, with a high quality of evidence. These results are supported by evidence of efficacy from 6 pilot/exploratory RCTs and several meta-analyses showing statistically significant and clinically relevant effect sizes. SMO efficacy appears to be (most) effective in the high-severity population and in trials with a longer treatment duration.

6. Safety of SMO in the treatment of AD

In a recent review, the safety profile of SMO in the treatment of AD has been analyzed based on data from 46 clinical studies, including the clinical studies previously discussed in previous sections, which included a sample population of 3067 participants exposed to SMO, and a pharmacovigilance database notified for SMO (Alcover® in Italy and Austria) with 299,013 exposed subjects (Addolorato et al., 2020). In this review, safety data are reported separately for the three recent large RCTs (GATE 1, GATE 2, and SMO032) and the 43 early clinical trials (Addolorato et al., 2020).

6.1 Adverse events reported in recent large RCTs

Route of administration of SMO in the three recent large RCTs was oral. Patients were treated in an inpatient/daycare setting for the treatment of AWS and in an outpatient setting in the treatment directed at maintenance of abstinence. Medication was dispensed in a quantity covering 7 days to one month of treatment in the two studies in the maintenance of abstinence.

These three recent large RCTs reported 194 common (non-serious) Treatment Emergent Adverse Events (TEAEs) in a total of 520 patients exposed to SMO 50 mg/kg divided into three administrations per day (Table 5). These TEAEs included dizziness (10.6% of patients), vertigo (4.8%), headache (4.2%), fatigue (3.5%), somnolence (3.1%) and nausea (2.9%) (Addolorato et al., 2020).

Table 5 Safety of SMO 50 mg/kg in alcohol-dependent patients.

	Number of patients (%)
Patients exposed to SMO	520 (100%)
Patients who experienced TEAEs	194 (37.3%)
Dizziness	55 (10.6%)
Vertigo	25 (4.8%)
Headache	22 (4.2%)
Fatigue	18 (3.5%)
Somnolence	16 (3.1%)
Nausea	15 (2.9%)
Hypersomnia	8 (1.5%)
Insomnia	8 (1.5%)
Euphoric Mood	7 (1.4%)
Vomiting	6 (1.2%)
Diarrhea	6 (1.2%)
Anxiety	6 (1.2%)
Patients with serious TEAEs	3 (0.6%)
Epilepsy	1 (0.2%)
Overdose	1 (0.2%)
Depression (suicidal depression)	1 (0.2%)

SMO, sodium oxybate; *TEAE*, treatment emergent adverse event.
Combined data from GATE 1, GATE 2, and SMO032 - from Addolorato, G., Lesch, O.-M., Maremmani, I., Walter, H., Nava, F., Raffaillac, Q., Caputo, F., 2020. Post-marketing and clinical safety experience with sodium oxybate for the treatment of alcohol withdrawal syndrome and maintenance of abstinence in alcohol-dependent subjects. *Expert Opinion on Drug Safety, 19*, 159–166. https://doi.org/10.1080/14740338.2020.1709821.

A higher incidence of TEAE was associated with higher SMO doses (Addolorato et al., 2020; Guiraud et al., 2021). No significant group differences between the SMO group and the control group were found in the incidence of adverse events (AEs) in GATE 1, GATE, 2, and SMO032 (Caputo et al., 2014; Guiraud et al., 2021, 2022). The most frequently experienced AE leading to discontinuation in the SMO group was nausea

(Guiraud et al., 2022). In SMO032 and GATE 2 studies, a total of 7 patients experienced 8 treatment emergent serious adverse events (SAEs) that were considered by the investigator to be related to study medication: six SAEs in the SMO groups (one overdose, one suicidal depression, one toxicity of various agents, one epilepsy, one loss of consciousness, and one discomfort), and two in the placebo groups (one drug toxicity and one joint dislocation) (Guiraud et al., 2021, 2022). In GATE 1, one subject experienced delirium tremens as an AE after treatment with SMO, and the AE was classified as severe (Caputo et al., 2014).

6.2 Adverse events reported in early clinical trials

A total of 43 clinical studies (excluding the above 3 large RCTs) with SMO have been found in 44 articles or reports of the biomedical literature and were included in the analysis. Route of administration in the trials was oral except in one study where SMO was administered intravenously. Maintenance of abstinence was conducted in an outpatient setting and SMO was dispensed to patients in bottles containing approximately four to seven days of treatment. Treatment of AWS was conducted essentially in an inpatient/daycare setting (Addolorato et al., 2020).

The 43 studies included a total of 2547 eligible patients, who were exposed to several dose regimens of SMO (Table 6). Overall, these studies

Table 6 Safety of SMO in alcohol-dependent patients combined data from 43 early clinical trials.

	SMO ≤50 mg/kg/day (% of patients)	SMO > 50 mg/kg/day (% of patients)
Patients exposed to SMO	1150 (100%)	1397 (100%)
Number of TEAEs	138 (11.5%)	499 (35.7%)
Vertigo/dizziness/drowsiness	110 (9.2%)	185 (13.2%)
Vomiting	5 (0.4%)	42 (3.0%)
Diarrhea	4 (0.3%)	30 (2.1%)
Nausea	1 (0.1%)	25 (1.8%)
Patients with serious TEAEs	0 (0.0%)	0 (0.0%)

SMO, sodium oxybate; *TEAE*, treatment emergent adverse event.
From Addolorato, G., Lesch, O.-M., Maremmani, I., Walter, H., Nava, F., Raffaillac, Q., Caputo, F., 2020. Post-marketing and clinical safety experience with sodium oxybate for the treatment of alcohol withdrawal syndrome and maintenance of abstinence in alcohol-dependent subjects. *Expert Opinion on Drug Safety, 19*, 159–166. https://doi.org/10.1080/14740338.2020.1709821.

reported 637 common (non-serious) TEAEs. Of these, 138 TEAEs were reported in 1150 patients exposed to SMO ≤ 50 mg/kg/day and 499 TEAEs were reported in 1397 patients exposed to SMO >50 mg/kg/day. Dizziness, vertigo and/or drowsiness were the most consistently reported adverse effects (9–14% of patients). Vomiting, diarrhea or nausea were observed in 1–3% of patients. In general, these TEAEs did not require treatment, since dizziness disappeared spontaneously after the first doses, and more than 50% of TEAEs disappeared after the first month of treatment. No serious TEAEs have been reported in the 2547 patients exposed to oral SMO (Table 6) (Addolorato et al., 2020).

6.3 Fatal SAEs in clinical trials

No deaths attributable to the use of SMO have been reported in the 46 clinical studies, including the clinical studies previously discussed, which included a sample population of 3067 participants exposed to SMO, and in the pharmacovigilance database notified for Alcover® with 299,013 exposed subjects (Addolorato et al., 2020). In GATE 2, one death was reported in the SMO group but not attributed to SMO: the patient was murdered while consuming alcohol (Guiraud et al., 2022).

6.4 Safety from pharmacovigilance

The pharmacovigilance database for Alcover® in Italy and Austria included 299,013 patients who were exposed to SMO for the treatment of AWS and/or for maintenance of abstinence from 1992 to October 2017. Common, non-serious AEs were reported in 17 patients (0.3%; mainly vertigo, nausea, and diarrhea) in Austria and 68 patients (0.02%; mainly vertigo, nausea, and sopor) in Italy. No fatal AE was reported in these pharmacovigilance databases (Addolorato et al., 2020). SAEs reported in the pharmacovigilance data are discussed in the next section.

6.5 Important identified risks

Important identified risks in the approved Risk Management Plans for SMO in the treatment of narcolepsy within EU are the following: overdose; respiratory depression; CNS depression; depression/suicidality; diversion/criminal misuse; alcohol interaction; psychosis; dependence/withdrawal; misuse/abuse; and convulsion (CMDh, 2021). These important identified risks have been analyzed and discussed for SMO in the treatment of AD based on data from the literature review.

6.5.1 Overdose

An SMO overdose expresses itself especially as a dampening effect on the CNS, possibly accompanied by confusion and shortness of breath. Treatment is symptomatic: artificial respiration, forced diuresis, if necessary, a short infusion of physostigmine to antagonize the sedation. In mild cases, the above-mentioned symptoms recede spontaneously, gastric lavage may be helpful (D&A Pharma, 2018).

In the 46 clinical studies which included a sample population of 3067 participants exposed to SMO, two cases (0.07%) of overdose were reported. The pharmacovigilance database for Alcover®, including 299,013 patients exposed to SMO, reports seven overdose cases (0.002%) (Addolorato et al., 2020).

These data are consistent with toxicity studies. Laborit found that sleep could be induced in the rat with 0.5 g/kg SMO (i.p.) and in rabbits and dogs using 1 g/kg (i.v.) (McCusker, Paget-Wilkes, Chronister, Goldberger, & ElSohly, 1999). Using the FDA methodology (FDA, 2005), the Human Equivalent Dose (HED) would be approximately 5.7 g/70 kg which is consistent with Xyrem posology (up to 4.5 g twice a night). Loss of gross motor skills and coma have been observed further to GHB overdose (Abanades et al., 2007; Carter, Richards, Mintzer, & Griffiths, 2006). This is in agreement with clinical practice as SMO has been widely used as an intravenous anesthetic (60–70 mg/kg) in certain obstetric and pediatric procedures (World Health Organization, 2012). In the mouse, LD50 values of SMO 2960 – 3700 mg/kg following i.p. injection were identified. In the rat, LD50 values were 9990 mg/kg following p.o. administration, and 1700 mg/kg following i.p. injection. The cause of death was reported to be respiratory depression. In the rabbit and dog, LD50 values in excess of 1000 mg/kg were reported following i.v. administration, which could be increased to over 7000 mg/kg with artificial respiration without lethality (European Medicines Agency, 2006; World Health Organization, 2012). Based on these pre-clinical data, the oral SMO LD50 HED would be 113 g/70 kg, i.e., around 100-fold the approved SMO posology for AD. Non-fatal respiratory depression was observed in rats with administration of SMO 600 mg/kg intravenously (Morse, Chadha, Felmlee, Follman, & Morris, 2017) which represents approximately 1/3 of LD 50.

These high thresholds may explain the lack of serious safety issues even in the rare cases of overdose with large doses of pharmaceutical SMO.

6.5.2 Respiratory depression

No cases of respiratory depression were reported by Addolorato et al. (2020) in their review of SMO safety data.

6.5.3 Psychosis

No cases of psychosis were reported by Addolorato et al. (2020) in their review of SMO safety data.

6.5.4 Convulsion

In the SMO032 study (Guiraud et al., 2021), one AD patient with chronic hypercholesterolemia experienced an epileptic seizure while treated with SMO 1.25 g t.i.d. The SAE did not recur, the patient recovered, and the event was considered resolved.

6.5.5 Alcohol interaction

One patient experienced loss of consciousness in the SMO032 study. This event was reported in a 45-year-old female patient treated with SMO 2.25 g t.i.d. (corresponding to 130 mg/kg/day) who relapsed to alcohol consumption (15 drinks) on the day of the event (Guiraud et al., 2021). In other clinical studies for the maintenance of abstinence, no side effects due to the combination of alcohol (mean daily alcohol intake ranged from 3.8 to 5.8 drinks) and oral SMO 50 mg/kg were found in those patients who were still drinking during the treatment. This finding is consistent with results from pharmacodynamics studies where no interaction between alcohol (administration of 38–53 g of pure alcohol for a man with body-weight of 75 kg) and SMO (administration of 1 g, 2 g, 2.25 g or 50 mg/kg) was detected in healthy volunteers (Addolorato et al., 2020). Thus, a consensus report by 50 Italian experts in the field of alcohol use disorders recommends not to wait for the patient to stop alcohol drinking before starting SMO therapy (Caputo et al., 2015). However, it is recommended to suspend or discontinue the treatment with SMO in case of relapse to heavy drinking (Guiraud et al., 2021).

6.5.6 Diversion

Among the 3067 participants exposed to SMO in clinical trials, not a single case of diversion/criminal misuse was reported (Addolorato et al., 2019). The pharmacovigilance database includes one case of diversion/criminal misuse (Addolorato et al., 2020).

6.5.7 Suicidal depression

In clinical trials, one patient experienced suicidal depression (Addolorato et al., 2020; Guiraud et al., 2022).

6.5.8 Dependence/withdrawal

Four cases (0.13%) of dependence/withdrawal were reported in clinical studies compared to two cases in the pharmacovigilance data base (Addolorato et al., 2020; Guiraud et al., 2022).

These results are consistent with preclinical data. None of the studies conducted in animals showed a development of a withdrawal syndrome or dependence at low SMO doses (e.g., therapeutic doses) (Bania, Ashar, Press, & Carey, 2003; Carai et al., 2005; Cook, Aceto, Coop, & Beardsley, 2002; Quang et al., 2006; Weerts et al., 2005). The development of tolerance to SMO's sedative effect has been observed in rodents after chronic and long-term exposure to sedative/ high/supra therapeutic SMO doses (Bania et al., 2003; Itzhak & Ali, 2002; Raybon & Boje, 2007; Van Sassenbroeck, De Paepe, Belpaire, & Buylaert, 2003).

Even at very high doses, studies on the development of a withdrawal syndrome in laboratory animals have yielded mixed results. A reproducible withdrawal syndrome in Sardinian alcohol-dependent rats was reported further to chronic oral administration of SMO doses close to LD50 (HED 86 g/70 kg) (Carai et al., 2005; Quang et al., 2006). In contrast, no evidence of physical dependence or a withdrawal syndrome has been found in Long Evan or Wistar rats administered similar SMO oral doses (HED 52–92 g/70 kg) (Cook et al., 2002; Quang et al., 2006). Results with Sprague-Dawley rats were more conflicting with mild withdrawal syndrome appearing but only after a prolonged exposure with extremely high SMO doses (HED 86–186 g/70 kg per o.s.) (Bania et al., 2003; Quang et al., 2006). Similarly, mild to moderate withdrawal syndromes were observed in baboons only further to chronic administration of very high SMO doses (HED 29 g/70 kg) (Weerts et al., 2005).

These data are consistent with the results in human studies. SMO physical dependence and withdrawal syndrome were reported, only in illicit GHB users with very high daily doses, markedly exceeding therapeutic dose regimen of pharmaceutical SMO: 10 g to 312 g/day (Kamal, 2016). The average self-administered dose reported in GHB-dependent patients ranges from 32–67.2 g/day to a maximum of 144 g/day at intervals of 45 min to 2.5 h resulting in the occurrence of a withdrawal syndrome at drug discontinuation (Kamal, 2016).

Therefore, as for other sedative/hypnotics (e.g., benzodiazepines), the severity of SMO physical dependence is likely to be a function of the dose and duration of use (Kamal, 2016).

Even though dependence on pharmaceutical SMO is rare, it is important that treatment options have been investigated for the SMO/GHB withdrawal syndrome. Two patients dependent on illicit GHB were successfully treated with diazepam (Addolorato, Caputo et al., 1999; Price, 2000). Other options have been described, including treatment with baclofen. De Jong et al. (2012) published a study of 23 patients dependent to illicit GHB successfully detoxified by pharmaceutical SMO titration followed by tapering (de Jong, Kamal, Dijkstra, & de Haan, 2012). Kamal (2016) treated 229 GHB-dependent patients using this titration and tapering protocol with a success rate of 85%, the remaining 15% voluntarily discontinued the detoxification due to different reasons.

In conclusion, evidence converges from both animal and human studies as well as from pharmacovigilance data that SMO dependence only develops in the case of chronic exposure to doses largely exceeding the therapeutic dose regimen and can result in a SMO withdrawal syndrome. In contrast, evidence supports a low risk of dependence and withdrawal syndrome at SMO posology and treatment duration for AD.

6.5.9 Abuse/misuse

Of the 1423 patients without severe psychiatric comorbidities or poly-addiction in the clinical studies, only 11 patients (0.8%) were involved in abuse/misuse. In addition, the pharmacovigilance database for Alcover® reported 10 cases of abuse/misuse out of 299,013 patients exposed to SMO (Addolorato et al., 2020).

These data are in line with abuse liability studies which showed that the abuse potential of SMO is dose dependent (Abanades et al., 2007; Johnson & Griffiths, 2013). At low doses, no reinforcing effect has been detected in animals (Beardsley, Balster, & Harris, 1996; Goodwin, Kaminski, Griffiths, Ator, & Weerts, 2011; Woolverton et al., 1999). Alcohol preferring rats self-administered SMO at low doses and did not increase SMO intake despite a free access to the drug (Colombo et al., 1998). In humans, doses of 1–2 g/70 kg (doses close to SMO posology for AD) showed an abuse potential comparable to doses of ethanol of 12–24 g/70 kg in subjects with a history of sedative (including alcohol) abuse (Johnson & Griffiths, 2013). Reinforcing effects in baboons have been detected only at very high oral doses corresponding to a HED of 70 g/70 kg/day (Goodwin et al., 2011).

In humans, dose-dependent mixed stimulant-sedative effects were observed after single oral administration of 40–70 mg/kg (corresponding to up to 1.8–4.2 folds the single doses of Alcover® syrup) (Abanades et al., 2006). Single oral administration of 6, 8, and 10 g/70 kg SMO showed similar abuse liability to alcohol doses of 72, 96, and 120 g/70 kg respectively in patients with history of sedative (including alcohol) abuse (Johnson & Griffiths, 2013). Carter et al. reported that the abuse liability of SMO (2–8 g/70 kg – 1.7 to 6.9 fold SMO posology for AD) is intermediate to triazolam (0.5 and 1 mg/70 kg) and pentobarbital (200 and 400 mg/70 kg) (Carter et al., 2006).

Thus, abuse/misuse of SMO is rare and only occurs in the case of very high doses. However, a higher incidence of abuse/misuse was found in AD patients with severe psychiatric comorbidities or poly-addiction with 64 cases out of a total of 552 patients (11.6%) in clinical studies. Nevertheless, cases of abuse did not result in SAEs (Caputo et al., 2009, 2011) which is in line with data from the literature. In a study examining the abuse potential of alcohol and SMO among individuals with histories of sedative abuse, including alcohol, it was found that participants who could tolerate single doses of alcohol up to 120 g/70 kg displayed a comparable tolerance to single doses of SMO up to 10 g/70 kg. AEs following a single oral administration of 6, 8, and 10 g/70 kg SMO included significant behavioral impairment and nausea/vomiting (Johnson & Griffiths, 2013). Another study showed that SMO effects (48.3 and 72.4 mg/kg) were close to those induced by flunitrazepam (1.25 mg) and ethanol (0.7 g/kg) in club drug users (Abanades et al., 2007). The SMO doses in this study were between 2.9 and 4.3 times as high as the approved therapeutic dose for the treatment of AD. These findings suggest that individuals who are tolerant to alcohol, such as AD patients, may exhibit better (cross)tolerance to higher doses of SMO.

6.6 SMO controlled access for the treatment of AD

SMO access has been well-controlled both in clinical trials and in real-life use in Italy and Austria. In clinical trials, patients were enrolled and followed by specialists in the treatment of addictions from dedicated addiction facilities or from hospitals. In real-life, Alcover® oral solutions are prescribed by addiction specialists from specialized addiction facilities (Italy) or hospitals (Austria) (Addolorato et al., 2020). Treatment of the maintenance of abstinence in clinical trials was always conducted in an out-patient setting, whereas in the treatment of AWS, the detoxification program included in-patient treatment or daily control (setting of a day-clinic). Patients treated for the maintenance of abstinence, self-administered the drug at home without healthcare professional supervision. In GATE 2, the administration was supervised by a responsible relative/friend/caregiver whereas no supervision was required in the SMO032 study. The quantity that patients were allowed to take home was covering 7 days of treatment in SMO032 and 30 days in GATE 2 (Guiraud et al., 2021, 2022). In a retrospective non-interventional survey investigating the therapeutic use of Alcover® in Italy which included 485 patients, almost all patients pursued long-term SMO treatment in an outpatient setting (Addolorato et al., 2020; Manzato, Nava, Manzato, Rosa, & Caputo, 2017). The mean dose used in real-life was ≈75 mg/kg/day (Manzato et al., 2017), indicating that a bottle of Alcover® syrup covered around four days of treatment.

6.7 SMO safety profile different from illicit GHB

Importantly, the SMO safety profile differs from the safety profile of illicit GHB (Carter et al., 2009). Illicit GHB use is not controlled resulting in very high doses of self-administered GHB and many safety concerns (Kamal, 2016). In addition, GHB can be manufactured at home, from inexpensive ingredients and recipes obtained from the Internet with purity issues (Carter et al., 2009; World Health Organization, 2012). Last, concerns are increasing about the direct consumption of GHB's precursor chemicals, i.e., GBL and 1,4-BD (World Health Organization, 2012). 1,4-BD and GBL share a higher potency compared to GHB (Carai et al., 2002). Sedation/coma of comparable length was obtained in rats with 500 mg/kg 1,4-BD and 1 g/kg SMO (Carai et al., 2002). In baboons, GBL appeared to be the most potent of the three compounds, followed by 1,4-BD and then GHB. In addition, GBL and 1,4-BD shared a quicker onset and a longer duration of action when compared to GHB (Goodwin et al., 2009). GBL has a relatively low acute toxicity in that the oral LD50 of GBL is in the range of 1540–1800 mg/kg in rats, 800–1720 mg/kg in mice, and 500–1690 mg/kg in guinea pigs with few other clinical signs than central nervous system depression (WHO, 2014). Using the FDA methodology (FDA, 2005), the HED of the GBL LD50 in rats would be 17 g/70 kg. Less than 2cl of pure GBL solution would be necessary to reach such predicted oral HED LD50. This quantity is relatively easy to drink by a recreational drug user and may explain the serious safety concerns reported with illicit GHB as opposed to pharmaceutical SMO.

7. Benefit-risk ratio of SMO in the treatment of AD

AD is a serious condition which affects 4.1 and 3.7% of the general population aged 15+ years in the region of the Americas and Europe, respectively (World Health Organization, 2018). About 12% of the overall mortality in these continents is due to AD (Rehm et al., 2012) and people with AUD there have an average life expectancy of 47–65 years and die 12–28 years earlier than people without AUD in the general population (Schwarzinger et al., 2017; Westman et al., 2015). Existing interventions for AD treatment only show modest efficacy with mixed-results and with many patients not responding to these treatments (European Medicines Agency, 2010a; Litten et al., 2012; van den Brink et al., 2018). Given the burden caused by AD and the modest (and inconsistent) efficacy of existing interventions, the WHO considers the development of new medicines for the treatment of AUD to be a priority for Europe and the World (Kaplan et al., 2013). Therefore, additional treatment options are needed, preferably with a different mechanism of action than currently available medications (Guiraud, 2023).

One very successful treatment approach for addictive behaviors is substitution therapy which is most effective for the treatment of opioid and nicotine dependence. SMO has been reported to be probably the closest to a substitution drug for AD (Guiraud et al., 2024). SMO is the sodium salt of GHB, which derives from GABA, the primary inhibitory neurotransmitter in the brain. SMO appears to be pharmacologically related to alcohol as data suggest that it inhibits GABAergic neurons, increases dopamine release, reduces glutamate, decreases serotonin, and is possibly accompanied by the release of endogenous opioids (Guiraud et al., 2024). SMO has been approved as an oral solution at low doses for the treatment of AWS and for the maintenance of abstinence in AD patients in Italy and Austria since 1991 and 1999, respectively (Guiraud, 2023).

Efficacy of SMO in the treatment of AWS and in the maintenance of abstinence in AD patients is characterized by data from 7 and 12 published trials, respectively, among which 12 RCTs with three phase 3 and one phase 2b trials. Studies included a total of 2231 patients (Guiraud, 2023). Five (network) meta-analyses with one conducted by the Cochrane Collaboration were performed.

In the treatment of AWS, SMO 50–70 mg/kg/day was more effective than placebo in a double-blind RCT and at least as effective as benzodiazepines and clomethiazole in reducing alcohol withdrawal symptoms in four RCTs, including two phase 3 equivalence studies. The active-controlled studies included patient populations which do not improve without effective treatment. Treatment durations were in accordance with standard clinical practice. Efficacy was measured with the CIWA-Ar. No rebound effects during the untreated follow-up periods were observed. Efficacy data from RCTs was supported by results from observational studies which showed effectiveness of SMO in the treatment of complicated AWS and in patients with comorbidities. However, no RCTs were available that tested the efficacy of SMO in the treatment of alcohol withdrawal complications such as seizures and delirium tremens.

In the maintenance of abstinence in AD patients, the phase 3 international double-blind placebo controlled RCT GATE 2 was positive in the primary analysis of the primary endpoint: SMO 50 mg/kg/day showed a statistically significantly and clinically relevant higher CAD after 6 months treatment ($p = 0.001$). Generalizability of these results was supported by a secondary analysis using a random effect model. A sensitivity analysis using multiple imputation also supported SMO efficacy in CAD, suggesting a low risk of bias related to missing data. SMO efficacy in CAD was sustained during the 6-month untreated follow-up. CAD was the recommended primary endpoint in European guidelines until the implementation of EMA guidelines in 2010. However, CAD measures two components among which the abstinence rate, the primary endpoint according to the EMA guideline (Guiraud, 2023). The statistically significant beneficial effect of SMO in CAD is explained by a numerically higher abstinence rate and a longer abstinence duration in relapsing patients. GATE 2 results are supported by five smaller RCTs which corroborated SMO efficacy in the maintenance of abstinence vs. placebo or naltrexone in their (co-)primary endpoint. The (network) meta-analyses of abstinence rate showed a significant and clinically relevant effect of SMO vs. placebo and vs. naltrexone. Heterogeneity of SMO effect was investigated through a subgroup analysis conducted in accordance with EMA guidelines and a network meta-regression analysis. Baseline severity and treatment duration appeared to be effect modifiers of SMO efficacy: SMO consistently showed large effect sizes in the high-severity population within and across trials, whereas treatment effect was smaller in mild-severity patients. At population level, SMO treatment effect was larger in studies with a longer treatment duration. In the high-severity group, SMO showed benefit in the abstinence rate after 3 months treatment duration whereas in the mild-severity group, SMO showed benefit only in PDA for longer (i.e., 6-month) treatment duration. The high-severity population includes very heavy drinking AD patients. Despite affecting "only" 0.74–0.85% of the general population aged 15 to 64 years, very heavy drinkers are responsible for a substantial part of the alcohol-attributable morbidities and have a life expectancy of only 47–61 years (Rehm et al., 2018). Therefore, demonstration of efficacy in this patient population is particularly relevant for a medicinal product as it can provide substantial clinical benefits for individual patients and for society as a whole.

Nevertheless, although the available data demonstrate an effect of SMO in the maintenance of abstinence, the precision of SMO effect size in each population severity group could be further improved and the causal relationship between population severity and SMO efficacy could be further established with the conduct of additional RCTs, e.g., with randomization stratified by population severity (Guiraud, 2023).

Safety data from 3 recent large randomized clinical studies (520 participants) and 43 earlier clinical studies (2547 participants) showed that SMO has a good safety profile in AD patients. The safety profile was confirmed by pharmacovigilance data resulting from 299,013 patients exposed to Alcover® in Austria and Italy. Main AEs were transient dizziness and vertigo. SAEs were rare. No deaths attributable to SMO have been reported. Risks of abuse or dependence are low in patients without psychiatric comorbidities or poly-drug use. The AEs of SMO were transient and did not require treatment discontinuation. The good safety profile can be explained by the posology which reflects low SMO dosages and by the controlled prescription and dispensing schemes in Austria and in Italy (Addolorato et al., 2020).

Overall, SMO showed efficacy in the treatment of uncomplicated AWS and in the maintenance of abstinence in AD patients with large effect sizes in high severity populations. In AWS, the main uncertainties are related to the efficacy of SMO in the treatment of seizures and delirium tremens and thus, SMO should not be used for these indications. In the maintenance of abstinence, the precision of SMO effect size in each population severity group could be further improved. However, since SMO was well tolerated in clinical trials and in 25+ years of real life therapeutic use, these limitations should not prevent the continued use of SMO in the maintenance of abstinence in the countries where the treatment is registered for this indication, i.e. Austria and Italy. In this context, a European group of alcohol researchers and clinicians reviewed SMO efficacy and safety data in the maintenance of abstinence in AD patients and concluded that SMO has an excellent risk-benefit ratio in this indication (van den Brink et al., 2018). Similarly, the EMA is reviewing Alcover® syrup efficacy and safety in the treatment of

AWS and in the maintenance of abstinence each year since 2016. Except for a variation to include an adverse event in the product information in 2018, no changes in Alcover® marketing authorization have been requested by EMA, indicating that no safety concerns and no lack of efficacy were identified, and that EMA considered the benefit risk of Alcover® syrup for the approved indications in Italy and Austria to be positive (European Medicines Agency, 2016, 2017a, 2018, 2019b, 2020b, 2021). Surprisingly, the EMA issued two negative opinions for the granules formulation SMO (Hopveus®/Alcover® granules) in the treatment of AWS and in the maintenance of abstinence in AD patients in 2017 and 2020, and considered that the efficacy was not sufficiently substantiated (European Medicines Agency 2017b, 2020a). The EMA highlighted that these opinions only concerned the granules formulation and do not affect Alcover® syrup (European Medicines Agency, 2017c). Limitations identified by the EMA for the demonstration of efficacy for the granules formulation were reviewed in a PhD thesis and were not considered as serious (Guiraud, 2023). Importantly, due to breaches vitiating the EMA assessment procedure, the 2020 EMA negative conclusions about Hopveus® was canceled by the European Court of Justice in March 2024 (Court of Justice of the European Union, 2024). Since procedural breaches identified by the European Court of Justice also occurred during the 2017 assessment procedure, a demand for retraction of the 2017 EMA scientific assessment has been submitted to the EMA in 2024.

In conclusion, SMO can be considered as a medicinal product with evidence of efficacy and acceptable safety and with a well-established use in the treatment of uncomplicated AWS and in the maintenance of abstinence in (severe) AD patients.

Conflict of interest statement

Julien Guiraud is employed by Vergio, Clichy, France and was employed by D&A Pharma until March 2022. Wim van den Brink reports personal fees from D&A Pharma, Kinnov Therapeutics, Bioproject, Lundbeck, Novartis, Indivior, Angelini, Mundipharma, Takeda, Opiant Inc, Recordati, Camurus, Novo Nordisk, and Clearmind Medicine.

Funding

Financial support for this work was provided by D&A Pharma.

References

Abanades, S., Farré, M., Barral, D., Torrens, M., Closas, N., Langohr, K., ... de la Torre, R. (2007). Relative abuse liability of γ-hydroxybutyric acid, flunitrazepam, and ethanol in club drug users. *Journal of Clinical Psychopharmacology, 27*, 625–638.

Abanades, S., Farre, M., Segura, M., Pichini, S., Barral, D., Pacifici, R., ... De La Torre, R. (2006). Hydroxybutyrate (GHB) in humans: Pharmacodynamics and pharmacokinetics. *Annals of the New York Academy of Sciences, 1074*, 559–576.

Absalom, N., Eghorn, L. F., Villumsen, I. S., Karim, N., Bay, T., Olsen, J. V., ... Wellendorph, P. (2012). 4 GABAA receptors are high-affinity targets for -hydroxybutyric acid (GHB). *Proceedings of the National Academy of Sciences, 109*, 13404–13409. https://doi.org/10.1073/pnas.1204376109.

Addolorato, G., Balducci, G., Capristo, E., Attilia, M. L., Taggi, F., Gasbarrini, G., & Ceccanti, M. (1999). Gamma-hydroxybutyric acid (GHB) in the treatment of alcohol withdrawal syndrome: A randomized comparative study versus benzodiazepine. *Alcoholism, Clinical and Experimental Research, 23*, 1596–1604 (a).

Addolorato, G., Caputo, F., Capristo, E., Bernardi, M., Stefanini, G. F., & Gasbarrini, G. (1999). A case of gamma-hydroxybutyric acid withdrawal syndrome during alcohol addiction treatment: Utility of diazepam administration. *Clinical Neuropharmacology, 22*, 60–62 (c).

Addolorato, G., Castelli, E., Stefanini, G. F., Casella, G., Caputo, F., Marsigli, L., ... Gasbarrini, G. (1996). An open multicentric study evaluating 4-hydroxybutyric acid sodium salt in the medium-term treatment of 179 alcohol dependent subjects. GHB Study Group. *Alcohol and Alcoholism (Oxford, Oxfordshire), 31*, 341–345.

Addolorato, G., Cibin, M., Caputo, F., Capristo, E., Gessa, G. L., Stefanini, G. F., & Gasbarrini, G. (1998). Gamma-hydroxybutyric acid in the treatment of alcoholism: Dosage fractioning utility in non-responder alcoholic patients. *Drug and Alcohol Dependence, 53*, 7–10 (b).

Addolorato, G., Lesch, O.-M., Maremmani, I., Walter, H., Nava, F., Raffaillac, Q., & Caputo, F. (2020). Post-marketing and clinical safety experience with sodium oxybate for the treatment of alcohol withdrawal syndrome and maintenance of abstinence in alcohol-dependent subjects. *Expert Opinion on Drug Safety, 19*, 159–166. https://doi.org/10.1080/14740338.2020.1709821.

Adinoff, B. (1994). Double-blind study of alprazolam, diazepam, clonidine, and placebo in the alcohol withdrawal syndrome: Preliminary findings. *Alcoholism, Clinical and Experimental Research, 18*, 873–878.

Aithal, G. P., & Grove, J. I. (2015). Mechanisms of alcohol toxicity. In J. Neuberger, & A. DiMartini (Eds.). *Alcohol abuse and liver disease* (pp. 55–64) Chichester, UK: John Wiley & Sons, Ltd. https://doi.org/10.1002/9781118887318.ch6.

Amato, L., Minozzi, S., Vecchi, S., & Davoli, M. (2010). Benzodiazepines for alcohol withdrawal. *Cochrane Database of Systematic Reviews*. https://doi.org/10.1002/14651858.CD005063.pub3.

Bahji, A., Bach, P., Danilewitz, M., Crockford, D., Devoe, D. J., El-Guebaly, N., & Saitz, R. (2022). Pharmacotherapies for adults with alcohol use disorders: A systematic review and network meta-analysis. *Journal of Addiction Medicine, 16*, 630–638. https://doi.org/10.1097/ADM.0000000000000992.

Bahji, A., Bach, P., Danilewitz, M., Crockford, D., El-Guebaly, N., Devoe, D. J., & Saitz, R. (2022). Comparative efficacy and safety of pharmacotherapies for alcohol withdrawal: A systematic review and network meta-analysis. *Addiction*, 15853. https://doi.org/10.1111/add.15853.

Bania, T. C., Ashar, T., Press, G., & Carey, P. M. (2003). Gamma-hydroxybutyric acid tolerance and withdrawal in a rat model. *10*, 9.

Barker, E. C., Puchowicz, M., Letterio, J., Higgins, K., & Sharkey, K. M. (2017). GHB levels in breast milk of women with narcolepsy with cataplexy treated with sodium oxybate. *Sleep Medicine, 36*, 172–177. https://doi.org/10.1016/j.sleep.2017.05.008.

Bay, T., Eghorn, L. F., Klein, A. B., & Wellendorph, P. (2014). GHB receptor targets in the CNS: Focus on high-affinity binding sites. *Biochemical Pharmacology, 87*, 220–228. https://doi.org/10.1016/j.bcp.2013.10.028.

Beardsley, P. M., Balster, R. L., & Harris, L. S. (1996). Evaluation of the discriminative stimulus and reinforcing effects of gammahydroxybutyrate (GHB). *Psychopharmacology, 127*, 315–322.

Borgen, L. A., Okerholm, R., Morrison, D., & Lai, A. (2003). The influence of gender and food on the pharmacokinetics of sodium oxybate oral solution in healthy subjects. *Journal of Clinical Pharmacology, 43*, 59–65.

Busardò, F., Kyriakou, C., Napoletano, S., Marinelli, E., & Zaami, S. (2015). Clinical applications of sodium oxybate (GHB): From narcolepsy to alcohol withdrawal syndrome. *European Review for Medical and Pharmacological Sciences, 19*, 4654–4663.

Busardò, F. P., Bertol, E., Mannocchi, G., Tittarelli, R., Pantano, F., Vaiano, F., ... Marinelli, E. (2016). Determination of GHB levels in breast milk and correlation with blood concentrations. *Forensic Science International, 265*, 172–181. https://doi.org/10.1016/j.forsciint.2016.02.020.

Busardò, F. P., & Jones, A. W. (2015). GHB pharmacology and toxicology: Acute intoxication, concentrations in blood and urine in forensic cases and treatment of the withdrawal syndrome. *Current Neuropharmacology, 13*, 47–70.

Caputo, F., Addolorato, G., Lorenzini, F., Domenicali, M., Greco, G., del RE, A., ... Bernardi, M. (2003). Gamma-hydroxybutyric acid versus naltrexone in maintaining alcohol abstinence: An open randomized comparative study. *Drug and Alcohol Dependence, 70*, 85–91.

Caputo, F., Addolorato, G., Stoppo, M., Francini, S., Vignoli, T., Lorenzini, F., ... Alcohol Treatment Study Group (2007). Comparing and combining gamma-hydroxybutyric acid (GHB) and naltrexone in maintaining abstinence from alcohol: An open randomised comparative study. *European Neuropsychopharmacology: The Journal of the European College of Neuropsychopharmacology, 17*, 781–789. https://doi.org/10.1016/j.euroneuro.2007.04.008.

Caputo, F., Francini, S., Brambilla, R., Vigna-Taglianti, F., Stoppo, M., Del Re, A., ... Bernardi, M. (2011). Sodium oxybate in maintaining alcohol abstinence in alcoholic patients with and without psychiatric comorbidity. *European Neuropsychopharmacology, 21*, 450–456. https://doi.org/10.1016/j.euroneuro.2010.12.005.

Caputo, F., Francini, S., Stoppo, M., Lorenzini, F., Vignoli, T., Del Re, A., ... Bernardi, M. (2009). Incidence of craving for and abuse of gamma-hydroxybutyric acid (GHB) in different populations of treated alcoholics: An open comparative study. *Journal of Psychopharmacology (Oxford), 23*, 883–890. https://doi.org/10.1177/0269881108094620.

Caputo, F., Mirijello, A., Cibin, M., Mosti, A., Ceccanti, M., Domenicali, M., ... Addolorato, G. (2015). Novel strategies to treat alcohol dependence with sodium oxybate according to clinical practice. *European Review for Medical and Pharmacological Sciences, 19*, 1315–1320.

Caputo, F., Skala, K., Mirijello, A., Ferrulli, A., Walter, H., Lesch, O., & Addolorato, G. (2014). Sodium oxybate in the treatment of alcohol withdrawal syndrome: A randomized double-blind comparative study versus oxazepam. The GATE 1 trial. *CNS Drugs, 28*, 743–752. https://doi.org/10.1007/s40263-014-0183-1.

Carai, M. A. M., Colombo, G., Reali, R., Serra, S., Mocci, I., Castelli, M. P., ... Gessa, G. L. (2002). Central effects of 1,4-butanediol are mediated by GABA(B) receptors via its conversion into gamma-hydroxybutyric acid. *European Journal of Pharmacology, 441*, 157–163.

Carai, M. A. M., Quang, L. S., Atzeri, S., Lobina, C., Maccioni, P., Orrù, A., ... Colombo, G. (2005). Withdrawal syndrome from γ-hydroxybutyric acid (GHB) and 1,4-butanediol (1,4-BD) in Sardinian alcohol-preferring rats. *Brain Research Protocols, 15*, 75–78. https://doi.org/10.1016/j.brainresprot.2005.04.001.

Carter, L. P., Pardi, D., Gorsline, J., & Griffiths, R. R. (2009). Illicit gamma-hydroxybutyrate (GHB) and pharmaceutical sodium oxybate (Xyrem®): Differences in characteristics and misuse. *Drug and Alcohol Dependence, 104*, 1–10. https://doi.org/10.1016/j.drugalcdep.2009.04.012.

Carter, L. P., Richards, B. D., Mintzer, M. Z., & Griffiths, R. R. (2006). Relative abuse liability of GHB in humans: A comparison of psychomotor, subjective, and cognitive effects of supratherapeutic doses of triazolam, pentobarbital, and GHB. *Neuropsychopharmacology: Official Publication of the American College of Neuropsychopharmacology, 31*, 2537–2551. https://doi.org/10.1038/sj.npp.1301146.

Cheng, H.-Y., McGuinness, L. A., Elbers, R. G., MacArthur, G. J., Taylor, A., McAleenan, A., ... Kessler, D. (2020). Treatment interventions to maintain abstinence from alcohol in primary care: Systematic review and network meta-analysis. *BMJ (Clinical Research ed.)*, m3934. https://doi.org/10.1136/bmj.m3934.

CMDh. (2021). List of safety concerns per approved Risk Management Plan (RMP) of active substances per product. Retrieved on May, 4, 2022 https://www.hma.eu/human-medicines/cmdh/pharmacovigilance/rmp.html.

Colombo, G., Agabio, R., Diaz, G., Fa, M., Lobina, C., Reali, R., & Gessa, G. L. (1998). *Gamma-hydroxybutyric acid intake in ethanol-preferring sP and -nonpreferring sNP rats*, 197–202.

Connelly, W. M., Errington, A. C., & Crunelli, V. (2013). γ-Hydroxybutyric acid (GHB) is not an agonist of extrasynaptic GABAA receptors. *PLoS One, 8*, e79062. https://doi.org/10.1371/journal.pone.0079062.

Cook, C., Aceto, M., Coop, A., & Beardsley, P. (2002). Effects of the putative antagonist NCS382 on the behavioral pharmacological actions of gammahydroxybutyrate in mice. *Psychopharmacology, 160*, 99–106. https://doi.org/10.1007/s00213-001-0951-8.

Court of Justice of the European Union. (2024). Judgment of the Court in Case C-291/22 P | D & A Pharma v Commission and EMA. Available at https://curia.europa.eu/jcms/upload/docs/application/pdf/2024–03/cp240046en.pdf accessed on April 21, 2024.

D&A Pharma. (2018). Alcover-sirup product information. Available at https://www.da-pharma.com/wp-content/uploads/2022/02/Alcover_Sirup_summery-characteristics.pdf accessed on April 15, 2024.

de Jong, C. A. J., Kamal, R., Dijkstra, B. A. G., & de Haan, H. A. (2012). Gamma-hydroxybutyrate detoxification by titration and tapering. *European Addiction Research, 18*, 40–45. https://doi.org/10.1159/000333022.

Di Bello, M., Gambassi, F., Mugnai, L., Masini, E., & Mannaioni, P. (1995). Gamma-hydroxybutyric acid induced suppression and prevention of alcohol withdrawal syndrome and relief of craving in alcohol-dependent patients. *Alcologia, 7*, 111–118.

Eller, M., Wang, Y., Wesnes, K., Alvarez-Horine, S., Benson, B., & Black, J. (2013). Evaluation of drug–drug interactions of sodium oxybate with divalproex: Results from a pharmacokinetic/pharmacodynamic study. *Sleep Medicine, 14*, e302–e303. https://doi.org/10.1016/j.sleep.2013.11.741.

European Medicines Agency. (2006). EPAR Xyrem - Scientific discussion. Retrieved on May 5, 2022 from https://www.ema.europa.eu/en/documents/scientific-discussion/xyrem-epar-scientific-discussion_en.pdf.

European Medicines Agency. (2007). Reflection Paper on Methodological Issues in Confirmatory Clinical Trials Planned with an Adaptive Design. CHMP/EWP/2459/02. Available at: https://www.ema.europa.eu/en/documents/scientific-guideline/reflection-paper-methodological-issues-confirmatory-clinical-trials-planned-adaptive-design_en.pdf Accessed March 5, 2021, b January 2021.

European Medicines Agency. (2010a). Guideline on the development of medicinal products for the treatment of alcohol dependence. EMA/CHMP/EWP/20097/2008. Available at: https://www.ema.europa.eu/en/documents/scientific-guideline/guideline-development-medicinal-products-treatment-alcohol-dependence_en.pdf Accessed March 4, 2021, a.

European Medicines Agency. (2010b). Overview of comments received on the draft guideline on the development of medicinal products for the treatment of alcohol dependence (CHMP/EWP/20097/08). Retrieved on May 6, 2022 from https://www.ema.europa.eu/en/documents/other/overview-comments-received-draft-guideline-development-medicinal-products-treatment-alcohol/ewp/20097/08_en.pdf.

European Medicines Agency. (2016). Pharmacovigilance Risk Assessment Committee (PRAC) Minutes of PRAC meeting on 10–13 May 2016. Retrieved on May 5, 2022 from https://www.ema.europa.eu/en/documents/minutes/minutes-prac-meeting-10–13-may-2016_en.pdf.

European Medicines Agency. (2017a). Pharmacovigilance Risk Assessment Committee (PRAC) Minutes of the meeting on 6–9 June 2017. Retrieved on May 5, 2022 from https://www.ema.europa.eu/en/documents/minutes/minutes-prac-meeting-6–9-june-2017_en.pdf.

European Medicines Agency. (2017b). Assessment report - Alcover (granules in sachet) - Procedure EMEA/H/A-29(4)/1451. d.

European Medicines Agency. (2017c). Questions and answers on Alcover granules (sodium oxybate 750, 1250 and 1750 mg) - EMA/833763/2017. Available at https://www.ema.europa.eu/en/documents/referral/alcover-article-294-referral-questions-and-answers-alcover-granules-sodium-oxybate-750–1250-and-1750-mg_en.pdf Last Access on April 22, 2024.

European Medicines Agency. (2018). Pharmacovigilance Risk Assessment Committee (PRAC) Minutes of PRAC meeting on 14–17 May 2018. Retrieved on May 5, 2022 from https://www.ema.europa.eu/en/documents/minutes/minutes-prac-meeting-14–17-may-2018_en.pdf.

European Medicines Agency. (2019a). Guideline on the investigation of subgroups in confirmatory clinical trials. EMA/CHMP/539146/2013. Available at: https://www.ema.europa.eu/en/documents/scientific-guideline/guideline-investigation-subgroups-confirmatory-clinical-trials_en.pdf Accessed March 5, 2021.

European Medicines Agency. (2019b). Pharmacovigilance Risk Assessment Committee (PRAC). Minutes of PRAC meeting on 13–16 May 2019. Retrieved on May 5, 2022 from https://www.ema.europa.eu/en/documents/minutes/minutes-prac-meeting-13–16-may-2019_en.pdf.

European Medicines Agency. (2020a). Assessment report - Hopveus - Procedure No. EMEA/H/C/004962/0000. Available at https://www.ema.europa.eu/en/documents/assessment-report/hopveus-epar-refusal-public-assessment-report_en.pdf accessed on April 30, 2024.

European Medicines Agency. (2020b). Pharmacovigilance Risk Assessment Committee (PRAC). Minutes of the meeting on 11 – 14 May 2020. Retrieved on May 5, 2022 from https://www.ema.europa.eu/en/documents/minutes/minutes-prac-meeting-11–14-may-2020_en.pdf.

European Medicines Agency. (2021). Pharmacovigilance Risk Assessment Committee (PRAC). Minutes for the meeting on 03–06 May 2021. Retrieved on May 5, 2022 from https://www.ema.europa.eu/en/documents/minutes/minutes-prac-meeting-3–6-may-2021_.pdf.

FDA. (2005). Estimating the maximum starting dose in initial clinical trials for therapeutics in adult healthy volunteers. Retrieved on May 5, 2022 from https://www.fda.gov/regulatory-information/search-fda-guidance-documents/estimating-maximum-safe-starting-dose-initial-clinical-trials-therapeutics-adult-healthy-volunteers 30.

FDA. (2015). Alcoholism: Developing Drugs for Treatment Guidance for Industry. Available at: https://www.fda.gov/files/drugs/published/Alcoholism—Developing-Drugs-for-Treatment.pdf Accessed March 5, 2021 17.

Ferrara, S. D., Tedeschi, L., Frison, G., Orlando, R., Mazzo, M., Zordan, R., ... Palatini, P. (1996). Effect of moderate or severe liver dysfunction on the pharmacokinetics of gamma-hydroxybutyric acid. *European Journal of Clinical Pharmacology, 50*, 305–310.

Ferrara, S. D., Zotti, S., Tedeschi, L., Frison, G., Castagna, F., Gallimberti, L., ... Palatini, P. (1992). Pharmacokinetics of gamma-hydroxybutyric acid in alcohol dependent patients after single and repeated oral doses. *British Journal of Clinical Pharmacology, 34*, 231–235.

Förstera, B., Castro, P. A., Moraga-Cid, G., & Aguayo, L. G. (2016). Potentiation of gamma aminobutyric acid receptors (GABAAR) by ethanol: How are inhibitory receptors affected? *Frontiers in Cellular Neuroscience, 10*. https://doi.org/10.3389/fncel.2016.00114.

Gallimberti, L., Canton, G., Gentile, N., Ferri, M., Cibin, M., Ferrara, S. D., ... Gessa, G. L. (1989). Gamma-hydroxybutyric acid for treatment of alcohol withdrawal syndrome. *Lancet 2*, 787–789.

Gallimberti, L., Cibin, M., Pagnin, P., Sabbion, R., Pani, P. P., Pirastu, R., ... Gessa, G. L. (1993). Gamma-hydroxybutyric acid for treatment of opiate withdrawal syndrome. *Neuropsychopharmacology, 9*, 77–81. https://doi.org/10.1038/npp.1993.45.

Gallimberti, L., Ferri, M., Ferrara, S. D., Fadda, F., & Gessa, G. L. (1992). Gamma-hydroxybutyric acid in the treatment of alcohol dependence: A double-blind study. *Alcoholism: Clinical and Experimental Research, 16*, 673–676. https://doi.org/10.1111/j.1530-0277.1992.tb00658.x.

Giorgetti, A., Busardò, F. P., & Giorgetti, R. (2022). Toxicological characterization of GHB as a performance-enhancing drug. *Frontiers in Psychiatry, 13*, 846983. https://doi.org/10.3389/fpsyt.2022.846983.

Goldstein, R. B., Chou, S. P., Smith, S. M., Jung, J., Zhang, H., Saha, T. D., ... Grant, B. F. (2015). Nosologic comparisons of DSM-IV and DSM-5 alcohol and drug use disorders: Results from the national epidemiologic survey on alcohol and related conditions–III. *Journal of Studies on Alcohol and Drugs, 76*, 378–388. https://doi.org/10.15288/jsad.2015.76.378.

Goodwin, A. K., Brown, P. R., Jansen, E. E. W., Jakobs, C., Gibson, K. M., & Weerts, E. M. (2009). Behavioral effects and pharmacokinetics of gamma-hydroxybutyrate (GHB) precursors gamma-butyrolactone (GBL) and 1,4-butanediol (1,4-BD) in baboons. *Psychopharmacology, 204*, 465–476. https://doi.org/10.1007/s00213-009-1477-8.

Goodwin, A. K., Kaminski, B. J., Griffiths, R. R., Ator, N. A., & Weerts, E. M. (2011). Intravenous self-administration of gamma-hydroxybutyrate (GHB) in baboons. *Drug and Alcohol Dependence*. https://doi.org/10.1016/j.drugalcdep.2010.10.006.

Groenman, A. P., Janssen, T. W. P., & Oosterlaan, J. (2017). Childhood psychiatric disorders as risk factor for subsequent substance abuse: A meta-analysis. *Journal of the American Academy of Child and Adolescent Psychiatry, 56*, 556–569. https://doi.org/10.1016/j.jaac.2017.05.004.

Gual, A., He, Y., Torup, L., van den Brink, W., & Mann, K. (2013). A randomised, double-blind, placebo-controlled, efficacy study of nalmefene, as-needed use, in patients with alcohol dependence. *European Neuropsychopharmacology, 23*, 1432–1442. https://doi.org/10.1016/j.euroneuro.2013.02.006.

Gueorguieva, R., Wu, R., Donovan, D., Rounsaville, B. J., Couper, D., Krystal, J. H., & O'Malley, S. S. (2012). Baseline trajectories of heavy drinking and their effects on postrandomization drinking in the COMBINE Study: Empirically derived predictors of drinking outcomes during treatment. *Alcohol (Fayetteville, N. Y.), 46*, 121–131. https://doi.org/10.1016/j.alcohol.2011.08.008.

Gueorguieva, R., Wu, R., Donovan, D., Rounsaville, B. J., Couper, D., Krystal, J. H., & O'Malley, S. S. (2011). Baseline trajectories of drinking moderate acamprosate and naltrexone effects in the combine study: Baseline trajectories of drinking in combine. *Alcoholism: Clinical and Experimental Research, 35*, 523–531. https://doi.org/10.1111/j.1530-0277.2010.01369.x.

Guerzoni, S., Pellesi, L., Pini, L. A., & Caputo, F. (2018). Drug-drug interactions in the treatment for alcohol use disorders: A comprehensive review. *Pharmacological Research, 133*, 65–76. https://doi.org/10.1016/j.phrs.2018.04.024.

Guiraud, J., 2023. Sodium oxybate for the treatment of alcohol dependence, University of Amsterdam. ed, Available at https://hdl.handle.net/11245.1/ad0b0a9e-e28c-432d-81a9-ccdf39b190f8, last access on December 21, 2023.

Guiraud, J., Addolorato, G., Antonelli, M., Aubin, H.-J., Bejczy, A., Benyamina, A., ... van den Brink, W. (2022). Sodium oxybate for the maintenance of abstinence in alcohol dependent patients: An international, multicenter, randomized, double-blind, placebo-controlled trial. *Journal of Psychopharmacology*.

Guiraud, J., Addolorato, G., Aubin, H.-J., Bachelot, S., Batel, P., de Bejczy, A., ... van den Brink, W. (2023). Sodium oxybate for alcohol dependence: A network meta-regression analysis considering population severity at baseline and treatment duration. *Alcohol and Alcoholism agac070*. https://doi.org/10.1093/alcalc/agac070.

Guiraud, J., Addolorato, G., Aubin, H.-J., Batel, P., de Bejczy, A., Caputo, F., ... SMO032 study group (2021). Treating alcohol dependence with an abuse and misuse deterrent formulation of sodium oxybate: Results of a randomised, double-blind, placebo-controlled study. *European Neuropsychopharmacology: The Journal of the European College of Neuropsychopharmacology, 52*, 18–30. https://doi.org/10.1016/j.euroneuro.2021.06.003.

Guiraud, J., Spanagel, R., & van den Brink, W. (2024). Substitution therapy for patients with alcohol dependence: Mechanisms of action and efficacy. *International Review of Neurobiology 175*, 187–239. https://doi.org/10.1016/bs.irn.2024.03.005.

ICH. (1998). International conference on harmonisation; guidance on statistical principles for clinical trials; availability–FDA. Notice. Fed Regist 63, 49583–49598.

ICH. (2016). M4E(R2) - Common technical document for the registration of pharmaceuticals for human use – Efficacy (EMA/CPMP/ICH/2887/1999).

Itzhak, Y., & Ali, S. F. (2002). Repeated administration of gamma-hydroxybutyric acid (GHB) to mice: Assessment of the sedative and rewarding effects of GHB. *Annals of the New York Academy of Sciences, 965*, 451–460.

Johnson, M. W., & Griffiths, R. R. (2013). Comparative abuse liability of GHB and ethanol in humans. *Experimental and Clinical Psychopharmacology, 21*, 112–123. https://doi.org/10.1037/a0031692.

Kamal, R. M. (2016). The detoxification approach for patients with GHB dependence. [S.l.: s.n.] - Retrieved on May 5, 2022 from https://repository.ubn.ru.nl/bitstream/handle/2066/151486/151486.pdf?sequence=1.

Kamal, R. M., van Noorden, M. S., Franzek, E., Dijkstra, B. A. G., Loonen, A. J. M., & De Jong, C. A. J. (2016). The neurobiological mechanisms of gamma-hydroxybutyrate dependence and withdrawal and their clinical relevance: A review. *Neuropsychobiology, 73*, 65–80. https://doi.org/10.1159/000443173.

Kaplan, W., Wirtz, V.J., Mantel-Teeuwisse, A., Stolk, P., Duthey, B., & Laing, R.. (2013). Priority medicines for Europe and the World - 2013 Update.

Keating, G. M. (2014). Sodium oxybate: A review of its use in alcohol withdrawal syndrome and in the maintenance of abstinence in alcohol dependence. *Clinical Drug Investigation, 34*, 63–80. https://doi.org/10.1007/s40261-013-0158-x.

Korninger, C., Roller, R. E., & Lesch, O. M. (2003). Gamma-hydroxybutyric acid in the treatment of alcohol withdrawal syndrome in patients admitted to hospital. *Acta Medica Austriaca, 30*, 83–86.

Krupitsky, E. M., Rudenko, A. A., Burakov, A. M., Slavina, T. Y., Grinenko, A. A., Pittman, B., ... Krystal, J. H. (2007). Antiglutamatergic strategies for ethanol detoxification: Comparison with placebo and diazepam. *Alcoholism: Clinical and Experimental Research 0, 070212174136008-???*. https://doi.org/10.1111/j.1530-0277.2007.00344.x.

Laborit, H. (1964). Sodium 4-hydroxybutyrate. *International Journal of Neuropharmacology, 3*, 433–451. https://doi.org/10.1016/0028-3908(64)90074-7.

Lale, R., Sklar, M., Wooldridge, J., & Sarkin, A. (2014). Gender congruence moderates beliefs about the causes of alcohol dependence and major depression. *International Journal of Mental Health and Addiction, 12*, 395–405. https://doi.org/10.1007/s11469-013-9465-y.

Langeland, W., Draijer, N., & van den Brink, W. (2004). Psychiatric comorbidity in treatment-seeking alcoholics: The role of childhood trauma and perceived parental dysfunction. *Alcoholism, Clinical and Experimental Research, 28*, 441–447. https://doi.org/10.1097/01.alc.0000117831.17383.72.

Leggio, L., Kenna, G. A., & Swift, R. M. (2008). New developments for the pharmacological treatment of alcohol withdrawal syndrome. A focus on non-benzodiazepine GABAergic medications. *Progress in Neuro-Psychopharmacology and Biological Psychiatry, 32*, 1106–1117. https://doi.org/10.1016/j.pnpbp.2007.09.021.

Leone, M. A., Vigna-Taglianti, F., Avanzi, G., Brambilla, R., & Faggiano, F. (2010). Gamma-hydroxybutyrate (GHB) for treatment of alcohol withdrawal and prevention of relapses. *Cochrane Database Syst Rev CD006266*. https://doi.org/10.1002/14651858.CD006266.pub2.

Liechti, M. E., Quednow, B. B., Liakoni, E., Dornbierer, D., von Rotz, R., Gachet, M. S., ... Bosch, O. G. (2016). Pharmacokinetics and pharmacodynamics of γ-hydroxybutyrate in healthy subjects: Effects of γ-hydroxybutyrate in healthy subjects. *British Journal of Clinical Pharmacology, 81*, 980–988. https://doi.org/10.1111/bcp.12863.

Litten, R. Z., Castle, I.-J. P., Falk, D., Ryan, M., Fertig, J., Chen, C. M., & Yi, H. (2013). The placebo effect in clinical trials for alcohol dependence: An exploratory Aanalysis of 51 naltrexone and acamprosate studies. *Alcoholism: Clinical and Experimental Research, 37*, 2128–2137. https://doi.org/10.1111/acer.12197.

Litten, R. Z., Egli, M., Heilig, M., Cui, C., Fertig, J. B., Ryan, M. L., ... Noronha, A. (2012). Medications development to treat alcohol dependence: A vision for the next decade: Medications development. *Addiction Biology, 17*, 513–527. https://doi.org/10.1111/j.1369-1600.2012.00454.x.

Liu, J., & Wang, L.-N. (2017). Baclofen for alcohol withdrawal. *Cochrane Database of Systematic Reviews*. https://doi.org/10.1002/14651858.CD008502.pub5.

Lyon, J. E., Khan, R. A., Gessert, C. E., Larson, P. M., & Renier, C. M. (2011). Treating alcohol withdrawal with oral baclofen: A randomized, double-blind, placebo-controlled trial. *Journal of Hospital Medicine, 6*, 469–474. https://doi.org/10.1002/jhm.928.

Mann, K., Batra, A., Fauth-Bühler, M., & Hoch, E. (2017). German guidelines on screening, diagnosis and treatment of alcohol use disorders (and the Guideline Group) *European Addiction Research, 23*, 45–60. https://doi.org/10.1159/000455841.

Mann, K., Lehert, P., & Morgan, M. Y. (2004). The efficacy of acamprosate in the maintenance of abstinence in alcohol-dependent individuals: Results of a meta-analysis. *Alcoholism: Clinical & Experimental Research, 28*, 51–63. https://doi.org/10.1097/01.ALC.0000108656.81563.05.

Mann, K., Torup, L., Sørensen, P., Gual, A., Swift, R., Walker, B., & van den Brink, W. (2016). Nalmefene for the management of alcohol dependence: Review on its pharmacology, mechanism of action and meta-analysis on its clinical efficacy. *European Neuropsychopharmacology, 26*, 1941–1949. https://doi.org/10.1016/j.euroneuro.2016.10.008.

Manzato, E., Nava, F., Manzato, G., Rosa, S., & Caputo, F. (2017). La terapia farmacologia dell'alcolismo: Il sodio oxibato (acido gamma-idrossibutirrico- GHB) farmaco anti-alcol approvato dall'organismo regolatorio italiano per la pratica clinica. Parte III. *Mission, 7*. https://doi.org/10.3280/mis46-2016oa4626.

Maremmani, I., Lamanna, F., & Tagliamonte, A. (2001). Long-term therapy using GHB (sodium gamma hydroxybutyrate) for treatment-resistant chronic alcoholics. *Journal of Psychoactive Drugs, 33*, 135–142. https://doi.org/10.1080/02791072.2001.10400478.

Mattila, M. J., Palva, E., Seppälä, T., & Ostrovskaya, R. U. (1978). Actions and interactions with alcohol of drugs on psychomotor skills: Comparison of diazepam and gamma-hydroxybutyric acid. *Archives Internationales de Pharmacodynamie et de Therapie, 234*, 236–246.

Mayo-Smith, M. F. (1997). Pharmacological management of alcohol withdrawal: A meta-analysis and evidence-based practice guideline. *JAMA: The Journal of the American Medical Association, 278*, 144. https://doi.org/10.1001/jama.1997.03550020076042.

McCusker, R. R., Paget-Wilkes, H., Chronister, C. W., Goldberger, B. A., & ElSohly, M. A. (1999). Analysis of gamma-hydroxybutyrate (GHB) in urine by gas chromatography-mass spectrometry. *Journal of Analytical Toxicology, 23*, 301–305. https://doi.org/10.1093/jat/23.5.301.

Minozzi, S., Amato, L., Vecchi, S., & Davoli, M. (2010). Anticonvulsants for alcohol withdrawal. *Cochrane Database of Systematic Reviews*. https://doi.org/10.1002/14651858.CD005064.pub3.

Moncini, M., Masini, E., Gambassi, F., & Mannaioni, P. F. (2000). Gamma-hydroxybutyric acid and alcohol-related syndromes. *Alcohol (Fayetteville, N. Y.), 20*, 285–291. https://doi.org/10.1016/S0741-8329(99)00093-2.

Morse, B. L., Chadha, G. S., Felmlee, M. A., Follman, K. E., & Morris, M. E. (2017). Effect of chronic γ-hydroxybutyrate (GHB) administration on GHB toxicokinetics and GHB-induced respiratory depression. *The American Journal of Drug and Alcohol Abuse, 43*, 686–693. https://doi.org/10.1080/00952990.2017.1339055.

National Institute on Alcohol Abuse and Alcoholism. (2021). *Alcohol use disorder: A comparison between DSM-IV and DSM-5.* Available at https://www.niaaa.nih.gov/sites/default/files/publications/AUD_A_Comparison.pdf Last Access 3 January 2023 2.

Nava, F., Premi, S., Manzato, E., Campagnola, W., Lucchini, A., & Gessa, G. L. (2007). Gamma-hydroxybutyrate reduces both withdrawal syndrome and hypercortisolism in severe abstinent alcoholics: An open study vs. diazepam. *The American Journal of Drug and Alcohol Abuse, 33*, 379–392. https://doi.org/10.1080/00952990701315046.

Nava, F., Premi, S., Manzato, E., & Lucchini, A. (2006). Comparing treatments of alcoholism on craving and biochemical measures of alcohol consumptions. *Journal of Psychoactive Drugs, 38*, 211–217. https://doi.org/10.1080/02791072.2006.10399846.

Nice. (2011). Alcohol-use disorders: Diagnosis, assessment and management of harmful drinking and alcohol dependence - Appendix 17d Pharmacological Interventions Forest Plots. Available at: https://www.nice.org.uk/guidance/cg115/evidence/appendix-17d-pharmacological-interventions-forest-plots-pdf-4911242584 Accessed October 10, 2021.

Nimmerrichter, A. A., Walter, H., Gutierrez-Lobos, K. E., & Lesch, O. M. (2002). Double-blind controlled trial of gamma-hydroxybutyrate and clomethiazole in the treatment of alcohol withdrawal. *Alcohol and Alcoholism (Oxford, Oxfordshire), 37*, 67–73.

Olsen, R. W., & Liang, J. (2017). Role of GABAA receptors in alcohol use disorders suggested by chronic intermittent ethanol (CIE) rodent model. *Molecular Brain, 10*, 45. https://doi.org/10.1186/s13041-017-0325-8.

Ormel, J. (1994). Common mental disorders and disability across cultures: Results from the WHO collaborative study on psychological problems in general health care. *JAMA: The Journal of the American Medical Association, 272*, 1741. https://doi.org/10.1001/jama.1994.03520220035028.

Palatini, P., Tedeschi, L., Frison, G., Padrini, R., Zordan, R., Orlando, R., ... Ferrara, S. D. (1993). Dose-dependent absorption and elimination of gamma-hydroxybutyric acid in healthy volunteers. *European Journal of Clinical Pharmacology, 45*, 353–356. https://doi.org/10.1007/BF00265954.

Perry, E. C. (2014). Inpatient management of acute alcohol withdrawal syndrome. *CNS Drugs, 28*, 401–410. https://doi.org/10.1007/s40263-014-0163-5.

Plinius Maior Society, 1994. Guidelines on evaluation of treatment of alcohol dependence.

Pribék, I. K., Kovács, I., Kádár, B. K., Kovács, C. S., Richman, M. J., Janka, Z., ... Lázár, B. A. (2021). Evaluation of the course and treatment of Alcohol Withdrawal Syndrome with the Clinical Institute Withdrawal Assessment for Alcohol - Revised: A systematic review-based meta-analysis. *Drug and Alcohol Dependence, 220*, 108536. https://doi.org/10.1016/j.drugalcdep.2021.108536.

Price, G. (2000). In-patient detoxification after GHB dependence. *The British Journal of Psychiatry, 177*, 181 –181.

Pross, N., Patat, A., Vivet, P., Bidaut, M., & Fauchoux, N. (2015). Pharmacodynamic interactions of a solid formulation of sodium oxybate and ethanol in healthy volunteers. *British Journal of Clinical Pharmacology, 80*, 480–492. https://doi.org/10.1111/bcp.12632.

Quang, L. S., Colombo, G., Lobina, C., Maccioni, P., Orru`, A., Gessa, G. L., ... Carai, M. A. M. (2006). Evaluation for the withdrawal syndrome from -hydroxybutyric acid (GHB), -butyrolactone (GBL), and 1,4-butanediol (1,4-BD) in different rat lines. *Annals of the New York Academy of Sciences, 1074*, 545–558. https://doi.org/10.1196/annals.1369.055.

Raybon, J. J., & Boje, K. M. K. (2007). Pharmacokinetics and pharmacodynamics of -hydroxybutyric acid during tolerance in rats: Effects on extracellular dopamine. *Journal of Pharmacology and Experimental Therapeutics, 320*, 1252–1260. https://doi.org/10.1124/jpet.106.113886.

Rehm, J., Baliunas, D., Borges, G. L. G., Graham, K., Irving, H., Kehoe, T., ... Taylor, B. (2010). The relation between different dimensions of alcohol consumption and burden of disease: An overview. *Addiction (Abingdon, England), 105*, 817–843. https://doi.org/10.1111/j.1360-0443.2010.02899.x.

Rehm, J., Guiraud, J., Poulnais, R., & Shield, K. D. (2018). Alcohol dependence and very high risk level of alcohol consumption: A life-threatening and debilitating disease. *Addiction Biology, 23*, 961–968. https://doi.org/10.1111/adb.12646.

Rehm, J., Shield, K. D., Rehm, M. X., Gmel, G., & Frick, U. (2012). Alcohol consumption, alcohol dependence and attributable burden of disease in Europe. *Centre for Addiction and Mental Health.*

Rösner, S., Hackl-Herrwerth, A., Leucht, S., Lehert, P., Vecchi, S., & Soyka, M. (2010). Acamprosate for alcohol dependence. *Cochrane Database of Systematic Reviews CD004332*. https://doi.org/10.1002/14651858.CD004332.pub2.

Rumgay, H., Murphy, N., Ferrari, P., & Soerjomataram, I. (2021). Alcohol and cancer: Epidemiology and biological mechanisms. *Nutrients, 13*, 3173. https://doi.org/10.3390/nu13093173.

Saitz, R., & O'Malley, S. S. (1997). Pharmacotherapies for alcohol abuse. *Medical Clinics, 81*, 881–907.

Samokhvalov, A. V., Irving, H., Mohapatra, S., & Rehm, J. (2010). Alcohol consumption, unprovoked seizures, and epilepsy: A systematic review and meta-analysis: Alcohol & epilepsy. *Epilepsia, 51*, 1177–1184. https://doi.org/10.1111/j.1528-1167.2009.02426.x.

Scherrer, B., Guiraud, J., Addolorato, G., Aubin, H., Bejczy, A., Benyamina, A., ... Spanagel, R. (2021). Baseline severity and the prediction of placebo response in clinical trials for alcohol dependence: A meta-regression analysis to develop an enrichment strategy. *Alcoholism, Clinical and Experimental Research*acer.14670. https://doi.org/10.1111/acer.14670.

Schwarzinger, M., Thiébaut, S. P., Baillot, S., Mallet, V., & Rehm, J. (2017). Alcohol use disorders and associated chronic disease – A national retrospective cohort study from France. *BMC Public Health, 18*. https://doi.org/10.1186/s12889-017-4587-y.

Shmulewitz, D., Aharonovich, E., Witkiewitz, K., Anton, R. F., Kranzler, H. R., Scodes, J., ... Alcohol Clinical Trials Initiative (ACTIVE Group) (2021). The World Health Organization risk drinking levels measure of alcohol consumption: Prevalence and health correlates in nationally representative surveys of U.S. adults, 2001-2002 and 2012-2013. *The American Journal of Psychiatry, 178*, 548–559. https://doi.org/10.1176/appi.ajp.2020.20050610.

Skala, K., Caputo, F., Mirijello, A., Vassallo, G., Antonelli, M., Ferrulli, A., ... Addolorato, G. (2014). Sodium oxybate in the treatment of alcohol dependence: From the alcohol withdrawal syndrome to the alcohol relapse prevention. *Expert Opinion on Pharmacotherapy, 15*, 245–257. https://doi.org/10.1517/14656566.2014.863278.

Spanagel, R. (2009). Alcoholism: A systems approach from molecular physiology to addictive behavior. *Physiological Reviews, 89*, 649–705. https://doi.org/10.1152/physrev.00013.2008.

Thai, D., Dyer, J. E., Benowitz, N. L., & Haller, C. A. (2006). GHB and ethanol effects and interactions in humans. *Journal of Clinical Psychopharmacology, 26*, 524–529. https://doi.org/10.1097/01.jcp.0000237944.57893.28.

UCB. (2022). Xyrem - Product information. Available at https://www.ema.europa.eu/en/documents/product-information/xyrem-epar-product-information_en.pdf accessed April 14, 2024.

van den Brink, W., Addolorato, G., Aubin, H.-J., Benyamina, A., Caputo, F., Dematteis, M., ... Spanagel, R. (2018). Efficacy and safety of sodium oxybate in alcohol-dependent patients with a very high drinking risk level. *Addiction Biology, 23*, 969–986. https://doi.org/10.1111/adb.12645.

van den Brink, W., Aubin, H.-J., Bladstrom, A., Torup, L., Gual, A., & Mann, K. (2013). Efficacy of as-needed nalmefene in alcohol-dependent patients with at least a high drinking risk level: Results from a subgroup analysis of two randomized controlled 6-month studies. *Alcohol and Alcoholism, 48*, 570–578. https://doi.org/10.1093/alcalc/agt061.

Van Sassenbroeck, D. K., De Paepe, P., Belpaire, F. M., & Buylaert, W. A. (2003). Characterization of the pharmacokinetic and pharmacodynamic interaction between gamma-hydroxybutyrate and ethanol in the rat. *Toxicological Sciences, 73*, 270–278. https://doi.org/10.1093/toxsci/kfg079.

Wang, Y. G., Swick, T. J., Carter, L. P., Thorpy, M. J., & Benowitz, N. L. (2009). Safety overview of post-marketing and clinical experience of sodium oxybate (Xyrem): Abuse, misuse, dependence, and diversion. *Journal of Clinical Sleep Medicine 5*, 365–371.

Weerts, E. M., Goodwin, A. K., Griffiths, R. R., Brown, P. R., Froestl, W., Jakobs, C., & Gibson, K. M. (2005). Spontaneous and precipitated withdrawal after chronic intragastric administration of gamma-hydroxybutyrate (GHB) in baboons. *Psychopharmacology, 179*, 678–687. https://doi.org/10.1007/s00213-004-2079-0.

Westman, J., Wahlbeck, K., Laursen, T. M., Gissler, M., Nordentoft, M., Hällgren, J., ... Ösby, U. (2015). Mortality and life expectancy of people with alcohol use disorder in Denmark, Finland and Sweden. *Acta Psychiatrica Scandinavica, 131*, 297–306. https://doi.org/10.1111/acps.12330.

Wetterling, T., Kanitz, R.-D., Besters, B., Fischer, D., Zerfass, B., John, U., ... Driessen, M. (1997). A new rating scale for the assessment of the alcohol-withdrawal syndrome (AWS scale). *Alcohol and Alcoholism, 32*, 753–760. https://doi.org/10.1093/oxfordjournals.alcalc.a008326.

WHO, 2014. Gamma-butyrolactone (GBL) - Critical Review Report.

Witkiewitz, K., Litten, R. Z., & Leggio, L. (2019). Advances in the science and treatment of alcohol use disorder. *Science Advances, 5*, eaax4043. https://doi.org/10.1126/sciadv.aax4043.

Woolverton, W. L., Rowlett, J. K., Winger, G., Woods, J. H., Gerak, L. R., & France, C. P. (1999). Evaluation of the reinforcing and discriminative stimulus effects of k-hydroxybutyrate in rhesus monkeys. *Drug and Alcohol Dependence* 7.

World Health Organization. (2000). *International guide for monitoring alcohol consumption and related harm*. Retrieved on May 5, 2022 from https://apps.who.int/iris/handle/10665/66529.

World Health Organization. (2012). *Gamma-hydroxybutyric acid (GHB) - Critical review report.* Retrieved on May 6, 2022 from https://kipdf.com/gamma-hydroxybutyric-acid-ghb-critical-review-report-expert-committee-on-drug-de_5ac4e7c91723dd210d0a315e.html.
World Health Organization. (2018). *Global status report on alcohol and health 2018.* Geneva: World Health Organization; 2018. Licence: CC BY-NC-SA 3.0 IGO. Retrieved on May 5, 2022 from http://www.who.int/substance_abuse/publications/global_alcohol_report/en/.
World Health Organization. (2019). *ICD-11: International classification of diseases* (11th revision). Retrieved on May 5, 2022 from https://icd.who.int/.

CHAPTER EIGHT

Ayahuasca for the treatment of alcohol use disorder

Eduardo A.V. Marinho[a,*], Yasmim A. Serra[a], Alexandre J. Oliveira-Lima[a], Tânia Marcourakis[b], and Laís F. Berro[a,c]
[a]Department of Health Sciences, Universidade Estadual de Santa Cruz, Ilhéus, BA, Brazil
[b]Laboratory of Neurotoxicology, Department of Clinical and Toxicological Analysis, School of Pharmaceutical Sciences, Universidade de São Paulo, São Paulo, SP, Brazil
[c]Department of Psychiatry and Human Behavior, Center for Innovation and Discovery in Addictions, University of Mississippi Medical Center, Jackson, MS, United States
*Corresponding author. e-mail address: eavmarinho@uesc.br

Contents

Abstract

For decades, psychedelics have been investigated for the treatment of psychiatric disorders. Specifically, evidence suggests that psychedelics may have therapeutic potential for the treatment of alcohol use disorder. Several studies with classic psychedelics, including LSD and psilocybin, show promising results, with psychedelics decreasing alcohol drinking and promoting abstinence in individuals with alcohol use disorder. In the last two decades, ayahuasca has emerged as another psychedelic with therapeutic potential for alcohol use disorder. Although its use by indigenous people from South America has been reported for thousands of years, ayahuasca, an Amazonian brewed beverage used in rituals, has gained attention in recent decades due to its reported effects in the central nervous system. Ayahuasca is a hallucinogenic beverage produced from the decoction of *Banisteriopsis caapi* and *Psychotria viridis*, plants that contain β-carbolines and N,N-dimethyltryptamine (DMT), respectively. The majority of clinical studies investigating ayahuasca for the treatment of alcohol use disorder are retrospective, and all show a significant decrease in alcohol use among ayahuasca users. Corroborating the clinical evidence, pre-clinical studies also have demonstrated that ayahuasca can block several of the abuse-related effects of alcohol. This chapter reviews the accumulating evidence from clinical and pre-clinical studies suggesting that ayahuasca may be a promising new pharmacotherapy for the treatment of alcohol use disorders, and discusses the potential mechanisms involved in these and other effects of ayahuasca.

International Review of Neurobiology, Volume 178
ISSN 0074-7742, https://doi.org/10.1016/bs.irn.2024.07.007

Substances that act on the central nervous system have been used since prehistoric times. The World Health Organization (WHO) defines psychoactive drugs as substances that affect mental processes, such as perception, consciousness, cognition, mood or emotions (World Health Organization WHO, 2020). Archaeological findings suggest that psychotropic substances, including alcohol and medicinal plants that induce altered states, have been used by humans for centuries for recreational, religious or healing purposes. Originally, their use was well accepted, as it was commonly associated with cultural values. However, the use of certain drugs for recreational purposes, to obtain pleasure and euphoria, or to relieve negative states, such as anxiety and pain, has increased steadily in society, and chronic use contributed to the development of substance use disorders (SUD) (Di Chiara, 1999; Scofield et al., 2016).

While psychoactive drugs may contribute to the development of SUD, some of these substances also have been studied for their therapeutic potential. For instance, psychedelics have been used anecdotally to treat alcohol use disorder (AUD) since the 1940s. While studies attempted to investigate these effects under controlled conditions, they came prematurely to a halt due to the classification of psychedelics as Schedule I drugs (i.e. "drugs with no currently accepted medical use and a high potential for abuse") by the Comprehensive Prevention and Control of Drug Abuse Act of 1970 in the United States. After decades of being banned from human psychiatric research, SUD-related psychedelic research resumed in the late 1990s, when new evidence indicated that mammals, including humans, can produce N, N-dimethyltryptamine (DMT), one of the main active components of psychedelic substances. DMT is endogenously synthesized from the essential amino acid L-tryptophan, and has been measured in rat cortex at levels similar to other neurotransmitters, such as serotonin (Aixalà, Dos Santos, Hallak, & Bouso, 2018; Belouin & Henningfeld, 2018; Calderon, Hunt, & Klein, 2018; DiVito & Leger, 2020; Nichols, Ronesi, Pratt, & Sanders-Bush, 2002).

When psychedelic research resumed, studies on SUD quickly reemerged, including the proposal of psilocybin, a serotonergic psychedelic, for the treatment of AUD (Benvenuti, Colombo, & Soverchia, 2023; Carhart-Harris & Goodwin, 2017). Since then, much progress has been made on the understanding of the mechanisms underlying the therapeutic effects of psychedelics and how they mediate AUD-related mechanisms, and other psychedelic drugs have garnered increasing interest. In particular, accumulating evidence from clinical and pre-clinical studies suggests that ayahuasca may have therapeutic potential for the treatment of AUD, which will be reviewed in this chapter.

1. Psychedelics

Psychedelics (also known as hallucinogens) are a class of psychoactive substances that produce changes in perception, mood and cognitive processes (Nichols, 2016). They may represent the oldest class of psychoactive agents known to man. Archaeological research suggests that humans used psychedelic plants more than 10,000 years ago in ritual ceremonies. Evidence suggests that psychedelics were important in the development of human society, as well as in the evolution of consciousness, mainly due to their cultural use and consumption in magical rituals for thousands of years (Grob, 1998).

One of the currently accepted classifications of psychedelics includes classical psychedelics and atypical (non-conventional) psychedelics. Classic psychedelics are: *phenethylamines*, such as 3,4,5-trimethoxy-phenethylamine (mescaline, derived from peyote); *tryptamines*, such as N,N-dimethyltryptamine (DMT, found in ayahuasca) and psilocybin (found in certain types of mushrooms such as *Psilocybe*); and *ergolines*, such as lysergic acid diethylamide (LSD) (Barsuglia et al., 2018; Murnane, 2018). Atypical psychedelics can be subdivided into: *dissociative psychedelics*, including N-methyl-D-aspartate (NMDA) receptor antagonists, such as phencyclidine (PCP) and ketamine (surgical anesthetic), and cannabinoid agonists, such as Δ9-tetrahydrocannabinol (Δ9-THC); and *deliriant hallucinogens*, including muscarinic receptor antagonists, such as scopolamine and entactogens (for example, 3,4-methylendioxymethamphetamine - MDMA - "ecstasy") (Calvey & Howells, 2018).

Classic psychedelics are known to act mainly as agonists or partial agonists of serotonin 5-HT_{2A} receptors, which play an important role in brain functions due to their interaction with other important neurotransmitters, such as dopamine (DA), glutamate, and gamma-aminobutyric acid (GABA) (Barsuglia et al., 2018; De Gregorio, Enns, Nun˜ez, Posa, & Gobbi, 2018) (Fig. 1). Serotonin has 14 receptors subtypes grouped in 7 families (5-HT_1 through 5-HT_7), which, alongside serotonin transporters (SERT), have frequently been target for the development of therapies for psychiatric disorders (Kurrasch-Orbaugh, Watts, Barker, & Nichols, 2003). Serotonin neurons are localized in the raphe nucleus and in the midbrain. They innervate several brain regions, modulating the activity of other brain systems both via inhibitory and excitatory pathways, depending on the various serotonin receptor subtypes. Of note, serotonin neurons densely innervate the ventral tegmental area (VTA), the nucleus accumbens (NAc), and the prefrontal cortex (PFC), all of which are highly involved in AUD.

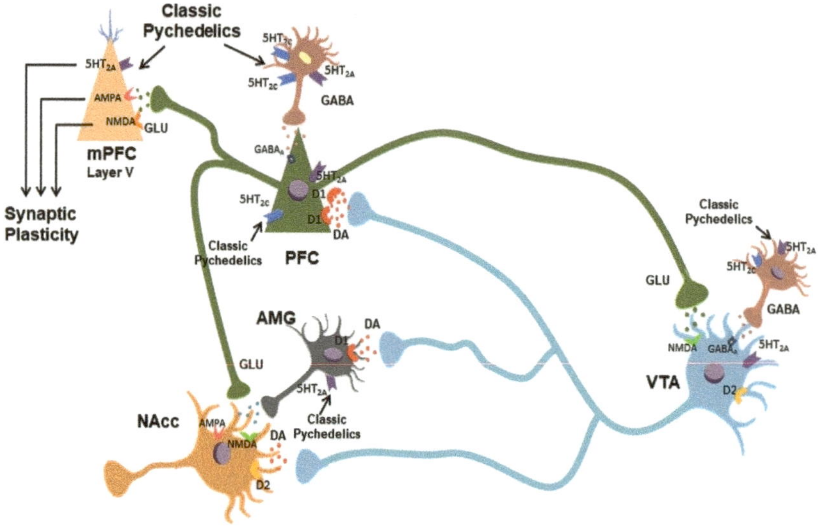

Fig. 1 **Mechanisms of action of classic psychedelics.** Psychedelics act primarily on serotonin 5-HT_{2A} and 5-HT_{2C} receptors. 5-HT_{2A} receptors are largely expressed in several brain regions, and studies suggest that these receptors mediate several of the effects of classic psychedelics. 5-HT_{2A} receptors are highly expressed in dopamine (DA) neurons of the ventral tegmental area (VTA) and in prefrontal cortex (PFC) glutamate (GLU) neurons, in addition to being expressed at lower levels in GABAergic interneurons in the VTA, PFC, nucleus accumbens (NAc) and amygdala (AMG). Of note, psychedelics bind to excitatory 5-HT_{2A} receptors on pyramidal cells in deep cortical layers, which leads to increased extracellular glutamate levels in the PFC. These project to the medial prefrontal cortex (mPFC), where glutamate release activates AMPA and NMDA receptors on cortical pyramidal neurons, thus promoting synaptic plasticity.

Specifically, increased activity of raphe nuclei serotonin neurons leads to changes in the firing rates of dopaminergic neurons in the VTA (Nichols, 2016).

The activation of 5-HT_{2A} receptors by classic psychedelics results in PFC hyperactivity that leads to a change in the functioning of the cortico-striatum–thalamus-cortical axis. This effect promotes impairment in information processing and distorted perception of stimuli, which results in the characteristic hallucinations reported with psychedelic use (Calvey & Howells, 2018). Classic psychedelics also act as agonists at other receptors, such as serotonin 5-HT_{1A} and 5-HT_{2C} receptors and dopamine D_2 receptors, all of which are present in various CNS structures. The combined action at these receptors also seems to contribute to the occurrence of psychedelic-induced hallucinations (Rossi et al., 2022).

Although $5\text{-}HT_{2A}$ receptors are described as fundamental in producing the hallucinogenic effects induced by classic psychedelics, not all $5\text{-}HT_{2A}$ receptor agonists produce similar effects. This has been explained by downstream effects of $5\text{-}HT_2$ receptor activation by different psychedelics. Receptors of the serotonin $5\text{-}HT_2$ family are Gq protein coupled and activate phospholipase C, which catalyzes membrane phospholipids into inositol triphosphate (IP3) and diaglyglycerol (DAG). Studies show that different signal transduction mechanisms can be activated at the same receptor depending on the conformation assumed by the agonist/receptor binding. This phenomenon is known as biased agonism, and explains how different agonists of the same receptor can result in different physiological and behavioral responses. LSD, for example, is an agonist at $5\text{-}HT_{2A}$ receptors and does not effectively stimulate the phospholipase C pathway. However, other signaling pathways are being implicated in the hallucinogenic effects of LSD and other psychedelics, such as the activation of different G protein isoforms. Specifically, interactions with the Gi/o and G12/13 proteins have been reported, both resulting in the activation of phospholipase A2 (Domínguez-Clavé et al., 2016; Kurrasch-Orbaugh et al., 2003).

Pharmacological manipulation of serotonin receptors can alter NAc DA levels, an effect caused by several substances with abuse potential, including alcohol. Therefore, the serotonergic neurotransmission exerts a regulatory mechanism in the limbic-corticostriatal circuit linked to the development and maintenance of AUD. The expression of serotonin receptors in both VTA DA and GABAergic neurons is particularly important for AUD-related behaviors. VTA GABAergic neurons act by inhibiting the activity of DA neurons originating from the VTA to the NAc. Excitatory $5\text{-}HT_{2A}$ receptors are expressed throughout the brain. $5\text{-}HT_{2A}$ receptors are highly expressed in DA neurons of the VTA and in PFC glutamate neurons, in addition to being expressed at lower levels in GABAergic interneurons in the VTA, PFC, NAc, amygdala and hippocampus. Therefore, because of their high expression at excitatory DA and glutamate neurons, activation of $5\text{-}HT_{2A}$ receptors directly increases DA levels. $5\text{-}HT_{2C}$ receptors, on the other hand, are highly expressed in GABAergic interneurons, with lower expression levels in DA and glutamate neurons. Therefore, activation of $5\text{-}HT_{2C}$ receptors leads to the activation of GABAergic interneurons and indirectly inhibits the activity of VTA DA neurons (Howell & Cunningham, 2015).

Chronic exposure to serotonin agonists or antagonists alters synaptic transmissions and causes neuroadaptations in serotonin receptors. Therefore, because serotonin signaling is related to the etiology of AUD, modulation of serotonin receptors by psychedelics may alter the desire for alcohol and other drugs. In fact, studies have shown that administration of serotonin agonists and antagonists can reduce alcohol drinking in both humans and animal models, although this effect is entirely dependent on the receptor target and pharmacological agent (Domanegg, Sommer, & Meinhardt, 2023). Based on these findings, psychedelics have been proposed for the treatment of AUD.

2. Psychedelics for the treatment of AUD

Clinical and pre-clinical evidence shows that drugs that target the serotonin system can influence alcohol drinking, suggesting that serotonin is involved in AUD (Bubar & Cunningham, 2006; 2008). One of the theories explaining the therapeutic effects of psychedelics on AUD involves the mesolimbic pathway. As described previously, drugs of abuse alter the activity of mesolimbic structures associated with drug craving, compulsive drug use, and negative affect observed during abstinence. It is possible that the mesolimbic dopaminergic circuit mediates compulsive behaviors, being the initial change that leads to alterations in other CNS systems (Calder & Hasler, 2023).

Specifically, during initial drug use, an increase in extracellular dopamine levels is observed in the mesolimbic system, specifically in the NAc. With chronic use, during AUD abstinence, there is a reduction in mesolimbic dopaminergic activity, with an important effect of serotonin agonists in attenuating the neuronal and behavioral responses associated with this pathway. Therefore, studies suggest that the therapeutic effects of psychedelics on AUD are possibly mediated by modulation of mesolimbic neurons via serotonin mechanisms, particularly mechanisms associated with psychedelic activity at 5-HT_{2A} receptors. In fact, studies have shown that 5-HT_{2A} receptor agonists decrease alcohol drinking in rats (Berquist & Fantegrossi, 2021; Maurel, De Vry, & Schreiber, 1999).

Serotonin 5HT_{2A} receptors are highly expressed in glutamatergic pyramidal neurons of the PFC that project to regions such as the NAc, amygdala and VTA (Bubar & Cunningham, 2006; Fletcher, Lê, & Higgins, 2008). Thus, studies suggest that plasticity induced by psychedelics would be initiated by the activation of 5HT_{2A} receptors, producing depolarization of prefrontal pyramidal neurons and increased glutamate release in the PFC. This would promote the activation of AMPA-type glutamatergic receptors and production of brain-derived neurotrophic factor (BDNF).

BDNF is a neurotrophin that is crucial for the promotion of neuroplasticity. BDNF binds to the TrkB receptor, resulting in the positive regulation of BDNF synthesis in the dendrites (Calder & Hasler, 2023). This creates a positive feedback loop that promotes prolonged periods of plasticity, which would counteract the neuronal atrophy observed in AUD, improving executive functioning and affective regulation.

In addition to neuroplasticity, reframing also has been proposed as an important mechanism by which psychedelics promote long-lasting therapeutic effects. Psychedelics appear to establish a new meaning to previously formed memories associated with alcohol use. This reframing appears to be mediated by the activation of signaling pathways that stimulate neuroplasticity, synaptogenesis and neurogenesis, which when increased promote a reduction in alcohol use (Grieco et al., 2022).

Based on these shared mechanisms, several studies have investigated the therapeutic effects of psychedelics in AUD. The majority of these studies, some of which date back to the 1960s, investigated classic psychedelics such as LSD and psilocybin. In addition to those early studies, over the past decade a growing number of clinical trials have been initiated exploring the therapeutic potential of these drugs for AUD. Generally, those studies demonstrate that controlled administration of psychedelics show promise as potential pharmacotherapies for AUD, particularly when administered in conjunction with psychotherapy (Calleja-Conde et al., 2022). Retrospective studies investigating the effects of psychedelic microdosing, a growing practice among recreational psychedelic users, show that 42–58 % of participants who microdose report reduced alcohol consumption (Anderson et al., 2019; Lea et al., 2020). According to an online survey, 83 % of participants reported no longer meeting criteria for SUD after psychedelic treatment (Garcia-Romeu et al., 2020).

LSD is the psychedelic that has been studied most extensively in the context of AUD. In addition to several pre-clinical studies showing decreased alcohol-related behaviors, including alcohol drinking, with LSD treatment (for review see Hilal et al., 2024), clinical studies from as early as 1950s show at least partially positive effects with LSD treatment (Chwelos, Blewett, Smith, & Hoffer, 1959; Smith, 1958). Clinical trials from the 1960s and 1970s show that LSD treatment associated with psychotherapy or group therapy significantly promoted abstinence among individuals with AUD (Kurland, Savage, & Pahnke, 1971; Maclean, Macdonald, Byrne, & Hubbard, 1961; O'reilly & Reich, 1962). However, several other clinical trials showed no significant effects of LSD treatment on AUD-related measures (Cheek et al., 1966; Denson & Sydiaha, 1970; Dusen, Wilson, Miners, & Hook, 1967; Johnson, 1969). Prospective studies also show that, based on past psychedelic experiences, more than 85 % of participants reported that LSD blocked the subjective effects of alcohol (Barrett, Archambault, Engelberg, & Pihl, 2000).

In addition to LSD, several studies have investigated the potential of psilocybin as a treatment for AUD. Psilocybin has been shown to decrease alcohol drinking (Alper, Cange, & Sah, 2023) and cue-induced reinstatement of alcohol drinking (Meinhardt, Pfarr, & Fouquet, 2021) in rodents. Similarly to LSD, early clinical studies with psilocybin showed positive effects for alcohol abstinence (Rydzyński & Gruszczyński, 1978; Rydzyński, Cwynar, Grzelak, & Jagiello, 1968). In an open-label clinical study associating long-term psychosocial treatment with two spaced out weeks of psilocybin treatment showed a significant increase in abstinence from alcohol after psilocybin treatment, an effect that was maintained even more than 20 weeks after its administration, at follow-up (Bogenschutz, Forcehimes, & Pommy, 2015). Of note, participants reported that specific subjective effects of psilocybin, including alterations in self-perception and in the person's relationship with alcohol, were potential reasons for prolonged abstinence after psilocybin treatment (Bogenschutz & Ross, 2018). In a recent randomized controlled clinical study, two administrations of psilocybin associated with motivational enhancement therapy and cognitive-behavioral therapy significantly decreased mean daily alcohol intake and heavy alcohol drinking days (Bogenschutz, Ross, & Bhatt, 2022).

Together, these studies demonstrate (1) the potential of LSD and psilocybin for the treatment of AUD, and (2) the growing interest in the study of psychedelics. In addition to LSD and psilocybin, several other psychedelics have been proposed as potential treatments for AUD, with ayahuasca emerging as a novel therapeutic based on evidence from decades of use by Native South American tribes.

3. Ayahuasca

Although its use by indigenous people from South America has been reported for thousands of years, ayahuasca, an Amazonian brewed beverage used in rituals, has gained attention in recent decades due to its reported CNS effects (Labate & Cavnar, 2010). Its religious and/or research use is approved in several countries, including Brazil, Australia, South Africa,

Canada, France, Spain, Germany, Japan and the Netherlands (Blainey, 2014; Labate & Feeney, 2012). Archaeological evidence of the earliest use of ayahuasca by indigenous people dates back more than four thousand years. This evidence includes ceramic vases and tubes that indicate that the beverage was used in the Ecuadorian Amazon as early as 1500 to 2000 BC. The word ayahuasca is of indigenous origin in the Quechua language, and it can be translated as "wine of the dead" (Labate, Dos Santos, Anderson, Mercante, & Barbosa, 2010).

Ayahuasca is considered a sacred beverage by numerous Amazonian indigenous tribes, and a medicinal beverage by mestizo healers from various parts of South America. It is a psychoactive drink used for religious purposes, and its traditional preparation comprises an aqueous decoction in the form of a tea consumed by indigenous people and, more recently, by urban populations in religious rituals and retreats. Ayahuasca is prepared from the decoction of two plants: *Psychotria viridis* and *Banisteriopsis caapi.* (Cata-Preta, Serra, Moreira-Junior, & Reis, 2018; Oliveira-Lima et al., 2015). The leaves of the shrub from the Rubiaceae family, *P. viridis*, contain indoalkylamine DMT as its main alkaloid, and to a lesser extent *N*-methyltryptamine (MMT). The bark and stems of the liana from the Malphighiaceae family, *B. caapi*, contain β-carboline derivatives of harmine, tetrahydroharmine and harmaline as its main alkaloids (Hamill, Hallak, Dursun, & Baker, 2019; Serra et al., 2022).

After oral administration of ayahuasca, DMT is rapidly inactivated due to the extensive first-pass metabolism through oxidative deamination by monoamine oxidase A (MAO-A) in the liver and gut. However, β-carboline alkaloids, such as those present in ayahuasca, are potent monoamine oxidase inhibitors (MAOIs) (Rossi et al., 2022). MAOIs are a variety of naturally and synthetically produced compounds used primarily in Western medicine to treat Parkinson's disease and depressive disorders. They exert their therapeutic effects by blocking the activity of monoamine oxidases, enzymes that catalyze the oxidation of endogenous and exogenous monoamines. Monoamines are a group of chemically similar molecules that serve as neurotransmitters and neuromodulators. They include catecholamines (dopamine, norepinephrine and epinephrine), tryptamines (serotonin, melatonin and DMT) and trace amines (tyramine, histamine and thyroamines) (Liester & Prickett, 2012). When ingested, monoamines are generally degraded by monoamine oxidase found in the gastrointestinal tract. However, in the presence of MAOIs, these monoamines are not metabolized in the gastrointestinal tract, which allows them to be absorbed by the organism. Therefore, the absorption of DMT in ayahuasca is possible due to the activity of the β-carbolines present in ayahuasca as MAOIs, which prevent DMT degradation and allow it to be absorbed in the gastrointestinal tract (Rossi et al., 2022).

Riba, McIlhenny, Valle, Bouso, and Barker (2012) evaluated DMT metabolism after ayahuasca administration. After oral administration of DMT, no psychoactive effects were observed due to the fast metabolism by MAO-A and the production of 3-indole-aceturic acid (3-IAA) in the urine. However, psychoactive effects were present when DMT was smoked, with a decrease in 3-IAA and an increase of DMT-N-oxide (DMT-NO) production, a process that is dependent on cytochrome P-450 (CYP450). The metabolic pathway observed with smoked DMT is similar to that observed after ayahuasca administration, demonstrating a shift int the metabolic route. Of note, DMT-NO can also undergo cyclization, producing 2-methyl-1,2,3,4-tetrahydro-beta-carboline (2-MTHBC). *N*-demethylation of DMT produces *N*-methyltryptamine (NMT), which also can undergo cyclization, producing 2-MTHBC. All of those metabolites can be found in urine after DMT ingestion in ayahuasca. Other pathways of DMT metabolism involve peroxidase, like kynureninase, which can form kynuramine metabolites (Carbonaro & Gatch, 2016; Gomes et al., 2014; Tourino et al., 2013).

DMT is structurally similar to serotonin, and it has partial or full agonist activity at several serotonin receptors, although with different affinities, including 5-HT_{1A}, 5-HT_{1B}, 5-HT_{1D}, 5-HT_{2A}, 5-HT_{2B}, 5-HT_{2C}, 5-HT_{5A}, 5-HT_6 and 5-HT_7 receptors (Carbonaro & Gatch, 2016). Despite the agonist binding profile of DMT, it has higher affinity for 5-HT_{2A} receptors, which has been proposed as the main receptor involved in its hallucinogenic effects. Therefore, DMT is generally considered the main psychoactive ingredient in ayahuasca and responsible for ayahuasca-induced hallucinations, while β-carbolines are proposed to promote DMT's effects via their activity as MAOIs. However, evidence suggests that β-carboline alkaloids also have psychoactive activity and therapeutic effects that may be relevant to AUD. For instance, tetrahydroharmine, the second main β-carboline in ayahuasca, also acts as a weak serotonin reuptake inhibitor, resulting in high CNS serotonin levels. β-carbolines also have been shown to directly increase circulating levels of monoamines, including serotonin, dopamine, norepinephrine and epinephrine. MAOIs are a class of antidepressants, and the effects of β-carbolines on monoamine oxidase might contribute to the overall therapeutic effect of ayahuasca (Carbonaro & Gatch, 2016, Rossi et al., 2022). Furthermore, these alkaloids also have been shown to bind to 5-HT_{2A} and 5-HT_{1A} receptors (Moura et al., 2006).

The psychoactive effects of ayahuasca are qualitatively similar to those of other drugs in the same pharmacological class, such as LSD and psilocybin. However, they are also phenomenologically unique. In humans, after ingestion ayahuasca's effects generally begin within 30–40 min, peak in ~2 h, and subside within 4–6 h (Cakic, Potkonyak, & Marshall, 2010; Callaway et al., 1999; James, Keppleer, Robertshaw, & Sessa, 2022). The main subjective effects described by ayahuasca users is colored eyes-closed visual hallucinations, complex thought processes and a general state of heightened awareness (Callaway et al., 1999). Commonly reported adverse effects include nausea, vomiting and diarrhea. According to Domínguez-Clavé et al. (2016), the duration of the subjective effects of ayahuasca correlate with the pharmacokinetics of DMT, but not of its β-carboline, particularly harmaline and tetrahydrohamanine. These alkaloids show peak plasma concentration after the onset of ayahuasca's subjective effects (Callaway et al., 1999).

Although DMT exerts psychedelic effects primarily through activation of 5-HT_{2A} receptors, recent and current studies have proposed that other non-serotonergic mechanisms might also be involved in the therapeutic effects of psychedelics. For instance, DMT has affinity for the sigma-1 opioid receptor, producing hyperactivity in mice independently of dopaminergic and serotonergic effects. It also binds to glutamatergic receptors type 2 and 3 (mGluR2/3), *N*-methyl-D-aspartate (NMDA) ionotropic receptors and trace amine associated receptor 1 (TAAR1) (for review see Carbonaro & Gatch, 2016). However, 5-HT_{2A} receptor activation has been implicated in many of the therapeutic effects of ayahuasca. 5-HT_{2A} receptors can be found in many brain regions, such as cortex, striatum, hippocampus, and amygdala (for review see Zhang & Stackman, 2015). They are metabotropic receptors, coupled to G-protein which activates phospholipase C (PLC) promoting the hydrolysis of the membrane phospholipid, phosphatidylinositol 4,5-bisphosphate (PIP2), generating inositol trisphosphate (IP3) and diacylglycerol (DAG). IP3 promotes the mobilization of intracellular calcium from the endoplasmic reticulum, increasing the activation of many genes involved in memory, learning and neuroplasticity, such as BDNF, egr-1 and c-fos (Domínguez-Clavé et al., 2016).

In fact, imaging studies show that ayahuasca can induce brain neuroplasticity, although the specific mechanisms underlying these effects remain poorly understood (Calder & Hasler, 2023). Pre-clinical studies show that DMT increases dendritic spine density, the expression of genes related to synaptic plasticity, immediate early genes and BDNF, especially in the neocortex, which is rich in 5-HT_{2A} receptors (Zhang & Stackman, 2015). 5-HT_{2A} receptors have been implicated in ayahuasca-induced neuroplasticity (Calder & Hasler, 2023). Of note, mGluR2/3 are modulated by 5-HT_{2A} agonists, leading to the formation of the heterodimer receptor, 5-HT_{2A}-mGluR2. The interaction between the glutamate and serotonin systems increases glutamatergic tone, which can increase BDNF production and, consequently, the beneficial long-lasting effects of ayahuasca (for review see Carbonaro & Gatch, 2016; James et al., 2022).

Sigma-1 receptors, which can also be activated by DMT in ayahuasca, are also involved in neurogenesis and neuroprotection, and are widely distributed in the CNS (Brierley & Davidson, 2012). Studies suggest that sigma-1 receptors in the brain can modulate neuronal differentiation and inflammation, increase BDNF and inhibit apoptosis (James et al., 2022). Morales-Garcia et al. (2020) showed that sigma-1 activation is involved in the neurogenic effects of DMT, and DMT-induced increased neurogenesis is correlated with improvement in spatial learning and memory tasks. These findings suggest that the cognitive enhancement observed with ayahuasca user may be the result of its enhanced neuroplasticity. Importantly, the β-carbolines present in ayahuasca also have been shown to exert beneficial effects on cognition (Moura et al., 2006), an effect that might be associated with β-carboline-induced neurogenesis (Morales-García et al., 2017).

Therefore, while the prevailing theory is that the effects of ayahuasca and other psychedelics are mediated by serotonergic mechanisms, future studies might unveil new and yet unknown mechanisms of action that may be relevant to the pathophysiology of AUD. While many of these mechanisms remain unknown, accumulating evidence shows that ayahuasca may be a potential therapeutic target for the treatment of AUD.

4. Ayahuasca for the treatment of AUD

Corroborating evidence showing that ayahuasca-assisted therapy decreases drug use and craving in drug-dependent individuals (Argento, Capler, Thomas, Lucas, & Tupper, 2019; Loizaga-Velder & Verres, 2014; Thomas, Lucas, Capler, Tupper, & Martin, 2013), several retrospective studies have shown that individuals who use ayahuasca report lower alcohol drinking compared to controls. In a study with members of an ayahuasca church, despite a higher prevalence of AUD compared to the general population before joining the church, 12 months of religious ayahuasca use led to a significant decrease in the prevalence of AUD in this population compared to the national norm (Barbosa et al., 2018). These findings are in agreement with studies showing that individuals who use ayahuasca regularly show less recent alcohol consumption (Barbosa et al., 2016) and lower scores on an alcohol use disorder scale compared to controls (Fábregas et al., 2010). Use of ayahuasca at retreats also has been associated with decreased alcohol use at a 6-month follow up (Thomas et al., 2013). Similarly, ayahuasca use in naturalistic settings has been shown to decrease self-reported current alcohol use, even after adjusting for religious or social group effects (Perkins et al., 2021).

Pre-clinical studies have been fundamental in determining whether the pharmacological properties of ayahuasca modulate the abuse-related effects of alcohol, or if the ritualistic context in which ayahuasca is predominantly consumed plays a major role in its therapeutic effects. Several pre-clinical studies have shown that ayahuasca blocks the abuse-related behavioral effects of alcohol. In pre-clinical studies, ayahuasca blocks the development and expression of behavioral sensitization to alcohol (Oliveira-Lima et al., 2015). Treatment with ayahuasca also blocked the reinstatement of alcohol-induced conditioned place preference in mice, a model of the rewarding effects of drugs (Cata-Preta et al., 2018). Ayahuasca also decreases alcohol drinking and preference in mice, an effect that is mediated by 5-HT_{2A} receptor mechanisms (Serra et al., 2022).

Together, these findings clearly demonstrate that ayahuasca has therapeutic effects in several behaviors associated with the abuse-related effects of alcohol in animal models. Clinical studies also show promising results, with ayahuasca consumption generally decreasing alcohol use and promoting abstinence. Future studies are required to further understand the importance of ayahuasca use under controlled settings, and whether the results obtained in naturalistic settings, such as in religious rituals or retreats, would be replicated under controlled laboratory or psychotherapy conditions. Finally, because ayahuasca is a psychedelic and its use is restricted, further large-scale studies are also needed to investigate its safety for use in individuals with AUD. Ayahuasca seems to have a mild side effect profile, and it does not seem to have the negative psychosocial effects observed with other drugs of abuse, including tolerance and dependence (for review, see Hamill et al., 2019). However, pre-clinical studies have shown significant reinforcing effects for DMT (Fantegrossi, Woods, & Winger, 2004), and ayahuasca has been shown to induce conditioned place preference in mice (Cata-Preta et al., 2018). In these studies, DMT had weak reinforcing effectiveness, and the rewarding effects of ayahuasca in mice were dose-dependent. Therefore, pre-clinical and clinical studies in which the doses of ayahuasca and, particularly, of its components (and their combinations) are controlled and thoroughly investigated are critical for understanding of optimal doses that may have therapeutic effects without potential unwanted side effects.

References

Aixalà, M., Dos Santos, R. G., Hallak, J. E. C., & Bouso, J. C. (2018). Psychedelics and personality. *ACS Chemical Neuroscience, 9*(10), 2304–2306. https://doi.org/10.1021/acschemneuro.8b00237 Epub 2018 Jun 4.PMID: 29863323.

Alper, K., Cange, J., Sah, R., et al. (2023). Psilocybin sex-dependentlyreduces alcohol consumption in C57BL/6J mice. *Frontiers in Pharmacology*. https://doi.org/10.3389/fphar.2022.1074633.

Anderson, T., Petranker, R., Rosenbaum, D., Weissman, C. R., Dinh-Williams, L. A., Hui, K., ... Farb, N. A. S. (2019). Microdosing psychedelics: Personality, mental health, and creativity differences in microdosers. *Psychopharmacology, 236*(2), 731–740. https://doi.org/10.1007/s00213-018-5106-2 Epub 2019 Jan 2. PMID: 30604183.

Argento, E., Capler, R., Thomas, G., Lucas, P., & Tupper, K. W. (2019). Exploring ayahuasca-assisted therapy for addiction: A qualitative analysis of preliminary findings among an Indigenous community in Canada. *Drug and Alcohol Review, 38*(7), 781–789.

Barbosa, P. C., Strassman, R. J., da Silveira, D. X., Areco, K., Hoy, R., Pommy, J., ... Bogenschutz, M. (2016). Psychological and neuropsychological assessment of regular hoasca users. *Comprehensive Psychiatry, 71*, 95–105.

Barbosa, P. C. R., Tófoli, L. F., Bogenschutz, M. P., Hoy, R., Berro, L. F., Marinho, E. A. V., ... Winkelman, M. J. (2018). Assessment of alcohol and tobacco use disorders among religious users of ayahuasca. *Frontiers in Psychiatry, 9*, 136.

Barrett, S. P., Archambault, J., Engelberg, M. J., & Pihl, R. O. (2000). Hallucinogenicdrugs attenuate the subjective response to alcohol in humans. *Human Psychopharmacology, 15*, 559–565. https://doi.org/10.1002/1099-1077(200010)15:73.0.CO;2-J.

Barsuglia, J. P., Polanco, M., Palmer, R., Malcolm, B. J., Kelmendi, B., & Calvey, T. (2018). A case report SPECT study and theoretical rationale for the sequential administration of ibogaine and 5-MeO-DMT in the treatment of alcohol use disorder. *Progress in Brain Research, 242*, 121–158. https://doi.org/10.1016/bs.pbr.2018.08.002 Epub 2018 Oct 25.PMID: 30471678.

Belouin, S. J., & Henningfeld, J. E. (2018). Psychedelics: Where we are now, why we got here, what we must do. *Neuropharmacology, 142*, 7–19.

Benvenuti, F., Colombo, D., Soverchia, L., et al. (2023). Psilocybin prevents reinstatement of alcohol seeking by disrupting the reconsolidation of alcohol-related memories. *Psychopharmacology, 240*, 1521–1530. https://doi.org/10.1007/s00213-023-06384-w.

Brierley, D. I., & Davidson, C. (2012). Developments in harmine pharmacology – implications for ayahuasca use and drug-dependence treatment. *Progress in Neuro-Psychopharmacology & Biological Psychiatry, 39*(2), 263–272.

Berquist, M. D., & Fantegrossi, W. E. (2021). Effects of 5-HT2A receptor agonist 2,5-dimethoxy-4-iodoamphetamine on alcohol consumption in Long-Evans rats. *Behavioural Pharmacology, 32*(5), 382–391. https://doi.org/10.1097/FBP.0000000000000628 PMID: 33595958; PMCID: PMC8266736.

Blainey, M. G. (2014). Forbidden therapies: Santo daime, ayahuasca, and the prohibition of entheogens in western onkeys. *Journal of Religion and Health, 54*(1), 287–302.

Bogenschutz, M. P., & Ross, S. (2018). Therapeutic applications of classic hallucinogens. *Current Topics in Behavioral Neurosciences, 36*, 361–391. https://doi.org/10.1007/7854_2016_464.
Bogenschutz, M. P., Forcehimes, A. A., Pommy, J. A., et al. (2015). Psilocybinassistedtreatment for alcohol dependence: A proof-of-concept study. *Journal of Psychopharmacology (Oxford, England), 29*, 289–299. https://doi.org/10.1177/0269881114565144.
Bogenschutz, M. P., Ross, S., Bhatt, S., et al. (2022). Percentage of heavy drinking days following psilocybin assisted psychotherapy vs placebo in the treatment of adult patients with alcoholuse disorder: A randomized clinical trial. *JAMA Psychiatry, 9*, 953–962. https://doi.org/10.1001/JamapSychiAtry.2022.2096.
Bubar, M. J., & Cunningham, K. A. (2006). Serotonin 5-Ht2a and 5-Ht2c receptors as potential targets for modulation of psychostimulant use and dependence. *Current Topics in Medicinal Chemistry, 6*(P), 1971–1985.
Bubar, M. J., & Cunningham, K. A. (2008). Prospects for onkeys N 5-Ht2r pharmacotherapy in psychostimulant abuse. *Progress in Brain Research, 172*, 319–346.
Cakic, V., Potkonyak, J., & Marshall, A. (2010). Dimethyltryptamine (DMT): Subjective effects and patterns of use among Australian recreational users. *Drug and Alcohol Dependence, 111*(1–2), 30–37. https://doi.org/10.1016/j.drugalcdep.2010.03.015.
Calder, A. E., & Hasler, G. (2023). Towards na understanding of psychedelic-induced neuroplasticity. *Neuropsychopharmacology, 48*, 104–112. https://doi.org/10.1038/s41386-022-01389-z.
Calderon, S. N., Hunt, J., & Klein, M. (2018). A regulatory perspective on the evaluation of hallucinogen drugs for human use. *Neuropharmacology, 142*, 135–142.
Calleja-Conde, J., Morales-García, J. A., Echeverry-Alzate, V., Bühler, K. M., Giné, E., & López-Moreno, J. A. (2022). Classic psychedelics and alcohol use disorders: A systematic review of human and animal studies. *Addiction Biology, 27*(6), e13229. https://doi.org/10.1111/adb.13229 PMID: 36301215; PMCID: PMC9541961.
Calvey, T., & Howells (2018). An introduction to psychedelic neuroscience. *Progress in Brain Research, 2018*(242), 1–23. https://doi.org/10.1016/bs.pbr.2018.09.013 Epub Nov 15. PMID: 30471677.
Callaway, J. C., McKenna, D. J., Grob, C. S., Brito, G. S., Raymon, L. P., Poland, R. E., ... Mash, D. C. (1999). Pharmacokinetics of hosca alkaloids in health humans. *Journal of Ethnopharmacology, 65*, 243–246.
Carbonaro, T. M., & Gatch, M. B. (2016). Neuropharmacology of N,N,-dimethyltryptamine. *Brain Research Bulletin, 126*(Pt1), 74–88. https://doi.org/10.1016/j.brainresbull.2016.04.016.
Carhart-Harris, R. L., & Goodwin, G. M. (2017). The therapeutic potential of psychedelic drugs: Past, present, and future. *Neuropsychopharmacology: Official Publication of the American College of Neuropsychopharmacology, 42*(11), 2105–2113. https://doi.org/10.1038/npp.2017.84 Epub 2017 Apr 26. PMID: 28443617; PMCID: PMC5603818.
Cata-Preta, E. G., Serra, Y. A., Moreira-Junior, E. C., Reis, H. S., et al. (2018). Ayahuasca and its DMT- and ß-carbolines- containonkeysn.ents block onkeysn.siononkeyanol-induced conditioned place preference in mice: Role of the treatment environment. *Frontiers in Pharmacology, 9*(May), 561.
Cheek, F. E., Osmond, H., et al. (1966). Observations regarding the use of LSD-25 in the treatment of alcoholism. *Journal of Psychopharmacology, 1*, 56–74.
Chwelos, N., Blewett, D. B., Smith, C. M., & Hoffer, A. (1959). Use of d-lysergic acid diethylamide in the treatment of alcoholism. *Quarterly Journal of Studies on Alcohol, 20*, 577–590.
De Gregorio, D., Enns, J. P., Nuñez, N. A., Posa, L., & Gobbi, G. (2018). D-Lysergic acid diethylamide, psilocybinonke other classic hallucinogens: Mechanism of action and potential therapeutic applications in mood disorders. *Progress in Brain Research,* 242.

Denson, R., & Sydiaha, D. (1970). A controlled study of LSD treatment in alcoholism and neurosis. *The British Journal of Psychiatry, 116*, 443–445. https://doi.org/10.1192/Bjp.116.533.443.

Di Chiara, G. (1999). Drug addiction as dopamine-dependent associative learning disorder. *European Journal of Pharmacology, 375*(1–3), 13–30.

DiVito, A. J., & Leger, R. F. (2020). Psychedelina as an emerging novel intervention in the treatment of substance use disorder: A review. *Molecular Biology Reports, 47*(12), 9791–9799. https://doi.org/10.1007/s11033-020-06009-x Epub 2020 Nov 24. PMID: 33231817.

Domanegg, K., Sommer, W. H., & Meinhardt, M. W. (2023). Psychedelic targeting of metabotropic glutamate receptor 2 and its implications for the treatment of alcoholism. *Cells, 12*, 963. https://doi.org/10.3390/cells12060963.

Domínguez-Clavé, E., Soler, J., Elices, M., Pascual, J. C., Álvarez, E., Revenga, M. F., ... Riba, J. (2016). Ayahuasca: Pharmacology, neuroscience and therapeutic potential. *Brain Research Bulletin, 126*, 89–101. https://doi.org/10.1016/j.brainresbull.2016.03.002.

Dusen, W. V., Wilson, W., Miners, W., & Hook, H. (1967). Treatment of alcoholism with lysergide. *Quarterly Journal of Studies on Alcohol, 28*, 295–304. https://doi.org/10.15288/qjsa.28.295.

Fábregas, J. M., González, D., Fondevila, S., Cutchet, M., Fernández, X., Barbosa, P. C., ... Bouso, J. C. (2010). Assessment of addiction severity among ritual users of ayahuasca. *Drug Alcohol Depend, 111*(3), 257–261. https://doi.org/10.1016/J.Drugalcdep.2010.03.024 Epub 2010 Jun 15. Pmid: 20554400.

Fantegrossi, W., Woods, J. H., & Winger, G. (2004). Transient reinforcing effects of phenylisopropylamine and indolealkylamine hallucinogens in onkeys monkeys. *Behavioural Pharmacology, 15*(2), 149–157. https://doi.org/10.1097/00008877-200403000-00007 Pmid: 15096915.

Fletcher, P. J., Lê, A. D., & Higgins, C. A. (2008). Serotonin receptors as potential targets for modulation of nicotine use and dependence. *Progress in Brain Research, v. 172*, 361–383.

Garcia-Romeu, A., Davis, A. K., Erowid, E., Erowid, F., Griffiths, R. R., & Johnson, M. W. (2020). Persisting reductions in cannabis opioid, and stimulant misuse after naturalistic psychedelic use: An online survey. *Frontiers in Psychiatry, 10*, 955. https://doi.org/10.3389/fpsyt.2019.00955 PMID: 32038317; PMCID: PMC6987443.

Gomes, M. M., Coimbra, J. B., Clara, R. O., Dorr, F. A., Moreno, A. C. R., Chagas, J. R., ... Campa, A. (2014). Biosynthesis of N,N-dimethyltryptamine (DMT) in a melanoma cell line and its metabolization by peroxidases. *Biochemical Pharmacology, 88*, 393–401.

Grieco, S. F., Castrén, E., Knudsen, G. M., Kwan, A. C., Olson, D. E., Zuo, Y., ... Xu, X. (2022). Psychedelics and neural plasticity: Therapeutic implications. *The Journal of Neuroscience, 42*(45), 8439–8449. https://doi.org/10.1523/JNEUROSCI.1121-22.2022 PMID: 36351821; PMCID: PMC9665925.

Grob, C. (1998). 2. Psychiatric research with hallucinogens: What have we learned? *The Heffter Review of Psychedelic Research.*

Hamill, J., Hallak, J., Dursun, S. M., & Baker, G. (2019). Ayahuasca: Psychological and physiologic effects, pharmacology and potential uses in addiction and mental illness. *Current Neuropharmacology, 17*(2), 108–128.

Hilal, F. F., Jeanblanc, J., Deschamps, C., Naassila, M., Pierrefiche, O., & Ben Hamida, S. (2024). Epigenetic drugs and psychedelics as emerging therapies for alcohol use disorder: Insights from preclinical studies. *Journal of Neural Transmission (Vienna), 131*(5), 525–561. https://doi.org/10.1007/s00702-024-02757-3 Epub 2024 Mar 30. PMID: 38554193.

Howell, L. L., & Cunningham, K. A. (2015). Serotonin 5-Ht2 receptor interactions with dopamine function: implications for therapeutics in cocaine use disorder. *Pharmacological Reviews, v. 67*, 176–197.

James, E., Keppleer, J., Robertshaw, T., & Sessa, B. (2022). N,N-dimethyltryptamine and Amazonian ayahuasca plant medicine. *Human Psychopharmacology: Clinical and Experimental, 37*, e2835. https://doi.org/10.1002/hup.2835.

Johnson, F. G. (1969). LSD in the treatment of alcoholism. *AJP, 126*, 481–487. https://doi.org/10.1176/ajp.126.4.481.

Kurland, A., Savage, C., Pahnke, W. N., et al. (1971). LSD in the treatment of alcoholics. *Pharmacopsychiatry, 4*, 83–94. https://doi.org/10.1055/s-0028-1094301.

Kurrasch-Orbaugh, D. M., Watts, V. J., Barker, E. L., & Nichols, D. E. (2003). Serotonin 5-hydroxytryptamine 2A receptor-coupled phospholipase C and phospholipase A2 signaling pathways have different receptor reserves. *The Journal of Pharmacology and Experimental Therapeutics, 304*(1), 229–237. https://doi.org/10.1124/jpet.102.042184 PMID: 12490596.

Labate, B. C., & Cavnar, C. (2010). The expansion of the field of research on ayahuasca: Some reflections about the ayahuasca track at the MAPS "Psychedelic Science in the 21st Century" conference. *The International Journal on Drug Policy, 22*(2), 174–178. https://doi.org/10.1016/j.drugpo.2010.09.002 Epub 2010 Nov 3. PMID: 21051213.

Labate, B. C., Dos Santos, R. G., Anderson, B., Mercante, M., & Barbosa, P. C. R. (2010). The treatment and handling of substance dependence with ayahuasca: Reflections on urrent and future research. *Ayahuasca, Ritual and Religion in Brazil, N. January*, 205–227.

Labate, B. C., & Feeney, K. (2012). Ayahuasca and the process of regulation in brazil and internationally: Implications and challenges. *International Journal of Drug Policy, v. 23*(n. 2), 154–161.

Lea, T., Amada, N., Jungaberle, H., Schecke, H., Scherbaum, N., & Klein, M. (2020). Perceived outcomes of psychedelic microdosing as self-managed therapies for mental and substance use disorders. *Psychopharmacology, 237*(5), 1521–1532. https://doi.org/10.1007/s00213-020-05477-0 Epub 2020 Feb 11. PMID: 32043165.

Liester, M. B., & Prickett, J. I. (2012). Hypotheses regarding the mechanisms of ayahuasca in the treatment of addictions. *Journal of Psychoactive Drugs, 44*(3), 200–208.

Loizaga-Velder, A., & Verres, R. (2014). Therapeutic effects of ritual ayahuasca use in the treatment of substance dependence–qualitative results. *Journal of Psychoactive Drugs, 46*(1), 63–72.

Maclean, J. R., Macdonald, D. C., Byrne, U. P., & Hubbard, A. M. (1961). The use of LSD-25 in the treatment of alcoholism and other psychiatric problems. *Quarterly Journal of Studies on Alcohol, 22*, 34–45.

Maurel, S., De Vry, J., & Schreiber, R. (1999). 5-HT receptor ligands differentially affect operant oral self-administration of ethanol in the rat. *European Journal of Pharmacology, 370*, 217–223.

Meinhardt, M. W., Pfarr, S., Fouquet, G., et al. (2021). Psilocybin targets a common molecular mechanism for cognitive impairment and increased craving in alcoholism. *Science Advances, 7*, eabh2399. https://doi.org/10.1126/sciadv.abh2399.

Moura, D. J., Rorig, C., Vieira, D. L., Henriques, J. A. P., Roesler, R., Saffi, J., & Boeira, J. M. (2006). Effects of β-carboline alkaloids on the object recognition task in mice. *Life Sciences, 79*, 2099–2104. https://doi.org/10.1016/j.lfs.2006.07.004.

Morales-Garcia, J. A., Calleja-Conde, J., Lopez-Moreno, J. Á., Alonso-Gil, S., Sanz-Cristobal, M., Riba, J., & Perez-Castillo, A. (2020). *Translational Psychiatry, 10*, 331. https://doi.org/10.1038/s41398-020-01011-0.

Morales-García, J. A., de la Fuente Revenga, M., Alonso-Gil, S., Rodríguez-Franco, M. I., Feilding, A., Perez-Castillo, A., & Riba, J. (2017). The alkaloids of *Banisteriopsis caapi*, the plant source of the Amazonian hallucinogen Ayahuasca, stimulate adult neurogenesis in vitro. *Scientific Reports, 7*, 5309. https://doi.org/10.1038/s41598-017-05407-9.

Murnane, K. S. (2018). The renaissance in psychedelic research: What do preclinical models have to offer. *Progress in Brain Research, 2018*(242), 25–67. https://doi.org/10.1016/bs.pbr.2018.08.003 Epub Oct 5. PMID: 30471682.

Nichols, C. D., Ronesi, J., Pratt, W., & Sanders-Bush, E. (2002). Hallucinogens and Drosophila: Linking serotonin receptor activation to behavior. *Neuroscience, 115*, 979–984.

Nichols, D. E. (2016). Psychedelics | Pharmacological Reviews. *68*(2), 264–355. https://doi.org/10.1124/pr.115.011478. Erratum in: Pharmacological Reviews (2016). *68*(2), 356. PMID: 26841800; PMCID: PMC4813425.

O'reilly, P. O., & Reich, G. (1962). Lysergic acid and the alcoholic. *Diseases of the Nervous System, 23*, 331–334.

Oliveira-Lima, A. J., et al. (2015). Effects of ayahuasca on the development of ethanol-induced behavioral sensitization and on a post-sensitization treatment in mice. *Physiology and Behavior, v. 142*, 28–36.

Perkins, D., Opaleye, E. S., Simonova, H., Bouso, J. C., Tófoli, L. F., GalvÃo-Coelho, N. L., ... Sarris, J. (2021). Associations between ayahuasca consumption in naturalistic settings and current alcohol and drug use: Results of a large international cross-sectional survey. *Drug and Alcohol Review*.

Riba, J., McIlhenny, E. H., Valle, M., Bouso, J. C., & Barker, S. A. (2012). Metabolism and disposition of N,N-dimethyltryptamine and harmala alkaloids after oral administration of ayahuasca. *Drug Testing and Analysis, 4*(7–8), 610–616. https://doi.org/10.1002/dta.1344.

Rossi, G. N., Guerra, L. T. L., Baker, G. B., Dursun, S. M., Saiz, J. C. B., Hallak, J. E. C., & dos Santos, R. G. (2022). Molecular pathways of the therapeutic effects of ayahuasca, a botanical psychedelic and potential rapid-acting antidepressant. *Biomolecules, 12*, 1618. https://doi.org/10.3390/biom12111618.

Rydzyński, Z., & Gruszczyński, W. (1978). Treatment of alcoholism withpsychotomimetic drugs. A follow-up study. *Activitas Nervosa Superior, 20*, 81–82.

Rydzyński, Z., Cwynar, S., Grzelak, L., & Jagiello, W. (1968). Prelminaryreport on the experience with psychosomimetic drugs in the treatment of alcobolism. *Activitas Nervosa Superior, 10*, 273.

Scofield, M. D., Heinsbroek, J. A., Gipson, C. D., Kupchik, Y. M., Spencer, S., Smith, A. C. W., ... Kalivas, P. W. (2016). The nucleus accumbens: Mechanisms of addiction across drug classes reflect the importance of glutamate homeostasis. *Pharmacological Reviews, 68*(3), 816–871.

Serra, Y. A., Barros-Santos, T., Anjos-Santos, A., Kisaki, N. D., Jovita- Farias, C., Leite, J. P. C., ... Marinho, E. A. V. (2022). Role of 5-Ht2a receptors in the effects of ayahuasca on ethanol self-administration using a two-bottle choice paradigm in male mice. *Psychopharmacology, 239*(6), 1679–1687.

Smith, C. M. (1958). A new adjunct to the treatment of alcoholism: The hallucinogenic drugs. *Quarterly Journal of Studies on Alcohol, 19*, 406–417.

Thomas, G., Lucas, P., Capler, N. R., Tupper, K. W., & Martin, G. (2013). Ayahuasca-assisted therapy for addiction: Results from a preliminary observational study in Canada. *Current Drug Abuse Reviews, 6*(1), 30–42.

Tourino, M. C., de Oliveira, E. M., Belle, L. P., Knebel, F. H., Albuquerque, R. C., Dorr, F. A., ... Campa, A. (2013). Tryptamine and dimethyltryptamine inhibit indoleamine 2,3 dioxygenase and increase the tumor-reactive effect of peripheral blood mononuclear cells. *Cell Biochemistry and Function, 31*(5), 361–364.

World Health Organization (WHO). (2020). Psychoactive drugs. Global status report on alcohol and health. https://www.who.int/health-topics/drugs-psychoactive#tab=tab_1.

Zhang, G., & Stackman, R. W., Jr (2015). The role of serotonin 5-HT2A receptors in memory and cognition. *Frontiers in Pharmacology, 6:article*, 225. https://doi.org/10.3389/fphar.2015.00225.

CHAPTER NINE

Cannabinoid for alcohol use disorder

Júlia Dalfovo Marquez[a], Talissa Dezanetti[a], Roger Walz[a,b], and Cristiane Ribeiro de Carvalho[a,c,*]

[a]Programa de Pós-Graduação em Neurociências, Universidade Federal de Santa Catarina (UFSC), Florianópolis, Brazil
[b]Departamento de Ciências Clínica Médica, Universidade Federal de Santa Catarina (UFSC), Florianópolis, Brazil
[c]Departamento de Patologia, Universidade Federal de Santa Catarina (UFSC), Florianópolis, Brazil
*Corresponding author. e-mail address: cristiane.carvalho@ufsc.br

Contents

Abstract

Several pieces of evidence have implicated the endocannabinoid system on dopaminergic mesolimbic brain reward, as well as the potential role of cannabinoid receptors CB1 and CB2 on modulation of reinforced properties of drug abuse and consequently to the treatment of substance use disorder, including alcoholism. Moreover, growing evidence has been proposed that cannabis or cannabinoid compounds may be helpful to treat alcohol use disorder (AUD). Cannabis is prevalent among individuals who also consume alcohol. While some authors reported that cannabis may be a promising candidate as a substitute medication for AUD, some studies have demonstrated that concomitant use of alcohol and cannabis may increase the risk of adverse outcomes. Considering that advances in the legalization and decriminalization movements regarding cannabis have led to increased availability worldwide, the current chapter aims to provide evidence on the benefits and risks of combining alcohol and cannabis, as well as the potential therapeutic use of cannabinoid compounds in treating AUD.

International Review of Neurobiology, Volume 178
ISSN 0074-7742, https://doi.org/10.1016/bs.irn.2024.08.005

1. Introduction

Cannabis is the most commonly used drug by people who also consume alcohol (SAMHSA, 2017). Cannabis use among people who report drinking in the past month or past year (i.e., concurrent use among drinkers) remains relatively low overall, with approximately 15% of drinkers reporting cannabis use in the same month or past year (Terry-McElrath et al., 2013) However, concurrent use of alcohol among cannabis users is significantly higher, with more than 75% of cannabis users reporting alcohol use within the same 30-day period (Midanik et al., 2007; Butterworth et al., 2014; Haas et al., 2015). Studies reveal that among those who abuse alcohol, 20%–50% of them also report using cannabis at the same time. This practice is known to increase the risk of adverse outcomes, including comorbid psychiatric disorders, poor clinical treatment outcomes, and risky behaviors such as heavy drinking and driving while intoxicated (Subbaraman et al., 2017). However, not all studies show these negative effects (Mallett et al., 2019).

Additionally some studies suggest a link between cannabis use and alcohol use disorder (AUD). It has been estimated that over 85% of individuals diagnosed with a cannabis use disorder will also meet the criteria for a lifetime AUD (Agrawal et al., 2007; Weinberger et al., 2016).

Still, evidence regarding the impact of cannabis use on alcohol consumption, specifically among those in treatment for AUD, remains sparse. Co-occurring drug use (including drugs other than cannabis) and alcohol use more broadly can negatively impact the treatment of substance use disorders (Schuckit et al., 2014). There is evidence indicating that using alcohol and cannabis together can have adverse effects. For example, some research shows that cannabis use is a risk factor for underage drinking, excessive drinking, and crash fatalities involving alcohol. However, not all studies show these effects (Mallett et al., 2019). On the other hand, some evidence has been shown that cannabis may be helpful as a substitute for alcohol to reduce excessive drinking and alcohol-related harms (Reiman, 2009; Subbaraman, 2014; Subbaraman et al., 2017, 2018).

Movements to legalize and decriminalize cannabis use have changed the legal and political landscape, leading to an increased availability of cannabis and cannabis-related products around the world, possibly affecting the alcohol and cannabis co-use patterns as well as AUD treatment. Therefore, in the current chapter, we will present some evidence regarding the benefits and risks of a combination of alcohol and cannabis, as well as the potential therapeutic of cannabinoid compounds to treat AUD.

1.1 Cannabinoids and cannabis-based products

Cannabis plants contain a number of biologically active substances named phytocannabinoid, the most studied are delta-9-tetrahydrocannabinol (Δ9-THC) and cannabidiol (CBD) and many other compounds. In the present chapter, cannabis term is used to describe the psychoactive broad preparations of the plant *Cannabis sp.* for medicinal cannabis and for recreational use, also known as marijuana or ganja, including leaves or other parts of plant or material (i.e. hashish, a resin extracted from the flower clusters and top leaves) of the *Cannabis sp.* while the term "cannabis-based products" is used to describe the cannabis related products for medical use, since standardized composition until products with unspecified composition of phytocannabinoids, including herbal cannabis, extract made from the resinous flowers and small leaves of the cannabis plant, magistral preparation, pharmaceutical products with marketing authorization such as nabilone, dronabinol, nabiximols and epidiolex (Table 1).

Both medical and recreational cannabis products available can contain different composition and concentrations of phytocannabinoid, particularly THC and CBD. Both, THC and CBD have different pharmacological properties and have different mechanisms of action within the central and peripheral nervous systems. Notably, CBD does not demonstrate associations with the adverse cognitive or subjective effects of THC. Studies using co-administration of CBD and THC reveal that CBD counteracts some of the intoxicating effects of THC (Bidwell et al., 2018), although other studies have failed to replicate this effect (Haney et al., 2016; Solowij et al., 2019). In addition, there appears to be consistent evidence that CBD may mitigate the psychotomimetic and anxiogenic effects that are sometimes associated with THC (Niesink & Van Laar, 2013).

Cannabis and cannabis-related products, either smoked or taken orally, produce many pharmacological effects that may vary according to the route of administration, emotional state, dose level, and individual vulnerability to specific effects. Other effects, such as euphoria and tachycardia (Fogel et al., 2017), as well as memory and cognitive impairment, are also reported. Several publications showed that the use of cannabinoids, especially marijuana, results in an enhanced risk of causing road accidents (Masten & Guenzburger, 2014). Furthermore, prolonged exposure to plant-derived, synthetic, or endogenous cannabinoid agonists leads to the development of pharmacological tolerance (González et al., 2005).

Table 1 General aspects of different cannabis-related products available for medicinal and non-medicinal use.

Type of product	Examples	Statements of use these products	Status around the world
Pharmaceutical products with marketing authorization issued by a drug regulatory authority	Nabilone (oral capsule containing synthetic cannabinoid analog to THC); Dronabinol (capsule or solution containing synthetic cannabinoid analog to THC); Nabiximols (oromucosal spray containing an extract from *Cannabis sativa* with 2.7 mg THC and 2.5 CBD per 100 mL); Epidiolex (oral solution containing purified *Cannabis sativa* 100 mg/mL of CBD);	Prescription medicines; Over-the-couther medicines	Several countries have regulate and allowed the use of Pharmaceutical products with marketing authorization with clearly dosage and condition for use.
Cannabis-based products with marketing authorization	Standardized cannabis-based medical products or preparations with known compositions of THC/CBD ratios; Herbal medical cannabis products (i.e. Bedrocan®, bediol® bedrilite®); Magistral preparations from any part of the plant; Cannabis extract from resinous flowers or small leaves of the plant; Cannabis-based products with unspecified composition;	Prescription herbal medicines; Over-the-couther medicines; Dietary supplements	There are considerable disparities regarding regulatory compliance and availability of these products around the world.
Non-medicinal products from cannabis	*Miscellaneous of products including:* Marijuana (dried leaves and flowers of cannabis); Hashish (dried resin from cannabis); Hash oil extracted from Hashish; Cannabis edibles;	No medicinal purpose	Some countries have legalized and/or decriminalized non-medical use.

Source: Herbal cannabis for medical use: a spectrum of regulatory approaches. World Drug Report. (2023). *Herbal cannabis for medical use: a spectrum of regulatory.*

1.2 The endocannabinoid system and alcohol addiction

The endocannabinoid system is comprised of cannabinoid CB1 and CB2 receptors and endogenous ligands of these receptors—so-called endocannabinoids—and the processes playing a role in the biosynthesis, release, transport, and metabolism of these endogenous lipid mediators. Endocannabinoids such as anandamide and 2-arachidonoylglycerol (2-AG) are highly lipophilic compounds not stored in vesicles after production. After their release on demand from depolarized postsynaptic neurons, endocannabinoids act as retrograde messengers, activating CB1 receptors on presynaptic terminals and modulating short and long-term synaptic plasticity. Depending on the presynaptic inhibition of GABA or glutamatergic transmission, their overall effect is either excitatory or inhibitory. This decisive modulatory action on synaptic transmission of the central transmitter systems has significant functional implications on many physiological functions, including reward processing target on dopaminergic mesolimbic pathway (Fig. 1).

The endocannabinoid system modulates emotional responses, memory, learning, and reward systems. Notably, there are many indications that this system is involved in alcohol-induced impairments; namely, CB1 receptor function is implicated in functional modulation of the mesocorticolimbic dopaminergic pathway and motivation for drug seeking (Kleczkowska et al., 2016). Several biochemical studies revealed changes within the components of the endocannabinoid system after acute and chronic alcohol intake (Basavarajappa et al., 199, 2003; Gonzalez et al., 2002).

The primary research provides the rationale that pharmacological or genetic interference with the endocannabinoid system may reduce drug cravings and subsequent relapse in addicted patients. Unfortunately, the interest of major pharmaceutical industries in the clinical development of new compounds targeting the endocannabinoid system has been critically damped by the worldwide withdrawal of the already approved antiobesity medication rimonabant—CB1 antagonist—due to serious psychiatric side effects. CB1 agonists dose-dependently increase alcohol consumption in rats, and these effects are prevented by pre-treatment with CB1 antagonists (Colombo et al., 2002). Conversely, rimonabant and other CB1 antagonists suppress rodent alcohol consumption (Colombo et al., 1998; Femenía et al., 2010), possibly by blunting alcohol-induced dopamine release (Hungund et al., 2003). Rimonabant may also have antagonist activity at the mu-opioid receptor (Seely et al., 2012), which calls into question whether the suppression of alcohol consumption is exclusively related to its

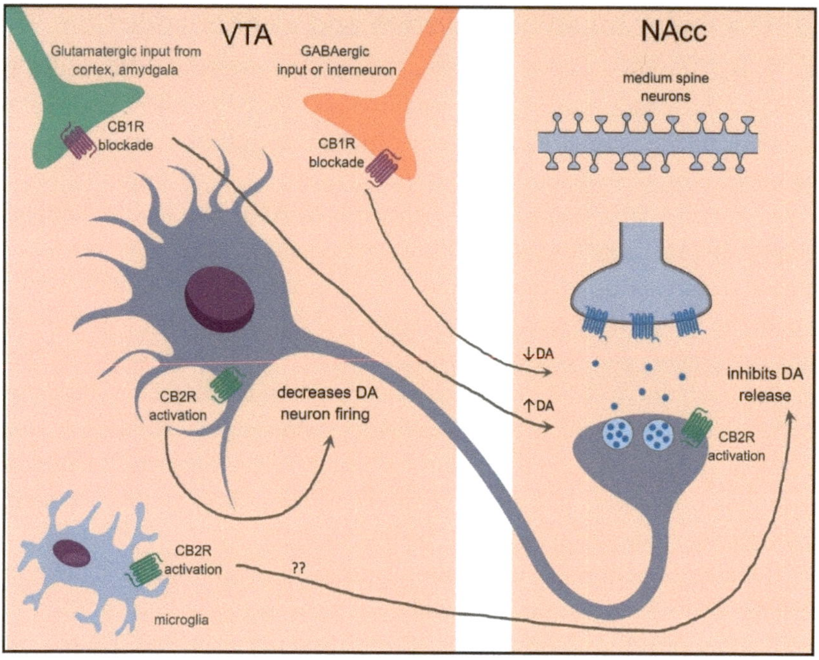

Fig. 1 Schematic representation of the endocannabinoid-mediated control of dopamine (DA) release within brain reward circuits. CB1 receptors are present in VTA GABAergic and glutamatergic neurons and their incoming connections to the VTA DA neurons. Both excitatory (glutamatergic) and inhibitory (GABAergic) projections affect dopaminergic signaling. CB2 receptors, on the other hand, are found in VTA DA neurons and microglia. For instance, cannabinoid agonists like Δ9-THC may produce rewarding effects by binding CB1 receptors on VTA GABAergic interneurons and their connections, decreasing GABA release. This, in turn, increases DA neuronal firing and enhances DA release in the NAc. Conversely, Δ9-THC may also have aversive effects by activating CB1 receptors on glutamatergic neurons in the VTA, leading to a decrease in excitatory glutamate input on VTA DA neurons. Additionally, the activation of CB2 receptors located in the cell body of VTA DA neurons and at the terminal of these neurons in the NAcc decreases DA neuron firing and DA release, respectively. Alternatively, the activation of CB2 receptors in microglia may also impact neuronal dopamine signaling. Thus, the subjective effects of Δ9-THC, as well as the regulation of DA release by cannabinoid receptors, may depend on the balance of both CB1 and CB2 actions. This figure was created on the "Mind the Graph" platform (www.mindthegraph.com).

CB1 activity. Nevertheless, the preclinical literature proves that the endocannabinoid system influences rodent alcohol intake.

Based on this preclinical evidence, two studies have addressed the effects of rimonabant on human alcohol consumption (Table 2). The only clinical

Table 2 Summary of how cannabis satisfies the seven criteria for substitute therapies.

Criterion	**Satisfied?**
1. Reduces alcohol use and related harms	Mixed evidences
2. Less harmful than alcohol	Yes
3. Misuse is less than that of alcohol	Mixed evidences
4. Can substitute for alcohol	Mixed evidences
5. Is safer in overdose than alcohol	Yes
6. Does not potentiate the effects of alcohol especially if either drug is taken in overdose	Mixed evidences
7. Offers significant health economic benefits	No

Adapted from Subbaraman, M. S. (2014). Can cannabis be considered a substitute medication for alcohol? *Alcohol and Alcoholism, 49*, 292–298.

trial in treatment-seeking alcohol-dependent subjects found no significant effect of rimonabant versus placebo on return to any drinking (Soyka et al., 2008). However, there was a non-significant reduction in relapse to heavy drinking in the rimonabant versus the placebo group (26.0% and 32.5%, respectively). Therefore, conducting another trial in a larger sample would be necessary before concluding that rimonabant is ineffective. The other study, a small between-subject human laboratory investigation, found no effect of rimonabant on alcohol self-administration (George et al., 2010). Thus, the limited evidence from human studies suggests that CB1 inverse agonists may not be a valuable treatment for AUD. However, the effect on relapse to heavy drinking cannot be ruled out.

Since that publication describes that under normal conditions, the gene encoding for CB2 receptors is expressed in different mouse brain structures involved in the regulation of emotional response (amygdala, hippocampus, and raphe nuclei) and brain rewards such as nucleus accumbens (NAcc) and ventral tegmental área (VTA) (García-Gutiérrez et al., 2010) highlighting the possible modulation of the reinforcing properties of drugs of abuse, there was a growing study demonstrating the role of the CB2 receptors on the regulation of dopamine neurotransmission and implicating this receptor as a new therapeutic target for drug addiction management.

Subsequently, Xi et al. (2011) demonstrated that CB2 receptors may also modulate dopaminergic neurotransmission in the NAcc. Both systemic and intra-NAcc administration of a selective and potent CB2 receptor agonist attenuated the increase in dopamine levels into the NAcc induced by cocaine self-administration in mice. The authors hypothesized that this effect may be due to the following possibilities: first, activating CB2 receptors in mesolimbic dopaminergic neurons of the VTA would lead to an inhibitory effect on dopamine release in the NAcc. Second, activation of CB2 receptors in microglia or astrocytes in the mesolimbic pathway could indirectly inhibit dopamine release in the NAcc by inflammatory mediators. Indeed, studies by Zhang confirmed that the CB2 receptor is expressed in VTA dopaminergic neurons projecting to the NAcc in rodents (Zhang et al., 2014; Zhang & Xi, 2017), and CB2 receptors activation of CB2 receptors decreases the firing rate and dopamine release in VTA dopaminergic neurons (Fig. 1).

Furthermore, growing evidence suggests that functional regulation of the CB2 receptors can modulate alcohol's reinforcing and motivational effects in rodents. The first report about the role of CB2 receptors in drug addiction was described by Emmanuel Onaivi's group (Ishiguro et al., 2006; Onaivi et al., 2008). Using voluntary alcohol consumption, when the animal can choose two bottles, one containing water and the other containing a progressively increased alcohol concentration, it was reported that mice that consumed a higher amount of alcohol presented a lower gene expression of CB2 receptors (CNR2) in the brain. Besides, the authors demonstrate that the administration of a selective CB2 agonist decreases alcohol consumption in stressed mice. This study also reported an association between CNR2 polymorphism, Q63R, and alcoholism in a population of Japanese patients. Pharmacological studies have been carried out to assess how the activation or blockade of CB2 receptors modulates alcohol consumption, motivation, and relapse. These studies revealed that acute CB2 receptor agonist administration induced a marked increase in alcohol intake in mice (Xi et al., 2011).

On the other hand, Al Mansouri et al. (2014) demonstrated that β-Caryophyllene, a selective CB2 receptor agonist, significantly reduced voluntary alcohol consumption and conditioning place preference in mice, and these effects were prevented by the antagonist CB2 antagonist, supporting the role of CB2 receptors mediate the effects of β-Caryophyllene. Likewise, an oral ethanol self-administration study analyzes the effects of AM630-induced blockade of the CB2 receptor. In this paradigm, alcohol consumption significantly increased, whereas JWH133 treatment induced the opposite effect. Interestingly, the administration of AM630 significantly increased tyrosine hydroxylase gene expression in the VTA and mu-opioid receptor CB1 receptor and gene expression in the NAcc. In contrast, the administration of CB2 agonists decreased the gene expression of these targets (Canseco-Alba et al., 2018).

In contrast, Powers et al. (2015) failed to observe any effect on alcohol consumption and CPP in HS/Ibg mice treated with the same CB2 agonist or antagonist. Discrepancies regarding the effects of pharmacological regulation of CB2 receptors on alcohol consumption may be related to some experimental variable, such as using different mouse strains (HS/Ibg or C57BL/6J) or dose, route, or schedule of administration of cannabinoid compounds. Another experimental paradigm revealed that the CB2 receptor agonist JWH133 increased astrocytes and microglia in the hippocampus of rats exposed for two weeks to an ethanol diet (11% v/v), indicating which CB2 receptors from astrocyte and microglial may be engaged following specific neurotoxic stimuli (Canseco-Alba et al., 2018) Altogether, the majority of the pharmacological studies carried out to date revealed that the agonism of the CB2 receptor can reduce ethanol consumption, preference, and motivation to drink in several animal models. Moreover, the CB2 receptor activation is proposed to induce a neuroprotective effect against ethanol neurotoxicity; such effects would further substantiate the potential therapeutic utility of pharmacological manipulation of CB2 receptors in AUD, mainly because this target seems to be devoid of psychiatry side effects, unlike CB1 receptor blockade.

1.3 Cannabis and alcohol use: Clinical studies

There is conflicting evidence about the effects of cannabis on alcohol consumption. While some studies suggest that cannabis use may replace alcohol consumption, others indicate that using both substances together may increase the risk of harm. For example, an experimental study conducted on 16 male concurrent alcohol/cannabis users showed that when given the options of alcohol only, cannabis only, or both alcohol and cannabis, the participants tended to consume less alcohol when cannabis was also available, in comparison to situations where cannabis was not available (Mello et al., 1978). A study was conducted with healthy volunteers to examine the effects of combinations of alcohol and Δ9-THC on physiological and subjective measures. The study found that there were no synergistic effects of acute combinations of very-low-dose ethanol and THC on subjective or physiologic response (heart rate, blood pressure, and blood alcohol concentration) or cognitive performance (Ballard & De Wit, 2011). On the other hand, according to a review, the combination of cannabis and alcohol can have more significant effects on impaired driving tasks, subjective sensations, and physiological measures like heart rate when compared to using either substance alone (Fares et al., 2022).

Additionally, it has been shown that both alcohol and cannabis use leads to notable changes in brain areas as well as abnormal brain functioning, which were more severe than those made by each substance separately. When damaged, these changes primarily affect brain areas such as the hippocampus, which may lead to deficits in memory and cognition, visuospatial functioning, reasoning, response perseveration, and reaction speed (Medina et al., 2007; Lisdahl et al., 2013). Besides, it has been shown that teenagers who used marijuana and alcohol together had impaired cognitive functioning and poor performance in specific cognitive domains. The differences in verbal recall and cognitive flexibility were still observed even after a month of abstinence from this concomitant use compared to the control group (Winward et al., 2014). The concomitant use of drugs that are frequently abused is known to enhance reaction time and increase the number of incorrect responses. Studies have shown that the combined effects of marijuana and alcohol can be additive or synergistic, resulting in a higher risk of driving accidents (Bramness et al., 2010; Dubois et al., 2015). Drivers who tested positive for both substances were found to have a greater likelihood of making errors than those who tested positive for alcohol or cannabis alone (Dubois et al., 2015). According to a study by Ronen et al. (2010), drinking alcohol and smoking THC together can result in higher heart rate, fatigue, and sleepiness during driving tests. Furthermore, this study also showed that healthy volunteers also presented impairment in non-driving tasks, such as arithmetic calculations, after this combination. The study suggests that the effect of these substances on driving performance also depends on the doses of each and the individual experienced use of these drugs.

A study conducted on 80 recreational users of cannabis and alcohol revealed that driving performance was worse for those who consumed both substances compared to those who consumed only one substance. The study also found that THC-blood levels were higher in participants who had consumed both (Downey et al., 2013). Another, the combined use of cannabis and alcohol among heavy cannabis users was found to negatively affect their ability to focus on multiple tasks compared to alcohol use alone. However, it is worth noting that heavy cannabis users may have built a tolerance to cannabis that could help them perform better on these tasks than lighter users (Ramaekers et al., 2011). Although the above driving simulation studies use double-blind, placebo-controlled designs, they were all relatively small samples and not necessarily generalizable to heavy alcohol users. Considering a retrospective cross-sectional forensic database study of drivers, a larger sample of 5042 individuals were caught driving under the influence of alcohol. The study found that those who tested positive for both THC and alcohol had a higher likelihood of being assessed as impaired. The risk of being judged as impaired increased significantly when alcohol and cannabis were detected (Bramness et al., 2010).

On the topic of driving impairment, various studies have produced differing conclusions. For instance, one study that examined the impact of combining alcohol and cannabis on driving impairment, body sway, and mood found no discernible effects of taking both substances (Liguori et al., 2002). Similarly, a more recent study that involved 25 experienced and 22 inexperienced drivers also failed to find any synergistic effect when combining alcohol and cannabis (Lenné et al., 2010). Besides, the combined use of cannabis and alcohol did not have a more significant effect on working memory, psychomotor ability, simple reaction time, or subjective mood than using either substance alone in healthy participants. Therefore, it can be assumed that cannabis does not enhance the effects of alcohol. However, it is essential to note that these studies were conducted on a small sample size and cannot be generalized to heavy drinkers (Ballard & De Wit, 2011).

Despite evidence that cannabis use adversely impacts alcohol use severity, there are several contradictory findings, particularly from case reports and correlational studies that pointed out cannabis as a substitute for alcohol by decreasing drinking behavior. Some medical cannabis patients and doctors have reported that a significant number of people are substituting cannabis for alcohol (Mikuriya, 2004; Lucas et al., 2013). However, retrospective self-report and selection bias limit the data on this topic. Nonetheless, a 30-year study on alcohol abusers indicated that almost half of them remained alcohol-free for a year, with some claiming to use cannabis as a substitute for alcohol (Vaillant & Schnurr, 1988). Some experts suggest that medical cannabis could be beneficial in reducing the amount of alcohol consumed by people with AUD (Mikuriya, 2004; Lucas et al., 2013).

Contrary to this point of view, marijuana use among individuals with alcohol or other drug dependence is associated with lower odds of achieving abstinence from drug and heavy alcohol use (Mojarrad et al., 2014). Furthermore, in a large randomized control trial, it was found that using cannabis during treatment for AUD was associated with fewer days of abstinence from alcohol compared to not using cannabis. In fact, for each additional day of cannabis use, there were approximately four fewer days of alcohol abstinence. When the levels of cannabis use were analyzed and grouped into quartiles, it was found that only those in the second quartile of cannabis use had significantly fewer alcohol-abstinent days as compared to cannabis abstainers. On average, individuals who used cannabis once or twice a month during AUD treatment had nearly ten fewer alcohol-abstinent days at the end of the treatment and 56 fewer abstinent days one-year post-treatment. However, cannabis use does not appear to be associated with the frequency of heavy drinking post-treatment (Subbaraman et al., 2017).

Future research must address alcohol and cannabis consumption patterns to help determine the impact of cannabis consumption alone and in combination with alcohol on alcohol treatment outcomes. These findings highlight the need for healthcare professionals to understand better how patients consume alcohol and cannabis together or separately, especially given the recent legalization of cannabis in many countries, as well as the discrepancy regarding possible risks and benefits of combining these substances.

1.4 Cannabidiol to treat substance use disorder

CBD is a "multi-target" drug, demonstrating various molecular targets (Ibeas Bih et al., 2015). Unlike Δ9-THC, CBD is not an agonist of the cannabinoid CB1 and cannabinoid CB2 receptors but is an inverse agonist in the low micromolar concentration level (Pertwee, 2008). Currently, CBD has a good safety profile, is well tolerated, and is approved by the U.S. Food and Drug Administration to treat resistant epileptic syndromes. The therapeutic effects of CBD are still being explored in clinical trials for many diseases. Several shreds of evidence demonstrate that CBD holds therapeutic potential for a variety of psychiatric and neurological disorders as a result of its anxiolytic, anticonvulsant, antidepressant, anti-inflammatory, antipsychotic, and neuroprotective effects (Campos et al., 2016; Ibeas Bih et al., 2015).

Growing evidence from preclinical, human laboratory and clinical studies demonstrates that CBD may also hold therapeutic potential in the area of substance abuse and addiction. For example, several preclinical studies have shown that CBD reduces opiate intoxication and reward and may prevent opioid (Hurd et al., 2015; De Carvalho & Takahashi, 2017) and stimulant relapse (Prud'homme et al., 2025; De Carvalho & Takahashi, 2017). CBD also appears to be a promising intervention for other addictive drugs, reducing the number of cigarettes smoked by tobacco smokers (Morgan et al., 2013). CBD also decreases cocaine-related problems, such as withdrawal symptoms, craving, impulsivity, and paranoia (Calpe-López et al., 2019; Hurd et al., 2019; Meneses-Gaya et al., 2021; Mongeau-Pérusse et al., 2021). Furthermore, CBD may be effective in tapering off daily marijuana use in individuals dependent on cannabis (Shannon & Opila-Lehman, 2015) and ameliorating symptoms of cannabis withdrawal in a case report (Crippa et al., 2013).

Many preclinical rodent studies have shown promising effects of CBD on alcohol reinforcement, motivation, and reinstatement (Gonzalez-Cuevas et al., 2018; Viudez-Martínez et al., 2018; Viudez-Martínez et al., 2020). In preclinical studies, CBD decreases alcohol-related outcomes, including some parameters of alcohol dependence, such as (i) alcohol consumption and motivation to use, (ii) relapse and withdrawal, and (iii) may protect against the deleterious effects of alcohol on the liver and brain in rodents (Nona et al., 2019; Turna et al., 2019).

CBD decreases the reinforcing properties of alcohol and decreases drinking motivation and consumption in mice (Viudez-Martínez et al., 2018; Viudez-Martínez et al., 2020). CBD has also been shown to attenuate cue significantly- and stress-induced alcohol-seeking, reinstatement, anxiety-like behavior, and high impulsivity in a rat model of AUD (Gonzalez-Cuevas et al., 2018). It has also been found to lessen anxiety, impulsivity and the urge to seek alcohol in rats. For instance, in one study, transdermal CBD gel decreased the tendency of rats to seek alcohol even after repeated treatment (Gonzalez-Cuevas et al., 2018). Another study found that male and female mice given CBD acutely had a lower alcohol intake than those in the control group (Viudez-Martínezet al., 2020). In addition, long-term of CBD treatment reduced alcohol preference and intake in a two-bottle choice paradigm (Viudez-Martínez et al., 2018).

Another study conducted on Sardinian alcohol-preferring rats has shown a decrease in alcohol self-administration when treated with CBD (Maccioni et al., 2022). However, the results were different in male baboons. In this case, acute oral administration of CBD did not have any effect on alcohol-seeking or drinking behaviors, including patterns of intake. Furthermore, chronic administration of CBD did not have any impact on these outcomes (Moore et al., 2023).

Altogether, these discrepancies reported in rodents may be related to species differences, the route of administration, and the self-administration protocol applied within each study. While studies using intraperitoneal administration demonstrate effective proof-of-concept designs, these designs often fail to translate clinically as this route of administration increases drug bioavailability, which may not be clinically meaningful (Al Shoyaib et al., 2019).

Studies utilizing clinically relevant routes of administration (e.g., oral ingestion) are better suited for evaluating therapeutic effects that may be altered by drug formulations, bioavailability, and pharmacokinetics that are meaningful for clinical translation. Before extrapolating these findings from animal studies to clinical approaches, these issues must be evaluated.

It is important to note that plant-based CBD contains not only CBD but also small amounts of other components of the Cannabis plant, such as terpenes and minor cannabinoids. These components influence the effects of CBD and THC in the brain and body, commonly referred to as the "entourage effect" (Russo, 2019). Therefore, previous studies that have tested the effects of synthetic CBD on alcohol use or other psychological and physical outcomes do not necessarily inform our understanding of the effects of plant-based CBD that people typically use when they purchase CBD products on the legal recreational or medical markets or buy hemp-derived CBD. Future research is needed to explore the effects of CBD on the subjective effects of alcohol and its potential to treat AUD. Currently CBD is under investigation as a pharmacotherapy for AUD patients in three clinical trials registered on clinicaltrials.gov (ClinicalTrials.gov identifier: NCT05613608, NCT04873453, NCT06105138). Much work remains to be done if we are to elucidate whether CBD is a viable treatment target for AUD.

An important concern about these studies conducted on CBD have used a wide variety of formulations and dosages. As a result, it's unclear how each CBD product is absorbed by the body due to the lack of pharmacokinetic data. This makes it difficult to compare the effectiveness of different CBD formulations. Currently, the only FDA-approved CBD product is Epidiolex, which is used to treat rare childhood seizure disorders. Given the unreliability of commercially available CBD products that are not FDA regulated, it should be consider using standatizated formulatations treat AUD.

1.5 Cannabis as substitute medication for alcohol

Substitution or maintenance therapy has been successfully applied to the treatment of drug addiction for decades. For example, when methadone was prescribed for heroin addicts and tobacco smoking was suppressed by making a substitution for nicotine in the form of patches or gum. Although AUD has not been seen as suitable for substitution therapy, benzodiazepines—positive allosteric modulators on the GABAA receptor enhance or facilitate the action of gamma amino butyric acid neurotransmission—could be as favorable substitutes for alcohol. However, clinicians have warned against prescribing benzodiazepines to patients with a history of substance abuse disorder because of their high potential to induce addiction due to reinforced properties. Besides, these drugs also produce side effects and are not safe if alcohol is consumed concomitantly. Substitute medicines are usually effective in preventing withdrawal from drug abuse and may also decrease craving and relapse. Frequently, they need to be used for the long term, and there seems to be value in reducing harm has been established, for example, for opioid dependence. Growing studies have indicated that medicinal cannabis could be reasoned as a substitute therapy for alcohol, especially among individuals who are trying to reduce drinking behavior. Based on these premises, medicinal cannabis might be safer and also produce less social harms, for this reason some studies have been pointed as a good candidate for substitute medication for alcohol.

Notably some studies have been supported among users of medicinal cannabis, as 40% report using cannabis to decrease alcohol consumption (Reiman, 2009); in states where medical cannabis is legal, alcohol consumption has also been observed to decrease (Anderson et al., 2013). Medical cannabis users have been found to drink less and experience fewer alcohol-related problems than non-medical cannabis users (Subbaraman & Kerr, 2018). Additionally, among medical-cannabis-using veterans, those who use cannabis more frequently as a substitute for alcohol tend to drink less on days when they use cannabis compared to days when they don't use it (Gunn et al., 2019).

According to the literature, there are some criteria that medication substitutes for drug abuse should accomplish for any substitution treatment: (1) It should reduce drug use and related harms; (2) It should ideally be without harm, or at least less harmful than drug abuse; (3) Misuse should be less than drug abuse; (4) It should be shown that it can substitute for drug a nor be used along with the drug; (5) It should be safer to overdose than drug abuse; (6) It should, under the best of circumstances, not potentiate the effects of drug abuse, mainly if either drug is consumed in overdose; (7) It should offer significant health and economic benefits.

An exciting review study from Subbaraman (2014) investigated whether Cannabis can be considered a substitute medication for alcohol based on evidence in the literature that corroborates or refutes each criterion of a medication substitute for alcohol. Table 1 summarizes the results from Subbaraman's review. It was reported in this review that mixed results were found regarding criterion number one. There are many studies suggesting that cannabis use can diminish alcohol drinking and related harms. However, these studies mostly have retrospective designs or selection bias, and few studies were carried out with AUD participants. Cannabis satisfies the second criterion because many studies have demonstrated that cannabis is safer than alcohol.

It is important to note that cannabis is not exempted from harm. Cannabis use may produce undesirable side effects, toxicity, and withdrawal syndrome, which are much less severe than alcohol. Besides, there is common sense that cannabis is a "soft drug" with less social damage than alcohol. Criteria number 3 has reported mixed results. For example, epidemiological studies suggest that AUD has higher rates than cannabis use disorder. However, the prevalence of cannabis use disorder among individuals with AUD is much higher than among those without AUD. On the other hand, there are no prospective studies evaluating whether cannabis as a substitute therapy for alcohol may lead to decreased harm or improved quality of life for patients. In addition, several studies reinforced that cannabis and alcohol are complements in that concomitant use of both may lead to more substance use compared with the use of either alone. Thus, criterion number four is partially proved. Regarding criteria number five, according to the Substance Abuse and Mental Administration (SAMHSA, 2017), "cannabis is rarely the only drug involved in a drug abuse death." The proportion of deaths attributed to cannabis alone has been documented as nearly zero, as reported by SAMHSA, implying that it cannot be taken in overdose itself. Besides, overdose, death would be more likely due to alcohol intoxication.

As mentioned before, concomitant use of cannabis with alcohol does not lead to higher blood or alcohol levels. Smoking cannabis may slow the absorption of alcohol, which decreases the psychoactive effects of alcohol (Lukas et al., 1992). Cannabis and alcohol, when combined, seem not to act additively or synergistically on any physiological and subjective measures in terms of impairment. However, it is essential to note that the literature on the impact of cannabis and alcohol combined use is limited and has mixed results. Some studies have shown that combined cannabis and alcohol leads to more significant impairment than either alone. Others have shown no differences. These divergent results may be due to differences in study designs (e.g., measures of impairment differed across studies). In light of the mixed evidence, cannabis only partially satisfies the sixth criterion. To this date, there are no studies that have compared individual health economic outcomes (seventh criterion) for those who substitute cannabis versus those who do not. So, prospective studies of long-term drinking patterns post-substitution on alcohol-related damage and risks, including changes in quality of life, are needed since the health and economic benefits can only be evaluated relative to these issues.

While more research and improved study designs are needed to better identify the extent and impact of cannabis substitution on those affected by AUD, cannabis does appear to be a potential substitute for alcohol. Perhaps more importantly, cannabis may be considered both safer and potentially less addictive than benzodiazepines and other pharmaceuticals that have been evaluated as substitutes for alcohol. But, there is no clear pattern of outcomes related to cannabis substitution. Most importantly, the recommendation to prescribe medicinal cannabis for alcohol-dependent individuals to help reduce drinking is premature. Future studies should use longitudinal data to better understand the consequences of cannabis substitution.

2. Concluding remarks

The last decades have seen a significant gain in an understanding of the action of cannabinoid compounds and the endocannabinoid system in reward processing and the development of addictive behavior; this basic science provides the rationale that pharmacological intervention of endocannabinoid system, but while CB1 receptor blockade rimonabant is not suitable due to serious psychiatric side effects. In contrast, some studies revealing the role of CB2 receptors in modulating the addictive properties of alcohol have also opened up a promising new avenue for the clinical development of novel therapeutic approaches. Ongoing future medication development, massive investments are being made in medical cannabis products, including CBD, however, whether substitution therapy with medicinal cannabis as proposed for AUD is a promising approach seems questionable; the same is true of CBD-based products, despite its good safety profile, preclinical findings are mixed and clinical evidence is scarce. Large randomized clinical trials will give us more conclusive insights into the potential therapeutic of CBD for AUD. Besides considering the miscellaneous cannabis-related products available in some countries around the world, further studies are needed to evaluate the possible risks and benefits of concomitant use with alcohol.

Despite the distinct goals of medical and recreational cannabis users, patients seeking therapeutic benefits from herbal cannabis products are, therefore, also at substantial risk of unwanted side effects, including developing cannabis use disorder. Finally, more studies are needed to inform patients about the risks associated with high-potency cannabis and the interaction between the major compounds and to help shape approaches that promote better therapeutic options.

References

Agrawal, A., Lynskey, M. T., Madden, P. A. F., Bucholz, K. K., & Heath, A. C. (2007). A latent class analysis of illicit drug abuse/dependence: Results from the national epidemiological survey on alcohol and related conditions. *Addiction, 102*(1), 94–104.

Al Mansouri, S., Ojha, S., Al Maamari, E., Al Ameri, M., Nurulain, S. M., & Bahi, A. (2014). The cannabinoid receptor 2 agonist, β-caryophyllene, reduced voluntary alcohol intake and attenuated ethanol-induced place preference and sensitivity in mice. *Pharmacology, Biochemistry, and Behavior, 124*, 260–268.

Al Shoyaib, A., Archie, S. R., & Karamyan, V. T. (2019). Intraperitoneal route of drug administration: Should it be used in experimental animal studies? *Pharmaceutical Research, 37*(1), 12.

Anderson, D. M., Hansen, B., & Rees, D. I. (2013). Medical marijuana laws, traffic fatalities, and alcohol consumption. *The Journal of Law & Economics, 56*(2), 333–369.

Ballard, M. E., & De Wit, H. (2011). Combined effects of acute, very-low-dose ethanol and delta(9)-tetrahydrocannabinol in healthy human volunteers. *Pharmacology, Biochemistry, and Behavior, 97*, 627–631.

Basavarajappa, B. S., Cooper, T. B., & Hungund, B. L. (1998). Chronic ethanol administration down-regulates cannabinoid receptors in mouse brain synaptic plasma membrane. *Brain Research, 793*, 212–218.

Basavarajappa, B. S., Saito, M., Cooper, T. B., & Hungund, B. L. (2003). Chronic ethanol inhibits the anandamide transport and increases extracellular anandamide levels in cerebellar granule neurons. *European Journal of Pharmacology, 466*, 73–83.

Bidwell, L. C., Mueller, R., YorkWilliams, S. L., Hagerty, S., Bryan, A. D., & Hutchison, K. E. (2018). A novel observational method for assessing acute responses to cannabis: Preliminary validation using legal market strains. *Cannabis and Cannabinoid Research, 3*(1), 35–44.

Bramness, J. G., Khiabani, H. Z., & Mørland, J. (2010). Impairment due to cannabis and ethanol: Clinical signs and additive effects. *Addiction, 105*(6), 1080–1087.

Butterworth, P., Slade, T., & Degenhardt, L. (2014). Factors associated with the timing and onset of cannabis use and cannabis use disorder: Results from the 2007 Australian National Survey of Mental Health and Well-Being. *Drug and Alcohol Review, 33*(5), 555–564.

Calpe-López, C., García-Pardo, M. P., & Aguilar, M. A. (2019). Cannabidiol treatment might promote resilience to cocaine and methamphetamine use disorders: A review of possible mechanisms. *Molecules, 24*(14), 2583.

Campos, A. C., Fogaça, MV, Sonego, A. B., & Guimaraes, F. S. (2016). Cannabidiol, neuroprotection and neuropsychiatric disorders. *Pharmacological Research: The Official Journal of the Italian Pharmacological Society, 112*, 119–127.

Canseco-Alba, A., Schanz, N., Ishiguro, H., Liu, Q. R., & Onaivi, E. S. (2018). Behavioral evaluation of seeking and preference of alcohol in mice subjected to stress. *Bio Protocol, 8*(20), e3061.

Colombo, G., Agabio, R., Fa, M., Guano, L., Lobina, C., Loche, A., ... Gessa, G. L. (1998). Reduction of voluntary ethanol intake in ethanol-preferring sP rats by the cannabinoid antagonist SR-141716. *Alcohol Alcoholism, 33*(2), 126–130.

Colombo, G., Serra, S., Brunetti, G., Gomez, R., Melis, S., Vacca, G., ... Gessa, L. (2002). Stimulation of voluntary ethanol intake by cannabinoid receptor agonists in ethanol-preferring sP rats. *Psychopharmacology, 159*(2), 181–187.

Crippa, J. A., Hallak, J. E., Machado-de-Sousa, J. P., Queiroz, R. H., Bergamaschi, M., Chagas, M. H., & Zuardi, A. W. (2013). Cannabidiol for the treatment of cannabis withdrawal syndrome: A case report. *Journal of Clinical Pharmacy and Therapeutics, 38*(2), 162–164.

De Carvalho, C. R., & Takahashi, R. N. (2017). Cannabidiol disrupts the reconsolidation of contextual drug-associated memories in Wistar rats. *Addiction Biology, 22*, 742–751 2017.

Downey, L. A., King, R., Papafotiou, K., Swann, P., Ogden, E., Boorman, M., et al. (2013). The effects of cannabis and alcohol on simulated driving: Influences of dose and experience. *Accident: Analysis and Prevention, 50*, 870–886.

Dubois, S., Mullen, N., Weaver, B., & Bédard, M. (2015). The combined effects of alcohol and cannabis on driving: Impact on crash risk. *Forensic Science International, 248*, 94–100.

Fares, A., Wright, M., Matheson, J., Mann, R. E., Stoduto, G., Le Foll, B., ... Di Ciano, P. (2022). Effects of combining alcohol and cannabis on driving, breath alcohol level, blood THC, cognition, and subjective effects: A narrative review. *Experimental and Clinical Psychopharmacology, 30*(6), 1036–1049.

Femenía, T., García-Gutiérrez, M. S., & Manzanares, J. (2010). CB1 receptor blockade decreases ethanol intake and associated neurochemical changes in fawn-hooded rats. *Alcoholism, Clinical and Experimental Research, 34*(1), 131–141.

Fogel, J. S., Kelly, T. H., Westgate, P. M., & Lile, J. A. (2017). Sex differences in the subjective effects of oral Delta(9)-THC in cannabis users. *Pharmacology Biochemistry and Behavior, 152*, 44–51.

García-Gutiérrez, M., Pérez-Ortiz, J., Gutiérrez-Adán, A., & Manzanares, J. (2010). Depression-resistant endophenotype in mice overexpressing cannabinoid CB2 receptors. *British Journal of Pharmacology, 160*, 1773–1784.

George, D. T., Herion, D. W., Jones, C. L., Phillips, M. J., Hersh, J., Hill, D., ... Singley, E. D., O.'Malley, S. S., Bishai, R., Rawlings, R. R., Kunos, G. (2010). Rimonabant (SR141716) has no effect on alcohol self-administration or endocrine measures in nontreatment-seeking heavy alcohol drinkers. *Psychopharmacology, 208*(1), 37–44.

Gonzalez, S., Fernandez-Ruiz, J., Sparpaglione, V., Parolaro, D., & Ramos, J. A. (2002). Chronic exposure to morphine, cocaine or ethanol in rats produced different effects in brain cannabinoid CB(1) receptor binding and mRNA levels. *Drug and Alcohol Dependence, 66*, 77–84.

González, S., Cebeira, M., & Fernández-Ruiz, J. (2005). Cannabinoid tolerance and dependence: A review of studies in laboratory animals. *Pharmacology, Biochemistry, and Behavior, 81*(2), 300–318.

Gonzalez-Cuevas, G., Martin-Fardon, R., Kerr, T. M., Stouffer, D. G., Parsons, L. H., Hammell, D. C., ... Weiss, F. (2018). Unique treatment potential of cannabidiol for the prevention of relapse to drug use: Preclinical proof of principle. *Neuropsychopharmacology: Official Publication of the American College of Neuropsychopharmacology, 43*, 2036–2045.

Gunn, R., Jackson, K., Borsari, B., & Metrik, J. (2019). A longitudinal examination of daily patterns of cannabis and alcohol co-use among medicinal and recreational veteran cannabis users. *Drug and Alcohol Dependence, 205*, 107661. https://doi.org/10.1016/j.drugalcdep.2019.107661 Dec 1.

Haas, A. L., Wickham, R., Macia, K., Shields, M., Macher, R., & Schulte, T. (2015). Identifying classes of conjoint alcohol and marijuana use in entering freshmen. *Psychology of Addictive Behaviors: Journal of the Society of Psychologists in Addictive Behaviors, 29*(3), 620–626.

Haney, M., Malcolm, R. J., Babalonis, S., Nuzzo, P. A., Cooper, Z. D., Bedi, G., ... Walsh, S. L. (2016). Oral cannabidiol does not alter the subjective, reinforcing or cardiovascular effects of smoked cannabis. *Neuropsychopharmacology: Official Publication of the American College of Neuropsychopharmacology, 41*(8), 1974–1982.

Hungund, B. L., Szakall, I., Adam, A., Basavarajappa, B. S., & Vadasz, C. (2003). Cannabinoid CB1 receptor knockout mice exhibit markedly reduced voluntary alcohol consumption and lack alcohol-induced dopamine release in the nucleus accumbens. *Journal of Neurochemistry, 84*(4), 698–704.

Hurd, Y. L., Spriggs, S., Alishayev, J., et al. (2019). Cannabidiol for the reduction of cue-induced craving and anxiety in drug-abstinent individuals with heroin use disorder: A double-blind randomized placebo-controlled trial. *The American Journal of Psychiatry, 176*(11), 911–922.

Hurd, Y. L., Yoon, M., Manini, A. F., Hernandez, S., Olmedo, R., Ostman, M., & Jutras-Aswad, D. (2015). Early phase in the development of cannabidiol as a treatment for addiction: Opioid relapse takes initial center stage. *Neurotherapeutics: The Journal of the American Society for Experimental NeuroTherapeutics, 12*(4), 807–815.

Ibeas Bih, C., Chen, T., Nunn, A. V., Bazelot, M., Dallas, M., & Whalley, B. J. (2015). Molecular targets of cannabidiol in neurological disorders. *Neurotherapeutics: The Journal of the American Society for Experimental NeuroTherapeutics, 12*(4), 699–730.

Ishiguro, H., Iwasaki, S., Teasenfitz, L., Higuchi, S., Horiuchi, Y., Saito, T., ... Onaivi, E. S. (2006). Involvement of cannabinoid CB2 receptor in alcohol preference in mice and alcoholism in humans. *Pharmacogenomics Journal, 7*, 380–385.

Kleczkowska, P., Smaga, I., Filip, M., et al. (2016). Cannabinoid ligands and alcohol addiction: A promising therapeutic tool or a humbug? *Neurotoxicity Research, 29*, 173–196.

Lenné, M. G., Dietze, P. M., Triggs, T. J., Walmsley, S., Murphy, B., & Redman, J. R. (2010). The effects of cannabis and alcohol on simulated arterial driving: Influences of driving experience and task demand. *Accident; Analysis and Prevention, 42*(3), 859–866.

Liguori, A., Gatto, C. P., & Jarrett, D. B. (2002). Separate and combined effects of marijuana and alcohol on mood, equilibrium and simulated driving. *Psychopharmacology, 163*(3–4), 399–405.

Lisdahl, K. M., Gilbart, E. R., Wright, N. E., & Shollenbarger, S. (2013). Dare to delay? The impacts of adolescent alcohol and marijuana use onset on cognition, brain structure, and function. *Frontiers in Psychiatry, 4*, 53.

Lucas, P., Reiman, A., Earleywine, M., Mcgowan, S. K., Oleson, M., Coward, M. P., et al. (2013). Cannabis as a substitute for alcohol and other drugs: A dispensary-based survey of substitution effect in Canadian medical cannabis patients. *Addiction Research & Theory, 21*, 435–442.

Lukas, S. E., Benedikt, R., Mendelson, J. H., Kouri, E., Sholar, M., & Amass, L. (1992). Marihuana attenuates the rise in plasma ethanol levels in human subjects. *Neuropsychopharmacology, 7*(1), 77–81 Aug; PMID: 1326277.

Maccioni, P., Bratzu, J., Carai, M. A. M., Colombo, G., & Gessa, G. L. (2022). Reducing effect of cannabidiol on alcohol self-administration in sardinian alcohol-preferring rats. *Cannabis and Cannabinoid Research, 7*(2), 161–169.

Mallett, K. A., Turrisi, R., Reavy, R., Russell, M., Cleveland, M. J., Hultgren, B., ... Hospital, M. (2019). An examination of parental permissiveness of alcohol use and monitoring, and their association with emerging adult drinking outcomes across college. *Alcoholism, Clinical and Experimental Research, 43*(4), 758–766.

Masten, S. V., & Guenzburger, G. V. (2014). Changes in driver cannabinoid prevalence in 12 U.S. states after implementing medical marijuana laws. *Journal of Safety Research, 50*, 35–52.

Medina, K. L., Schweinsburg, A. D., Cohen-Zion, M., Nagel, B. J., & Tapert, S. F. (2007). Effects of alcohol and combined marijuana and alcohol use during adolescence on hippocampal volume and asymmetry. *Neurotoxicology and Teratology, 29*(1), 141–152.

Mello, N. K., Mendelson, J. H., Kuehnle, J. C., & Sellers, M. L. (1978). Human polydrug use: Marihuana and alcohol. *The Journal of Pharmacology and Experimental Therapeutics, 207*, 922–935.

Meneses-Gaya, C., Crippa, J. A., Hallak, J. E., Miguel, A. Q., Laranjeira, R., Bressan, R. A., ... Lacerda, A. L. (2021). Cannabidiol for the treatment of crack-cocaine craving: An exploratory double-blind study Brazilian. *Journal of Psychiatry, 43*(5), 467–476.

Midanik, L. T., Tam, T. W., & Weisner, C. (2007). Concurrent and simultaneous drug and alcohol use: Results of the 2000 National Alcohol Survey. *Drug and Alcohol Dependence, 90*, 72–80.

Mikuriya, T. (2004). Cannabis as a substitute for alcohol: A harm- reduction approach. *Journal of Cannabis Therapeutics, 4*, 79–93.
Mojarrad, M., Samet, J. H., Cheng, D. M., Winter, M. R., & Saitz, R. (2014). Marijuana use and achievement of abstinence from alcohol and other drugs among people with substance dependence: A prospective cohort study. *Drug and Alcohol Dependence, 142*, 91–97.
Mongeau-Pérusse, V., Brissette, S., Bruneau, J., Conrod, P., Dubreucq, S., Gazil, G., ... Jutras-Aswad, D. (2021). Cannabidiol as a treatment for craving and relapse in individuals with cocaine use disorder: A randomized placebo-controlled trial. *Addiction, 116*(9), 2431–2442.
Moore, C. F., Zamarripa, C. A., & Weerts, E. M. (2023). Oral cannabidiol does not alter alcohol seeking and self-administration in baboons. *Drug and Alcohol Dependence, 245*, 109829.
Morgan, C. J., Das, R. K., Joye, A., Curran, H. V., & Kamboj, S. K. (2013). Cannabidiol reduces cigarette consumption in tobacco smokers: Preliminary findings. *Addictive Behaviors, 38*(9), 2433–2436.
Niesink, R. J., & Van Laar, M. W. (2013). Does cannabidiol protect against adverse psychological effects of THC? *Front Psychiatry, 4*, 130.
Nona, C. N., Hendershot, C. S., & Le Foll, B. (2019). Effects of cannabidiol on alcohol-related outcomes: A review of preclinical and human research. *Experimental and Clinical Psychopharmacology, 27*(4), 359–369.
Onaivi, E. S., Ishiguro, H., Gong, J. P., Patel, S., Meozzi, P. A., Myers, L., ... Gardner, E. (2008). Brain neuronal CB2 cannabinoid receptors in drug abuse and depression: From mice to human subjects. *PLoS One, 3*, e1640.
Pertwee, R. G. (2008). The diverse CB1 and CB2 receptor pharmacology of three plant cannabinoids: Delta9-tetrahydrocannabinol, cannabidiol and delta9-tetrahydrocannabivarin. *British Journal of Pharmacology, 153*(2), 199–215.
Powers, M. S., Breit, K. R., & Chester, J. A. (2015). Genetic versus pharmacological assessment of the role of cannabinoid type 2 receptors in alcohol reward-related behaviors. *Alcoholism, Clinical and Experimental Research, 39*, 2438–2446.
Prud'homme, M., Cata, R., & Jutras-Aswad, D. (2025). Cannabidiol as an intervention for addictive behaviors: A systematic review of the evidence. *Substance Abuse: Research and Treatment, 9* SART.S25081.
Ramaekers, J. G., Theunissen, E. L., de Brouwer, M., Toennes, S. W., Moeller, M. R., & Kauert, G. (2011). Tolerance and cross-tolerance to neurocognitive effects of THC and alcohol in heavy cannabis users. *Psychopharmacology, 214*(2), 391–401.
Reiman, A. (2009). Cannabis as a substitute for alcohol and other drugs. *Harm Reduction Journal, 6*, 35.
Ronen, A., Chassidim, H. S., Gershon, P., Parmet, Y., Rabinovich, A., Bar-Hamburger, R., et al. (2010). The effect of alcohol, THC and their combination on perceived effects, willingness to drive and performance of driving and non-driving tasks. *Accident; Analysis and Prevention, 42*, 1855–1865.
Russo, E. B. (2019). The case for the entourage effect and conventional breeding of clinical cannabis: No "strain," no gain. *Frontiers in Plant Science, 9*, 1969.
SAMHSA. (2017). Center for Behavioral Health Statistics and Quality. (2018). *2017 National survey on drug use and health: Detailed tables* (Rockville, MD: Substance Abuse and Mental Health Services Administration).
Schuckit, M. A., Smith, T. L., & Kalmijn, J. A. (2014). The patterns of drug and alcohol use and associated problems over 30 years in 397 men. *Alcoholism, Clinical and Experimental Research, 38*(1), 227–234.
Seely, K. A., Brents, L. K., Franks, L. N., Rajasekaran, M., Zimmerman, S. M., Fantegrossi, W. E., & Prather, P. L. (2012). AM-251 and rimonabant act as direct antagonists at mu-opioid receptors: Implications for opioid/cannabinoid interaction studies. *Neuropharmacology, 63*(5), 905–915.

Shannon, S., & Opila-Lehman, J. (2015). Cannabidiol oil for decreasing addictive use of marijuana: A case report. *Integrative Medicine (Encinitas), 14*(6), 31–35.
Solowij, N., Broyd, S., Greenwood, L. M., van Hell, H., Martelozzo, D., Rueb, K., ... Croft, R. (2019). A randomised controlled trial of vaporised Δ^9-tetrahydrocannabinol and cannabidiol alone and in combination in frequent and infrequent cannabis users: Acute intoxication effects. *European Archives of Psychiatry and Clinical Neuroscience, 269*(1), 17–35.
Soyka, M., Koller, G., Schmidt, P., Lesch, O. M., Leweke, M., Fehr, C., ... Mann, K. F. ACTOL Study Investigators. (2008). Cannabinoid receptor 1 blocker rimonabant (SR 141716) for treatment of alcohol dependence: Results from a placebo-controlled, double-blind trial. *Journal of Clinical Psychopharmacology, 28*(3), 317–324.
Subbaraman, M. S. (2014). Can cannabis be considered a substitute medication for alcohol? *Alcohol and Alcoholism, 49*, 292–298.
Subbaraman, M. S., & Kerr, W. C. (2018). Alcohol use and risk of related problems among cannabis users is lower among those with medical cannabis recommendations, though not due to health. *Journal of Studies on Alcohol and Drugs, 79*(6), 935–942.
Subbaraman, M. S., Metrik, J., Patterson, D., & Swift, R. (2017). Cannabis use during treatment for alcohol use disorders predicts alcohol treatment outcomes. *Addiction, 112*(4), 685–694.
Terry-McElrath, Y. M., O'Malley, P. M., & Johnston, L. D. (2013). Simultaneous alcohol and marijuana use among U.S. high school seniors from 1976 to 2011: Trends, reasons, and situations. *Drug and Alcohol Dependence, 133*(1), 71–79.
Turna, J., Syan, S. K., Frey, B. N., et al. (2019). Cannabidiol as a novel candidate alcohol use disorder pharmacotherapy: A systematic review. *Alcoholism, Clinical and Experimental Research, 43*(4), 550–563.
Vaillant, G. E., & Schnurr, P. (1988). What is a case? A 45-Year study of psychiatric impairment within a college sample selected for mental health. *Archives of General Psychiatry, 45*, 313–319.
Viudez-Martínez, A., García-Gutiérrez, M. S., & Manzanares, J. (2020). Gender differences in the effects of cannabidiol on ethanol binge drinking in mice. *Addiction Biology, 25*(3), e12765. https://doi.org/10.1111/adb.12765.
Viudez-Martínez, A., García-Gutiérrez, M. S., Navarrón, C. M., Morales-Calero, M. I., Navarrete, F., Torres-Suárez, A. I., & Manzanares, J. (2018). Cannabidiol reduces ethanol consumption, motivation and relapse in mice. *Addiction Biology, 23*(1), 154–164. https://doi.org/10.1111/adb.12495.
Weinberger, A. H., Platt, J., & Goodwin, R. D. (2016). Is cannabis use associated with an increased risk of onset and persistence of alcohol use disorders? A three-year prospective study among adults in the United States. *Drug and Alcohol Dependence, 161*, 363–367.
Winward, J. L., Hanson, K. L., Tapert, S. F., & Brown, S. A. (2014). Heavy alcohol use, marijuana use, and concomitant use by adolescents are associated with unique and shared cognitive decrements. *Journal of the International Neuropsychological Society: JINS, 20*(8), 784–795.
Xi, Z. X., Peng, X., Li, X., Song, R., Zhang, H., Liu, Q. R., ... Gardner, E. L. (2011). Brain cannabinoid CB2 receptors modulate cocaine's actions in mice. *Nature Neuroscience, 14*, 1160–1166.
Zhang, H., Gao, M., Liu, Q. R., Bi, G. H., Li, X., Yang, H. J., ... Xi, Z. X. (2014). Cannabinoid CB2 receptors modulate midbrain dopamine neuronal activity and dopamine-related behavior in mice. *Proceedings of the National Academy of Sciences of the United States of America, 111*, E5007–E5015.
Zhang, H. Y., & Xi, Z. X. (2017). *The neuroscience of cocaine mechanisms and treatment.* Cambridge, MA: Academic Press, Chapter 70—Cannabinoid CB2 receptor: A new target for treatment of cocaine addiction.

CHAPTER TEN

Environmental enrichment and complementary clinical interventions as therapeutic approaches for alcohol use disorder in animal models and humans

Rosana Camarini[a,*], Priscila Marianno[a], Beatriz Yamada Costa[a], Paola Palombo[b], and Ana Regina Noto[b]

[a]Department of Pharmacology, Institute of Biomedical Sciences, Universidade de Sao Paulo, São Paulo, Brazil
[b]Department of Psychobiology, Universidade Federal de Sao Paulo, São Paulo, Brazil
*Corresponding author. e-mail address: camarini@icb.usp.br

Contents

Abstract

Alcohol use disorder (AUD) is a multifactorial disorder arising from a complex interplay of various genetic, environmental, psychological, and social factors. Environmental factors influence alcohol misuse and can lead to AUD. While stress plays a crucial role in the onset and progression of this disorder, environmental enrichment (EE) also influences ethanol-induced behavioral and neurobiological responses. These alterations include reduced ethanol consumption, diminished operant self-administration,

International Review of Neurobiology, Volume 178
ISSN 0074-7742, https://doi.org/10.1016/bs.irn.2024.08.003

attenuated behavioral sensitization, and enhanced conditioned place preference. EE exerts modulatory effects on multiple neurobiological processes, such as the brain-derived neurotrophic factor/TrkB signaling pathway, the oxytocinergic system, and the hypothalamic–pituitary–adrenal axis. EE, which includes stimulating activities to counteract ethanol effects in animal studies, has parallels in human intervention that have shown potential benefits. Physical activity, cognitive behavioral therapy, and meditation, alongside techniques involving cognitive stimulation, social interaction, and recreational activities, may lead to more effective therapeutic outcomes in treatments of AUD.

1. Alcohol use disorder: a multifactorial disorder

Alcohol is a widely consumed drug, but its misuse poses significant health risks and can culminate in alcohol use disorder (AUD), characterized by a pathological pattern of excessive alcohol consumption and difficulties in controlling alcohol consumption, despite significant distress. Alcohol and/or other substance misuse is considered one of the greatest risk factors for illness, morbidity and mortality and, thus, a major contributor to the global burden of disease, with significant social and economic costs (American Psychiatric Association, 2013; Ezzati, Lopez, Rodgers, Vander Hoorn, & Murray, 2002; Winstock et al., 2021; World Health Organization, 2023).

AUD is a multifactorial condition that results from an interaction of genetic, environmental, psychological, and social factors. Genes contribute approximately 50% of the risk for AUD. Factors such as stress, trauma, accessibility and availability of the drug, and coping mechanisms influence the onset and progression of the disorder. AUD often co-occurs with other mental health disorders, and vice-versa. Social factors also contribute to the etiology of AUD. These include family dynamics, peer pressure, environment and community infrastructure, cultural attitudes toward drinking, and availability of support systems (Chartier, Karriker-Jaffe, Cummings, & Kendler, 2017). As a multifaceted disorder, addressing the diverse contributing factors of AUD is essential to improve prevention, diagnosis, and treatment.

Elements of social control, including both protective and risk environments, influence opportunities to use alcohol and thereby affect the genetic contribution to AUD. Adverse environmental conditions can act as social stressors that trigger or amplify genetic effects related to AUD. On the other hand, elements of social control, such as social norms, laws, regulations, and institutional policies, can limit or discourage behaviors considered harmful by society, like heavy drinking (Chartier et al., 2017).

The pharmacological treatments currently approved for AUD—disulfiram, naltrexone, acamprosate, nalmefene, and baclofen—assist in reducing cravings, decreasing alcohol consumption, and maintaining abstinence (Witkiewitz, Litten, & Leggio, 2019). However, their effectiveness varies among individuals, and not all patients benefit equally, as a significant percentage of patients struggle to maintain abstinence and may relapse into alcohol use during treatment (Heilig, Augier, Pfarr, & Sommer, 2019).

Treating AUD effectively requires not only pharmacological approaches but also interventions that mitigate the impact of environmental influences. For instance, promoting healthy social activities, enhancing support networks, and cultivating supportive living environments to enhance well-being can aid in reducing alcohol consumption and establishing long-term recovery. In this context, cognitive behavioral therapy, meditation, physical activity, and other forms of cognitive stimulation can act as supportive treatments for the patients.

2. Environmental enrichment: history, concepts, and neurobiological effects

Environmental enrichment (EE) traces its origins back to Donald O. Hebb's pioneering experiment in 1947, which demonstrated the profound impact of early and stimulating experiences on rat learning. Although Hebb had presented his findings as an abstract at an American Psychological Association meeting, they were not published in full; yet, they sparked subsequent research into the effects of environmental alterations on brain structure and function (Brown, 2020; Hebb, 1947).

EE represents an experimental paradigm in which laboratory animals are housed in enriched environments, offering a diverse array of sensory-motor stimuli and increased opportunities for social interaction, surpassing the confines of standard laboratory housing conditions (Simpson & Kelly, 2011). These environments are characterized by the incorporation of various objects, such as toys, tunnels, shelters, climbing structures, and running wheels, which are periodically rotated or altered to maintain novelty (Fig. 1). The cages used in EE experiments are typically larger than standard cages to accommodate these objects, with the aim of enhancing sensory, cognitive, and motor abilities in animals by promoting greater social interaction,

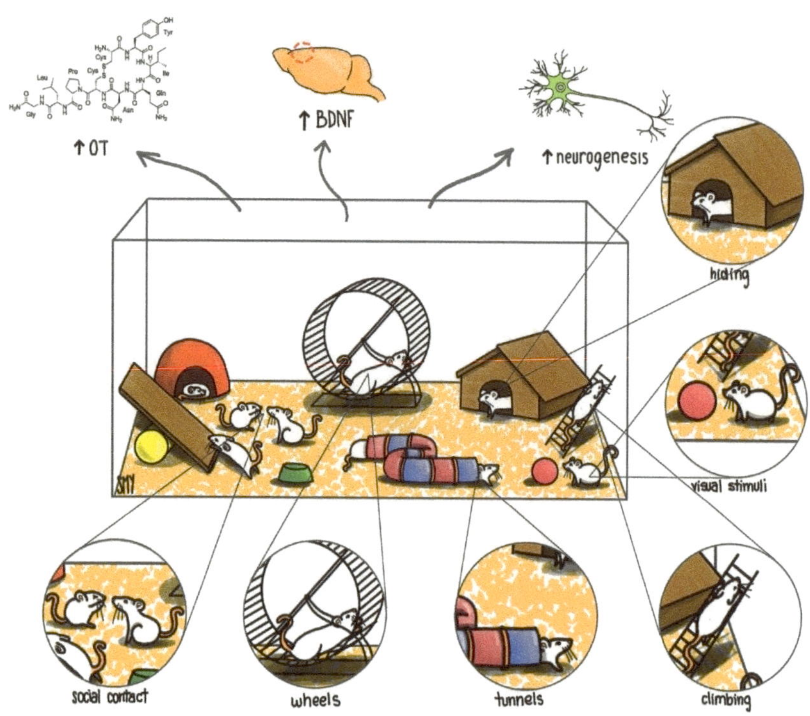

Fig. 1 Environmental enrichment comprises a diversity of objects that offer social and non-social stimulation, encouraging exploratory behavior, social interaction, and opportunities to exercise, such as climbing or running. As part of the protocol, the enriched environment may incorporate running wheels, tunnels, shelters, climbing structures, and toys of varying size and shapes. At a molecular level, EE increases hypothalamic oxytocin (OT) levels and mRNA, brain-derived neurotrophic factor (BDNF) levels and mRNA, and promotes neurogenesis.

environmental exploration, and voluntary exercise (Nithianantharajah & Hannan, 2006; Simpson & Kelly, 2011). EE is recognized for its role in enhancing the well-being of laboratory animals by incorporating social and nonsocial elements, thereby exerting positive effects on cognitive functions such as learning and memory (Bennett, McRae, Levy, & Frick, 2006; Young, Lawlor, Leone, Dragunow, & During, 1999), or emotional states, including depression and anxiety (Kimura et al., 2022; Takuma, Ago, & Matsuda, 2011), among other benefits.

The diversity of EE protocols documented in the literature may contribute to a significant variability in the results, making it challenging to compare findings across studies. These differences encompass various factors, including the presence or absence of elements like running wheels, the duration of exposure at different life stages, the frequency of object rotation, the number of conspecifics present, and comparisons to isolated or group-housed control groups. For instance, isolation rearing, often viewed as a form of chronic stress, should not be seen merely as a control but rather as an impoverished group, given that isolation can trigger a heightened stress response due to the lack of stimulation. This emphasizes the importance of evaluating EE outcomes with appropriate control groups (Simpson & Kelly, 2011).

Despite the variability in protocols, some crucial aspects are maintained across EE studies. These include the complexity of the environment, with the presence of diverse stimuli, and the novelty generated by the rearrangement and replacement of objects and food, fostering the formation of spatial maps and learning (Nithianantharajah & Hannan, 2006; van Praag, Kempermann, & Gage, 2000). Physical enrichment strategies involve structural modifications to the environment, such as increased space and resources for exercise, play, and exploration. Sensory stimulation, nutritional supplementation, and feeding are also considered forms of EE. Sensory stimulation involves providing innocuous stimuli aimed at activating one or more of an animal's senses, such as auditory, olfactory, and visual stimulation, in a controlled environment. Nutritional supplementation refers to dietary interventions used to address behavioral deficits, while feeding refers to the provision of novel food items or novel ways of feed provision. The social component of EE has also proved to be important for the model, as EE animals show heightened social interest to other animals present in the cage or in tests of social interaction (Kempermann, 2019; Rae et al., 2024; Ratuski & Weary, 2022; van Praag, Kempermann, & Gage, 2000).

Adding shelters or nest boxes should also be taken into account in EE models, as they can affect mouse behaviors. Studies have evaluated various materials, including paper strips, tissue, cotton string, wood wool, and wood shavings, each exerting different effects on behavior. For instance, materials like corn husks have been shown to reduce aggressive behavior, while others may inadvertently increase fighting (e.g. wood wool) or pose specific risks such as entangling around the limbs, tails or bodies of neonates (e.g. cotton wool, wood wool or shredded paper). Similarly, the inclusion of shelters or nest boxes has been investigated to reduce fear responses in mice. The material and design of the shelters can influence their effectiveness, with various results depending on factors like mouse strain and the nesting material (shredded paper versus wood wool). Providing nest boxes results in positive outcomes, including increased longevity and weaning rates in certain mouse strains (Bayne, 2018; Jennings et al., 1998). However, the inclusion of shelter may induce aggression, especially in mice. In laboratory conditions, male mice housed socially may exhibit some level of aggression due to the established hierarchy. Hierarchies are typically stable, but there are exceptions. For some mouse strains, such as CD-1 mice, EE has been associated with increased aggression levels among males, suggesting that it may adversely affect their well-being (Howerton, Garner, & Mench, 2008; van Loo et al., 2002). EE for group-housed BALB/cByJ mice was also associated with increased stress levels (McQuaid, Audet, Jacobson-Pick, & Anisman, 2013).

Hebb´s research laid the groundwork for subsequent studies investigating the effects of EE on behavioral, anatomical, and molecular aspects (Brown, 2020). Subsequent research has reported morphological changes in various cerebral areas of rats exposed to enriched environments, indicating that raising rats in environments with toys and mazes led to an increase in the weight and thickness of cortical tissue and hippocampus (Bennett, Diamond, Krech, & Rosenzweig, 1964; Rosenzweig, Krech, Bennett, & Diamond, 1962), as well as a higher number of newly formed glial cells (Altman & Das, 1964; Diamond et al.,1966).

EE plays an important role in shaping neural structures and functions in adult organisms, enhancing the adult neurogenesis in the dentate gyrus and decreasing spontaneous apoptosis in the hypothalamus. It impacts various stages of adult neurogenesis, influencing cell survival, precursor cell proliferation, and dendrite and spine formation. Moreover, EE promotes the survival of adult-born neurons without significantly affecting precursor cell proliferation. Physical activity—such as access to a running wheel—, considered an important element of EE, can also influence precursor cell proliferation, potentially masking the distinction between EE and physical activity (Kempermann, 2019; Segovia, Yagüe, García-Verdugo, & Mora, 2006; Young et al., 1999). Furthermore, exposure during adulthood to a challenging and complex environment increased spine density on CA1 hippocampal pyramidal neurons, indicating an increased number of synapses per cell, which can lead to alterations in cell connectivity, ultimately enhancing spatial learning (Moser, Trommald, & Andersen, 1994). EE facilitated long-term potentiation, a mechanism of synaptic strengthening, in this brain region, attributed to alterations in the threshold for activity-dependent recruitment of the cAMP-PKA signaling pathway (Duffy, Craddock, Abel, & Nguyen, 2001). These observed changes are consistent with the enrichment-induced modifications in gene expression involved in synaptic function and cellular plasticity in various brain regions, such as the cerebral cortex and hippocampus.

Rodents raised in EE are known to exhibit elevated levels of neurotrophins, including brain-derived neurotrophic factor (BDNF), glial-derived neurotrophic factor, nerve growth factor, and neurotrophin-3 (Ickes et al., 2000; Young et al., 1999). A study aimed to analyze the transcriptional and epigenetic regulatory effects of EE on BDNF in the prefrontal cortex of male mice showed upregulation of Bdnf exons II, IV, VI, and IX mRNA expression, alongside reduced methylation levels at two CpG sites of exon IV in the PFC of EE mice. These findings suggest an EE-induced epigenetic control of BDNF exon expression, potentially involving exon IV methylation (Costa et al., 2023). These factors contribute to neuronal function and survival, as well as synaptic plasticity (Huang & Reichardt, 2001), therefore, suggesting that EE can confer several benefits to development and survival of the brain.

The functional outcomes associated with these changes induced by EE include enhanced behavioral performances in animals, such as cognitive and learning improvement. EE improves cognitive function and mitigates the learning and memory deficits observed in age-related memory decline or neurodegenerative diseases, potentially by increasing BDNF expression and modulating synaptic plasticity (Bennett et al., 2006; Young et al., 1999). While the initial characterization and conceptualization of EE were rooted in its effects on cognitive functions like learning and memory, EE has shown promising potential in the context of psychiatric disorders, including anxiety and depression, as well as comorbid pain (Kimura et al., 2022; Takuma et al., 2011). Chourbaji et al. (2012) demonstrated that exposure to EE decreased anxiety-like behavior in BDNF+/− mice, which are genetically engineered to exhibit reduced levels of BDNF and display heightened anxiety.

EE has been linked to enhancements in emotional regulation, stress resilience, cognitive function, promotion of prosocial behavior, increase of social dominance, and decrease in motivations for appetitive taste stimuli, as evidenced by both animal models and human studies. Additionally, EE has been explored as a potential intervention for mitigating symptoms of schizophrenia, depression, and neurodevelopmental disorders like autism spectrum disorder in humans and animal models (Aronoff, Hillyer, & Leon, 2016; Benaroya-Milshtein et al., 2004; Gubert & Hannan, 2019; Kimura et al., 2022; Pang et al., 2013).

Various studies have convincingly demonstrated that EE elicits responses in animals that are opposite to those induced by stress. However, the proposed inoculation stress hypothesis offers a potential explanation for this phenomenon. According to this hypothesis, EE represents a state of chronic mild stress, which triggers adaptive changes in neuronal and neuroendocrine systems, eventually enhancing the resilience of enriched animals to subsequent stressors. Exposure to novelty, social interaction, and physical activity, have been shown to elicit stress-like responses in various studies (Crofton, Zhang, & Green, 2015).

Indeed, EE can influence both anxiety responses and plasma corticosterone levels. While EE has been shown to decrease the anxiety induced by acute stress (Novaes et al., 2017) and predictable chronic stress (Zanca et al., 2015), its combination with chronic unpredictable stressors seems to reduce the stress threshold. Whether this makes it more challenging for the animal to cope with stress, or if the stress-inoculated mice may exhibit more active defense behavior, further studies are needed to determine. By instance, chronic unpredictable stress induced anxiety-like behavior and elevated ACTH in EE mice (Belo-Silva et al., 2024). Regarding corticosterone levels, no alterations (Leal-Galicia, Saldívar-González, Morimoto, & Arias, 2007) higher (Marashi, Barnekow, Ossendorf, & Sachser, 2003; Moncek, Duncko, Johansson, & Jezova, 2004) or lower (Grégoire, Bonenfant, Le Nguyen, Aumont, & Fernandes, 2014; Konkle, Kentner, Baker, Stewart, & Bielajew, 2010) corticosterone levels have been reported in EE mice compared to controls. These findings, along with the inoculation stress hypothesis, underscore the intricate interplay among EE, stress, and the hypothalamic–pituitary–adrenal (HPA) axis in regulating anxiety.

In a review by Malone, Shaykin, Stairs, and Bardo (2022), the consequences of EE on neurobehavioral effects of drugs of abuse were reported across several levels, including cellular, neurochemical, neuroendocrinal, and neuroimmune systems. The review highlighted significant findings, such as alterations in dopamine systems, particularly in dopamine transporter function, effects on the neural immune system, the HPA axis, BDNF levels, transcriptional changes, involvement of transcription factors, and exosomal microRNA profile. Most of studies investigating the protective effects of EE on addiction-like behaviors have primarily focused on psychostimulants. Further research is needed to elucidate the mechanisms through which EE influences ethanol-induced behaviors.

As an environmental intervention capable of inducing neurobiological changes and influencing multiple behaviors, EE offers a new approach to mental health that complements traditional therapeutic interventions. This marks a growing trend in its application within substance use disorder (SUD) and AUD treatment.

3. The impact of environmental enrichment on multiple alcohol-induced behaviors and its mechanisms

3.1 Behavioral effects

EE has shown multiple beneficial effects across various models of addiction-like behaviors of drugs (for a review, see Malone et al., 2022). EE rats exhibited reduced cocaine intravenous self-administration compared to control rats, yet they displayed enhanced cocaine conditioned place preference (CPP) (Green et al., 2010). Gipson, Beckmann, El-Maraghi, Marusich, and Bardo (2011) showed that EE prevented the escalation of cocaine self-administration, particularly at lower doses, with social interaction emerging as the main contributing factor to this protective effect over novelty. Additionally, this protective effect of EE at lower doses was further evidenced in studies involving methylphenidate. It facilitated abstinence from cocaine (Ewing, Ranaldi, 2018) and from heroin in rats (Barrera et al., 2023; Peck, Galaj, Eshak, Newman, & Ranaldi, 2015), reduced the influence of cocaine-associated cues on drug-related responses (Ranaldi, Kest, Zellner, & Hachimine-Semprebom, 2011), and provided protective effects against the acquisition of opiate self-administration (Hofford, Chow, Beckmann, & Bardo, 2017; Imperio et al., 2018; Barrera et al., 2023). Moreover, EE exhibited anti-craving effects in nicotine addiction (Gomez, Sun, Midde, Harrod, & Zhu, 2015; Venebra-Muñoz et al., 2014) and decreased sensitivity to the psychostimulant effects of nicotine in rats (Green, Cain, Thompson, & Bardo, 2003).

Several preclinical studies have addressed the effects of EE on alcohol-related effects, such as alcohol consumption using the two-bottle-choice paradigm, operant self-administration, CPP, behavioral sensitization, abstinence, and relapse models.

Research by Maccioni et al. (2022) demonstrated that Sardinian alcohol-preferring (sP) rats housed in EE exhibited lower levels of lever-responding for alcohol, reduced alcohol self-administration, and lower alcohol break-point compared to those in impoverished or standard environments, suggesting a decrease in the reinforcing and motivational effects of alcohol.

Similarly, findings from Deehan, Palmatier, Cain, and Kiefer (2011) using 24-h access to two-bottle choice and operant chambers indicate that alcohol-preferring male rats reared in EE exhibited reduced ethanol consumption, preference, and motivation compared to those raised in standard environments. The findings are specifically related to ethanol, as all rearing conditions consumed equivalent amounts of solutions containing high sucrose concentrations (10%). C57BL/6J mice, known for their propensity to consume large amounts of ethanol under the drinking-in-the-dark paradigm, exhibited decreased binge-like alcohol consumption when exposed to enriched environments. In addition, they also demonstrated lower levels of anxiety-like, compulsive-like, and novelty-seeking behaviors (Rodríguez-Ortega & Cubero, 2018). Females raised in EE from weaning to adulthood, within the context of SHR rats—a model of attention deficit hyperactivity disorder—demonstrated reduced novelty-induced locomotion, lower consumption of saccharin and ethanol under forced conditions, and exhibited decreased ethanol preference in a free-choice setting compared to control rats (de Carvalho, Pandolfo, Pamplona, & Takahashi, 2010).

Exposure to EE during periods of abstinence has been shown to mitigate the risk of relapse to alcohol-seeking behavior in rats (Campbell, Jin, & Lawrence, 2019) and to prevent depression-related behavioral phenotypes as well as dysregulated activity of the HPA axis associated with alcohol abstinence (Pang et al., 2013). Under conditions of stress, such as social and environmental deprivation, ethanol preference tends to escalate. However, this heightened preference can be effectively mitigated through the implementation of EE in both ethanol-naïve mice and those with prior ethanol experience (Holgate, Garcia, Chatterjee, & Bartlett, 2017). Similar results were demonstrated in female C57BL/6 mice (Pang et al., 2013). Moreover, a study by Marianno, Abrahao, and Camarini (2017) demonstrated that EE resulted in a reduction in alcohol consumption in the drinking-in-the-dark model in Swiss mice, especially under conditions of acute stress. Mice resilient to depressive-like behaviors did not exhibit increased ethanol intake induced by social defeat, unlike susceptible mice, which showed an increase in neuroinflammatory responses, characterized by increased levels of IL-6 in the prefrontal cortex and striatum compared to controls. When defeated mice were exposed to EE, they did not show an increase in neuroinflammatory markers and did not exhibit an increase in ethanol self-administration (Reguilón, Ferrer-Pérez, Manzanedo, Miñarro, & Rodríguez-Arias, 2021).

Transitioning adult mice from standard cage to EE resulted in a decrease in binge-like alcohol consumption, as seen in Holgate et al. (2017).

Conversely, discontinuing EE led to a gradual increase in binge-like alcohol intake, followed by an increase in anxiety. This implies in protective and therapeutic effects of EE on binge-like behavior and emphasizes the importance of continued EE for sustaining these beneficial effects.

However, conflicting findings have been reported, as demonstrated by larger ethanol consumption in rats reared in the EE compared to those reared individually. In contrast, when transferred to isolation after initial enrichment, animals exhibited reduced ethanol intake (Rockman, Gibson, & Benarroch, 1989). Additional studies have confirmed these discrepant results. Exposure to EE throughout adolescence elevated ethanol consumption in females, but not in male rats, as determined by 24-h two-bottle choice test (Suárez, Fabio, Bellia, Fernández, & Pautassi, 2020). Maccioni et al. (2022) observed an increase in binge-like alcohol drinking among Sardinian alcohol-preferring (sP) male rats housed in EE conditions, characterized by consumption exceeding 2 g/kg of ethanol and blood alcohol levels surpassing 100 mg% during daily 1-h drinking sessions. In this latter study, the increase in ethanol intake was attributed to the potentiation of alcohol consumption induced by the timing of drinking sessions, with intake nearly doubling in the final hours of the dark phase compared to the initial hour. This heightened sensitivity to alcohol intake in EE sP rats can be ascribed to increased emotional distress and anticipation of alcohol availability, leading to larger and more concentrated alcohol consumption. In the study by Berardo, Fabio, and Pautassi (2016), male Wistar EE rats, but not female, exhibited higher ethanol intake and preference compared to rats under standard conditions. It is worth noting that ethanol concentration provided in this later study was relatively low (7.5%) compared to the majority of studies (10%–15%), although EE sP rats had exhibited a preference for the highest ethanol concentration (30%) (Maccioni et al., 2022).

Despite these variations, the consensus remains that EE helps to reduce ethanol intake. Several factors can contribute to these discrepant findings, including ethanol concentration, regimen of ethanol exposure, sex, age, specie, and strain.

Studies about the effects of EE on ethanol-induced CPP revealed that despite increasing sociability and reducing anxiety-like behaviors, EE heightened the preference for ethanol in Swiss male mice (Rae et al., 2018) and led to a preference for ethanol reward over social reward, an effect mimicked by the oxytocin analogue carbetocin (Rae et al., 2024). Additionally, in a study comparing ethanol-induced CPP between adults and adolescents, adolescents displayed greater sensitivity to the enhancement of ethanol CPP induced by EE (Pautassi et al., 2017). In contrast, female SHR rats expressed a reduction in EtOH-induced CPP compared to standard condition rats (de Carvalho et al., 2010); and the reinstatement of ethanol-induced CPP was inhibited by EE in C57BL/6 male mice (Li, Meng, Huang, Wang, & Li, 2015). Although contentious, it appears that EE may not attenuate the rewarding effects of ethanol, as demonstrated in Pavlovian conditioning studies involving males.

In an investigation focused on neuroplasticity, EE emerged as a preventive measure against the development and expression of ethanol-induced behavioral sensitization (Rueda, Teixeira, Yonamine, & Camarini, 2012). The hypothesis is that EE decreases the incentive salience of ethanol, disrupting the association between environmental cues and the effects of drugs, thus attenuating neurochemical mechanisms responsible for sensitization.

These findings underscore the significant role of environmental factors, such as social and EE, in modulating alcohol consumption, relapse, locomotor activation, and other ethanol-related behaviors (Fig. 2).

3.2 Neurobiological effects

The effects of EE on oxytocinergic (OT) system, BDNF levels, and components of the HPA axis offer insights into how this environment modulates ethanol-induced effects.

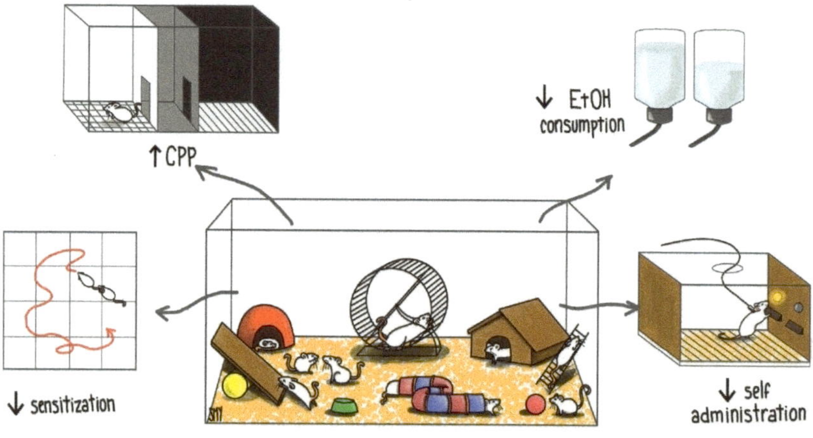

Fig. 2 The figure shows the effects of enrichment environment (EE) on ethanol-induced behavioral effects. EE modifies addiction-like behaviors associated with ethanol, including reduced ethanol consumption, decreased self-administration, prevention of behavioral sensitization, and heightened conditioned place preference.

The study by Rae et al. (2018) investigated the stimulatory impact of EE exposure on alcohol reward, as evidenced by enhanced ethanol CPP, possibly mediated by an OT-dependent mechanism. EE induced neuroadaptations in the OT system, such as increased hypothalamic oxytocin (OT) levels and mRNA levels, decreased oxytocin receptor (OTR) binding in specific brain regions, and diminished OTR signaling activity in the striatum, evidenced by lower phospholipase C in the striatum (Rae et al., 2018, 2024). This implies a possible compensatory mechanism that entails receptor internalization to counterbalance the increased availability of ligands. Administration of the OT analogue carbetocin mimicked EE effects on ethanol CPP and induced similar alterations in OTR binding. Conversely, administration of the OTR antagonist L,369–899 during EE reversed the pronounced ethanol rewarding effect (Rae et al., 2018). Additionally, EE elevated D1 receptor expression in the nucleus accumbens, which was attenuated by the oxytocin antagonist L-368–889 (Rae et al., 2024), suggesting an oxytocin-dopamine interaction underlying the ethanol pro-rewarding effects of EE. These discoveries offer insights into the neurobiological mechanisms through which EE enhances the rewarding effects of ethanol in socially enriched environment, emphasizing the role of OT-dopaminergic pathways.

Dysregulation of BDNF, its receptor TrkB, and associated signaling pathways has been noted in various brain regions following ethanol exposure, with conflicting reports of upregulation or downregulation in several areas (Angelucci et al. 2009; Davis, 2008). EE modulates BDNF levels, including in ethanol-treated mice (Pautassi et al., 2017; Rueda et al., 2012), with results that vary from decrease to increase or no change. EE decreased BDNF levels in the prefrontal cortex but not in the hippocampus (Rueda et al., 2012). Additionally, Pautassi et al. (2017) observed that ethanol significantly reduced BDNF levels in the prefrontal cortex in adolescents reared under standard housing conditions, but not in adult mice nor adolescents housed in EE conditions.

Evidence suggests that BDNF regulates ethanol intake by triggering MAP kinase signaling and producing downstream gene products like preprodynorphin (Logrip, Janak, & Ron, 2008). In this context, prenatal ethanol exposure (PEE) induced significant alterations in the transcriptional regulation of various opioid genes within the brain mesocorticolimbic regions in infant or adolescent offspring, resulting in an anxiety-like phenotype, which were partially mitigated by EE (Wille-Bille et al., 2020). Indeed, PEE causes extensive harm to the fetus, potentially resulting in fetal alcohol spectrum disorders (FASD), which include a range of physical, cognitive, and behavioral impairments (Berman & Hannigan, 2000). EE has been investigated as a potential intervention to mitigate these effects, including reducing the risk of addiction to psychostimulants in prenatal ethanol exposed rats (Wang et al., 2018). While the precise mechanisms behind the protective effects of EE in PEE are not fully understood, they may involve anatomical and neurochemical changes, such as enhanced neuron size and synaptic development, increased expression of growth factors, reduction of neuroinflammation, and changes in hippocampal network function (Aghaie et al., 2020; Hannigan & Berman, 2000). For instance, in 8-week-old animals exposed to PEE, significant changes in excitatory transmission were observed in newly generated dentate granule cells of EE mice (Kajimoto et al., 2016). These findings suggest that EE interventions could benefit FASD-related issues.

Following the findings that EE during alcohol abstinence reverted abnormalities in HPA axis function (Pang et al., 2013), glucocorticoid receptor (GR), mineralocorticoid receptor (MR), Crh, pomc1 (pro-opiomelanocortin, the precursor of ACTH), and dopamine 2 receptor (drd2) mRNA levels were measured. In the hypothalamus, GR gene expression was higher in alcohol-abstinent C57BL/6J mice housed in standard conditions, but not in abstinent mice housed in EE. No significant differences were found in MR gene expression between the groups. Hypothalamic Crh gene expression decreased in both abstinent groups (control and enriched) compared to water controls. In the pituitary, up-regulation of GR and MR expression in abstinent mice was maintained in EE mice. EE abstinent mice showed higher pituitary pomc1 and drd2 gene expression compared to water controls. Altogether, these findings provide evidence of alterations in gene expression during ethanol abstinence in EE mice (Pang et al., 2013).

Overall, EE exerts multiple effects on ethanol-induced behavioral and neurobiological effects.

4. Physical activity as environmental enrichment

EE setups may incorporate elements that promote physical activity, such as running wheels or toys for exploration, although their primary aim is to facilitate natural, voluntary movement and engagement with the environment. In contrast, exercise typically refers to structured, planned, and repetitive physical activity (Caspersen, Powell, & Christenson, 1985). Physical activity has been found to notably diminish ethanol consumption and preference across various paradigms, including two-bottle choice, drinking-in-the-dark and operant oral self-administration (Darlington et al., 2016; Gallego, Cox, Funk, Foster, & Ehringer, 2015; Reguilón, Ferrer-Pérez, Ballestín, Miñarro, & Rodríguez-Arias, 2020), suggesting a reduction in the rewarding effects of ethanol.

The effects of voluntary running exercise, whether undertaken prior to, concurrently with ethanol exposure, or during ethanol withdrawal, have been explored in rodents. In adolescent C57BL/6 male and female mice, concurrent voluntary wheel running resulted in decreased ethanol consumption and preference, with notable sex differences observed—females exhibited higher alcohol preference and showed stronger responses to exercise-induced reduction in consumption. These effects were associated with changes in hippocampal BDNF expression (Gallego et al., 2015). Voluntary wheel running also demonstrated protective effects against ethanol withdrawal-induced seizure susceptibility in male CR rats (McCulley et al., 2012). Centanni, Conley, Luchsinger, Lantier, and Winder (2022) examined the impact of intermittent running wheel access on alcohol consumption patterns and negative affective symptoms during ethanol abstinence in female C57BL/6J mice. Wheel access modified drinking behavior, with increased consumption when the wheel was locked and decreased consumption when unlocked. Unlimited wheel access mitigated abstinence-induced negative affect-like behavior, such as anhedonia and depression (Centanni et al., 2022). Furthermore, ethanol consumption remained stable during alternating periods of wheel running, whereas wheel running increased in the absence of ethanol and decreased upon its reintroduction (Ozburn, Harris, & Blednov, 2008), supporting the theory of hedonic substitution.

Research has explored how physical activity influences the escalation of ethanol consumption prompted by stress, yielding favorable results. Specifically, C57BL/6J mice exposed to voluntary wheel running during adolescence and later subjected to social defeat did not show an increase in ethanol self-administration behavior, particularly among those susceptible to depressive-like behavior induced by social defeat. Susceptible mice exhibited decreased BDNF levels in hippocampus and striatum following ethanol self-administration, regardless access to running wheel (Reguilón et al., 2020; Reguilón, Ferrer-Pérez, Manzanedo, Miñarro, & Rodríguez-Arias, 2023). Voluntary wheel running during adolescence improved anxiety-like behavior induced by stress, but increased ethanol intake/preference in adulthood in Long Evans male rats. It is likely that the presence of a running wheel acted as a reinforcement for the rats engaged in running, thereby sensitizing their neural circuitry to future rewards, which resulted in increased drinking upon discontinuation of access to the wheel. This could also be interpreted as a form of hedonic substitution (Lynch, Porter, & Butler, 2019).

Thus, incorporating physical activity into treatment for AUD could potentially reduce alcohol consumption, improve negative affect during abstinence, and help prevent the development of AUD in individuals at high risk. The mechanisms underlying these effects may involve several factors, including enhanced clearance or other physiological adaptations, modulation of BDNF signaling, reversal of ethanol-induced neuroinflammation by reducing the expression of chemokines associated with neuroinflammation, adaptations in the HPA axis response to stressors contributing to reduced severity of ethanol withdrawal, and changes in gene expression related to ethanol preference quantitative trait loci, such as potassium channel genes, Oprm1, Prkcg, Stxbp1, Crhr1, Gabra3, Slc6a13, Stx1b, Pomc, Rassf5, Polr2a, and Camta2 (Darlington et al., 2016; Gallego et al., 2015; McCulley et al., 2012; Reguilón et al., 2020).

These findings suggest that there may be a hedonic substitution effect between wheel running and ethanol consumption.

5. Translational implications: complementary clinical interventions as environmental enrichment in AUD

Among patients diagnosed with AUD, there is a challenge in controlling consumption, associated with reactive decision-making in response to environmental and/or emotional triggers that drive drug seeking, which often is accompanied by subsequent regret, self-judgment, guilt, and shame (Witkiewitz, Marlatt, & Walker, 2005).

Anxiety, depression, and other negative affect states are among the main predictors of substance use, relapse, and treatment outcomes, followed by interpersonal conflicts and social pressure. This is a dynamic relationship, where negative emotional states increase the likelihood of substance use and vice versa. In AUD, the presence of mental comorbidities is frequent, particularly anxiety and depression. A systematic review of data from different countries indicated that approximately 40% of individuals diagnosed with AUD also have major depressive disorder and anxiety disorder.

Within this population, individuals engage in harmful drug use to seek pleasure, escape from reality, and avoid the discomfort of abstinence (Witkiewitz et al., 2019).

The environment significantly impacts alcohol consumption, often taking place in social settings, where individuals may be influenced by peer pressure or social norms. Additionally, social factors such as stress, loneliness, and social isolation can contribute to alcohol misuse. EE strategies may address the social aspects of AUD by promoting positive social interactions, providing alternative activities to drinking, and fostering supportive social networks, thereby potentially reducing alcohol-related harm and promoting healthier behaviors.

Because of its chronic nature and failure to maintain abstinence, frequent relapses, and the presence of other mental comorbidities, developing an effective treatment for AUD is challenging. Behavioral interventions, combined with pharmacological therapies, may better assist patients in dealing with the challenges of this disorder.

Social interaction, sensory stimulation, and physical activity are some of the components of EE. In preclinical and clinical studies, as there is no standardization of EE, diverse settings and stimuli aimed at promoting well-being and sociability emerge as a promising recommendation for individuals who suffer from SUD or AUD.

In light of favorable results regarding EE as an adjunct in the treatment of AUD in animal models, some studies are exploring the translation of a controlled, multimodal protocol into humans (Barillot et al., 2023). In this chapter, we explore certain forms of EE that could be implemented to aid in the treatment of AUD and to improve patient adherence to pharmacotherapy.

5.1 Cognitive behavioral therapy (CBT)

The cognitive-behavioral approach focuses on how the individual perceives the situation itself, which can interfere with motivational, affective, and behavioral reactions. It is one of the most used in the treatment of SUD because, through it, it is possible to analyze and alter automatic thoughts and distorted beliefs that can lead to dysfunctional emotions and behaviors (Carroll & Kiluk, 2017).

Literature data classify SUD as a type of learned behavior; thus, using behavior modification techniques can help patients cope with everyday situations that lead to drug use. The main goal of CBT in this scenario is to assist in the reconstruction of dysfunctional cognitions in order to achieve flexibility in evaluating specific situations and developing functional strategies related to prevention of use (Carroll & Kiluk, 2017).

In this approach, techniques are used to identify dysfunctional thoughts that can lead to negative feelings and maladaptive behavior, to control bad cognitions and beliefs, to develop adaptive and functional thoughts, minimize the dysfunctional ones, and to help have a real view of oneself and the future (Ceci et al., 2022).

Performed in approximately 12 sessions, CBT identifies high-risk situations and the reasons that drive individuals to drink abusively, consumption patterns, and their own perceptions related to alcohol use. Additionally, it helps in developing response skills, relapse prevention, and social support, in order to make patients more responsible and with greater self-control (Carroll & Kiluk, 2017; Ceci et al., 2022).

Various literary reviews demonstrate the effectiveness of CBT for SUD (Carroll & Onken, 2005; Magill & Ray, 2009). Magill and Ray (2009), analyzing 53 controlled clinical trials for adults diagnosed with AUD, observed a significant treatment effect in patients who practiced CBT, with approximately 58% of patients performing better than those in the control group.

Recently, Peng et al. (2022) observed that individuals who underwent group CBT for 8 weeks showed significant improvement in self-management, problem-solving, and positive response, resulting in greater treatment adherence and reduced relapse. Similarly, a group of American researchers, treating female patients diagnosed with AUD, reported that both individual and group CBT significantly reduced the percentage of alcohol consumption days and coping skills maintained throughout the rest of the treatment (Epstein et al., 2018).

By combining this technique with other psychosocial treatments or pharmacotherapy, several studies demonstrate improved effects (Magill & Ray, 2009). Therapy combined with the drug naltrexone resulted in an increase in patients' abstinence time compared to the group that received only CBT (Walters, Connor, Feeney, & Young, 2009). Additionally, Feeney, Connor, Young, Tucker, and McPherson (2006) associated acamprosate and naltrexone with behavioral treatment and observed an increase in treatment frequency, abstinence, and number of days until the first relapse.

In the same vein, the literature has also shown that combining CBT with other psychotherapeutic techniques, such as motivational enhancement therapy (MET), yields even better results (Magill, Kiluk, & Ray, 2023). Through the application of motivational psychology to produce motivated changes within the individual, MET helps the patient find reasons for changes in addictive behavior and reduce treatment dropout (Madson, Schumacher, Baer, & Martino, 2016) and alcohol consumption (Martín-Pérez et al., 2019).

Therefore, as it is known that the individual's perception of a situation interferes with the behavioral, affective, and motivational reactions they may express, CBT is of utmost importance because by analyzing and modifying automatic thoughts and distorted beliefs, it is possible to obtain more balanced emotions and behaviors (Epstein et al., 2018).

5.2 Mindfulness

New behavioral interventions have shown potential benefits as adjunct tools in the treatment of AUD and other substances, among which mindfulness-based interventions (MBIs) stand out (Sancho et al., 2018). Literature data have elucidated positive outcomes of MBIs for addiction treatment (Korecki, Schwebel, Votaw, & Witkiewitz, 2020; Sancho et al., 2018), with a reduction in craving, frequency, and quantity of use (Korecki et al., 2020), a decrease in negative factors, and a mitigation of relapse incidence (Li, Howard, Garland, McGovern, & Lazar, 2017).

Mindfulness is a term that can be used to define the human ability to pay attention to the present moment with openness and curiosity, as well as it can refer to meditative practices or structured protocols to develop such ability (Kabat-Zinn, 2013). Based on Buddhist teachings, Jon Kabat-Zinn structured a pioneering protocol (mindfulness-based stress reduction—MBSR) for patients in a hospital setting. The promising initial scientific results of MBSR for stress management and quality of life have driven the dissemination of mindfulness in the Western healthcare field (Kabat-Zinn, 2013). Since then, new protocols have been developed, among which it is worth mentioning mindfulness-based cognitive therapy for depression and MBRP (mindfulness-based relapse prevention) for SUDs (Bowen & Enkema, 2014).

MBRP emerged from a range of studies from the Addictive Behaviors Research Center, aiming to promote gains from Marlatt's Relapse Prevention Program (Roos, Bowen, & Witkiewitz, 2017). Structured in eight weekly group sessions, combining cognitive-behavioral techniques, relapse prevention, and mindfulness meditation practices (Bowen & Enkema, 2014; Roos et al. 2017).

The protocol generally includes a sequence of meditation practices, teachings, and group discussions, with 2-h sessions each. The sessions are led by meditation practitioners who have undergone rigorous training. The offered practices involve attention to direct experiences: physical sensations, emotions, and thoughts. During the practices, participants are invited to feel, acknowledge, and accept their experiences of the present moment, aiming to accept and compassionately embrace experiences, pleasant or unpleasant, such as pain, restlessness, anger, or self-depreciative thoughts (Davidson & Kaszniak, 2015). Additionally, there is a collective discussion about acceptance, courage, wisdom, and awareness of the transient nature of life, among other attitudes.

As an adjunct technique in AUD treatment, mindfulness brings benefits because as there is stimulation of full attention to the present moment, including the perception of sensations, emotions, and thoughts, it is possible to control trigger situations that lead to alcohol consumption (Roos et al. 2017). The practice of open attention and acceptance of physical and emotional discomfort provides a pause of awareness between trigger and consumption behavior, increasing the possibility of more conscious choices. Additionally, there is also better acceptance of discomfort, expanding perspectives on immediate craving management, as well as reframing guilt and self-judgment post-use (Bowen et al., 2014).

Mindfulness extends the opportunity to respond to triggers more skillfully and consciously, as opposed to conditioned reactive behavior (Bowen et al., 2014).

A randomized clinical trial conducted with 286 inpatients for SUD in non-profit private facilities demonstrated that patients diagnosed with severe AUD who participated in the MBRP protocol presented a significantly lower risk of relapse to use. Additionally, a drastic reduction in excessive alcohol consumption was observed at 6- and 12-month follow-up after treatment completion (Bowen et al., 2014). In addition to this, Davis, Clinton, Akil, and Becker (2008) observed that patients in residential treatment associated with self-help meetings (Alcoholics Anonymous), undergoing mindfulness protocol, decreased substance craving and use, besides a drastic improvement in stress associated with use, compared to the control group.

Witkiewitz et al. (2019) demonstrated in an outpatient study with adult women who underwent the same protocol, a reduction in the time and dose of daily drinks consumed, in the percentage of days of excessive consumption, in reactivity to alcohol stimulus, and craving for it.

Studies have also shown that increased drug craving in patients with AUD may be related to hypersensitivity to alcohol cues in attentional brain networks. Brown et al. (2020) exposed dependent patients to a series of substance-related images and evaluated the level of craving evoked when exposed to them. After eight mindfulness sessions, a decrease in alcohol craving was observed when the same images associated with use were shown. This change may be related to the fact that this meditation can act by controlling attention.

In line with these findings, Korecki et al. (2020), in a systematic review, gathered the latest publications on the use of MBI as an adjunct in SUD treatment. It was observed that this technique seems to be as effective as existing evidence-based treatments for SUDs in reducing the frequency and quantity of alcohol use, substance-related problems, substance craving, and increasing abstinence rates.

Thus, it is believed that the benefits found from this technique for SUD are related to the development of non-judgmental observation of thoughts and behavioral impulses, that is, by understanding the feelings associated with craving, an individual may be able to "overcome" them without engaging in substance use. Additionally, by working on improving the automaticity of action, it is possible to mitigate automatic behaviors and bring purposeful attention to the present moment, which can help reduce the frequency and quantity of substance consumption and related problems (Korecki et al., 2020).

5.3 Physical exercise

Behavioral therapies, such as those mentioned above, show positive results during patient treatment; however, literature data demonstrate that in the year following its completion, there is a high relapse rate of substance use (Moriarty, Bourbeau, & Zuhl, 2020). Because of this, researchers are studying various approaches that can assist and complement the treatment for AUD.

In this regard, for years, various researchers have been examining the role of physical activities in improving symptoms associated with alcohol abuse. Physical activity can be understood as "planned, structured, and repetitive body movement aimed at improving or maintaining one or more components of physical fitness" (Caspersen et al. 1985), and it is believed to be beneficial because it is a healthy alternative that can improve mood and psychological well-being (factors that can lead to relapse) and also because it is the only one available to people without access to other forms of treatment, such as medication or psychotherapy (Manthou et al., 2016).

Health institutes have been resistant to consider this intervention as treatment, referring to it only as a recreational activity. French, English, and Australian clinical guidelines considered physical exercise as a beneficial recreational activity for daily activities and as a simple way to relax (Haber, 2009; National Collaborating Centre for Mental Health NCCMH, 2011; Société Française d'Alcoologie SFA, 2015). Recently, the updated American guidelines from NIDA considered physical activity as a promising adjunct treatment for SUD (National Institute on Drug Abuse NIDA, 2012).

The practice of physical exercise can be considered a viable and safe approach for patients undergoing treatment. Its results show improvements in physical health problems and sedentary behavior reduction (Hallgren, Vancampfort, Schuch, Lundin, & Stubbs, 2017), improvement in somatic problems (cardiovascular diseases, diabetes, metabolic syndrome), and mental disorders (sleep, anxiety, and depression) (Giesen, Deimel, & Bloch, 2015).

Regarding the consumption of substances of abuse, several studies have shown positive effects. Positive results were found by implementing an aerobic exercise program (Brown, Prince, Minami, & Abrantes, 2016), high-intensity interval training (Flemmen, Unhjem, & Wang, 2014), and combining both (Giesen et al. 2015; Muller & Clausen, 2015). However, researchers explain that the observed effects are related to the type of exercise performed and its structural parameters, such as intensity and frequency (Giménez-Meseguer, Tortosa-Martínez, & Cortell-Tormo, 2020).

Aerobic activity is the most commonly used type for patients diagnosed with AUD (Giménez-Meseguer et al. 2020). Sinyor, Brown, Rostant, and Seraganian (1982) observed a significant increase in the abstinence rate and alcohol consumption over three months after engaging in physical conditioning activities compared to the control group. Similarly, a randomized study observed a decrease in the number of days of alcohol consumption, doses consumed, greater treatment adherence, and increased abstinence when comparing patients who engaged in moderate aerobic exercises for 12 weeks with a control group that received only counseling (Brown et al., 2014). This same research group also observed that the treatment group, post-exercise sessions, experienced a decrease in anxiety, alcohol craving, and mood improvement compared to the initial time and also to the control group (Brown et al., 2016).

The same type of activity, practiced by patients for 8 weeks, resulted in a decrease in weekly alcohol consumption, desire, and also in the score on the Alcohol Use Disorders Identification Test (Georgakouli et al., 2017). In addition, recent literature reviews have shown that moderate to severe intensity physical exercise may be an adjunctive and non-pharmacological treatment for patients diagnosed with AUD (Giesen et al. 2015; Hallgren et al., 2017).

When analyzing the intensity of the exercises to be performed, the literature revealed better efficacy in those of moderate to severe intensity. Ussher, Sampuran, Doshi, West, and Drummond (2004) demonstrated that these intensities provide short-term relief from alcohol cravings compared to light intensities. It was also observed that compared to low levels of physical activity, moderate intensity assisted patients in excessive alcohol use, as a "dose-response" effect, i.e., the amount consumed by the intervention group decreased by 4% for each day of increased exercise (Roessler et al., 2017).

This same moderate intensity in acute sessions also yielded beneficial results. Hallgren et al. (2021) observed that with a short session of this type of exercise, a reduction in alcohol craving and mood improvement in adults diagnosed with AUD is possible.

5.4 Other techniques and final considerations

In addition to physical activity, therapy, and meditation, some researchers believe there are benefits in techniques such as cognitive stimulation, social interaction, recreational activities, and virtual reality (Barillot et al., 2023; Galaj, Barrera, & Ranaldi, 2020). For example, Bennett, Cardone, and Jarczyk (1998) observed that a weekend camping trip resulted in decreased alcohol relapse compared to the group that did not take the trip. Individuals exposed to video game showed reduced substance abuse (Cutter et al., 2014). A new intervention model called Motivational Chess, which integrates chess and motivational interviewing, shows promise in improving executive functions, particularly working memory, in cocaine-dependent individuals during the initial period of abstinence. The study suggests that this type of intervention may accelerate cognitive recovery in this population (Gonçalves et al., 2014).

In Gold, Baron, Bowirrat, and Blum (2020), the authors explore the interactions among serotonin, opioid peptides, cannabinoids, GABA, glutamate, acetylcholine, and dopamine and their associated second messengers in the processing of rewards. They report the repercussions of deficiencies or dysregulations in these neurochemical systems in terms of susceptibility to addiction. Emphasizing the critical role of balanced neurotransmission for holistic well-being, the study sheds light on the intricate mechanisms of endorphinergic and dopaminergic pathways within the brain reward cascade. It has been intriguingly discussed that EE might help balance out pathological manifestations in psychiatric disorders (Kimura et al., 2022).

Several researchers argue that engaging in recreational activities can stimulate the mesocorticolimbic system and ncrease dopamine concentration, leading to improved well-being, increased optimism, and reduced stress and depression, thus assisting in the treatment of patients (for a review, see Alexander et al., 2021). Various human activities can stimulate the production of endorphins or enkephalins, such as laughter, consuming palatable food, engaging in sexual activity, practicing yoga, meditating, listening to music, and exercising (Gold et al., 2020; Kami, Tajima, & Senba, 2017). As a result, individuals may experience a positive mood, reduced stress, and an overall sense of reward or satisfaction.

The promising preclinical results regarding the effects of EE on alcohol-related behaviors, combined with the findings in clinical studies, suggest the potential for this intervention to produce more effective therapeutic effects for the treatment of AUD.

References

Aghaie, C. I., Hausknecht, K. A., Wang, R., Dezfuli, P. H., Haj-Dahmane, S., Kane, C. J. M., et al. (2020). Prenatal ethanol exposure and postnatal environmental intervention alter dopaminergic neuron and microglia morphology in the ventral tegmental area during adulthood. *Alcoholism: Clinical and Experimental Research, 44*, 435–444.

Alexander, R., Aragón, O. R., Bookwala, J., Cherbuin, N., Gatt, J. M., Kahrilas, I. J., et al. (2021). The neuroscience of positive emotions and affect: Implications for cultivating happiness and wellbeing. *Neuroscience & Biobehavioral Reviews, 121*, 220–249.

Altman, J., & Das, G. D. (1964). Autoradiographic examination of the effects of enriched environment on the rate of glial multiplication in the adult rat brain. *Nature, 204*, 1161–1163.

American Psychiatric Association (APA). (2013). *Diagnostic and statistical manual of mental disorders. DSM-5*. APA.

Angelucci, F., De Bartolo, P., Gelfo, F., Foti, F., Cutuli, D., & Bossù, P. (2009). Increased concentrations of nerve growth factor and brain-derived neurotrophic factor in the rat cerebellum after exposure to environmental enrichment. *Cerebellum (London, England), 8*, 499–506.

Aronoff, E., Hillyer, R., & Leon, M. (2016). Environmental enrichment therapy for autism: Outcomes with increased access. *Neural Plasticity, 2016*, 2734915.

Barillot, L., Chauvet, C., Besnier, M., Jaafari, N., Solinas, M., & Chatard, A. (2023). Effect of environmental enrichment on relapse rates in patients with severe alcohol use disorder: Protocol for a randomised controlled trial. *BMJ Open, 13*, e069249.

Barrera, E. D., Timken, P. D., Lee, E., Persaud, K. R. S., Goldstein, H., Parasram, D. N., et al. (2023). Environmental enrichment facilitates electric barrier induced heroin abstinence after incubation of craving in male and female rats. *Drug and Alcohol Dependence, 244*, 109799.

Bayne, K. (2018). Environmental enrichment and mouse models: Current perspectives. *Animal Models and Experimental Medicine, 1*, 82–90.

Belo-Silva, A. E., de Gusmão Taveiros Silva, N. K., Marianno, P., de Araújo Costa, G., da Rovare, V. P., Bailey, A., et al. (2024). Effects of the combination of chronic unpredictable stress and environmental enrichment on anxiety-like behavior assessed using the elevated plus maze in Swiss male mice: Hypothalamus-Pituitary-Adrenal Axis-mediated mechanisms. *Hormones and Behavior, 3*, 105538.

Benaroya-Milshtein, N., Hollander, N., Apter, A., Kukulansky, T., Raz, N., Wilf, A., et al. (2004). Environmental enrichment in mice decreases anxiety, attenuates stress responses and enhances natural killer cell activity. *European Journal of Neuroscience, 20*, 1341–1347.

Bennett, L. W., Cardone, S., & Jarczyk, J. (1998). Effects of a therapeutic camping program on addiction recovery. *Journal of Substance Abuse Treatment, 15*, 469–474.

Bennett, E. L., Diamond, M. C., Krech, D., & Rosenzweig, M. R. (1964). Chemical and anatomical plasticity of brain. *Science (New York, N.Y.), 146*, 610–619.

Bennett, J. C., McRae, P. A., Levy, L. J., & Frick, K. M. (2006). Long-term continuous, but not daily, environmental enrichment reduces spatial memory decline in aged male mice. *Neurobiology of Learning and Memory, 85*, 139–152.

Berardo, L. R., Fabio, M. C., & Pautassi, R. M. (2016). Post-weaning environmental enrichment, but not chronic maternal isolation, enhanced ethanol intake during periadolescence and early adulthood. *Frontiers in Behavioral Neuroscience, 10*, 195.

Berman, R. F., & Hannigan, J. H. (2000). Effects of prenatal alcohol exposure on the hippocampus: Spatial behavior, electrophysiology, and neuroanatomy. *Hippocampus, 10*, 94–110.

Bowen, S., & Enkema, M. C. (2014). Relationship between dispositional mindfulness and substance use: Findings from a clinical sample. *Addictive Behaviors, 39*, 532–537.

Bowen, S., Witkiewitz, K., Clifasefi, S. L., Grow, J., Chawla, N., Hsu, S. H., et al. (2014). Relative efficacy of mindfulness-based relapse prevention, standard relapse prevention, and treatment as usual for substance use disorders. *JAMA Psychiatry, 71*, 547.

Brown, R. E. (2020). Donald O. Hebb and the Organization of Behavior: 17 years in the writing. *Molecular Brain, 13*, 55.

Brown, D. R., Jackson, T. C. J., Claus, E. D., Votaw, V. R., Stein, E. R., Robinson, C. S. H., et al. (2020). Decreases in the late positive potential to alcohol images among alcohol treatment seekers following mindfulness-based relapse prevention. *Alcohol and Alcoholism, 55*, 78–85.

Brown, R. A., Abrantes, A. M., Minami, H., Read, J. P., Marcus, B. H., Jakicic, J. M., et al. (2014). A preliminary, randomized trial of aerobic exercise for alcohol dependence. *Journal of Substance Abuse Treatment, 47*, 1–9.

Brown, R. A., Prince, M. A., Minami, H., & Abrantes, A. M. (2016). An exploratory analysis of changes in mood, anxiety and craving from pre- to post-single sessions of exercise, over 12 weeks, among patients with alcohol dependence. *Mental Health and Physical Activity, 11*, 1–6.

Campbell, E. J., Jin, S., & Lawrence, A. J. (2019). Environmental enrichment reduces the propensity to relapse following punishment-imposed abstinence of alcohol seeking. *Physiology & Behavior, 210*, 112638.

Carroll, K. M., & Kiluk, B. D. (2017). Cognitive behavioral interventions for alcohol and drug use disorders: Through the stage model and back again. *Psychology of Addictive Behaviors, 31*, 847–861.

Carroll, K. M., & Onken, L. S. (2005). Behavioral therapies for drug abuse. *American Journal of Psychiatry, 162*, 1452–1460.

Caspersen, C. J., Powell, K. E., & Christenson, G. M. (1985). Physical activity, exercise, and physical fitness: Definitions and distinctions for health-related research. *Public Health Reports, 100*, 126–131.

Ceci, F. M., Francati, S., Ferraguti, G., Coriale, G., Ciccarelli, R., Minni, A., et al. (2022). Behavioral dysregulations by chronic alcohol abuse. Motivational enhancement therapy and cognitive behavioral therapy outcomes. *Rivista di psichiatria, 57*, 1–9.

Centanni, S. W., Conley, S. Y., Luchsinger, J. R., Lantier, L., & Winder, D. G. (2022). The impact of intermittent exercise on mouse ethanol drinking and abstinence-associated affective behavior and physiology. *Alcohol, Clinical and Experimental Research, 46*, 114–128.

Chartier, K. G., Karriker-Jaffe, K. J., Cummings, C. R., & Kendler, K. S. (2017). Review: Environmental influences on alcohol use: Informing research on the joint effects of genes and the environment in diverse U.S. populations. *The American Journal on Addictions, 26*, 446–460.

Chourbaji, S., Hörtnagl, H., Molteni, R., Riva, M. A., Gass, P., & Hellweg, R. (2012). The impact of environmental enrichment on sex-specific neurochemical circuitries—Effects on brain-derived neurotrophic factor and the serotonergic system. *Neuroscience, 220*, 267–276.

Costa, G. A., de Gusmão Taveiros Silva, N. K., Marianno, P., Chivers, P., Bailey, A., & Camarini, R. (2023). Environmental enrichment increased BDNF transcripts in the prefrontal cortex: Implications for an epigenetically controlled mechanism. *Neuroscience, 526*, 277–289.

Crofton, E. J., Zhang, Y., & Green, T. A. (2015). Inoculation stress hypothesis of environmental enrichment. *Neuroscience & Biobehavioral Reviews, 49*, 19–31.

Cutter, C. J., Schottenfeld, R. S., Moore, B. A., Ball, S. A., Beitel, M., Savant, J. D., et al. (2014). A pilot trial of a videogame-based exercise program for methadone maintained patients. *Journal of Substance Abuse Treatment, 47*, 299–305.

Darlington, T. M., McCarthy, R. D., Cox, R. J., Miyamoto-Ditmon, J., Gallego, X., & Ehringer, M. A. (2016). Voluntary wheel running reduces voluntary consumption of ethanol in mice: Identification of candidate genes through striatal gene expression profiling. *Genes, Brain, and Behavior, 15*, 474–490.

Davidson, R. J., & Kaszniak, A. W. (2015). Conceptual and methodological issues in research on mindfulness and meditation. *American Psychologist, 70*, 581–592.

Davis, B. A., Clinton, S. M., Akil, H., & Becker, J. B. (2008). The effects of novelty-seeking phenotypes and sex differences on acquisition of cocaine self-administration in selectively bred high-responder and low-responder rats. *Pharmacology Biochemistry & Behavior, 90*, 331–338.

Davis, M. I. (2008). Ethanol-BDNF interactions: Still more questions than answers. *Pharmacology & Therapeutics, 118*, 36–57.

de Carvalho, C. R., Pandolfo, P., Pamplona, F. A., & Takahashi, R. N. (2010). Environmental enrichment reduces the impact of novelty and motivational properties of ethanol in spontaneously hypertensive rats. *Behavioural Brain Research, 208*, 231–236.

Deehan, G. A., Jr., Palmatier, M. I., Cain, M. E., & Kiefer, S. W. (2011). Differential rearing conditions and alcohol-preferring rats: Consumption of and operant responding for ethanol. *Behavioral Neuroscience, 125*, 184–193.

Diamond, M. C., Law, F., Rhodes, H., Lindner, B., Rosenzweig, M. R., Krech, D., et al. (1966). Increases in cortical depth and glia numbers in rats subjected to enriched environment. *Journal of Comparative Neurology, 128*, 117–126.

Duffy, S. N., Craddock, K. J., Abel, T., & Nguyen, P. V. (2001). Environmental enrichment modifies the PKA-dependence of hippocampal LTP and improves hippocampus-dependent memory. *Learning and Memory, 8*, 26–34.

Epstein, E. E., McCrady, B. S., Hallgren, K. A., Gaba, A., Cook, S., Jensen, N., et al. (2018). Individual versus group female-specific cognitive behavior therapy for alcohol use disorder. *Journal of Substance Abuse Treatment, 88*, 27–43.

Ewing, S., & Ranaldi, R. (2018). Environmental enrichment facilitates cocaine abstinence in an animal conflict model. *Pharmacology Biochemistry & Behavior, 166*, 35–41.
Ezzati, M., Lopez, A. D., Rodgers, A., Vander Hoorn, S., & Murray, C. J. (2002). Selected major risk factors and global and regional burden of disease. *The Lancet, 360*(9343).
Feeney, G. F. X., Connor, J. P., Young, R., McD., Tucker, J., & McPherson, A. (2006). Combined acamprosate and naltrexone, with cognitive behavioural therapy is superior to either medication alone for alcohol abstinence: A single centres' experience with pharmacotherapy. *Alcohol and Alcoholism, 41*, 321–327.
Flemmen, G., Unhjem, R., & Wang, E. (2014). High-intensity interval training in patients with substance use disorder. *BioMed Research International, 2014*, 616935 2014.
Galaj, E., Barrera, E. D., & Ranaldi, R. (2020). Therapeutic efficacy of environmental enrichment for substance use disorders. *Pharmacology Biochemistry & Behavior, 188*, 172829.
Gallego, X., Cox, R. J., Funk, E., Foster, R. A., & Ehringer, M. A. (2015). Voluntary exercise decreases ethanol preference and consumption in C57BL/6 adolescent mice: Sex differences and hippocampal BDNF expression. *Physiology & Behavior, 138*, 28–36.
Georgakouli, K., Manthou, E., Georgoulias, P., Ziaka, A., Fatouros, I. G., Mastorakos, G., et al. (2017). Exercise training reduces alcohol consumption but does not affect HPA-axis activity in heavy drinkers. *Physiology & Behavior, 179*, 276–283.
Giesen, E. S., Deimel, H., & Bloch, W. (2015). Clinical exercise interventions in alcohol use disorders: A systematic review. *Journal of Substance Abuse Treatment, 52*, 1–9.
Giménez-Meseguer, J., Tortosa-Martínez, J., & Cortell-Tormo, J. M. (2020). The benefits of physical exercise on mental disorders and quality of life in substance use disorders patients. Systematic review and meta-analysis. *International Journal of Environmental Research and Public Health, 17*, 3680.
Gipson, C. D., Beckmann, J. S., El-Maraghi, S., Marusich, J. A., & Bardo, M. T. (2011). Effect of environmental enrichment on escalation of cocaine self-administration in rats. *Psychopharmacology (Berl), 214*, 557–566.
Gold, M. S., Baron, D., Bowirrat, A., & Blum, K. (2020). Neurological correlates of brain reward circuitry linked to opioid use disorder (OUD): Do Homo sapiens acquire or have a reward deficiency syndrome? *Journal of the Neurological Sciences, 418*, 117137.
Gomez, A. M., Sun, W. L., Midde, N. M., Harrod, S. B., & Zhu, J. (2015). Effects of environmental enrichment on ERK1/2 phosphorylation in the rat prefrontal cortex following nicotine-induced sensitization or nicotine self-administration. *European Journal of Neuroscience, 41*, 109–119.
Gonçalves, P. D., Ometto, M., Bechara, A., Malbergier, A., Amaral, R., Nicastri, S., et al. (2014). Motivational interviewing combined with chess accelerates improvement in executive functions in cocaine dependent patients: A one-month prospective study. *Drug and Alcohol Dependence, 141*, 79–84.
Green, T. A., Alibhai, I. N., Roybal, C. N., Winstanley, C. A., Theobald, D. E., Birnbaum, S. G., et al. (2010). Environmental enrichment produces a behavioral phenotype mediated by low cyclic adenosine monophosphate response element binding (CREB) activity in the nucleus accumbens. *Biological Psychiatry, 67*, 28–35.
Green, T. A., Cain, M. E., Thompson, M., & Bardo, M. T. (2003). Environmental enrichment decreases nicotine-induced hyperactivity in rats. *Psychopharmacology (Berl), 170*, 235–241.
Grégoire, C. A., Bonenfant, D., Le Nguyen, A., Aumont, A., & Fernandes, K. J. (2014). Untangling the influences of voluntary running, environmental complexity, social housing and stress on adult hippocampal neurogenesis. *PLoS One, 9*, e86237.
Gubert, C., & Hannan, A. J. (2019). Environmental enrichment as an experience-dependent modulator of social plasticity and cognition. *Brain Research, 1717*, 1–14.

Haber, P. (2009). *Guidelines for the treatment of alcohol problems.* Australian Government Department of Health and Ageing.

Hallgren, M., Vancampfort, D., Hoang, M. T., Andersson, V., Ekblom, Ö., Andreasson, S., et al. (2021). Effects of acute exercise on craving, mood and anxiety in non-treatment seeking adults with alcohol use disorder: An exploratory study. *Drug and Alcohol Dependence, 220*, 108506.

Hallgren, M., Vancampfort, D., Schuch, F., Lundin, A., & Stubbs, B. (2017). More reasons to move: Exercise in the treatment of alcohol use disorders. *Frontiers in Psychiatry, 28*, 8.

Hannigan, J. H., & Berman, R. F. (2000). Amelioration of fetal alcohol-related neurodevelopmental disorders in rats: Exploring pharmacological and environmental treatments. *Neurotoxicology and Teratology, 22*, 103–111.

Hebb, D. O. (1947). The effects of early experience on problem-solving at maturity. *American Psychologist, 2*, 306–307.

Heilig, M., Augier, E., Pfarr, S., & Sommer, W. H. (2019). Developing neuroscience-based treatments for alcohol addiction: A matter of choice? *Translational Psychiatry, 9*, 255.

Hofford, R. S., Chow, J. J., Beckmann, J. S., & Bardo, M. T. (2017). Effects of environmental enrichment on self-administration of the short-acting opioid remifentanil in male rats. *Psychopharmacology (Berl), 234*, 3499–3506.

Holgate, J. Y., Garcia, H., Chatterjee, S., & Bartlett, S. E. (2017). Social and environmental enrichment has different effects on ethanol and sucrose consumption in mice. *Brain and Behavior, 7*, e00767.

Howerton, C. L., Garner, J. P., & Mench, J. A. (2008). Effects of a running wheel-igloo enrichment on aggression, hierarchy linearity, and stereotypy in group-housed male CD-1 (ICR) mice. *Applied Animals Behaviour Science, 115*, 90–103.

Huang, E. J., & Reichardt, L. F. (2001). Neurotrophins: Roles in neuronal development and function. *Annual Review of Neuroscience, 24*, 677–736.

Ickes, B. R., Pham, T. M., Sanders, L. A., Albeck, D. S., Mohammed, A. H., & Granholm, A. C. (2000). Long-term environmental enrichment leads to regional increases in neurotrophin levels in rat brain. *Experimental Neurology, 164*, 45–52.

Imperio, C. G., McFalls, A. J., Hadad, N., Blanco-Berdugo, L., Masser, D. R., Colechio, E. M., et al. (2018). Exposure to environmental enrichment attenuates addiction-like behavior and alters molecular effects of heroin self-administration in rats. *Neuropharmacology, 139*, 26–40.

Jennings, M., Batchelor, G. R., Brain, P. F., Dick, A., Elliott, H., Francis, R. J., et al. (1998). Refining rodent husbandry: The mouse. *Laboratory Animals, 32*, 233–259.

Kabat-Zinn, J. H. (2013). *Full catastrophe living, Revised Edition: How to cope with stress, pain and illness using mindfulness meditation.* UK: Little, Brown Book Group.

Kajimoto, K., Valenzuela, C. F., Allan, A. M., Ge, S., Gu, Y., & Cunningham, L. A. (2016). Prenatal alcohol exposure alters synaptic activity of adult hippocampal dentate granule cells under conditions of enriched environment. *Hippocampus, 26*, 1078–1087.

Kami, K., Tajima, F., & Senba, E. (2017). Exercise-induced hypoalgesia: Potential mechanisms in animal models of neuropathic pain. *Anatomical Science International, 92*, 79–90.

Kempermann, G. (2019). Environmental enrichment, new neurons and the neurobiology of individuality. *Nature Reviews. Neuroscience, 20*, 235–245.

Kimura, L. F., Novaes, L. S., Picolo, G., Munhoz, C. D., Cheung, C. W., & Camarini, R. (2022). How environmental enrichment balances out neuroinflammation in chronic pain and comorbid depression and anxiety disorders. *British Journal of Pharmacology, 179*, 1640–1660.

Konkle, A. T., Kentner, A. C., Baker, S. L., Stewart, A., & Bielajew, C. (2010). Environmental-enrichment-related variations in behavioral, biochemical, and physiologic responses of Sprague-Dawley and Long Evans rats. *Journal of the American Association for Laboratory Animal Science, 49*, 427–436.

Korecki, J. R., Schwebel, F. J., Votaw, V. R., & Witkiewitz, K. (2020). Mindfulness-based programs for substance use disorders: A systematic review of manualized treatments. *Substance Abuse Treatment, Prevention, and Policy, 15*, 51.

Leal-Galicia, P., Saldívar-González, A., Morimoto, S., & Arias, C. (2007). Exposure to environmental enrichment elicits differential hippocampal cell proliferation: Role of individual responsiveness to anxiety. *Developmental Neurobiology, 67*, 395–405.

Li, W., Howard, M. O., Garland, E. L., McGovern, P., & Lazar, M. (2017). Mindfulness treatment for substance misuse: A systematic review and meta-analysis. *Journal of Substance Abuse Treatment, 75*, 62–96.

Li, X., Meng, L., Huang, K., Wang, H., & Li, D. (2015). Environmental enrichment blocks reinstatement of ethanol-induced conditioned place preference in mice. *Neuroscience Letters, 599*, 92–96.

Logrip, M. L., Janak, P. H., & Ron, D. (2008). Dynorphin is a downstream effector of striatal BDNF regulation of ethanol intake. *FASEB Journal, 22*, 2393–2404.

Lynch, C. A., Porter, B., & Butler, T. R. (2019). Access to voluntary running wheel exercise: Prevention of anxiety-like behavior in chronically stressed rats, but potentiation of ethanol intake/preference. *Physiology & Behavior, 206*, 118–124.

Maccioni, P., Somenzi, L. R., Lobina, C., Carai, M. A. M., Gessa, G. L., & Colombo, G. (2022). Environmental enrichment augments binge-like alcohol drinking in Sardinian alcohol-preferring rats. *Alcohol (Fayetteville, N.Y.), 105*, 1–7.

Madson, M. B., Schumacher, J. A., Baer, J. S., & Martino, S. (2016). Motivational interviewing for substance use: Mapping out the next generation of research. *Journal of Substance Abuse Treatment, 65*, 1–5.

Magill, M., & Ray, L. A. (2009). Cognitive-behavioral treatment with adult alcohol and illicit drug users: A meta-analysis of randomized controlled trials. *Journal of Studies on Alcohol and Drugs, 70*, 516–527.

Magill, M., Kiluk, B. D., & Ray, L. A. (2023). Efficacy of cognitive behavioral therapy for alcohol and other drug use disorders: Is a one-size-fits-all approach appropriate? *Substance Abuse and Rehabilitation, 14*, 1–11.

Malone, S. G., Shaykin, J. D., Stairs, D. J., & Bardo, M. T. (2022). Neurobehavioral effects of environmental enrichment and drug abuse vulnerability: An updated review. *Pharmacology Biochemistry & Behavior, 221*, 173471.

Manthou, E., Georgakouli, K., Fatouros, I. G., Gianoulakis, C., Theodorakis, Y., & Jamurtas, A. Z. (2016). Role of exercise in the treatment of alcohol use disorders. *Biomedical Reports, 4*, 535–545.

Marashi, V., Barnekow, A., Ossendorf, E., & Sachser, N. (2003). Effects of different forms of environmental enrichment on behavioral, endocrinological, and immunological parameters in male mice. *Hormones and Behavior, 43*, 281–292.

Marianno, P., Abrahao, K. P., & Camarini, R. (2017). Environmental enrichment blunts ethanol consumption after restraint stress in C57BL/6 mice. *PLoS One, 12*, e0170 317.

Martín-Pérez, C., Navas, J. F., Perales, J. C., López-Martín, Á., Cordovilla-Guardia, S., Portillo, M., et al. (2019). Brief group-delivered motivational interviewing is equally effective as brief group-delivered cognitive-behavioral therapy at reducing alcohol use in risky college drinkers. *PLoS One, 14*, e0226271.

McCulley, W. D., 3rd, Walls, S. A., Khurana, R. C., Rosenwasser, A. M., & Devaud, L. L. (2012). Running wheel activity protects against increased seizure susceptibility in ethanol withdrawn male rats. *Pharmacology Biochemistry & Behavior, 100*, 485–489.

McQuaid, R. J., Audet, M. C., Jacobson-Pick, S., & Anisman, H. (2013). The differential impact of social defeat on mice living in isolation or groups in an enriched environment: Plasma corticosterone and monoamine variations. *The International Journal of Neuropsychopharmacology, 16*, 351–363.

Moncek, F., Duncko, R., Johansson, B. B., & Jezova, D. (2004). Effect of environmental enrichment on stress related systems in rats. *Journal of Neuroendocrinology, 16*, 423–431.

Moriarty, T., Bourbeau, K., & Zuhl, M. (2020). Exercise and neural adaptations: Designing a novel treatment for alcohol addiction. *Alternative Therapies in Health and Medicine, 26*, 48–57.

Moser, M. B., Trommald, M., & Andersen, P. (1994). An increase in dendritic spine density on hippocampal CA1 pyramidal cells following spatial learning in adult rats suggests the formation of new synapses. *Proceedings of the National Academy of Sciences of the United States of America, 91*, 12673–12675.

Muller, A. E., & Clausen, T. (2015). Group exercise to improve quality of life among substance use disorder patients. *Scandinavian Journal of Public Health, 43*, 146–152.

National Collaborating Centre for Mental Health (NCCMH). (2011). *Alcohol use disorders: The NICE guideline on the diagnosis, assessment and management of harmful drinking and alcohol dependence*. NCCMH.

National Institute on Drug Abuse (NIDA). (2012). *Principles of drug addiction treatment: A research-based guide* (3rd ed.). NIDA.

Nithianantharajah, J., & Hannan, A. J. (2006). Enriched environments, experience-dependent plasticity and disorders of the nervous system. *Nature Reviews. Neuroscience, 7*, 697–709.

Novaes, L. S., Dos Santos, N. B., Batalhote, R. F. P., Malta, M. B., Camarini, R., Scavone, C., et al. (2017). Environmental enrichment protects against stress-induced anxiety: Role of glucocorticoid receptor, ERK, and CREB signaling in the basolateral amygdala. *Neuropharmacology, 113*, 457–466.

Ozburn, A. R., Harris, R. A., & Blednov, Y. A. (2008). Wheel running, voluntary ethanol consumption, and hedonic substitution. *Alcohol (Fayetteville, N.Y.), 42*, 417–424.

Pang, T. Y., Du, X., Catchlove, W. A., Renoir, T., Lawrence, A. J., & Hannan, A. J. (2013). Positive environmental modification of depressive phenotype and abnormal hypothalamic-pituitary-adrenal axis activity in female C57BL/6J mice during abstinence from chronic ethanol consumption. *Frontiers in Pharmacology, 4*, 93.

Pautassi, R. M., Suárez, A. B., Hoffmann, L. B., Rueda, A. V., Rae, M., Marianno, P., et al. (2017). Effects of environmental enrichment upon ethanol-induced conditioned place preference and pre-frontal BDNF levels in adolescent and adult mice. *Scientific Reports, 7*, 8574.

Peck, J. A., Galaj, E., Eshak, S., Newman, K. L., & Ranaldi, R. (2015). Environmental enrichment induces early heroin abstinence in an animal conflict model. *Pharmacology Biochemistry & Behavior, 138*, 20–25.

Peng, W., Zhang, H., Yang, J., Wang, J., Kang, J., Zhu, R., et al. (2022). Group cognitive behavioral therapy as an effective approach for patients with alcohol dependence: A perspective study. *Medicine (Baltimore), 101*, e30459.

Rae, M., Gomes, I., Spelta, L. E. W., Bailey, A., Marcourakis, T., Devi, L., et al. (2024). Environmental enrichment enhances ethanol preference over social reward in male Swiss mice: Involvement of oxytocin-dopamine interactions. *Neuropharmacology, 253*, 109971.

Rae, M., Zanos, P., Georgiou, P., Chivers, P., Bailey, A., & Camarini, R. (2018). Environmental enrichment enhances conditioned place preference to ethanol via an oxytocinergic-dependent mechanism in male mice. *Neuropharmacology, 138*, 267–274.

Ranaldi, R., Kest, K., Zellner, M., & Hachimine-Semprebom, P. (2011). Environmental enrichment, administered after establishment of cocaine self-administration, reduces lever pressing in extinction and during a cocaine context renewal test. *Behavioural Pharmacology, 22*, 347–353.

Ratuski, A. S., & Weary, D. M. (2022). Environmental enrichment for rats and mice housed in laboratories: A metareview. *Animals (Basel), 12*, 414.

Reguilón, M. D., Ferrer-Pérez, C., Ballestín, R., Miñarro, J., & Rodríguez-Arias, M. (2020). Voluntary wheel running protects against the increase in ethanol consumption induced by social stress in mice. *Drug and Alcohol Dependence, 212*, 108004.
Reguilón, M. D., Ferrer-Pérez, C., Manzanedo, C., Miñarro, J., & Rodríguez-Arias, M. (2021). Ethanol intake in male mice exposed to social defeat: Environmental enrichment potentiates resilience. *Neurobiology of Stress, 15*, 100413.
Reguilón, M. D., Ferrer-Pérez, C., Manzanedo, C., Miñarro, J., & Rodríguez-Arias, M. (2023). Voluntary wheel running during adolescence prevents the increase in ethanol intake induced by social defeat in male mice. *Psychopharmacology*.
Rockman, G. E., Gibson, J. E., & Benarroch, A. (1989). Effects of environmental enrichment on voluntary ethanol intake in rats. *Pharmacology Biochemistry & Behavior, 34*, 487–490.
Rodríguez-Ortega, E., & Cubero, I. (2018). Environmental enrichment modulates drug addiction and binge-Like consumption of highly rewarding substances: A role for anxiety and compulsivity brain systems? *Frontiers in Behavioral Neuroscience, 12*, 295.
Roessler, K. K., Bilberg, R., Søgaard Nielsen, A., Jensen, K., Ekstrøm, C. T., & Sari, S. (2017). Exercise as adjunctive treatment for alcohol use disorder: A randomized controlled trial. *PLoS One, 12*, e0186076.
Roos, C. R., Bowen, S., & Witkiewitz, K. (2017). Baseline patterns of substance use disorder severity and depression and anxiety symptoms moderate the efficacy of mindfulness-based relapse prevention. *Journal of Consulting and Clinical Psychology, 85*, 1041–1051.
Rosenzweig, M. R., Krech, D., Bennett, E. L., & Diamond, M. C. (1962). Effects of environmental complexity and training on brain chemistry and anatomy: A replication and extension. *Journal of Comparative and Physiological Psychology, 55*, 429–437.
Rueda, A. V., Teixeira, A. M., Yonamine, M., & Camarini, R. (2012). Environmental enrichment blocks ethanol-induced locomotor sensitization and decreases BDNF levels in the prefrontal cortex in mice. *Addiction Biology, 17*, 736–745.
Sancho, M., De Gracia, M., Rodríguez, R. C., Mallorquí-Bagué, N., Sánchez-González, J., Trujols, J., et al. (2018). Mindfulness-based interventions for the treatment of substance and behavioral addictions: A systematic review. *Frontiers in Psychiatry, 9*, 95.
Segovia, G., Yagüe, A. G., García-Verdugo, J. M., & Mora, F. (2006). Environmental enrichment promotes neurogenesis and changes the extracellular concentrations of glutamate and GABA in the hippocampus of aged rats. *Brain Research Bulletin, 70*, 8–14.
Simpson, J., & Kelly, J. P. (2011). The impact of environmental enrichment in laboratory rats—Behavioural and neurochemical aspects. *Behavioural Brain Research, 222*, 246–264.
Sinyor, D., Brown, T., Rostant, L., & Seraganian, P. (1982). The role of a physical fitness program in the treatment of alcoholism. *Journal of Studies on Alcohol, 43*, 380–386.
Société Française d'Alcoologie (SFA). (2015). Mésusage de l'alcool, dépistage, diagnostic, et traitement. Recommandation de bonne pratique. *Alcoologie et Addictologie, 37*(1), 5–84.
Suárez, A., Fabio, M. C., Bellia, F., Fernández, M. S., & Pautassi, R. M. (2020). Environmental enrichment during adolescence heightens ethanol intake in female, but not male, adolescent rats that are selectively bred for high and low ethanol intake during adolescence. *American Journal of Drug and Alcohol Abuse, 46*, 553–564.
Takuma, K., Ago, Y., & Matsuda, T. (2011). Preventive effects of an enriched environment on rodent psychiatric disorder models. *Journal of Pharmacological Sciences, 117*, 71–76.
Ussher, M., Sampuran, A. K., Doshi, R., West, R., & Drummond, D. C. (2004). Acute effect of a brief bout of exercise on alcohol urges. *Addiction (Abingdon, England), 99*, 1542–1547.
van Loo, P. L. P., Kruitwagen, C. L. J. J., Koolhaas, J. M., Van de Weerd, H. A., Van Zutphen, L. F. M., & Baumans, V. (2002). Influence of cage enrichment on aggressive behaviour and physiological parameters in male mice. *Applied Animal Behaviour Science, 76*, 65–81.

van Praag, H., Kempermann, G., & Gage, F. H. (2000). Neural consequences of environmental enrichment. *Nature Reviews. Neuroscience, 1*, 191–198.
Venebra-Muñoz, A., Corona-Morales, A., Santiago-García, J., Melgarejo-Gutiérrez, M., Caba, M., & García-García, F. (2014). Enriched environment attenuates nicotine self-administration and induces changes in ΔFosB expression in the rat prefrontal cortex and nucleus accumbens. *Neuroreport, 25*, 688–692.
Walters, D., Connor, J. P., Feeney, G. F. X., & Young, R. M. (2009). The cost effectiveness of naltrexone added to cognitive-behavioral therapy in the treatment of alcohol dependence. *Journal of Addictive Diseases, 28*, 137–144.
Wang, R., Hausknecht, K. A., Shen, Y. L., Haj-Dahmane, S., Vezina, P., & Shen, R. Y. (2018). Environmental enrichment reverses increased addiction risk caused by prenatal ethanol exposure. *Drug and Alcohol Dependence, 191*, 343–347.
Wille-Bille, A., Bellia, F., Jiménez García, A. M., Miranda-Morales, R. S., D'Addario, C., & Pautassi, R. M. (2020). Early exposure to environmental enrichment modulates the effects of prenatal ethanol exposure upon opioid gene expression and adolescent ethanol intake. *Neuropharmacology, 165*, 107917.
Winstock, A. R., Maier, L. J., Zhuparris, A., Davies, E., Puljevic, C., Kuypers, K. P. C., et al. (2021). Global drug survey (GDS) 2021 key findings report.
Witkiewitz, K., Litten, R. Z., & Leggio, L. (2019). Advances in the science and treatment of alcohol use disorder. *Science Advances, 5*, eaax4043.
Witkiewitz, K., Marlatt, G. A., & Walker, D. (2005). Mindfulness-based relapse prevention for alcohol and substance use disorders. *Journal of Cognitive Psychotherapy, 19*, 211–228.
World Health Organization. (2023). *Global alcohol action plan 2022–2030 to strengthen implementation of the global strategy to reduce the harmful use of alcohol*. Geneva: World Health Organization Licence: CC BY-NC-SA 3.0 IGO.
Young, D., Lawlor, P. A., Leone, P., Dragunow, M., & During, M. J. (1999). Environmental enrichment inhibits spontaneous apoptosis, prevents seizures and is neuroprotective. *Nature Medicine, 5*, 448–453.
Zanca, R. M., Braren, S. H., Maloney, B., Schrott, L. M., Luine, V. N., & Serrano, P. A. (2015). Environmental enrichment increases glucocorticoid receptors and decreases GluA2 and Protein Kinase M Zeta (PKMζ) trafficking during chronic stress: A protective mechanism? *Frontiers in Behavioral Neuroscience, 9*, 303.

CHAPTER ELEVEN

Animal models for studying therapeutic targets and treatments for alcohol use disorder

Cristiane Aparecida Favoretto[a,b], Thamires Righi[a], Gustavo Juliate Damaceno Fernandes[a], Natalia Bonetti Bertagna[a], Ben Tagami Rodolpho[a], Nilma do Rocio Lara de Lima Janisset[a], Caio Jovita-Farias[a], Giovanna Victória Lopes Costa[a], Alexia dos Anjos-Santos[a], Fernando Bezerra Romualdo da Silva[c], Rodrigo Molini Leão[d], and Fábio Cardoso Cruz[a,*]

[a]Molecular and Behavioral Neuroscience Laboratory, Department of Pharmacology, Universidade Federal de São Paulo (UNIFESP), São Paulo, São Paulo, Brazil
[b]Department of Molecular Medicine, The Scripps Research Institute, La Jolla, CA, United States
[c]Laboratory of Experimental and Clinical Neuroscience, INSERM, Université de Poitiers, France
[d]Laboratory of Pharmacology, Biomedical Sciences Institute, Department of Pharmacology, Federal University of Uberlândia, Uberlândia, MG, Brazil
*Corresponding author. e-mail address: f.cruz@unifesp.br

Contents

International Review of Neurobiology, Volume 178
ISSN 0074-7742, https://doi.org/10.1016/bs.irn.2024.08.004

Abstract

Over the decades, preclinical models have been developed and refined to investigate the rewarding effects of addictive substances and the neurobiological underpinnings of alcohol and other drug use disorders. This chapter delves into the methodological foundations, advantages, and limitations of leading animal models used to study alcohol use disorders (AUDs). Some models focus on the early stages of alcohol use and abuse. For instance, conditioned place preference assesses associative learning between a specific context and the effects of the drug, while locomotor sensitization measures increased locomotor activity following repeated drug exposure. In contrast, contingent models such as operant and non-operant alcohol self-administration protocols gauge voluntary intake, preference, motivation, and seeking behavior for alcohol solutions among experimental subjects. Additionally, we discuss the chronic intermittent alcohol vapor model, extensively utilized to induce a phenotype resembling dependence through non-contingent inhalation of alcohol vapor, resulting in elevated blood alcohol concentrations. Given the focus on pharmacological treatments for AUDs, we explore how different animal models can be employed to evaluate potential therapies and extrapolate findings to alcohol-related behaviors in humans. This chapter aims to provide readers with a comprehensive understanding of various animal models for AUDs, aiding in the interpretation of preclinical studies and the selection of suitable models for future research endeavors.

1. Introduction

Substance Use Disorders (SUD), including Alcohol Use Disorders (AUD), are a substantial public health problem around the world (Manthey et al., 2019). The exact neurobiology involved in this disorder is not known. There are few medicines approved for treating SUD, and the efficacy of available treatments is low since a high percentage of people relapse during the treatment (Koob, 2024).

AUD are complex, multifactorial,and involve different behaviors and neurobiological mechanisms. Therefore, understanding AUD requires the use of complex strategies. Animal models have helpful in unveiling the mechanisms underlying AUD and for studying potential new therapeutical targets (Kwako et al., 2017; Nieto, Grodin, Aguirre, Izquierdo, & Ray, 2021).

Each experimental model has unique characteristics, as well as limitations and advantages to its use. Therefore, it is essential that researchers determine the central aims of the study and its specific characteristics, such as the stage of AUD to be explored, and the appropriate animal model (Nieto et al., 2021). By itself, no experimental model can answer whether a treatment is effective or not. However, it can help elucidate neurobiological aspects of the disorder and test for preventive interventions and treatments.

In this chapter, we present some of the leading preclinical, experimental models used to investigate the neural mechanisms underlying AUD in their different phases, or stages - including initial use, abuse, addiction, and relapse. We will focus on principles and methodological characteristics, advantages, limitations, and applications of some of the most used animal models in the AUD preclinical field, which include conditioned place preference and models of alcohol intake like alcohol operant and non-operant self-administration,and the "Drinking in the Dark" paradigm, and ethanol vapor chamber exposure (Goltseker, Hopf, & Barak, 2019; Kwako et al., 2017; Nieto et al., 2021).

2. Conditioned place preference

2.1 The rationale behind the model and its procedures

Conditioned place preference (CPP) has been established as a preclinical behavior model to study the rewarding effects of different drugs of abuse (McKendrick & Graziane, 2020; Prus, James, & Rosecrans, 2009). Based on associative learning processing, which Pavlov identified and tested in his classical conditioning experiments, the CPP model evaluates motivated behaviors induced by the rewarding effects of drugs or natural rewards (Huston, Silva, Topic, & Müller, 2013).

Rodents, especially rats and mice, possibly correspond to the most used species in alcohol and other drugs of abuse-related studies. However, it is worth mentioning that alcohol-CPP has been described and executed in different species, including flies (Kaun, Azanchi, Maung, Hirsh, & Heberlein, 2011), planaria (Raffa, 2008; Tallarida et al., 2014), zebrafish (Collier, Khan, Caramillo, Mohn, & Echevarria, 2014; Mathur, Lau, & Guo, 2011), and even humans (Childs & De Wit, 2016). This capability of being easily adapted to different species is an advantage of the CPP model.

One of the first studies using CPP protocols that resemble the currently known and used format (further described below) was carried out by Rossi and Reid in 1976, to assess the reinforcing properties of morphine in rats (Rossi & Reid, 1976). CPP measures associations formed between a contextual environment presenting specific characteristics (e.g., wall colors, patterns, and floor texture) and a rewarding stimulus (e.g., drug effects). It consists of repeated pairing sessions that will promote the association of a particular environment with the studied substance effects (drug-paired environment), while a sensorily different environment will be associated with the absence of the substance effects (vehicle-paired environment) (McKendrick & Graziane, 2020; Tzschentke, 2007). In other words, the drug effects initially considered an unconditioned stimulus can be paired or associated with neutral environmental cues. Following consecutive conditioning episodes or sessions, the environmental stimuli acquire motivational properties per se. Thus, exposure to the drug-paired environment and its cues, even in the absence of the drug, will induce behavior changes, expressed as seeking or preference behaviors, in which the experimental subjects will choose to spend more time in the drug-paired context relative to other(s) (Huston et al., 2013; McKendrick et al., 2020; Sanchis-Segura & Spanagel, 2006).

Furthermore, the CPP model can also assess whether a drug type or dose produces unpleasant effects. The subject will show avoidance-like behavior toward the environment previously associated with the drug, exhibiting conditioned place aversion (CPA). CPA is observed when experimental animals spend significantly more time in the vehicle-paired environment compared to the drug-paired environment (Prus et al., 2009; Sanchis-Segura, Zghoul, & Spanagel, 2006).

A typical apparatus used in CPP experiments consists of two or three-compartment units. In both cases, the larger outer compartments, where experimental individuals are confined during drug or vehicle conditioning sessions, are designed to have different characteristics (e.g., white or black walls, horizontal or cross-grid flooring, different scents placed under the flooring). The main difference between both units is the presence, in the three-compartment unit of a smaller central area with no unique characteristics, where the animals are not confined during any session.

When the experiment is conducted in the two-compartment unit, the experimental subject must choose between the drug-paired or vehicle-paired compartments. If the experiment is conducted in the three-compartment unit, the gates between those compartments are opened during the pre- or post-conditioning sessions (described below). In this case, the subject can roam freely between them, choosing to remain at the drug-paired or the vehicle-paired units, or at the central unpaired area (Prus et al., 2009; Yates, 2023).

The CPP paradigm offers a simple way of assessing the conditioned reward induced by diverse natural or drug stimuli and essentially consists of three phases (Cunningham, Gremel, & Groblewski, 2006): Preconditioning, Conditioning, and Postconditioning (Bardo & Bevins, 2000; Blanco-Gandía, Aguilar, Miñarro, & Rodríguez-Arias, 2018; McKendrick & Graziane, 2020; Prus et al., 2009; Sanchis-Segura et al., 2006; Yates, 2023). Each of the phases are described below:

2.1.1 Preconditioning

Also known as the pre-test, this session is intended to determine the subject's initial or natural preference for the apparatus compartments (baseline) (Cunningham, Ferree, & Howard, 2003; McKendrick & Graziane, 2020). During this session, animals are given free access to the entire apparatus, including its compartments, before returning to their home cages. Some studies include more than one (usually two or three) preconditioning sessions in their protocols. In those cases, the first preconditioning sessions can be called "habituation", while the last one will measure the initial preference for the apparatus sides. Habituation sessions reduce the novelty and stress factors associated with handling and the new environment. An intrinsic preference for one compartment over the other may also complicate the interpretation of place preference outcomes (Napier, Herrold, & De Wit, 2013). According to the animal's baseline preference, the experimenter may perform a biased or unbiased model design (Cunningham et al., 2003, 2006). An apparatus is classified as "biased" when untrained subjects consistently prefer one compartment of the apparatus over another without any prior conditioning. In contrast, an apparatus is "unbiased" when subjects show no preference between compartments. Additionally, the term "biased" can refer to assigning the conditioned stimulus (CS) based on each subject's initial preference identified in a pre-test, while "unbiased" refers to a random assignment that ignores these initial preferences (Cunningham et al., 2003, 2006).

2.1.2 Conditioning

The experimenter administers a vehicle or a drug to the animal and confines it in a distinct context. This procedure typically involves multiple trials, during which a drug (e.g., alcohol) is paired with one of the compartments in the apparatus. Conversely, a vehicle solution (e.g., saline) is paired with a different compartment during alternate trials. For alcohol conditioning, 4 to 10 sessions with either the drug or the vehicle are typically conducted. The duration of these trials varies, with sessions lasting 5, 15, or 30 min (Cunningham et al., 2006; Eaton, Dzhala, Robinson, Radevski, & Akins, 2023; Lorenz, Moye, & Leong, 2022; Morales, Varlinskaya, & Spear, 2012; Pati, Pina, & Kash, 2019).

2.1.3 Postconditioning (preference test)

Also known as post-test, in this phase, animals are again given free access to the test apparatus (all compartments) in a drug-free state, and the time spent in each of them is recorded (Eaton et al., 2023; Morales et al., 2012; Pati et al., 2019). The total time or percentage spent in the drug-paired compartment can be compared between preconditioning and postconditioning sessions. A significant increase in the time spent in the drug-paired compartment compared to the vehicle-paired compartment during postconditioning, or an increase in the drug-paired side during postconditioning relative to preconditioning, indicates the development of conditioned place preference (CPP), which may be associated with the rewarding properties of the drug (Cunningham et al., 2003, 2006; McKendrick & Graziane, 2020).

2.1.4 CPP and relapse

Since alcohol use disorders are characterized as chronic and relapsing conditions, craving and relapse remain significant challenges during treatment. Additionally, environmental stimuli associated with drug intake are critical triggers for craving and relapse (Gawel et al., 2016; Henriques et al., 2021; Palombo et al., 2021; Tavares, Bianchi, Yokoyama, Palombo, & Cruz, 2023). To further model behaviors associated with abstinence and relapse in preclinical substance use disorder studies, CPP protocols may include additional experimental phases after the completion of conditioned preference acquisition, known as extinction and reinstatement (Blanco-Gandía et al., 2018; Cunningham et al., 2003; Gawel et al., 2016), as detailed below.

2.1.5 Extinction and reinstatement phases

After the acquisition of conditioned place preference, the preference/seeking behaviors can be extinguished by successively exposing subjects to the previously drug-paired environment in the absence of the drug (in a drug-free state). This can be accomplished by (1) confining the experimental individuals in both apparatus' compartments after treatment with vehicle solution or (2) allowing free exploration of both compartments without drug or vehicle treatment. The extinction phase aims to weaken the associative memory between drugs and the environment (Aguilar, Rodríguez-Arias, & Miñarro, 2009; Liu, Le Foll, Liu, Wang, & Lu, 2008).

Treatment with a priming injection (i.e., a low dose of the testing drug) and exposure to stress can reinstate or reactivate preference for the previously drug-paired environment (Napier et al., 2013). Drug priming-induced and stress-induced reinstatement of alcohol CPP have been reported in numerous studies (Aguilar et al., 2009; Bhutada et al., 2012; Font, Miquel, & Aragon, 2008; Kuzmin, Sandin, Terenius, & Ögren, 2003; Martín-Sánchez et al., 2019; Okhuarobo, Igbe, Yahaya, & Sule, 2018; Yunusoğlu, 2021).

2.2 Model advantages and limitations

The CPP model offers several advantages. Firstly, it boasts experimental simplicity and short duration. Additionally, it enables the evaluation of both drug preference and aversion. The model is susceptible to low drug doses, enhancing its efficacy in detecting subtle effects. Notably, the postconditioning test is conducted in a drug-free state, mitigating the influence of other drug effects on the results. This helps to avoid confounding variables, such as locomotor consequences (Bardo & Bevins, 2000; Cunningham et al., 2006).

The CPP model has several limitations for translational analysis. Firstly, its non-contingent nature means subjects lack control over drug administration, affecting its relevance to human drug-use disorders (Cunningham et al., 2006; Mueller & De Wit, 2011). Additionally, the brief duration of the protocol contrasts with the chronic exposure typical in human drug dependence (Mueller & De Wit, 2011). Furthermore, the model's design makes it challenging to observe dose-response effects accurately, as testing different doses requires separate experimental groups and may only yield limited outcome variations (Bardo & Bevins, 2000).

3. Non-operant alcohol self-administration models

3.1 Two-bottle choice

3.1.1 The rationale behind the model and its procedures

The two-bottle choice test is a widely utilized method to investigate voluntary alcohol consumption in animals. This paradigm involves providing animals with access to two bottles continuously for several consecutive days: one bottle containing water and the other containing alcohol at concentrations ranging from 3% to 20% (Hilderbrand & Lasek, 2017). Alternatively, alcohol access can be provided using an intermittent protocol (Carnicella, Ron, & Barak, 2014; Vierkant, Xie, Wang, & Wang, 2023).

This method allows researchers to observe alcohol preference and measure intake levels (Bahi & Dreyer, 2019; Little et al., 1999). Moreover, exposure to alcohol in the two-bottle choice paradigm can serve as a period of habituation for subsequent alcohol operant self-administration studies (Faccidomo, Salling, Galunas, & Hodge, 2015).

In most cases, rat strains do not naturally consume alcohol unless sweeteners like sucrose or saccharin are added to encourage initial drinking in experiments. Researchers often utilize "fading" techniques, gradually reducing sweeteners while increasing alcohol concentrations, to facilitate higher alcohol intake in rodents (Hilderbrand & Lasek, 2017; Vierkant et al., 2023).

The progression of alcohol abuse involves a gradual increase from moderate to excessive intake, involving cycles of intoxication, withdrawal, craving, and relapse (Koob & Volkow, 2010; Koob, 2003). Two-bottle intermittent access protocols are crucial for promoting high levels of voluntary alcohol consumption and assessing preferences between alcohol solutions and water in animal models. In a typical intermittent access protocol, rodents are individually housed and undergo an acclimatization period where they are exposed to two water bottles. Subsequently, they are given three 24-h sessions per week (usually Monday, Wednesday, and Friday) with free access to one bottle of water and one containing 20% ethanol. Drinking sessions can begin during either the light or dark cycle, with fluid intake measured at intervals, typically 30–60 min or 24 h after the session begins (Carnicella et al., 2014; Vierkant et al., 2023).

Different forms of intermittent access to alcohol have been implemented in research. Early studies, such as those by Wayner et al. (1972) and Sinclair and Senter (1967), outlined protocols of 24 h of alcohol access followed by 48 h of abstinence or alternating one week of alcohol access with one week of withdrawal. The duration of exposure in these protocols varies widely from 28 days (Agoglia et al., 2022) to several months (Kimbrough, Kim, Cole, Brennan, & George, 2017; Newman et al., 2018; Perry et al., 2023).

3.1.2 *Model advantages and limitations*

With intermittent access to alcohol, animals typically start by consuming moderate amounts and often escalate to excessive alcohol intake over time. This progression makes the model valuable for studying the transition from social-like to problematic drinking behaviors (Becker & Ron, 2014; Carnicella et al., 2014). However, it is important to note that only 50–80% of rats typically escalate to excessive alcohol consumption, which depends on factors such as strain, breed, and laboratory conditions (Carnicella et al., 2014).

One disadvantage of the intermittent access to alcohol model is that animals are often single-housed, and this social isolation during the procedure can lead to increased ethanol consumption in rodents, regardless of their age at the time of isolation (Chappell, Carter, McCool, & Weiner, 2013). Another disadvantage of voluntary alcohol consumption is that this procedure generally does not result in sufficient blood ethanol concentrations to induce behavioral signs of intoxication. This is especially notable in outbred animals that have not been specifically selected for alcohol preference (Hilderbrand & Lasek, 2017).

3.2 Involuntary alcohol consumption

3.2.1 *The rationale behind the model and its procedures*

In involuntary consumption protocols, animals are exposed to alcohol without access to water (Jeanblanc, Rolland, Gierski, Martinetti, & Naassila, 2019). Typically, this involves housing animals individually or in their home cages, where their water bottles are replaced with one containing an alcohol solution (Cronise, Finn, Metten, & Crabbe, 2005).

Besides measuring alcohol consumption directly, the forced consumption of alcohol for a few hours daily can be used to habituate animals to the taste of the drug and facilitate the execution of other subsequent alcohol consumption protocols, such as operant self-administration (Jeanblanc et al., 2019). For example, Wise (1973) demonstrated in rats that forced consumption of 20% ethanol for 30 days increased subsequent consumption in these animals when given a choice in the two-bottle choice paradigm.

Similar to voluntary protocols, sweetened alcoholic solutions, often using sucrose or saccharin, can be employed to facilitate intake of this drug (Hilderbrand & Lasek, 2017).

3.2.2 Model advantages and limitations

The involuntary consumption model offers several advantages, including ease of implementation and the ability to achieve high blood alcohol concentrations, sometimes exceeding 7 g/kg alcohol perday (Aufrère, Bourhis, & Beaugé, 1997). It is a versatile protocol capable of analyzing various components related to AUD (Chen et al., 2015; Montesinos, Pascual, Rodríguez-Arias, Miñarro, & Guerri, 2016; Pascual, Blanco, Cauli, Miñarro, & Guerri, 2007; Pascual, Baliño, Aragón, & Guerri, 2015; Wilkin, Lalor, Parker, & Newsome, 2016). However, forced drinking can reduce the daily intake of fluids depending on the specific protocol and strain of animals used, potentially leading to dehydration and stress (Azarov & Woodward, 2014; Blank, Duncan, & Meadows, 1991).

3.3 The drinking in the dark (did)

3.3.1 The rationale behind the model and its procedures

Several preclinical animal models have been created to mimic binge-like alcohol drinking, as defined by the National Institute on Alcohol Abuse and Alcoholism (NIAAA) as drinking patterns resulting in blood ethanol concentrations (BECs) exceeding 0.08% (80 mg/dL) (National Institute on Alcohol Abuse and Alcoholism DHHS, 2007; Rhodes, Best, Belknap, Finn, & Crabbe, 2005; for more detail: Lopez et al., 2023). Rhodes and colleagues (2005) developed the Drinking in the Dark (DID) protocol based on findings that mice consume more alcohol during their active dark cycle (Dole & Gentry, 1984). This protocol involves housing individual mice and providing access to a bottle containing 20% alcohol solution for 2 to 4 h, starting 3 h into the dark cycle, effectively inducing high BECs (Lopez et al., 2023).

3.3.2 Model advantages and limitations

The DID procedure is distinguished by its ability to study alcohol consumption effectively without requiring extensive training or the use of sweet compounds to induce high levels of intake. This method proves valuable for investigating neurobiological aspects underlying binge-like alcohol consumption (Thiele & Navarro, 2014).

4. Operant alcohol self-administration

4.1 The rationale behind the model and its procedures

Operant self-administration is an animal model used to assess the rewarding properties of addictive drugs and the motivational aspects of addiction, considered a gold standard protocol (Panlilio & Goldberg, 2007; Sanchis-Segura & Spanagel, 2006; Spanagel, 2017). This model is based on operant conditioning principles established by Skinner in the 1930s (Skinner, 1938), where animals perform a specific task reinforced by the delivery of a substance, such as alcohol (Samson, 1986).

The apparatus used for operant self-administration procedures is an operant chamber equipped with at least one lever or nose poke hole for animal interaction. Typically, one lever/nose poke is designated as 'active,' triggering a pump that administers the drug via various routes (oral, intragastric, intravenous, or intracranial). Additionally, discriminative stimuli such as light and sound often accompany substance delivery (Planeta, 2013; Sanchis-Segura & Spanagel, 2006).

To acquire operant self-administration behavior, animals are initially trained to perform an action (e.g., pressing a lever) to receive a drug under a specific reinforcement schedule (Planeta, 2013). Once the acquisition is achieved, the maintenance phase begins to establish stable levels of response and reinforcement (Planeta, 2013; Sanchis-Segura & Spanagel, 2006). Typically, acquisition and maintenance phases employ a fixed ratio reinforcement schedule, where the reinforcer is delivered every time a predetermined number of responses are emitted (Sanchis-Segura & Spanagel, 2006).

One effective reinforcement schedule for evaluating drug-seeking motivation is the progressive ratio (PR). In this schedule, the number of responses required to obtain a reinforcement increases with each reinforcement obtained. PR schedules may utilize different scales (arithmetic or geometric) to increment the response requirement (e.g., 1, 1, 2, 2, 3, 3, etc.) (Killeen, Posadas-Sanchez, Johansen, & Thrailkill, 2009). PR sessions conclude when the animal fails to emit the required number of responses within a specified time interval to obtain a reinforcement. The 'breaking point' is a commonly used index of performance, representing the highest response requirement achieved to earn a single reinforcer (Sanchis-Segura & Spanagel, 2006).

Operant self-administration procedures also enable the assessment of reinstatement of drug-seeking behavior triggered by various stimuli, such as stress, drug priming (i.e., acute exposure to a lower dose of the drug), and environmental cues. In the reinstatement procedure, following the acquisition of operant responses, animals undergo non-reinforcing conditions until these responses are extinguished. Typically, animals are exposed repeatedly to conditions resembling the training phase but in the absence of the drug (Armstrong, Rosenthal, Stout, & Richard, 2023; Bianchi et al., 2018; Felipe et al., 2021; Palombo et al., 2017; Tavares et al., 2023). Consequently, the association between response and outcome weakens, leading to similar response rates on the active lever as on the inactive lever (Marchant, Campbell, Pelloux, Bossert, & Shaham, 2019; Millan, Marchant, & McNally, 2011). Recent studies have also incorporated contingent punishment (e.g., foot shocks) during the extinction phase to simulate the negative effects of drug withdrawal (Marchant, Khuc, Pickens, Bonci, & Shaham, 2013, 2016, 2018).

Studies utilize the "ABA renewal" animal model to investigate the role of contextual cues in triggering relapse to drug-seeking behaviors (Bianchi et al., 2018, 2024; Palombo et al., 2017; Tavares et al., 2023). In this paradigm, animals are initially trained to self-administer alcohol in Context A. Subsequently, alcohol self-administration is extinguished in a different environment, Context B. During testing, animals are reintroduced to either Context A (the original alcohol-associated context) or Context B (the extinction context) as a control condition. This experimental design enables researchers to determine whether the reintroduction of the drug-associated context reinstates drug-seeking behavior (Crombag, Bossert, Koya, & Shaham, 2008; Khoo, Sciascia, Brown, & Chaudhri, 2020).

Moreover, exposure to stressors such as foot shock, pharmacological stressors like yohimbine (an α2-adrenergic antagonist inducing anxious-like symptoms), and social defeat have been shown to promote the reinstatement of alcohol-seeking behavior after operant response extinction in rats (Funk et al., 2005; Lê & Shaham, 2002; Lê, Harding, Juzytsch, Funk, & Shaham, 2005).

These studies underscore the significant impact of both contextual and stress-related factors in triggering alcohol craving and relapse (Bianchi et al., 2018; Chaudhri, Sahuque, Schairer, & Janak, 2010; Felipe et al., 2021; Marchant et al., 2013; Palombo et al., 2017; Tavares et al., 2023; Funk et al., 2005; Lê & Shaham, 2002; Lê et al., 2005).

4.2 Model advantages and limitations

The operant self-administration model is highly valid as it replicates behaviors related to drug consumption, motivation, and seeking observed in humans. Additionally, the neural mechanisms underlying these behaviors show similarities between rodents and humans (Panlilio & Goldberg, 2007; Spanagel, 2017). However, despite its predictive capability, the model faces limitations. Depending on the route of administration, animals may require surgical procedures for substance delivery, which increases protocol complexity (Planeta, 2013). In alcohol studies, rodents typically do not consume alcohol willingly, necessitating adjustments such as addition of sweeteners, which complicates experimental design (Wise, 1975).

5. Chronic intermittent ethanol vapor exposure

5.1 The rationale behind the model and its procedures

Developed by Goldstein et al. in 1971, Chronic Intermittent Ethanol Vapor Exposure (CIEV) induces physical dependence and tolerance in rodents, predicting the neurobiological and behavioral effects seen in human alcohol addiction (Gilpin, Richardson, Cole, & Koob, 2008; Goldstein & Pal, 1971; Goldstein, 1975; Koob, 2003; Roberts, 2000; Rogers, Wiener, & Bloom, 1979). This method maintains alcohol in the animal's bloodstream at levels of 150–300 mg/dL, leading to physiological adaptations like those in AUD. CIEV is currently the gold standard for inducing alcohol dependence in rodents, providing an effective approach to studying the neurobiology of alcohol dependence (Vendruscolo et al., 2012).

Compared to other methods, CIEV offers significant advantages for studying ethanol addiction. It overcomes rodents' natural aversion to ethanol and allows for easy maintenance of consistent BECs (Gilpin et al., 2008).

5.2 Chronic intermittent ethanol vapor exposure in rats

In CIEV, animals are housed in standard rat cages placed into separate sealed clear acrylic chambers where they are exposed to a controlled amount of ethanol vapor. Ethanol vapor is adjusted to maintain BECs in the 150–200 mg/dL range (this range of blood alcohol levels causes mild-to-moderate withdrawal symptoms (Macey, Schulteis, Heinrichs, & Koob, 1996)). Animals are exposed to ethanol vapor inhalation daily for 14 h, followed by 10 h of withdrawal, over several weeks. Blood samples are collected weekly to measure BECs, while control animals are not exposed to ethanol vapor. To evaluate the escalation of alcohol intake, self-administration sessions are conducted 6 h after the beginning of the ethanol vapor abstinence cycle. Typically, the animals undergo regular 30-min sessions with 10% alcohol solution on an FR1 reinforcement schedule, conducted two to three times a week (see Gilpin et al., 2008, for review).

Rats exposed to ethanol vapor show increased alcohol consumption and an escalation in alcohol intake over self-administration sessions. These animals also exhibit increased response levels for alcohol on a PR schedule of reinforcement and maintain alcohol intake even when quinine is added to the solution, despite its aversive bitter taste. Overall, ethanol vapor exposure leads to an escalation of alcohol intake and to alcohol compulsive-like drinking behavior (Gilpin et al., 2008; Guglielmo et al., 2022; Vendruscolo et al., 2012).

5.3 Chronic intermittent ethanol vapor exposure in mice

Mice are invaluable in research due to the availability of diverse transgenic and inbred strains and the relative ease of generating specific mutations for mechanistic studies. While the CIEV protocol can be performed in mice using a similar system to that described for rats, some methodological adjustments are necessary. Currently, there are two primary patterns of ethanol vapor inhalation exposure that significantly increase voluntary alcohol intake in mice (Becker & Hale, 1993).

One pattern involves continuous exposure for 64 h (one cycle), and the other consists of an intermittent pattern that also totals 64 h but consists of four 16-h bouts of exposure separated by 8-h withdrawal periods. Both exposure patterns lead to significant increases in alcohol drinking, but the intermittent pattern results in more robust and sustained increases in voluntary consumption, particularly when repeated over multiple cycles (Lopez & Becker, 2005).

During ethanol vapor exposure, it is crucial to use pyrazole to inhibit alcohol metabolism in mice, as they metabolize alcohol rapidly. Pyrazole is essential for maintaining stable BECs during vapor inhalation. Recent findings demonstrated that pyrazole keeps BECs values stable throughout a 16-h exposure period, whereas omitting pyrazole led to significant declines to near zero by the end of the exposure (Griffin, Lopez, & Becker, 2009). In ethanol-dependent mice, pyrazole is administered alongside a loading dose of ethanol before they enter the vapor chamber. Control mice (non-dependent), which receive air exposure, also receive pyrazole injections but without the ethanol loading dose.

5.4 Model advantages and limitations

The CIEV model offers several advantages over other animal models of addiction, such as liquid diet and intragastric ethanol intubation. Unlike these methods, CIEV allows precise control over alcohol dose, duration, and exposure pattern. It also facilitates additional behavioral tests and provides flexibility in experimental protocols, making it adaptable across various laboratory settings and experiments (Rogers et. al., 1979; Gilpin et al., 2009).

This model demonstrates high reliability and predictive validity, crucial criteria for evaluating animal models (Edwards & Koob, 2013; Schulteis, Markou, Cole, & Koob, 1995). It reliably induces increased alcohol intake and compulsive-like drinking and manifests numerous physical and motivational symptoms akin to human alcohol dependence. Its predictive validity extends to making inferences about human conditions and identifying potentially therapeutic drugs (Vendruscolo & Roberts, 2014).

However, a notable criticism of the alcohol vapor exposure model is its limited face validity. Humans do not typically develop dependence through inhalation or forced consumption of alcohol. Nevertheless, this concern is partially mitigated by rodents' shorter lifespans, necessitating a compressed timeframe for achieving high BECs. Sustaining such levels in rodents without experimental intervention is challenging, except with alcohol-preferring rat strains.

Another limitation is the necessity of forced alcohol exposure to induce clinically relevant BECs and addiction-like behaviors. To address these limitations, researchers have developed a novel animal model involving voluntary induction of alcohol dependence through ethanol vapor self-administration. In this approach, animals are trained to self-administer alcohol vapor in 8-h sessions every other day. This method enables rats to attain BECs ranging from 150 to 300 mg/dL and exhibit addiction-like traits such as escalated intake, anxiety-like behavior, hyperalgesia, and somatic withdrawal signs (see Guglielmo, Kallupi, Cole, & George, 2017 for more details).

6. Others useful procedures

6.1 Behavioral sensitization

Behavioral sensitization in rodents is characterized by a gradual and progressive increase in locomotor activity after repeated drug administration

(Camarini, Marianno, & Rae, 2018; Cofresí, Bartholow, & Piasecki, 2019; Post & Contel, 1983; Robinson & Becker, 1986; Sanchis-Segura & Spanagel, 2006). This phenomenon involves neurochemical and molecular adaptations in the mesolimbic dopaminergic system (Nestler, 2001; Robinson & Becker, 1986). According to Robinson (1993), these adaptations are associated with an enhancement in the motivational salience of the stimulus (Planeta & Cruz, 2005). Behavioral sensitization has been demonstrated with many drugs, such as cocaine, amphetamine, nicotine, and opioids (Kalivas, Duffy, & Latimer, 1987; Miserendino & Nestler, 1995; Planeta & Marin, 2002; Robinson & Becker, 1986).

Alcohol administration can lead to behavioral sensitization (Camarini & Pautassi, 2016). One of the pioneering studies on alcohol-induced locomotor sensitization was conducted by Masur and Boerngen in 1980. In this research, female Swiss mice received alcohol at various doses (0, 1.0, 2.5, or 3.5 g/kg, intraperitoneally), and their locomotor activity was assessed for 60 min over 30 consecutive days. A progressive increase in locomotor activity was observed throughout the study period (Masur & Boerngen, 1980).

Stress is a crucial factor in AUD. Neuroadaptive processes triggered by stress lead to lasting changes in mesolimbic systems, which can manifest as heightened stimulatory effects of alcohol (referred to as locomotor cross-sensitization). This model is, therefore, valuable for exploring how stress affects the neurobiology of alcohol responses (Camarini & Pautassi, 2016).

However, studies show that locomotor sensitization is not simply a behavioral response to repeated substance administration or stress. Sensitization also involves aspects of associative learning. In this regard, research demonstrates that when the drug is repeatedly administered in the same context as the locomotor sensitization test, animals exhibit a more robust sensitization compared to those receiving the drug in a different environment (Anagnostaras & Robinson, 1996; Anagnostaras, Schallert, & Robinson, 2002; Koya et al., 2009; Marin et al., 2009; Mattson et al., 2008; Robinson, Browman, Crombag, & Badiani, 1998). Thus, the sensitization model is also important for investigating associative aspects related to the development of dependence.

6.2 Quinine adulteration test

One of the key characteristics of AUD is the persistent consumption of alcohol despite experiencing negative consequences (Koob & Volkow, 2010; American Psychiatric Association, 2013). To explore this resistance to adversity in alcohol intake, researchers have employed a method involving the addition of the bitter-tasting substance quinine to alcohol (Hopf et al., 2010; De Oliveira Sergio, Frasier, & Hopf, 2023). In this procedure, alcohol solutions are gradually mixed with increasing concentrations of quinine.

This approach aims to evaluate the extent to which animals continue to consume alcohol despite the aversive taste of quinine. Studies use this test to measure compulsive-like consumption behaviors (Vendruscolo et al., 2012).

6.3 The Lieber-DeCarli diet

Alcoholic liver disease (ALD) is a prevalent chronic liver disease in the world. To investigate ALD, Charles Lieber and Leonore M. DeCarli developed a method incorporating alcohol into fully liquid diets (Brandon-Warner, Schrum, Schmidt, & McKillop, 2012; Guo, Zheng, Benedé-Ubieto, Cubero, & Nevzorova, 2018; Lieber & DeCarli, 1989). This approach has become the preferred model for studying chronic alcohol consumption, leading to an increase in many liver processes metabolic and initial signs of hepatocyte damage (Bertola, Mathews, Ki, Wang, & Gao, 2013; Guo et al., 2018). The diet consists of 15% protein, 36% fat, and the rest from carbohydrates, which are isocalorically replaced by ethanol in the Lieber-DeCarli liquid diet (LDC). Additionally, the diet is enriched with essential vitamins (A, D, E, K, and B), minerals, and fibers (Lieber & DeCarli, 1989). The ethanol in the liquid diet is gradually increased from 1% to a final concentration of 5.07% w/v (6.4% v/v) over a 7-day period (Bertola et al., 2013; Guo et al., 2018). Animals on the LDC diet consume up to 14 to 16 g/kg/day of alcohol, reaching blood alcohol levels between 100 and 160 mg/dL (Guo et al., 2018). The duration of LDC feeding depends on the stage of chronic liver injury being studied. Early ALD signs can appear within 3–4 weeks of alcohol exposure, varying by species, strain, and gender (D'Souza et al., 2010). This model is straightforward, cost-effective, and provides highly reproducible outcomes (see Guo et al., 2018 for more details).

7. Concluding remarks

The experimental models discussed in this chapter are helpful for the investigation of neural mechanisms underlying AUD and the exploration of potential new treatments. The development of animal models and refinement of protocols are crucial for advancing our understanding of AUD and their associated behavioral, physiological, and neurochemical alterations. Without robust preclinical studies, progress in understanding SUD would be hindered, thereby limiting the development of new therapeutic strategies and treatments that could ultimately enhance the health and quality of life of individuals suffering from AUD.

References

Agoglia, A. E., Zhu, M., Quadir, S. G., Bluitt, M. N., Douglass, E., Hanback, T., ... Herman, M. A. (2022). Sex-specific plasticity in CRF regulation of inhibitory control in central amygdala CRF1 neurons after chronic voluntary alcohol drinking. *Addiction Biology, 27*(1), e13067. https://doi.org/10.1111/adb.13067.

Aguilar, M. A., Rodríguez-Arias, M., & Miñarro, J. (2009). Neurobiological mechanisms of the reinstatement of drug-conditioned place preference. *Brain Research Reviews, 59*(2), 253–277. https://doi.org/10.1016/j.brainresrev.2008.08.002.

American Psychiatric Association. (2013). Diagnostic and statistical manual of mental disorders (5th edn). American Psychiatric Association. https://doi.org/10.1176/appi.books.9780890425596.

Anagnostaras, S. G., & Robinson, T. E. (1996). Sensitization to the psychomotor stimulant effects of amphetamine: Modulation by associative learning. *Behavioral Neuroscience, 110*(6), 1397.

Anagnostaras, S. G., Schallert, T., & Robinson, T. E. (2002). Memory processes governing amphetamine-induced psychomotor sensitization. *Neuropsychopharmacology: Official Publication of the American College of Neuropsychopharmacology, 26*(6), 703–715.

Armstrong, A., Rosenthal, H., Stout, N., & Richard, J. M. (2023). Reinstatement of Pavlovian responses to alcohol cues by stress. *Psychopharmacology, 240*(3), 531–545. https://doi.org/10.1007/s00213-022-06255-w.

Aufrère, G., Bourhis, B. L., & Beaugé, F. (1997). Ethanol intake after chronic intoxication by inhalation of ethanol vapour in rats: Behavioural dependence. *Alcohol (Fayetteville, N. Y.), 14*(3), 247–253. https://doi.org/10.1016/S0741-8329(96)00175-9.

Azarov, A. V., & Woodward, D. J. (2014). Early ethanol and water intake: Choice mechanism and total fluid regulation operate in parallel in male alcohol preferring (P) and both Wistar and Sprague Dawley rats. *Physiology & Behavior, 123*, 11–19. https://doi.org/10.1016/j.physbeh.2013.09.011.

Bahi, A., & Dreyer, J.-L. (2019). No effect of sex on ethanol intake and preference after dopamine transporter (DAT) knockdown in adult mice. *Psychopharmacology, 236*(4), 1349–1365. https://doi.org/10.1007/s00213-018-5144-9.

Bardo, M. T., & Bevins, R. A. (2000). Conditioned place preference: What does it add to our preclinical understanding of drug reward? *Psychopharmacology, 153*(1), 31–43. https://doi.org/10.1007/s002130000569.

Becker, H. C., & Hale, R. L. (1993). Repeated episodes of ethanol withdrawal potentiate the severity of subsequent withdrawal seizures: An animal model of alcohol withdrawal "kindling". *Alcoholism, Clinical and Experimental Research, 17*(1), 94–98. https://doi.org/10.1111/j.1530-0277.1993.tb00731.x.

Becker, H. C., & Ron, D. (2014). Animal models of excessive alcohol consumption: Recent advances and future challenges. *Alcohol (Fayetteville, N. Y.), 48*(3), 205–208. https://doi.org/10.1016/j.alcohol.2014.04.001.

Bertola, A., Mathews, S., Ki, S. H., Wang, H., & Gao, B. (2013). Mouse model of chronic and binge ethanol feeding (the NIAAA model). *Nature Protocols, 8*(3), 627–637. https://doi.org/10.1038/nprot.2013.032.

Bhutada, P., Mundhada, Y., Ghodki, Y., Dixit, P., Umathe, S., & Jain, K. (2012). Acquisition, expression, and reinstatement of ethanol-induced conditioned place preference in mice: Effects of exposure to stress and modulation by mecamylamine. *Journal of Psychopharmacology, 26*(2), 315–323. https://doi.org/10.1177/0269881111431749.

Bianchi, P. C., Carneiro De Oliveira, P. E., Palombo, P., Leão, R. M., Cogo-Moreira, H., Planeta, C. D. S., & Cruz, F. C. (2018). Functional inactivation of the orbitofrontal cortex disrupts context-induced reinstatement of alcohol seeking in rats. *Drug and Alcohol Dependence, 186*, 102–112. https://doi.org/10.1016/j.drugalcdep.2017.12.045.

Bianchi, P. C., Palombo, P., Carneiro de Oliveira, P. E., Engi, S. A., Tavares, G. E. B., Anjos-Santos, A., ... Leão, R. M. (2024). Involvement of Pre-limbic Cortex-Nucleus accumbens projections in Context-Induced alcohol seeking. *Brain Research, 1841*. https://doi.org/10.1016/j.brainres.2024.149086.

Blanco-Gandía, M. C., Aguilar, M. A., Miñarro, J., & Rodríguez-Arias, M. (2018). Reinstatement of drug-seeking in mice using the conditioned place preference paradigm. *Journal of Visualized Experiments,* (136), 56983. https://doi.org/10.3791/56983.

Blank, S. E., Duncan, D. A., & Meadows, G. G. (1991). Suppression of natural killer cell activity by ethanol consumption and food restriction. *Alcoholism: Clinical and Experimental Research, 15*(1), 16–22. https://doi.org/10.1111/j.1530-0277.1991.tb00514.x.

Brandon-Warner, E., Schrum, L. W., Schmidt, C. M., & McKillop, I. H. (2012). Rodent models of alcoholic liver disease: Of mice and men. *Alcohol (Fayetteville, N. Y.), 46*(8), 715–725. https://doi.org/10.1016/j.alcohol.2012.08.004.

Camarini, R., Marianno, P., & Rae, M. (2018). *Social factors in ethanol sensitization. Em. International Review of Neurobiology, 140*, Elsevier, 53–80. https://doi.org/10.1016/bs.irn.2018.07.003.

Camarini, R., & Pautassi, R. M. (2016). Behavioral sensitization to ethanol: Neural basis and factors that influence its acquisition and expression. *Brain Research Bulletin, 125*, 53–78. https://doi.org/10.1016/j.brainresbull.2016.04.006.

Carnicella, S., Ron, D., & Barak, S. (2014). Intermittent ethanol access schedule in rats as a preclinical model of alcohol abuse. *Alcohol (Fayetteville, N. Y.), 48*(3), 243–252. https://doi.org/10.1016/j.alcohol.2014.01.006.

Chappell, A. M., Carter, E., McCool, B. A., & Weiner, J. L. (2013). Adolescent rearing conditions influence the relationship between initial anxiety-like behavior and ethanol drinking in male long evans rats. *Alcoholism: Clinical and Experimental Research, 37*(s1), https://doi.org/10.1111/j.1530-0277.2012.01926.x.

Chaudhri, N., Sahuque, L. L., Schairer, W. W., & Janak, P. H. (2010). Separable roles of the nucleus accumbens core and shell in context- and cue-induced alcohol-seeking. *Neuropsychopharmacology: Official Publication of the American College of Neuropsychopharmacology, 35*(3), 783–791. https://doi.org/10.1038/npp.2009.187.

Chen, P., Miyamoto, Y., Mazagova, M., Lee, K., Eckmann, L., & Schnabl, B. (2015). Microbiota protects mice against acute alcohol-induced liver injury. *Alcoholism: Clinical and Experimental Research, 39*(12), 2313–2323. https://doi.org/10.1111/acer.12900.

Childs, E., & De Wit, H. (2016). Alcohol-induced place conditioning in moderate social drinkers. *Addiction (Abingdon, England), 111*(12), 2157–2165. https://doi.org/10.1111/add.13540.

Cofresí, R. U., Bartholow, B. D., & Piasecki, T. M. (2019). Evidence for incentive salience sensitization as a pathway to alcohol use disorder. *Neuroscience & Biobehavioral Reviews, 107*, 897–926.

Collier, A. D., Khan, K. M., Caramillo, E. M., Mohn, R. S., & Echevarria, D. J. (2014). Zebrafish and conditioned place preference: A translational model of drug reward. *Progress in Neuro-Psychopharmacology and Biological Psychiatry, 55*, 16–25. https://doi.org/10.1016/j.pnpbp.2014.05.014.

Crombag, H. S., Bossert, J. M., Koya, E., & Shaham, Y. (2008). Context-induced relapse to drug seeking: A review. *Philosophical Transactions of the Royal Society B: Biological Sciences, 363*(1507), 3233–3243. https://doi.org/10.1098/rstb.2008.0090.

Cronise, K., Finn, D. A., Metten, P., & Crabbe, J. C. (2005). Scheduled access to ethanol results in motor impairment and tolerance in female C57BL/6J mice. *Pharmacology, Biochemistry, and Behavior, 81*(4), 943–953. https://doi.org/10.1016/j.pbb.2005.07.005.

Cunningham, C. L., Ferree, N. K., & Howard, M. A. (2003). Apparatus bias and place conditioning with ethanol in mice. *Psychopharmacology, 170*(4), 409–422. https://doi.org/10.1007/s00213-003-1559-y.

Cunningham, C. L., Gremel, C. M., & Groblewski, P. A. (2006). Drug-induced conditioned place preference and aversion in mice. *Nature Protocols, 1*(4), 1662–1670. https://doi.org/10.1038/nprot.2006.279.

D'Souza El-Guindy, Kovacs, E. J., De Witte, P., Spies, C., Littleton, J. M., ... Meadows (2010). Laboratory models available to study alcohol-induced organ damage and immune variations: Choosing the appropriate model. *Alcoholism: Clinical and Experimental Research, 34*(9), 1489–1511. https://doi.org/10.1111/j.1530-0277.2010.01234.x.

De Oliveira Sergio, T., Frasier, R. M., & Hopf, F. W. (2023). Animal models of compulsion alcohol drinking: Why we love quinine-resistant intake and what we learned from it. *Frontiers in Psychiatry, 14*, 1116901. https://doi.org/10.3389/fpsyt.2023.1116901.

Dole, V. P., & Gentry, R. T. (1984). Toward an analogue of alcoholism in mice: Scale factors in the model. *Proceedings of the National Academy of Sciences, 81*(11), 3543–3546.

Eaton, S. E., Dzhala, S., Robinson, L. E., Radevski, M. E., & Akins, C. K. (2023). Ethanol induces a dose-dependent conditioned place preference and conditioned place aversion in Japanese quail. *Experimental and Clinical Psychopharmacology, 31*(1), 1–6. https://doi.org/10.1037/pha0000548.

Edwards, S., & Koob, G. F. (2013). Escalation of drug self-administration as a hallmark of persistent addiction liability. *Behavioural Pharmacology, 24*(5 and 6), 356–362. https://doi.org/10.1097/FBP.0b013e3283644d15.

Faccidomo, S., Salling, M. C., Galunas, C., & Hodge, C. W. (2015). Operant ethanol self-administration increases extracellular-signal regulated protein kinase (ERK) phosphorylation in reward-related brain regions: Selective regulation of positive reinforcement in the prefrontal cortex of C57BL/6J mice. *Psychopharmacology, 232*(18), 3417–3430. https://doi.org/10.1007/s00213-015-3993-z.

Felipe, J. M., Palombo, P., Bianchi, P. C., Zaniboni, C. R., Anésio, A., Yokoyama, T. S., ... Cruz, F. C. (2021). Dorsal hippocampus plays a causal role in context-induced reinstatement of alcohol-seeking in rats. *Behavioural Brain Research, 398*, 112978. https://doi.org/10.1016/j.bbr.2020.112978.

Font, L., Miquel, M., & Aragon, C. M. G. (2008). Involvement of brain catalase activity in the acquisition of ethanol-induced conditioned place preference. *Physiology & Behavior, 93*(4–5), 733–741. https://doi.org/10.1016/j.physbeh.2007.11.026.

Funk, D, Harding, S, Juzytsch, W, & Lê, AD (2005). Effects of unconditioned and conditioned social defeat on alcohol self-administration and reinstatement of alcohol seeking in rats. *Psychopharmacology (Berl), 183*(3), 341–349. https://doi.org/10.1007/s00213-005-0194-1.

Gawel, K., Labuz, K., Gibula-Bruzda, E., Jenda, M., Marszalek-Grabska, M., Silberring, J., & Kotlinska, J. H. (2016). Acquisition and reinstatement of ethanol-induced conditioned place preference in rats: Effects of the cholinesterase inhibitors donepezil and rivastigmine. *Journal of Psychopharmacology, 30*(7), 676–687. https://doi.org/10.1177/0269881116642539.

Gilpin, N. W., Richardson, H. N., Cole, M., & Koob, G. F. (2008). Vapor inhalation of alcohol in rats. *Current Protocols in Neuroscience, 44*(1), https://doi.org/10.1002/0471142301.ns0929s44.
Gilpin, N. W., Smith, A. D., Cole, M., Weiss, F., Koob, G. F., & Richardson, H. N. (2009). Operant behavior and alcohol levels in blood and brain of alcohol-dependent rats. *Alcoholism: Clinical and Experimental Research, 33*(12), 2113–2123. https://doi.org/10.1111/j.1530-0277.2009.01051.x.
Goldstein, D. B. (1975). Physical dependence on alcohol in mice. *Federation Proceedings, 34*(10), 1953–1961.
Goldstein, D. B., & Pal, N. (1971). Alcohol dependence produced in mice by inhalation of ethanol: Grading the withdrawal reaction. *Science (New York, N. Y.), 172*(3980), 288–290. https://doi.org/10.1126/science.172.3980.288.
Goltseker, K., Hopf, F. W., & Barak, S. (2019). Advances in behavioral animal models of alcohol use disorder. *Alcohol (Fayetteville, N. Y.), 74*, 73–82. https://doi.org/10.1016/j.alcohol.2018.05.014.
Griffin, W. C., 3rd, Lopez, M. F., & Becker, H. C. (2009). Intensity and duration of chronic ethanol exposure is critical for subsequent escalation of voluntary ethanol drinking in mice. *Alcoholism, Clinical and Experimental Research, 33*(11), 1893–1900. https://doi.org/10.1111/j.1530-0277.2009.01027.x.
Guglielmo, G. de, Simpson, S., Kimbrough, A., Conlisk, D., Baker, R., Cantor, M., & George, O. (2022). Voluntary and forced exposure to ethanol vapor produces similar escalation of alcohol drinking but differential recruitment of brain regions related to stress, habit, and reward in male rats (p. 2022.05.11.491572). bioRxiv. https://doi.org/10.1101/2022.05.11.491572.
Guglielmo, G., Kallupi, M., Cole, M. D., & George, O. (2017). Voluntary induction and maintenance of alcohol dependence in rats using alcohol vapor self-administration. *Psychopharmacology, 234*(13), 2009–2018. https://doi.org/10.1007/s00213-017-4608-7.
Guo, F., Zheng, K., Benedé-Ubieto, R., Cubero, F. J., & Nevzorova, Y. A. (2018). The Lieber-DeCarli diet-A flagship model for experimental alcoholic liver disease. *Alcoholism, Clinical and Experimental Research, 42*(10), 1828–1840. https://doi.org/10.1111/acer.13840 Epub 2018 Aug 9. PMID: 30025151.
Henriques, G. M., Anjos-Santos, A., Rodrigues, I. R. S., Nascimento-Rocha, V., Reis, H. S., Libarino-Santos, M., ... Berro, L. F. (2021). Ibogaine blocks cue- and drug-induced reinstatement of conditioned place preference to ethanol in male mice. *Frontiers in Pharmacology, 12*, 739012. https://doi.org/10.3389/fphar.2021.739012.
Hilderbrand, E. R., & Lasek, A. W. (2017). Studying sex differences in animal models of addiction: An emphasis on alcohol-related behaviors. *ACS Chemical Neuroscience, 9*(8), 1907–1916. https://doi.org/10.1021/acschemneuro.7b00449.
Huston, J. P., Silva, M. A. D. S., Topic, B., & Müller, C. P. (2013). What's conditioned in conditioned place preference? *Trends in Pharmacological Sciences, 34*(3), 162–166. https://doi.org/10.1016/j.tips.2013.01.004.
Jeanblanc, J., Rolland, B., Gierski, F., Martinetti, M. P., & Naassila, M. (2019). Animal models of binge drinking, current challenges to improve face validity. *Neuroscience & Biobehavioral Reviews, 106*, 112–121. https://doi.org/10.1016/j.neubiorev.2018.05.002.
Kalivas, P. W., Duffy, P., & Latimer, L. G. (1987). Neurochemical and behavioral effects of corticotropin-releasing factor in the ventral tegmental area of the rat. *Journal of Pharmacology and Experimental Therapeutics, 242*(3), 757–763.
Kaun, K. R., Azanchi, R., Maung, Z., Hirsh, J., & Heberlein, U. (2011). A Drosophila model for alcohol reward. *Nature Neuroscience, 14*(5), 612–619. https://doi.org/10.1038/nn.2805.
Khoo, S. Y.-S., Sciascia, J. M., Brown, A., & Chaudhri, N. (2020). Comparing ABA, AAB, and ABC renewal of appetitive pavlovian conditioned responding in alcohol- and sucrose-trained male rats. *Frontiers in Behavioral Neuroscience, 14*, 5. https://doi.org/10.3389/fnbeh.2020.00005.

Killeen, P. R., Posadas-Sanchez, D., Johansen, E. B., & Thrailkill, E. A. (2009). Progressive ratio schedules of reinforcement. *Journal of Experimental Psychology: Animal Behavior Processes, 35*(1), 35–50. https://doi.org/10.1037/a0012497.

Kimbrough, A., Kim, S., Cole, M., Brennan, M., & George, O. (2017). Intermittent access to ethanol drinking facilitates the transition to excessive drinking after chronic intermittent ethanol vapor exposure. *Alcoholism: Clinical and Experimental Research, 41*(8), 1502–1509. https://doi.org/10.1111/acer.13434.

Koob, G. F. (2003). Alcoholism: Allostasis and beyond. *Alcoholism: Clinical and Experimental Research, 27*(2), 232–243. https://doi.org/10.1097/01.ALC.0000057122.36127.C2.

Koob, G. F. (2024). Alcohol use disorder treatment: Problems and solutions. *Annual Review of Pharmacology and Toxicology, 64*(1), 255–275. https://doi.org/10.1146/annurev-pharmtox-031323-115847.

Koob, G. F., & Volkow, N. D. (2010). Neurocircuitry of addiction. *Neuropsychopharmacology: Official Publication of the American College of Neuropsychopharmacology, 35*(1), 217–238. https://doi.org/10.1038/npp.2009.110.

Koya, E., Golden, S. A., Harvey, B. K., Guez-Barber, D. H., Berkow, A., Simmons, D. E., & Hope, B. T. (2009). Targeted disruption of cocaine-activated nucleus accumbens neurons prevents context-specific sensitization. *Nature Neuroscience, 12*(8), 1069–1073.

Kuzmin, A., Sandin, J., Terenius, L., & Ögren, S. O. (2003). Acquisition, expression, and reinstatement of ethanol-induced conditioned place preference in mice: Effects of opioid receptor-like 1 receptor agonists and naloxone. *Journal of Pharmacology and Experimental Therapeutics, 304*(1), 310–318. https://doi.org/10.1124/jpet.102.041350.

Kwako, L. E., Momenan, R., Grodin, E. N., Litten, R. Z., Koob, G. F., & Goldman, D. (2017). Addictions neuroclinical assessment: A reverse translational approach. *Neuropharmacology, 122*, 254–264. https://doi.org/10.1016/j.neuropharm.2017.03.006.

Lê, A. D., Harding, S., Juzytsch, W., Funk, D., & Shaham, Y. (2005). Role of alpha-2 adrenoceptors in stress-induced reinstatement of alcohol seeking and alcohol self-administration in rats. *Psychopharmacology, 179*(2), 366–373. https://doi.org/10.1007/s00213-004-2036-y.

Lê, A. D., & Shaham, Y. (2002). Neurobiology of relapse to alcohol in rats. *Pharmacology & Therapeutics, 94*(1–2), 137–156. https://doi.org/10.1016/S0163-7258(02)00200-0.

Lieber, C. S., & DeCarli, L. M. (1989). Liquid diet technique of ethanol administration: 1989 update. *Alcohol and Alcoholism (Oxford, Oxfordshire), 24*(3), 197–211.

Little, H. J., O'Callaghan, M. J., Butterworth, A. R., Wilson, J., Cole, J., & Watson, W. P. (1999). Low alcohol preference among the "high alcohol preference" C57 strain of mice; preference increased by saline injections. *Psychopharmacology, 147*(2), 182–189. https://doi.org/10.1007/s002130051159.

Liu, Y., Le Foll, B., Liu, Y., Wang, X., & Lu, L. (2008). Conditioned place preference induced by licit drugs: Establishment, extinction, and reinstatement. *The Scientific World JOURNAL, 8*, 1228–1245. https://doi.org/10.1100/tsw.2008.154.

Lopez, M. F., & Becker, H. C. (2005). Effect of pattern and number of chronic ethanol exposures on subsequent voluntary ethanol intake in C57BL/6J mice. *Psychopharmacology, 181*(4), 688–696. https://doi.org/10.1007/s00213-005-0026-3.

Lopez, M. F., Davis, E. C., Cucinello-Ragland, J. A., Regunathan, S., Edwards, S., & Becker, H. C. (2023). Agmatine reduces alcohol drinking and produces antinociceptive effects in rodent models of alcohol use disorder. *Alcohol (Fayetteville, N. Y.), 109*, 23–33. https://doi.org/10.1016/j.alcohol.2023.01.003.

Lorenz, E., Moye, C., & Leong, K.-C. (2022). Paired housing or a socially-paired context decreases ethanol conditioned place preference in male rats. *Brain Sciences, 12*(11), 1485. https://doi.org/10.3390/brainsci12111485.

Macey, D. J., Schulteis, G., Heinrichs, S. C., & Koob, G. F. (1996). Time-dependent quantifiable withdrawal from ethanol in the rat: Effect of method of dependence induction. *Alcohol (Fayetteville, N. Y.), 13*(2), 163–170. https://doi.org/10.1016/0741-8329(95)02030-6.

Manthey, J., Shield, K. D., Rylett, M., Hasan, O. S. M., Probst, C., & Rehm, J. (2019). Global alcohol exposure between 1990 and 2017 and forecasts until 2030: A modelling study. *The Lancet, 393*(10190), 2493–2502. https://doi.org/10.1016/S0140-6736(18)32744-2.

Marchant, N. J., Campbell, E. J., Pelloux, Y., Bossert, J. M., & Shaham, Y. (2019). Context-induced relapse after extinction versus punishment: Similarities and differences. *Psychopharmacology, 236*(1), 439–448. https://doi.org/10.1007/s00213-018-4929-1.

Marchant, N. J., Khuc, T. N., Pickens, C. L., Bonci, A., & Shaham, Y. (2013). Context-induced relapse to alcohol seeking after punishment in a rat model. *Biological Psychiatry, 73*(3), 256–262. https://doi.org/10.1016/j.biopsych.2012.07.007.

Marin, M. T., Berkow, A., Golden, S. A., Koya, E., Planeta, C. S., & Hope, B. T. (2009). Context-specific modulation of cocaine-induced locomotor sensitization and ERK and CREB phosphorylation in the rat nucleus accumbens. *European Journal of Neuroscience, 30*(10), 1931–1940.

Martín-Sánchez, A., Warnault, V., Montagud-Romero, S., Pastor, A., Mondragón, N., De La Torre, R., & Valverde, O. (2019). Alcohol-induced conditioned place preference is modulated by CB2 cannabinoid receptors and modifies levels of endocannabinoids in the mesocorticolimbic system. *Pharmacology, Biochemistry, and Behavior, 183*, 22–31. https://doi.org/10.1016/j.pbb.2019.06.007.

Masur, J., & Boerngen, R. (1980). The excitatory component of ethanol in mice: A chronic study. *Pharmacology, Biochemistry, and Behavior, 13*(6), 777–780. https://doi.org/10.1016/0091-3057(80)90206-3.

Mathur, P., Lau, B., & Guo, S. (2011). Conditioned place preference behavior in zebrafish. *Nature Protocols, 6*(3), 338–345. https://doi.org/10.1038/nprot.2010.201.

Mattson, B. J., Koya, E., Simmons, D. E., Mitchell, T. B., Berkow, A., Crombag, H. S., & Hope, B. T. (2008). Context-specific sensitization of cocaine-induced locomotor activity and associated neuronal ensembles in rat nucleus accumbens. *European Journal of Neuroscience, 27*(1), 202–212.

McKendrick, G., Garrett, H., Tanniru, S., Ballard, S., Sun, D., Silberman, Y., ... Graziane, N. M. (2020). A novel method to study reward-context associations and drug-seeking behaviors. *Journal of Neuroscience Methods, 343*, 108857. https://doi.org/10.1016/j.jneumeth.2020.108857.

McKendrick, G., & Graziane, N. M. (2020). Drug-induced conditioned place preference and its practical use in substance use disorder research. *Frontiers in Behavioral Neuroscience, 14*, 582147. https://doi.org/10.3389/fnbeh.2020.582147.

Millan, E. Z., Marchant, N. J., & McNally, G. P. (2011). Extinction of drug seeking. *Behavioural Brain Research, 217*(2), 454–462. https://doi.org/10.1016/j.bbr.2010.10.037.

Miserendino, M. J., & Nestler, E. J. (1995). Behavioral sensitization to cocaine: Modulation by the cyclic AMP system in the nucleus accumbens. *Brain Research, 674*(2), 299–306.

Montesinos, J., Pascual, M., Rodríguez-Arias, M., Miñarro, J., & Guerri, C. (2016). Involvement of TLR4 in the long-term epigenetic changes, rewarding and anxiety effects induced by intermittent ethanol treatment in adolescence. *Brain, Behavior, and Immunity, 53*, 159–171. https://doi.org/10.1016/j.bbi.2015.12.006.

Morales, M., Varlinskaya, E. I., & Spear, L. P. (2012). Evidence for conditioned place preference to a moderate dose of ethanol in adult male Sprague–Dawley rats. *Alcohol (Fayetteville, N. Y.), 46*(7), 643–648. https://doi.org/10.1016/j.alcohol.2012.06.001.

Mueller, D., & De Wit, H. (2011). *Conditioned place preference in rodents and humans. Em J. Raber (Org.), Animal Models of Behavioral Analysis, 50*, Humana Press, 133–152. ⟨https://doi.org/10.1007/978-1-60761-883-6_6⟩.

Napier, T. C., Herrold, A. A., & De Wit, H. (2013). Using conditioned place preference to identify relapse prevention medications. *Neuroscience & Biobehavioral Reviews, 37*(9), 2081–2086. https://doi.org/10.1016/j.neubiorev.2013.05.002.

National Institute on Alcohol Abuse and Alcoholism (DHHS). (2007). What colleges need to know now: An update on college drinking research. ERIC Clearinghouse.

Nestler, E. J. (2001). Molecular neurobiology of addiction. *American Journal on Addictions, 10*(3), 201–217.

Newman, E. L., Albrechet-Souza, L., Andrew, P. M., Auld, J. G., Burk, K. C., Hwa, L. S., ... Miczek, K. A. (2018). Persistent escalation of alcohol consumption by mice exposed to brief episodes of social defeat stress: Suppression by CRF-R1 antagonism. *Psychopharmacology, 235*(6), 1807–1820. https://doi.org/10.1007/s00213-018-4905-9.

Nieto, S. J., Grodin, E. N., Aguirre, C. G., Izquierdo, A., & Ray, L. A. (2021). Translational opportunities in animal and human models to study alcohol use disorder. *Translational Psychiatry, 11*(1), 496. https://doi.org/10.1038/s41398-021-01615-0.

Okhuarobo, A., Igbe, I., Yahaya, A., & Sule, Z. (2018). Effect of caffeine on alcohol consumption and alcohol-induced conditioned place preference in rodents. *Journal of Basic and Clinical Physiology and Pharmacology, 30*(1), 19–28. https://doi.org/10.1515/jbcpp-2018-0068.

Palombo, P., Engi, S. A., Yokoyama, T. S., Bezerra, A. G., Curado, D. F., Anésio, A., ... Galduróz, J. C. F. (2021). Effects of biperiden (cholinergic muscarinic m1/m4 receptor antagonist) on ethanol conditioned place preference in mice. *Neuroscience Letters, 745*, 135551. https://doi.org/10.1016/j.neulet.2020.135551.

Palombo, P., Leao, R. M., Bianchi, P. C., De Oliveira, P. E. C., Planeta, C. D. S., & Cruz, F. C. (2017). Inactivation of the prelimbic cortex impairs the context-induced reinstatement of ethanol seeking. *Frontiers in Pharmacology, 8*, 725. https://doi.org/10.3389/fphar.2017.00725.

Panlilio, L. V., & Goldberg, S. R. (2007). Self-administration of drugs in animals and humans as a model and an investigative tool. *Addiction (Abingdon, England), 102*(12), 1863–1870. https://doi.org/10.1111/j.1360-0443.2007.02011.x.

Pascual, M., Baliño, P., Aragón, C. M. G., & Guerri, C. (2015). Cytokines and chemokines as biomarkers of ethanol-induced neuroinflammation and anxiety-related behavior: Role of TLR4 and TLR2. *Neuropharmacology, 89*, 352–359. https://doi.org/10.1016/j.neuropharm.2014.10.014.

Pascual, M., Blanco, A. M., Cauli, O., Miñarro, J., & Guerri, C. (2007). Intermittent ethanol exposure induces inflammatory brain damage and causes long-term behavioural alterations in adolescent rats. *European Journal of Neuroscience, 25*(2), 541–550. https://doi.org/10.1111/j.1460-9568.2006.05298.x.

Pati, D., Pina, M. M., & Kash, T. L. (2019). Ethanol-induced conditioned place preference and aversion differentially alter plasticity in the bed nucleus of stria terminalis. *Neuropsychopharmacology: Official Publication of the American College of Neuropsychopharmacology, 44*(11), 1843–1854. https://doi.org/10.1038/s41386-019-0349-0.

Perry, T. W., Sneddon, E. A., Reichert, A. N., Schuh, K. M., Shand, N. A., Quinn, J. J., & Radke, A. K. (2023). *Sex, but not early life stress, effects on two-bottle choice alcohol drinking behaviors in mice*. https://doi.org/10.1101/2023.01.21.524642.

Planeta, C. S. (2013). Animal models of alcohol and drug dependence. *Revista Brasileira de Psiquiatria, 35*(suppl 2), S140–S146. https://doi.org/10.1590/1516-4446-2013-1149.

Planeta, C. S., & Cruz, F. C. (2005). Neurophysiological basis of tobacco dependence. *Archives of Clinical Psychiatry (São Paulo), 32*, 251–258.

Planeta, C. D. S., & Marin, M. T. (2002). Effect of cocaine on periadolescent rats with or without early maternal separation. *Brazilian Journal of Medical and Biological Research, 35*, 1367–1371.

Post, R. M., & Contel, N. R. (1983). Human and animal studies of cocaine: Implications for development of behavioral pathology. *Stimulants: Neurochemical, Behavioral and Clinical Perspectives, 1*, 169–203.

Prus, A. J., James, J. R., & Rosecrans, J. A. (2009). *Conditioned place preference. Em The ECPH Encyclopedia of Psychology (p. 1)*. Singapore: Springer Nature,−1. https://doi.org/10.1007/978-981-99-6000-2_178-1.

Raffa, R. B. (2008). *Planaria: A model for drug action and abuse* (0th edn.). CRC Press. https://doi.org/10.1201/9781498713597.

Rhodes, J. S., Best, K., Belknap, J. K., Finn, D. A., & Crabbe, J. C. (2005). Evaluation of a simple model of ethanol drinking to intoxication in C57BL/6J mice. *Physiology & Behavior, 84*(1), 53–63. https://doi.org/10.1016/j.physbeh.2004.10.007.

Roberts, A. (2000). Excessive ethanol drinking following a history of dependence animal model of allostasis. *Neuropsychopharmacology: Official Publication of the American College of Neuropsychopharmacology, 22*(6), 581–594. https://doi.org/10.1016/S0893-133X(99)00167-0.

Robinson, T. (1993). The neural basis of drug craving: An incentive-sensitization theory of addiction. *Brain Research Reviews, 18*(3), 247–291. https://doi.org/10.1016/0165-0173(93)90013-P.

Robinson, T. E., & Becker, J. B. (1986). Enduring changes in brain and behavior produced by chronic amphetamine administration: A review and evaluation of animal models of amphetamine psychosis. *Brain Research Reviews, 11*(2), 157–198.

Robinson, T. E., Browman, K. E., Crombag, H. S., & Badiani, A. (1998). Modulation of the induction or expression of psychostimulant sensitization by the circumstances surrounding drug administration. *Neuroscience & Biobehavioral Reviews, 22*(2), 347–354.

Rogers, J., Wiener, S. G., & Bloom, F. E. (1979). Long-term ethanol administration methods for rats: Advantages of inhalation over intubation or liquid diets. *Behavioral and Neural Biology, 27*(4), 466–486. https://doi.org/10.1016/S0163-1047(79)92061-2.

Rossi, N. A., & Reid, L. D. (1976). Affective states associated with morphine injections. *Physiological Psychology, 4*(3), 269–274. https://doi.org/10.3758/BF03332869.

Samson, H. H. (1986). Initiation of ethanol reinforcement using a sucrose-substitution procedure in food- and water-sated rats. *Alcoholism: Clinical and Experimental Research, 10*(4), 436–442. https://doi.org/10.1111/j.1530-0277.1986.tb05120.x.

Sanchis-Segura, C., Zghoul, T., & Spanagel, R. (2006). *Methods for behavioural assessment of drug-reinforcement and addictive features. Em M. Koch, Animal Models of Neuropsychiatric Diseases*. Published by Imperial College Press and Distributed by World Scientific Publishing Co., 181–222. https://doi.org/10.1142/9781860948022_0005.

Sanchis-Segura, C., & Spanagel, R. (2006). REVIEW: Behavioural assessment of drug reinforcement and addictive features in rodents: An overview. *Addiction Biology, 11*(1), 2–38. https://doi.org/10.1111/j.1369-1600.2006.00012.x.

Schulteis, G., Markou, A., Cole, M., & Koob, G. F. (1995). Decreased brain reward produced by ethanol withdrawal. *Proceedings of the National Academy of Sciences of the United States of America, 92*(13), 5880–5884. https://doi.org/10.1073/pnas.92.13.5880.

Sinclair, J. D., & Senter, R. J. (1967). Increased preference for ethanol in rats following alcohol deprivation. *Psychonomic Science, 8*(1), 11–12. https://doi.org/10.3758/BF03330642.

Skinner, B. F. (1938). The behavior of organisms: An experimental analysis. *BF Skinner Foundation.*

Spanagel, R. (2017). Animal models of addiction. *Dialogues in Clinical Neuroscience, 19*(3), 247–258. https://doi.org/10.31887/DCNS.2017.19.3/rspanagel.

Tallarida, C. S., Bires, K., Avershal, J., Tallarida, R. J., Seo, S., & Rawls, S. M. (2014). Ethanol and cocaine: Environmental place conditioning, stereotypy, and synergism in planarians. *Alcohol (Fayetteville, N. Y.), 48*(6), 579–586. https://doi.org/10.1016/j.alcohol.2014.07.006.

Tavares, G. E. B., Bianchi, P. C., Yokoyama, T. S., Palombo, P., & Cruz, F. C. (2023). Involvement of cortical projections to basolateral amygdala in context-induced reinstatement of ethanol-seeking in rats. *Behavioural Brain Research, 448*, 114435. https://doi.org/10.1016/j.bbr.2023.114435.

Thiele, T. E., & Navarro, M. (2014). "Drinking in the dark" (DID) procedures: A model of binge-like ethanol drinking in non-dependent mice. *Alcohol (Fayetteville, N. Y.), 48*(3), 235–241.

Tzschentke, T. M. (2007). REVIEW ON CPP: Measuring reward with the conditioned place preference (CPP) paradigm: update of the last decade. *Addiction Biology, 12*(3–4), 227–462. https://doi.org/10.1111/j.1369-1600.2007.00070.x.

Vendruscolo, L. F., Barbier, E., Schlosburg, J. E., Misra, K. K., Whitfield, T. W., Logrip, M. L., ... Koob, G. F. (2012). Corticosteroid-dependent plasticity mediates compulsive alcohol drinking in rats. *Journal of Neuroscience, 32*(22), 7563–7571. https://doi.org/10.1523/JNEUROSCI.0069-12.2012.

Vendruscolo, L. F., & Roberts, A. J. (2014). Operant alcohol self-administration in dependent rats: Focus on the vapor model. *Alcohol (Fayetteville, N. Y.), 48*(3), 277–286. https://doi.org/10.1016/j.alcohol.2013.08.006.

Vierkant, V., Xie, X., Wang, X., & Wang, J. (2023). Experimental models of alcohol use disorder and their application for pathophysiological investigations. *Current Protocols, 3*(6), e831. https://doi.org/10.1002/cpz1.831.

Wayner, M. J., Greenberg, I., Tartaglione, R., Nolley, D., Fraley, S., & Cott, A. (1972). A new factor affecting the consumption of ethyl alcohol and other sapid fluids. *Physiology & Behavior, 8*(2), 345–362. https://doi.org/10.1016/0031-9384(72)90383-6.

Wilkin, R. J. W., Lalor, P. F., Parker, R., & Newsome, P. N. (2016). Murine models of acute alcoholic hepatitis and their relevance to human disease. *The American Journal of Pathology, 186*(4), 748–760. https://doi.org/10.1016/j.ajpath.2015.12.003.

Wise, R. A. (1973). Voluntary ethanol intake in rats following exposure to ethanol on various schedules. *Psychopharmacologia, 29*(3), 203–210. https://doi.org/10.1007/BF00414034.

Wise, R. A. (1975). *Maximization of ethanol intake in the rat. Em M. M. Gross (Org.), Alcohol Intoxication and Withdrawal, 59*, Springer US, 279–294. https://doi.org/10.1007/978-1-4757-0632-1_19.

Yates, J. R. (2023). Quantifying conditioned place preference: A review of current analyses and a proposal for a novel approach. *Frontiers in Behavioral Neuroscience, 17*, 1256764. https://doi.org/10.3389/fnbeh.2023.1256764.

Yunusoğlu, O. (2021). Resveratrol impairs acquisition, reinstatement and precipitates extinction of alcohol-induced place preference in mice. *Neurological Research, 43*(12), 985–994. https://doi.org/10.1080/01616412.2021.1948749.

Further reading

Colombo, G., Serra, S., Brunetti, G., Gomez, R., Melis, S., Vacca, G., ... Gessa, G. (2002). Stimulation of voluntary ethanol intake by cannabinoid receptor agonists in ethanol-preferring sP rats. *Psychopharmacology, 159*(2), 181–187. https://doi.org/10.1007/s002130100887.

Cunningham, C. L., Bakner, L., Schuette, L. M., & Young, E. A. (2021). Morphine and ethanol pretreatment effects on expression and extinction of ethanol-induced conditioned place preference and aversion in mice. *Psychopharmacology, 238*(1), 55–66. https://doi.org/10.1007/s00213-020-05658-x.

Cunningham, C. L., Clemans, J. M., & Fidler, T. L. (2002). Injection timing determines whether intragastric ethanol produces conditioned place preference or aversion in mice. *Pharmacology, Biochemistry, and Behavior, 72*(3), 659–668. https://doi.org/10.1016/S0091-3057(02)00734-7.

Cunningham, C. L., & Noble, D. (1992). Conditioned activation induced by ethanol: Role in sensitization and conditioned place preference. *Pharmacology, Biochemistry, and Behavior, 43*(1), 307–313. https://doi.org/10.1016/0091-3057(92)90673-4.

Cunningham, C. L., Okorn, D. M., & Howard, C. E. (1997). Interstimulus interval determines whether ethanol produces conditioned place preference or aversion in mice. *Animal Learning & Behavior, 25*(1), 31–42. https://doi.org/10.3758/BF03199022.

Cunningham, C. L., Tull, L. E., Rindal, K. E., & Meyer, P. J. (2002). Distal and proximal pre-exposure to ethanol in the place conditioning task: Tolerance to aversive effect, sensitization to activating effect, but no change in rewarding effect. *Psychopharmacology, 160*(4), 414–424. https://doi.org/10.1007/s00213-001-0990-1.

Gage, G. A., Muench, M. A., Jee, C., Kearns, D. N., Chen, H., & Tunstall, B. J. (2023). Intermittent-access operant alcohol self-administration promotes binge-like drinking and drinking despite negative consequences in male and female heterogeneous stock rats. *Neuropharmacology, 235*, 109564. https://doi.org/10.1016/j.neuropharm.2023.109564.

Vena, A. A., Zandy, S. L., Cofresí, R. U., & Gonzales, R. A. (2020). Behavioral, neurobiological, and neurochemical mechanisms of ethanol self-administration: A translational review. *Pharmacology & Therapeutics, 212*, 107573. https://doi.org/10.1016/j.pharmthera.2020.107573.

CHAPTER TWELEVE

GABAergic compounds for the treatment of alcohol use disorder

Laís F. Berro*, **James K. Rowlett, and Donna M. Platt**
Department of Psychiatry and Human Behavior, Center for Innovation and Discovery in Addictions, University of Mississippi Medical Center, Jackson, MS, United States
*Corresponding author. e-mail address: lberro@umc.edu

Contents

Abstract

Decades of research have implicated the gamma-aminobutyric acid (GABA)ergic system as one of the main mediators of the behavioral effects of alcohol. Of importance, the addiction-related effects of alcohol also have been shown to be mediated in part by GABAergic systems, raising the possibility that pharmacotherapies targeting GABAergic receptors may be promising candidates for the treatment of alcohol use disorder (AUD). Alcohol modulates the activity of GAB_{AA} and GA_BA_B receptors, and studies show that compounds targeting some of those receptors may decrease the addiction-related behavioral effects of alcohol. Specifically, drugs that share similar pharmacological properties with alcohol, such as positive allosteric modulators (PAMs) of GAB_{AA} and GA_BA_B receptors, have been proposed as substitution therapies for AUD. Available evidence also suggests that negative allosteric modulators (NAMs) of GABAergic receptors may be potential therapeutics for AUD, although this effect is selective for specific receptor subtypes. Therefore, this Chapter reviews the available evidence on the use of GABAergic compounds for the treatment of AUD. Several GAB_{AA} and GA_BA_B ligands show promising results, with a particularly positive ther apeutic profile demonstrated for α5GAB_{AA} receptor NAMs, α4/6δGAB_{AA} receptor modulators (both positive and negative, including neurosteroids), and GA_BA_B receptor

International Review of Neurobiology, Volume 178
ISSN 0074-7742, https://doi.org/10.1016/bs.irn.2024.08.001

PAMs. As newer and better GABAergic compounds become available, future research should focus on understanding how these ligands can modulate different clinical symptoms of AUD, with potential new areas of research encompassing alcohol withdrawal syndrome and AUD-related insomnia.

Decades of research have implicated the central nervous system (CNS) gamma-aminobutyric acid (GABA)ergic system as one of the main mediators of the behavioral effects of alcohol. Alcohol promotes inhibition by potentiating the effects of GABA and enhancing its inhibitory effects (Uusi-Oukari & Korpi, 2024). The effects of alcohol on GABAergic receptors have been implicated in alcohol-induced sedation, anxiolysis, loss of inhibition and relaxation (Berro et al., 2019; Chandler et al., 2017; Porjesz & Rangaswamy, 2007). Of importance, the addiction-related effects of alcohol also have been shown to be mediated, in part, by GABAergic systems, raising the possibility that pharmacotherapies targeting GABAergic receptors may be promising candidates for the treatment of alcohol use disorder (AUD).

Alcohol enhances GABA neurotransmission via interactions with GABAergic receptors (Uusi-Oukari & Korpi, 2024). Alcohol modulates the activity of $GABA_A$ receptors, and studies show that $GABA_A$ receptor mechanism are involved in the addiction-related effects of alcohol in animals and humans (for review, see Chandler, Overton, Rüedi-Bettschen, & Platt, 2018). $GABA_A$ receptors in the central nervous system are pentamers constituted from structurally distinct proteins, with each protein family comprised of different subunits from at least 8 different families of distinct proteins (6 α, 3 β, 3 γ, 1 δ, 1 ε, 1 θ, 1 π, and 3 ρ subunits). However, the majority of $GABA_A$ receptors consist of α, β, and γ subunit families arranged in the sequence $\alpha\beta\alpha\beta\gamma$ (Fig. 1). $GABA_A$ receptors are widely distributed throughout the CNS, and their activation leads to chloride channel opening and consequent hyperpolarization of the cell (Lüddens, Seeburg, & Korpi, 1994). While the specifics of how alcohol interacts with $GABA_A$ receptors are a matter of debate, studies conclusively show that alcohol-induced potentiation of GABA-induced currents is due to an increase in the frequency and duration of $GABA_A$ receptor channel opening (Tatebayashi, Motomura, & Narahashi, 1998).

Alcohol also interacts with $GABA_B$ receptors, although the specific mechanisms underlying this interaction are poorly understood (Evenseth, Gabrielsen, & Sylte, 2020). Importantly, $GABA_B$ receptors have been shown to modulate alcohol-related effects (for review, see Maccioni & Colombo, 2009). The $GABA_B$ receptor is composed of $GABA_{B1}$ and

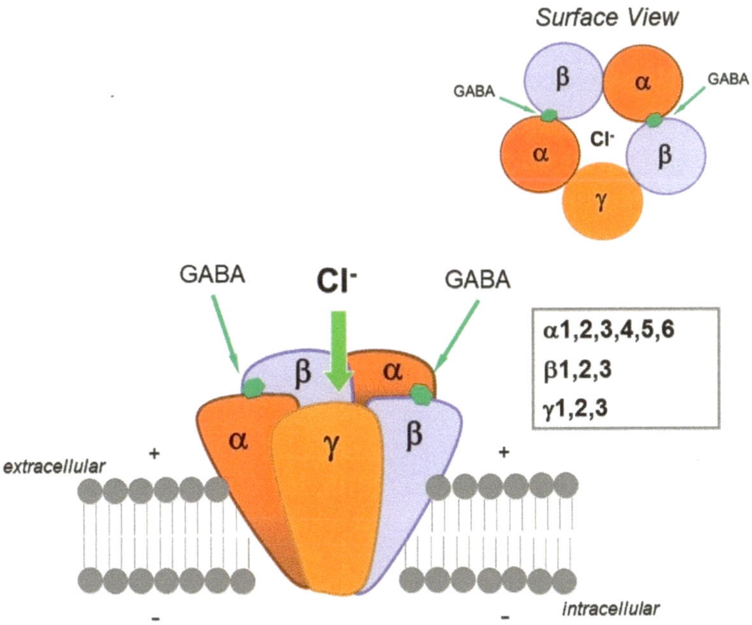

Fig. 1 $GABA_A$ receptor structure and α, β, and γ subunit family subtypes.

$GABA_{B2}$ subunits, and it is a G-protein coupled receptor (Munk et al., 2016). Multiple isoforms of the $GABA_{B1}$ subunit exist, and the most abundant are $GABA_{B1a}$ and $GABA_{B1b}$ (Kaupmann et al., 1997). Of note, the $GABA_B$ receptor contains both an orthosteric (GABA) and an allosteric binding sites (Evenseth et al., 2020), which has allowed for the development of allosteric modulators of this receptor.

Given the existence of multiple $GABA_A$ and $GABA_B$ receptor subtypes, pharmacological tools with selectivity for specific receptor subtypes have been used to tease apart GABAergic mechanisms in AUD (for review, see Chandler et al., 2018; Maccioni & Colombo, 2009). Of note, in addition to mechanistic behavioral pharmacology research, such tools also have been considered as potential pharmacotherapies for AUD. Specifically, drugs that share similar pharmacological properties with alcohol, such as positive allosteric modulators (PAMs) of $GABA_A$ and $GABA_B$ receptors, have been proposed as substitution therapies for AUD (Chick & Nutt, 2012). Available evidence also suggests that negative modulators of GABAergic receptors may be potential therapeutics for AUD, although this effect is selective for specific receptor subtypes. This Chapter will review the available evidence on the use of GABAergic compounds for the treatment of AUD.

1. $GABA_A$ compounds

1.1 Non-selective $GABA_A$ compounds

$GABA_A$ receptor ligands have been used for decades for the treatment of aspects of AUD, particularly alcohol withdrawal syndrome. The most largely accepted strategy to treat alcohol dependence consists of a preliminary detoxification period, with abstinence being maintained using PAMs of $GABA_A$ receptors as pharmacotherapies. Barbiturates, which act as $GABA_A$ receptor PAMs, were used extensively for the management of alcohol withdrawal, until they were replaced by the safer $GABA_A$ PAMs, the benzodiazepines (Erstad & Cotugno, 1995). Currently, conventional long-acting benzodiazepines, such as diazepam, lorazepam and chlordiazepoxide, are considered the first-line therapy for treating alcohol withdrawal syndrome (Amato, Minozzi, Vecchi, & Davoli, 2010). Benzodiazepines are effective at reducing seizures, delirium tremens, and withdrawal scores in AUD patients (Cohen, Alexander, & Holt, 2022).

Benzodiazepines also have been studied for the treatment of AUD outside of the context of withdrawal. In early studies, chlordiazepoxide showed the highest treatment retention rate compared to other pharmacotherapies in patients with AUD (Kissin, 1975). In a more recent study, a 30-day alcohol detoxification protocol with diazepam significantly reduced heavy alcohol drinking, alcohol craving, self-reported anxiety, and clinician-assessed anxiety in patients with AUD (Simioni, Labreuche, Behal, Cottencin, & Rolland, 2017), suggesting that long-term treatment with benzodiazepines past initial withdrawal may decrease drinking outcomes after alcohol cessation. These findings are corroborated by studies in nonhuman primates showing that pretreatment with the benzodiazepine triazolam significantly decreased blood alcohol levels at higher doses (Sawyer et al., 2014).

However, the extended use of benzodiazepines in AUD is limited due to (1) their potential for misuse, (2) their potential to induce tolerance and physical dependence, and (3) their interactions with alcohol to promote CNS depression (Schuster & Humphries, 1981). In fact, nonhuman primate studies have shown that low doses of the benzodiazepine triazolam increased alcohol drinking (Sawyer et al., 2014), a finding that is corroborated by clinical studies with other benzodiazepines (Zack, Poulos, & Woodford, 2006). Drug discrimination studies also show that benzodiazepines fully or partially substitute for alcohol in nonhuman primates (Grant et al., 2000; Berro et al., 2019; Platt et al., 2005) and humans

(Jackson, Stephens, & Duka, 2003; Zack, Toneatto, & MacLeod, 1999), emphasizing that non-selective $GABA_A$ PAMs share discriminative stimulus effects with alcohol. Of note, compelling evidence suggests that the discriminative stimulus effects of drugs in nonhuman animals can be similar to their subjective effects in humans (Schuster & Johanson, 1988). While the drug discrimination findings favor the idea of a benzodiazepine substitution therapy for AUD, they also raise concerns over potential dose-dependent drug interactions and misuse. Commensurate with these concerns, current guidelines recommend that detoxification treatment using a benzodiazepine should not exceed 2 weeks (Rolland et al., 2016).

Given the apparent problems observed with the use of non-selective $GABA_A$ PAMs on AUD-related outcomes, more recent studies have focused on the investigation of $GABA_A$ PAMs with selectivity for one or more α subunits. Distinct α subunits have been shown to not only determine the behavioral effects of benzodiazepines (Engin, 2023; Licata & Rowlett, 2008), but also to contribute differentially to the addiction-related effects of alcohol (for review, see Chandler et al., 2018). These studies have made significant advances in identifying $GABA_A$ receptor mechanisms involved in the addiction-related effects of alcohol, and also have identified potential pharmacological tools that might have therapeutic effects for AUD without the negative side effects of conventional benzodiazepines.

1.2 Compounds with selectivity for α1$GABA_A$ receptors

The most widely studied PAM with selectivity for α1-containing $GABA_A$ receptors (α1$GABA_A$ receptors), the sleep aid zolpidem, has been investigated both in pre-clinical and clinical alcohol studies. Zolpidem showed only partial substitution for alcohol in drug discrimination studies in rats (Bienkowski, Iwinska, Stefanski, & Kostowski, 1997; Sanger, 1997), squirrel monkeys (Platt et al., 2005) and macaques (Berro et al., 2019; Helms, Rogers, Waters, & Grant, 2008), indicating that it only partially shares discriminative stimulus effects with alcohol. In nonhuman primates, zolpidem had no effect on alcohol drinking at behaviorally-relevant doses (Sawyer et al., 2014). These findings are in agreement with a clinical study showing that zolpidem did not enhance the subject-rated "drug liking" and "drug strength perception" effects of alcohol (Wilkinson, 1998). Together, these studies suggest that alcohol and zolpidem do not fully share similar mechanisms and, therefore, zolpidem may not be effective as a substitution therapy for AUD.

Other experimental α1GABA$_A$-selective and clinically-available α1GABA$_A$-preferring PAMs have been investigated in alcohol studies, with similar findings. Drug discrimination studies show partial overlap between the discriminative stimulus effects of alcohol and the α1-preferring PAM zaleplon in rats (Bienkowski et al., 1997; Sanger, 1997) and monkeys (Platt et al., 2005). The study in monkeys also showed similar effects for the α1-selective partial PAM CL 218,872 (Platt et al., 2005). Of note, the α1-selective antagonist βCCt had no effects on alcohol drinking (Sawyer et al., 2014) and did not alter alcohol discrimination (Platt et al., 2005) in rhesus and squirrel monkeys, respectively. On the other hand, the α1-preferring antagonists 3-PBC and 3-ISOPBC decreased alcohol self-administration in baboons (Holtyn, Tiruveedhula, Stephen, Cook, & Weerts, 2017; Kaminski, Van Linn, Cook, Yin, & Weerts, 2013). Finally, a compound showing zero efficacy at α1GABA$_A$ receptors (L-838,417, selective efficacy at α2/3/5GABA$_A$ receptors) fully substituted for alcohol in nonhuman primate studies, confirming previous findings (e.g., Platt et al., 2005) showing that modulation of α1GABA$_A$ receptors is not necessary to reproduce the discriminative stimulus effects of alcohol. Together, these data clearly demonstrate a lack of a role for α1GABA$_A$ receptors in the addiction-related effects of alcohol.

While compounds with selectivity for α1GABA$_A$ receptors do not seem to be relevant therapeutic targets for AUD, the α1-selective drug zolpidem is commonly prescribed to treat insomnia (Foulds et al., 2018). Sleep impairment is highly prevalent among patients with AUD and represents a risk factor for relapse (Koob & Colrain, 2020). Alarmingly, combined zolpidem-alcohol misuse is associated with a high risk of intensive care unit admission in patients visiting the emergency department (Zosel, Osterberg, & Mycyk, 2011). In fact, zolpidem is among the most common drugs leading to emergency department visits (Geller, Zhou, Budnitz, Lovegrove, & Dal Pan, 2020), and long-term use of zolpidem is associated with tolerance and dependence (Moore & Mattison, 2018). Therefore, in addition to not being an effective medication for AUD, zolpidem-alcohol co-use may represent a significant public health concern, particularly in light of the recent increased prevalence of zolpidem use and misuse in the general population (Khan et al., 2022).

1.3 Compounds with selectivity for α2/3GABA$_A$ receptors

Because of the structural similarities between α2- and α3-containing GABA$_A$ receptors (α2GABA$_A$ and α3GABA$_A$ receptors, respectively), identifying compounds with selective affinity for either of these compounds has been a challenge (Whiting et al., 1999). Recent advances have allowed for the identification of compounds with selective efficacy ("functional" selectivity) for α2GABA$_A$ and/or α3GABA$_A$ receptors. These new pharmacological tools are of great importance given the evidence implicating an association between *GABRA2* and *GABRA3* – the genes encoding the α2 and α3 subunits, respectively – with an increased risk for developing AUD (for review see Koulentaki & Kouroumalis, 2018). Several preclinical and clinical studies have demonstrated a role for α2GABA$_A$ and α3GABA$_A$ receptors in the addiction-related effects of alcohol, and most of the evidence comes from human genetics studies or genetically altered mouse studies (for review see Chandler et al., 2018).

The first pharmacological investigation into the effects of α2GABA$_A$ and/or α3GABA$_A$ receptor ligands on the addiction-related effects of alcohol was conducted in rhesus monkeys. These studies demonstrated that, similar to non-selective compounds, the GABA$_A$ PAMs L-838,417 (selective efficacy at α2/3/5GABA$_A$ receptors), HZ-166 (selective efficacy at α2/3GABA$_A$ receptors) and YT-III-31 (selective efficacy at α3GABA$_A$ receptors) all shared discriminative stimulus effects with alcohol (Berro et al., 2019). Moreover, compounds with selective efficacy at α2/3GABA$_A$ or at α3GABA$_A$ receptors also increased alcohol drinking in monkeys (Berro et al., 2019). While these findings indicate that α2GABA$_A$ and α3GABA$_A$ receptors play an important role in key addiction-related effects of alcohol, they also show primarily a facilitatory role for α3GABA$_A$ and potentially α2GABA$_A$ receptors in several addiction-related effects of alcohol in monkeys. Unfortunately, no selective α2/3GABA$_A$ receptor antagonists have been identified and tested for their ability to mimic and/or modulate the effects of alcohol.

Importantly, studies have shown that α2/3GABA$_A$-selective PAMs have decreased reinforcing effectiveness (i.e., decrease potential for misuse) compared to conventional benzodiazepines (Shinday et al., 2013). Such compounds also retain the anxiolytic-like (Meng et al., 2020; Rowlett, Lelas, Tornatzky, & Licata, 2006) and anticonvulsant effects of GABA$_A$ PAMs (Knutson et al., 2020; Ping et al., 2023). Finally, because studies in nonhuman primates suggest that α2/3GABA$_A$ receptors do not seem to be involved in the physical dependence-like effects of benzodiazepines (Ator, Atack, Hargreaves, Burns, & Dawson, 2010; Duke et al., 2021), it is possible that compounds with selective efficacy at α2/3GABA$_A$ receptors may represent improved therapeutic tools for the treatment of alcohol withdrawal syndrome compared to non-selective benzodiazepines. To the best of our knowledge, no studies to date have investigated the efficacy of such compounds in AUD patients, and future research is needed to identify whether α2/3GABA$_A$-selective PAMs may be beneficial in the context of AUD.

1.4 Compounds with selectivity for α5GABA$_A$ receptors

Several compounds with selectivity for α5-containing GABA$_A$ (α5GABA$_A$) receptors exist and have been tested in alcohol studies, adding to the mounting evidence supporting a role for α5GABA$_A$ receptors mechanisms in the addiction-related effects of alcohol (for review, see Chandler et al., 2018). The α5GABA$_A$-selective PAM QH-ii-066 has been shown to increase alcohol drinking in rhesus monkeys (Rüedi-Bettschen et al., 2013), but not in rats (Chandler et al., 2019). QH-ii-066 fully substituted for alcohol in drug discrimination studies in squirrel monkeys (Platt et al., 2005) and rhesus monkeys (Berro et al., 2019), an effect that was also shown for the partial α5GABA$_A$-selective PAM panadiplon (Platt et al., 2005). Pretreatment with QH-ii-066 also enhanced alcohol-seeking behavior (Chandler et al., 2019) and relapse-like drinking behavior (Chandler et al., 2021) in rats.

While PAMs enhance the effects of alcohol in animal models, negative allosteric modulators (NAMs) with selectivity for α5GABA$_A$ receptors also have been tested in alcohol models, with promising results. Earlier studies investigated the effects of the α5-preferring NAM Ro 15-4513, showing that it attenuates the discriminative stimulus effects of alcohol in mice (Middaugh, Bao, Becker, & Daniel, 1991; Rees & Balster, 1988; Stephens, Pistovcakova, Worthing, Atack, & Dawson, 2005), rats (Gatto & Grant, 1997) and cynomologus monkeys (Helms, Rogers, & Grant, 2009). More recent studies investigated the effects of the experimental α5GABA$_A$ NAM L-655,708. In rats, L-655,708 attenuated cue-induced reinstatement of alcohol seeking and decreased alcohol drinking (Chandler et al., 2019) and relapse-like drinking behavior (Chandler et al., 2021). In nonhuman primates, L-655,708 selectively inhibited alcohol (but not sucrose) drinking in rhesus monkeys (Rüedi-Bettschen et al., 2013), and decreased the discriminative stimulus effects of alcohol in squirrel monkeys (Platt et al., 2005). In addition, the α5GABA$_A$ NAM RY-023 has been shown to decrease alcohol drinking in rats (Chandler et al., 2019; Cook et al., 2005) and to facilitate extinction of alcohol seeking and reduce reactivity to alcohol-associated cues in rhesus monkeys (Pareek et al., 2023). Finally, the α5GABA$_A$ NAM RY-024 decreased alcohol self-administration and blocked the sedative-motor effects of alcohol in rats (McKay et al., 2004).

While several experimental α5GABA$_A$ NAMs are available, the only compound that has been tested in humans is α5IA. In a clinical study in healthy volunteers, α5IA did not alter alcohol-induced intoxication and slowing of eye movements, and subject-rated alcohol liking (Nutt et al., 2007). However, α5IA significantly improved alcohol-induced learning impairment and partially reversed the sedative effects of alcohol (Nutt et al., 2007). Of note, the discrepancy between the limited clinical data and the preclinical findings on α5GABA$_A$ NAMs may be due to the receptor binding profile of α5IA (25% efficacy at α5GABA$_A$ receptors, Sternfeld et al., 2004) compared to the other experimental α5GABA$_A$ NAMs tested, particularly RY023 (55% efficacy, June et al., 2001) and L-655,708 (35% efficacy, Atack et al., 2006). Clinical studies are needed to determine whether selective NAMs with greater efficacy at α5GABA$_A$ receptors will show improved therapeutic potential in patients with AUD, as suggested by the mounting pre-clinical evidence.

1.5 Compounds with selectivity for α4/6δ GABA$_A$ receptors

The majority of α4- and α6-containing GABA$_A$ receptors are co-expressed with the δ subunit (α4/6δGABA$_A$ receptors), a receptor conformation that is sensitive to alcohol and has been shown to mediate some of its effects (for review, see Chandler et al., 2018). Specifically, regarding the addiction-related effects of alcohol, the partial agonist at α4/6δGABA$_A$ receptors gaboxadol (THIP) has been shown to enhance acquisition of voluntary alcohol intake (Smith, Robidoux, & Amit, 1992) and to increase alcohol drinking and preference (Boyle, Segal, Smith, & Amit, 1993) in rats. Mouse studies, on the other hand, show that gaboxadol decreases alcohol drinking (Melón, Nolan, Colar, Moore, & Boehm, 2017; Moore et al., 2007; Ramaker, Ford, Fretwell, & Finn, 2011). In drug discrimination studies, gaboxadol does not produce alcohol-like effects in mice (Shelton & Grant, 2002) or monkeys (Grant et al., 2000). Therefore, the potential of gaboxadol as a pharmacotherapy for AUD is still up for debate and further studies are needed to conclusively understand the extent to which it alters the addiction-related effects of alcohol.

Neuroactive steroids (neurosteroids), which act at α4/6δGABA$_A$ receptors (Lambert, Belelli, Peden, Vardy, & Peters, 2003), also have been studied in the context of AUD. Neurosteroids differ from conventional benzodiazepines because, in addition to being non-selective at the benzodiazepine GABA$_A$ receptor binding sites, they also act as α4/6δGABA$_A$ PAMs (Lambert et al., 2003). These compounds, including pregnanolone, allopregnanolone and its synthetic analogue ganaxolone, have been shown to have alcohol-like discriminative stimulus effects (Bowen, Purdy, & Grant, 1999; Grant, Azarov, Bowen, Mirkis, & Purdy, 1996; Grant, Azarov, Shively, & Purdy, 1997; Grant, Helms, Rogers, & Purdy, 2008) and increase alcohol seeking in rodents (Ramaker, Strong, Ford, & Finn, 2012). However, alcohol-induced decreases in allopregnanolone have been associated with increased alcohol craving (Pierucci-Lagha et al., 2006). In agreement, a recent study showed that treatment with pregnanolone decreased stress- and alcohol cue- induced craving and stress-induced anxiety in patients with AUD (Milivojevic et al., 2023). Of note, neuroactive steroids that act as $\alpha4/6\delta GABA_A$ NAMs, such as epipregnanolone and epiallopregnanolone, reduce the discriminative stimulus effects of alcohol (Helms, Rossi, & Grant, 2012).

Given the evidence showing interactions between alcohol use and neurosteroid levels, and research showing that neurosteroids can modulate the subjective and reinforcing effects of alcohol (for review, see Gatta, Camussi, Auta, Guidotti, & Pandey, 2022; Morrow, Boero, & Porcu, 2020), it is clear that neurosteroids should be further investigated as novel treatment options for patients with AUD. The recent FDA approval of synthetic (ganaxolone, zuranolone) and endogenous (brexanolone) neurosteroids for conditions other than AUD (i.e., seizure disorder, postpartum depression) emphasizes the growing interest in the use of these substances as pharmacotherapies, and may pave the way for future AUD clinical studies.

2. $GABA_B$ compounds

Over the last several decades, the use of $GABA_B$ receptor agonists and PAMs as therapeutics for AUD has been gaining interest, as the evidence supporting their therapeutic potential continues to grow. Several pre-clinical studies have shown that the $GABA_B$ receptor agonist, baclofen, decreases acquisition and maintenance of alcohol drinking, binge-like alcohol drinking, alcohol-seeking behavior, and the reinforcing, motivational, locomotor stimulant, and rewarding effects of alcohol in rodents and monkeys (for review, see Agabio & Colombo, 2014; Holtyn & Weerts, 2022). For instance, baclofen reduced alcohol drinking and seeking, as well as the motivational properties of alcohol, in baboons (Duke, Kaminski, & Weerts, 2014).

Importantly, baclofen also has been shown to decrease the symptoms of alcohol withdrawal syndrome in rodents (Colombo et al., 2000). These data corroborate preliminary clinical findings and randomized-controlled trials demonstrating the effectiveness of baclofen in the treatment of alcohol withdrawal syndrome in humans (for review, see Agabio & Colombo, 2014). While clinical studies investigating the effectiveness of baclofen in the treatment of AUD have yielded conflicting results (for review, see Augier, 2021), two recent randomized clinical trials showed encouraging effects. Treatment with baclofen significantly reduced relapse rates and promoted abstinence (Morley et al., 2018; Rigal et al., 2020), and also decreased daily alcohol intake (Rigal et al., 2020) in patients with AUD. However, the clinical use of baclofen is still limited due to the contradictory literature and due to its associated side effects (Rigal et al., 2020), including the development of tolerance that may lead to increased risk for lethal intoxication (Léger, Brunet, Le Roux, Lerolle, & Boels, 2017).

Given the limitations in the use of $GABA_B$ agonists such as baclofen due to potential side effects (see Augier, 2021), more recent studies have focused on the investigation of $GABA_B$ PAMs as alternative pharmacotherapies targeting similar mechanisms. These studies show that $GABA_B$ PAMs, such as CGP7930 and GS39783, reduced excessive alcohol drinking, relapse- and binge-like drinking, operant alcohol self-administration, reinstatement of alcohol seeking, and alcohol-induced conditioned place preference in rats and mice (for review, see Augier, 2021; Colombo, 2024). While these are newer compounds with fewer clinical studies, the $GABA_B$ PAM ASP8062 is currently in clinical trials, with Phase 1 studies showing that it was well tolerated, with no safety concerns or major side effects (Walzer, Marek, Wu, Nagata, & Han, 2020), including when administered in combination with alcohol (Ito et al., 2022). Therefore, $GABA_B$ PAMs provide a new and exciting area of research in AUD, and future clinical studies are needed to determine whether the pre-clinical data will translate to AUD patients.

3. Conclusions

GABAergic compounds have been investigated in the context of alcohol and AUD for decades. These studies have allowed for the understanding of the mechanisms underlying the different behavioral effects of alcohol, but they also have contributed to the identification of potential new pharmacotherapies for AUD. Based on the evidence reviewed in this Chapter, several $GABA_A$ and $GABA_B$ ligands show promising results, with particularly encouraging therapeutic profiles demonstrated for $\alpha5GABA_A$ receptor NAMs, $\alpha4/6\delta GABA_A$ receptor modulators (both positive and negative, including neurosteroids), and $GABA_B$ receptor PAMs. As newer GABAergic compounds become available, future research should focus on understanding how these ligands can modulate different clinical symptoms of AUD, with potential new areas of research encompassing alcohol withdrawal syndrome and AUD-related insomnia.

Acknowledgements

This work was supported by the National Institutes of Health (DA011792, DA043204, DA049886, DA058666 and AA029023).

References

Agabio, R., & Colombo, G. (2014). GABAB receptor ligands for the treatment of alcohol use disorder: preclinical and clinical evidence. *Frontiers in Neuroscience, 8*, 140.

Amato, L., Minozzi, S., Vecchi, S., & Davoli, M. (2010). Benzodiazepines for alcohol withdrawal. *Cochrane Database of Systematic Reviews (Online)*, (3), CD005063.

Atack, J. R., Bayley, P. J., Seabrook, G. R., Wafford, K. A., McKernan, R. M., & Dawson, G. R. (2006). L-655,708 enhances cognition in rats but is not proconvulsant at a dose selective for 5-containing GABAA receptors. *Neuropharmacology, 51*, 1023–1029.

Ator, N. A., Atack, J. R., Hargreaves, R. J., Burns, H. D., & Dawson, G. R. (2010). Reducing abuse liability of GABAA/benzodiazepine ligands via selective partial efficacy at alpha1 and alpha2/3 subtypes. *The Journal of Pharmacology and Experimental Therapeutics, 332*, 4–16.

Augier, E. (2021). Recent advances in the potential of positive allosteric modulators of the GABAB receptor to treat alcohol use disorder. *Alcohol and Alcoholism (Oxford, Oxfordshire), 56*(2), 139–148.

Berro, L. F., Rüedi-Bettschen, D., Cook, J. E., Golani, L. K., Li, G., Jahan, R., ... Platt, D. M. (2019). GABA(A) receptor subtypes and the abuse-related effects of ethanol in rhesus monkeys: Experiments with selective positive allosteric modulators. *Alcoholism, Clinical and Experimental Research, 43*(5), 791–802.

Bienkowski, P., Iwinska, K., Stefanski, R., & Kostowski, W. (1997). Discriminative stimulus properties of ethanol in the rat: Differential effects of selective and nonselective benzodiazepine receptor agonists. *Pharmacology, Biochemistry, and Behavior, 58*, 969–973.

Bowen, C. A., Purdy, R. H., & Grant, K. A. (1999). Ethanol-like discriminative stimulus effects of endogenous neuroactive steroids: Effects of ethanol training dose and dosing procedure. *The Journal of Pharmacology and Experimental Therapeutics, 289*, 405–411.

Boyle, A. E., Segal, R., Smith, B. R., & Amit, Z. (1993). Bidirectional effects of GABAergic agonists and antagonists on maintenance of voluntary ethanol intake in rats. *Pharmacology, Biochemistry, and Behavior, 46*, 179–182.

Chandler, C. M., Follett, M. E., Porter, N. J., Liang, K. Y., Vallender, E. J., Miller, G. M., ... Platt, D. M. (2017). Persistent negative effects of alcohol drinking on aspects of novelty-directed behavior in male rhesus macaques. *Alcohol (Fayetteville, N. Y.), 63*, 19–26.

Chandler, C. M., Overton, J. S., Rüedi-Bettschen, D., & Platt, D. M. (2018). GABAA receptor subtype mechanisms and the abuse-related effects of ethanol: Genetic and pharmacological evidence. *Handbook of Experimental Pharmacology, 248*, 3–27.

Chandler, C. M., Reeves-Darby, J., Jones, S. A., McDonald, J. A., Li, G., Rahman, M. T., ... Platt, D. M. (2019). α5GABAA subunit-containing receptors and sweetened alcohol cue-induced reinstatement and active sweetened alcohol self-administration in male rats. *Psychopharmacology, 236*(6), 1797–1806.

Chandler, C. M., Reeves-Darby, J., Jones, S. A., Li, G., Rahman, M. T., Cook, J. M., & Platt, D. M. (2021). Modulation of relapse-like drinking in male Sprague-Dawley rats by ligands targeting the α5GABAA receptor. *Neuropharmacology, 199*, 108785.

Chick, J., & Nutt, D. J. (2012). Substitution therapy for alcoholism: time for a reappraisal? *Journal of Psychopharmacology (Oxford, England), 26*, 205–212.

Cohen, S. M., Alexander, R. S., & Holt, S. R. (2022). The spectrum of alcohol use: Epidemiology, diagnosis, and treatment. *The Medical Clinics of North America, 106*(1), 43–60.

Colombo, G. (2024). Positive allosteric modulators of the GABAB receptor: A new class of ligands with therapeutic potential for alcohol use disorder. *Alcohol and Alcoholism (Oxford, Oxfordshire), 59*(3), agae018.

Colombo, G., Agabio, R., Carai, M. A., Lobina, C., Pani, M., Reali, R., ... Gessa, G. L. (2000). Ability of baclofen in reducing alcohol intake and withdrawal severity: I–Preclinical evidence. *Alcoholism, Clinical and Experimental Research, 24*(1), 58–66.

Cook, J. B., Foster, K. L., Eiler, W. J., 2nd, McKay, P. F., Woods, J., 2nd, Harvey, S. C., ... June, H. L. (2005). Selective GABAA alpha5 benzodiazepine inverse agonist antagonizes the neurobehavioral actions of alcohol. *Alcoholism, Clinical and Experimental Research, 29*(8), 1390–1401.

Duke, A. N., Kaminski, B. J., & Weerts, E. M. (2014). Baclofen effects on alcohol seeking, self-administration and extinction of seeking responses in a within-session design in baboons. *Addiction Biology, 19*(1), 16–26.

Duke, A. N., Tiruveedhula, V. V. N. P. B., Sharmin, D., Knutson, D. E., Cook, J. M., Platt, D. M., & Rowlett, J. K. (2021). Tolerance and dependence following chronic alprazolam treatment in rhesus monkeys: Role of GABA(A) receptor subtypes. *Drug and Alcohol Dependence, 228*, 108985.

Engin, E. (2023). GABAA receptor subtypes and benzodiazepine use, misuse, and abuse. *Frontiers in Psychiatry, 13*, 1060949.

Erstad, B. L., & Cotugno, C. L. (1995). Management of alcohol withdrawal. *American Journal of Health-System Pharmacy: AJHP: Official Journal of the American Society of Health-System Pharmacists, 52*, 697–709.

Evenseth, L. S. M., Gabrielsen, M., & Sylte, I. (2020). The GABA(B) receptor-structure, ligand binding and drug development. *Molecules (Basel, Switzerland), 25*(13), 3093.

Foulds, J. A., Manning, V., Garfield, J. B. B., Allsop, S. J., Lam, T., Arunogiri, S., & Lubman, D. I. (2018). Prescribed sedative and other psychotropic medication use among clients attending alcohol and other drug treatment. *Drug and Alcohol Review, 37*(6), 738–742.

Gatta, E., Camussi, D., Auta, J., Guidotti, A., & Pandey, S. C. (2022). Neurosteroids (allopregnanolone) and alcohol use disorder: From mechanisms to potential pharmacotherapy. *Pharmacology & Therapeutics, 240*, 108299.

Gatto, G. J., & Grant, K. A. (1997). Attenuation of the discriminative stimulus effects of ethanol by the benzodiazepine partial inverse agonist Ro 15-4513. *Behavioural Pharmacology, 8*, 139–146.

Geller, A. I., Zhou, E. H., Budnitz, D. S., Lovegrove, M. C., & Dal Pan, G. J. (2020). Changes in emergency department visits for zolpidem-attributed adverse drug reactions after FDA Drug Safety Communications. *Pharmacoepidemiology and Drug Safety, 29*(3), 352–356.

Grant, K. A., Azarov, A., Bowen, C. A., Mirkis, S., & Purdy, R. H. (1996). Ethanol-like discriminative stimulus effects of the neurosteroid 3a-hydroxy-5a-pregnan-20-one in female Macaca fascicularis monkeys. *Psychopharmacology, 124*, 340–346.

Grant, K. A., Azarov, A., Shively, C. A., & Purdy, R. H. (1997). Discriminative stimulus effects of ethanol and 3α-hydroxy-5α-pregnan-20-one in relation to menstrual cycle phase in cynomolgus monkeys (Macaca fascicularis). *Psychopharmacology, 130*, 59–68.

Grant, K. A., Helms, C. M., Rogers, L. S. M., & Purdy, R. H. (2008). Neuroactive steroid stereospecificity of ethanol-like discriminative stimulus effects in monkeys. *The Journal of Pharmacology and Experimental Therapeutics, 326*, 354–361.

Grant, K. A., Waters, C. A., Green-Jordan, K., Azarov, A., & Szeliga, K. T. (2000). Characterization of the discriminative stimulus effects of GABA(A) receptor ligands in Macaca fascicularis monkeys under different ethanol training conditions. *Psychopharmacology (Berl), 152*(2), 181–188.

Helms, C. M., Rogers, L. S., & Grant, K. A. (2009). Antagonism of the ethanol-like discriminative stimulus effects of ethanol, pentobarbital, and midazolam in cynomolgus monkeys reveals involvement of specific GABAA receptor subtypes. *The Journal of Pharmacology and Experimental Therapeutics, 331*, 142–152.

Helms, C. M., Rogers, L. S., Waters, C. A., & Grant, K. A. (2008). Zolpidem generalization and antagonism in male and female cynomolgus monkeys trained to discriminate 1.0 or 2.0 g/kg ethanol. *Alcoholism, Clinical and Experimental Research, 32*, 1197–1206.

Helms, C. M., Rossi, D. J., & Grant, K. A. (2012). Neurosteroid influences on sensitivity to ethanol. *Frontiers in Endocrinology (Lausanne), 3*, 10.

Holtyn, A. F., Tiruveedhula, V. V., Stephen, M. R., Cook, J. M., & Weerts, E. M. (2017). Effects of the benzodiazepine GABAA α1-preferring antagonist 3-isopropoxy-β-carboline hydrochloride (3-ISOPBC) on alcohol seeking and self-administration in baboons. *Drug and Alcohol Dependence, 170*, 25–31.

Holtyn, A. F., & Weerts, E. M. (2022). GABAB receptors and alcohol use disorders: Preclinical studies. *Current Topics in Behavioral Neurosciences, 52*, 157–194.

Ito, M., Spence, A., Blauwet, M. B., Heo, N., Goldwater, R., Maruff, P., & Marek, G. J. (2022). A phase 1 study to assess potential interaction between ASP8062 and alcohol in healthy adult subjects. *Journal of Psychopharmacology (Oxford, England), 36*(6), 756–767.

Jackson, A., Stephens, D. N., & Duka, T. (2003). Lorazepam substitutes for the alcohol stimulus in social drinkers. *Psychopharmacology, 166*, 181–187.

June, H. L., Harvey, S. C., Foster, K. L., McKay, P. F., Cummings, R., Garcia, M., ... Cook, J. M. (2001). GABAA receptors containing 5 subunits in the CA1 and CA3 hippocampal fields regulate ethanol-motivated behaviors: An extended ethanol reward circuitry. *The Journal of Neuroscience, 21*, 2166–2177.

Kaminski, B. J., Van Linn, M. L., Cook, J. M., Yin, W., & Weerts, E. M. (2013). Effects of the benzodiazepine GABAA alpha1-preferring ligand, 3-propoxy-beta-carboline hydrochloride (3-PBC), on alcohol seeking and self-administration in baboons. *Psychopharmacology, 227*(1), 127–136.

Kaupmann, K., Huggel, K., Heid, J., Flor, P. J., Bischoff, S., Mickel, S. J., ... Bettler, B. (1997). Expression cloning of GABA(B) receptors uncovers similarity to metabotropic glutamate receptors. *Nature, 386*(6622), 239–246.

Khan, H., Garg, A., Yasmeen, Agarwal, N. B., Yadav, D. K., Ashif Khan, M., & Hussain, S. (2022). Zolpidem use and risk of suicide: A systematic review and meta-analysis. *Psychiatry Research, 316*, 114777.

Kissin, B. (1975). The use of psychoactive drugs in the long-term treatment of chronic alcoholics. *252*, 385–395.

Knutson, D. E., Smith, J. L., Ping, X., Jin, X., Golani, L. K., Li, G., ... Witkin, J. M. (2020). Imidazodiazepine anticonvulsant, KRM-II-81, produces novel, non-diazepam-like antiseizure effects. *ACS Chemical Neuroscience, 11*(17), 2624–2637.

Koob, G. F., & Colrain, I. M. (2020). Alcohol use disorder and sleep disturbances: A feed-forward allostatic framework. *Neuropsychopharmacology: Official Publication of the American College of Neuropsychopharmacology, 45*(1), 141–165.

Koulentaki, M., & Kouroumalis, E. (2018). GABA(A) receptor polymorphisms in alcohol use disorder in the GWAS era. *Psychopharmacology, 235*(6), 1845–1865.

Lambert, J. J., Belelli, D., Peden, D. R., Vardy, A. W., & Peters, J. A. (2003). Neurosteroid modulation of GABAA receptors. *Progress in Neurobiology, 71*, 67–80.

Léger, M., Brunet, M., Le Roux, G., Lerolle, N., & Boels, D. (2017). Baclofen self-poisoning in the era of changing indication: Multicentric reports to a french poison control centre. *Alcohol and Alcoholism (Oxford, Oxfordshire), 52*(6), 665–670.

Licata, S. C., & Rowlett, J. K. (2008). Abuse and dependence liability of benzodiazepine-type drugs: GABA(A) receptor modulation and beyond. *Pharmacology, Biochemistry, and Behavior, 90*(1), 74–89.

Lüddens, H., Seeburg, P. H., & Korpi, E. R. (1994). Impact of beta and gamma variants on ligand-binding properties of gamma-aminobutyric acid type A receptors. *Molecular Pharmacology, 45*(5), 810–814.

Maccioni, P., & Colombo, G. (2009). Role of the GABA(B) receptor in alcohol-seeking and drinking behavior. *Alcohol (Fayetteville, N. Y.), 43*(7), 555–558.

McKay, P. F., Foster, K. L., Mason, D., Cummings, R., Garcia, M., Williams, L. S., ... June, H. L. (2004). A high affinity ligand for GABAA-receptor containing alpha5 subunit antagonizes ethanol's neurobehavioral effects in Long-Evans rats. *Psychopharmacology, 172*(4), 455–462.

Melón, L. C., Nolan, Z. T., Colar, D., Moore, E. M., & Boehm 2nd, S. L. (2017). Activation of extrasynaptic delta-GABA(A) receptors globally or within the posterior-VTA has estrous-dependent effects on consumption of alcohol and estrous-independent effects on locomotion. *Hormones and Behavior, 95*, 65–75.

Meng, Z., Berro, L. F., Sawyer, E. K., Rüedi-Bettschen, D., Cook, J. E., Li, G., ... Rowlett, J. K. (2020). Evaluation of the anti-conflict, reinforcing, and sedative effects of YT-III-31, a ligand functionally selective for alpha3 subunit-containing GABA(A) receptors. *Journal of Psychopharmacology (Oxford, England), 34*(3), 348–357.

Middaugh, L. D., Bao, K., Becker, H. C., & Daniel, S. S. (1991). Effects of Ro 15-4513 on ethanol discrimination in C57BL/6 mice. *Pharmacology, Biochemistry, and Behavior, 38*, 763–767.

Milivojevic, V., Sullivan, L., Tiber, J., Fogelman, N., Simpson, C., Hermes, G., et al. (2023). Pregnenolone effects on provoked alcohol craving, anxiety, HPA axis, and autonomic arousal in individuals with alcohol use disorder. *Psychopharmacology (Berl), 240*, 101–114.

Moore, T. J., & Mattison, D. R. (2018). Assessment of patterns of potentially unsafe use of zolpidem. *JAMA Internal Medicine, 178*(9), 1275–1277.

Moore, E. M., Serio, K. M., Goldfarb, K. J., Stepanovska, S., Linsenbardt, D. N., & Boehm, S. L. (2007). GABAergic modulation of binge-like ethanol intake in C57BL/6J mice. *Pharmacology, Biochemistry, and Behavior, 88*, 105–113.

Morley, K. C., Baillie, A., Fraser, I., Furneaux-Bate, A., Dore, G., Roberts, M., ... Haber, P. S. (2018). Baclofen in the treatment of alcohol dependence with or without liver disease: multisite, randomised, double-blind, placebo-controlled trial. *The British Journal of Psychiatry: the Journal of Mental Science, 212*(6), 362–369.

Morrow, A. L., Boero, G., & Porcu, P. (2020). A rationale for allopregnanolone treatment of alcohol use disorders: Basic and clinical studies. *Alcoholism, Clinical and Experimental Research, 44*(2), 320–339.

Munk, C., Isberg, V., Mordalski, S., Harpsøe, K., Rataj, K., Hauser, A. S., ... Gloriam, D. E. (2016). GPCRdb: The G protein-coupled receptor database—An introduction. *British Journal of Pharmacology, 173*(14), 2195–2207.

Nutt, D. J., Besson, M., Wilson, S. J., Dawson, G. R., & Lingford-Hughes, A. R. (2007). Blockade of alcohol's amnestic activity in humans by an alpha5 subtype benzodiazepine receptor inverse agonist. *Neuropharmacology, 53*, 810–820.

Pareek, T., Overton, J. S., Nguyen, L. T., Rahman, M. T., Sharmin, D., Cook, J. M., & Platt, D. M. (2023). Modeling cue-exposure therapy for alcohol use disorder in rhesus monkeys: Effects of putative cognitive enhancers. *Drug and Alcohol Dependence, 243*, 109735.

Pierucci-Lagha, A., Covault, J., Feinn, R., Khisti, R. T., Morrow, A. L., Marx, C. E., ... Kranzler, H. R. (2006). Subjective effects and changes in steroid hormone concentrations in humans following acute consumption of alcohol. *Psychopharmacology, 186*(3), 451–461.

Ping, X., Meyer, M. J., Zahn, N. M., Golani, L. K., Sharmin, D., Pandey, K. P., ... Witkin, J. M. (2023). Comparative anticonvulsant activity of the GABAkine KRM-II-81 and a deuterated analog. *Drug Development Research, 84*(3), 527–531.

Platt, D. M., Duggan, A., Spealman, R. D., Cook, J. M., Li, X., Yin, W., & Rowlett, J. K. (2005). Contribution of alpha1GABAA and alpha5GABAA receptor subtypes to the discriminative stimulus effects of ethanol in squirrel monkeys. *The Journal of Pharmacology and Experimental Therapeutics, 313*, 658–667.

Porjesz, B., & Rangaswamy, M. (2007). Neurophysiological endophenotypes, CNS disinhibition, and risk for alcohol dependence and related disorders. *The Scientific World Journal, 7*, 131–141.

Ramaker, M. J., Ford, M. M., Fretwell, A. M., & Finn, D. A. (2011). Alteration of ethanol drinking in mice via modulation of the GABAA receptor with ganaxolone, finasteride, and gaboxadol. *Alcoholism, Clinical and Experimental Research, 35*, 1994–2007.

Ramaker, M. J., Strong, M. N., Ford, M. M., & Finn, D. A. (2012). Effect of ganaxolone and THIP on operant and limited-access ethanol self-administration. *Neuropharmacology, 63*(4), 555–564.

Rees, D. C., & Balster, R. L. (1988). Attenuation of the discriminative stimulus properties of ethanol and oxazepam, but not of pentobarbital, by Ro 15-4513 in mice. *The Journal of Pharmacology and Experimental Therapeutics, 244*, 592–598.

Rigal, L., Sidorkiewicz, S., Tréluyer, J. M., Perrodeau, E., Le Jeunne, C., Porcher, R., & Jaury, P. (2020). Titrated baclofen for high-risk alcohol consumption: A randomized placebo-controlled trial in out-patients with 1-year follow-up. *Addiction (Abingdon, England), 115*(7), 1265–1276.

Rolland, B., Paille, F., Gillet, C., Rigaud, A., Moirand, R., Dano, C., ... Aubin, H. J. (2016). Pharmacotherapy for alcohol dependence: The 2015 recommendations of the French Alcohol Society, Issued in Partnership with the European Federation of Addiction Societies. *CNS Neuroscience & Therapeutics, 22*(1), 25–37.

Rowlett, J. K., Lelas, S., Tornatzky, W., & Licata, S. C. (2006). Anti-conflict effects of benzodiazepines in rhesus monkeys: Relationship with therapeutic doses in humans and role of GABAA receptors. *Psychopharmacology, 184*(2), 201–211.

Rüedi-Bettschen, D., Rowlett, J. K., Rallapalli, S., Clayton, T., Cook, J. M., & Platt, D. M. (2013). Modulation of α5 subunit-containing GABAA receptors alters alcohol drinking by rhesus monkeys. *Alcoholism, Clinical and Experimental Research, 37*(4), 624–634.

Sanger, D. J. (1997). The effects of new hypnotic drugs in rats trained to discriminate ethanol. *Behavioural Pharmacology, 8*, 287–292.

Sawyer, E. K., Moran, C., Sirbu, M. H., Szafir, M., Van Linn, M., Namjoshi, O., ... Platt, D. M. (2014). Little evidence of a role for the alpha1GABAA subunit-containing receptor in a rhesus monkey model of alcohol drinking. *Alcoholism, Clinical and Experimental Research, 38*(4), 1108–1117.

Schuster, C. L., & Humphries, R. H. (1981). Benzodiazepine dependency in alcoholics. *Connecticut Medicine, 45*, 11–13.

Schuster, C. R., & Johanson, C. E. (1988). Relationship between the discriminative stimulus properties and subjective effects of drugs. *Psychopharmacology Series, 4*, 161–175.

Shelton, K. L., & Grant, K. A. (2002). Discriminative stimulus effects of ethanol in C57BL/6J and DBA/2J inbred mice. *Alcoholism, Clinical and Experimental Research, 26*, 747–757.

Shinday, N. M., Sawyer, E. K., Fischer, B. D., Platt, D. M., Licata, S. C., Atack, J. R., ... Rowlett, J. K. (2013). Reinforcing effects of compounds lacking intrinsic efficacy at α1 subunit-containing GABAA receptor subtypes in midazolam- but not cocaine-experienced rhesus monkeys. *Neuropsychopharmacology: Official Publication of the American College of Neuropsychopharmacology, 38*(6), 1006–1014.

Simioni, N., Labreuche, J., Behal, H., Cottencin, O., & Rolland, B. (2017). Thirty- versus ten-day diazepam treatment for alcohol detoxification and a comparison of drinking patterns, craving, and anxiety for up to 12 weeks: A "proof-of-concept" open-label randomized controlled trial. *Journal of Clinical Psychopharmacology, 37*(6), 722–728.

Smith, B. R., Robidoux, J., & Amit, Z. (1992). GABAergic involvement in the acquisition of voluntary ethanol intake in laboratory rats. *Alcohol and Alcoholism (Oxford, Oxfordshire), 27*, 227–231.

Stephens, D. N., Pistovcakova, J., Worthing, L., Atack, J. R., & Dawson, G. R. (2005). Role of GABAA alpha5-containing receptors in ethanol reward: The effects of targeted gene deletion, and a selective inverse agonist. *European Journal of Pharmacology, 526*(1-3), 240–250.

Sternfeld, F., Carling, R. W., Jelley, R. A., Ladduwahetty, T., Merchant, K. J., Moore, K. W., ... MacLeod, A. M. (2004). Selective, orally active -aminobutyric acidA 5 receptor inverse agonists as cognition enhancers. *Journal of Medicinal Chemistry, 47*, 2176–2179.

Tatebayashi, H., Motomura, M., & Narahashi, T. (1998). Alcohol modulation of single GABAA receptor-channel kinetics. *Neuroreport, 9*, 1769–1775.

Uusi-Oukari, M., & Korpi, E. R. (2024). GABAergic mechanisms in alcohol dependence. *International Review of Neurobiology, 175*, 75–123.

Walzer, M., Marek, G. J., Wu, R., Nagata, M., & Han, D. (2020). Single- and multiple-dose safety, tolerability, and pharmacokinetic profiles of ASP8062: Results from 2 phase 1 studies. *Clinical Pharmacology in Drug Development, 9*(3), 297–306.

Whiting, P. J., Bonnert, T. P., McKernan, R. M., Farrar, S., Le, Bourdelles, B., ... Wafford, K. A. (1999). Molecular and functional diversity of the expanding GABAA receptor gene family. *Annals of the New York Academy of Sciences, 868*, 645–653.

Wilkinson, C. J. (1998). The abuse potential of zolpidem administered alone and with alcohol. *Pharmacology, Biochemistry, and Behavior, 60*, 193–202.

Zack, M., Poulos, C. X., & Woodford, T. M. (2006). Diazepam dose-dependently increases or decreases implicit priming of alcohol associations in problem drinkers. *Alcohol and Alcoholism (Oxford, Oxfordshire), 41*, 604–610.

Zack, M., Toneatto, T., & MacLeod, C. M. (1999). Clinical use of benzodiazepines and decreased memory activation in anxious problem drinkers. *Alcoholism, Clinical and Experimental Research, 23*, 174–182.

Zosel, A., Osterberg, E. C., & Mycyk, M. B. (2011). Zolpidem misuse with other medications or alcohol frequently results in intensive care unit admission. *American Journal of Therapeutics, 18*(4), 305–308.

CHAPTER THIRTEEN

Ghrelin system and GLP-1 as potential treatment targets for alcohol use disorder

Elisabet Jerlhag*
Institute of Neuroscience and Physiology, Department of Pharmacology, The Sahlgrenska Academy at the University of Gothenburg, Gothenburg, Sweden
*Corresponding author. e-mail address: Elisabet.Jerlhag@pharm.gu.se

Contents

Abstract

Peptides of the gut–brain axis have gained recent attention as potential treatment targets for addiction. While the number of gut–brain peptides is vast, ghrelin and glucagon-like peptide-1 (GLP-1) have been suggested as important players. Ghrelin is traditionally considered an orexigenic peptide, but recent studies found that it increases alcohol intake in rodents and craving for alcohol in humans. Additionally, suppression of the ghrelin receptor attenuates alcohol-related responses in animal models reflecting alcohol use disorder (AUD). For instance, a lower alcohol intake, suppressed motivation to consume alcohol, and attenuated reward from alcohol is observed after ghrelin receptor antagonism treatment. On a similar note, a partial ghrelin receptor agonist prevents hangover symptoms in humans. When it comes to the anorexigenic peptide GLP-1, agonists of its receptor are approved to treat diabetes type 2 and obesity. Extensive preclinical studies have revealed that these GLP-1 receptor agonists reduce alcohol intake, suppress the motivation to consume alcohol, and prevent relapse drink, with effects tentatively associated with a reduced alcohol-induced reward. These preclinical findings have to some extent been varied in humans, as GLP-1 receptor agonists decrease alcohol intake in overweight patients with AUD. Furthermore, genetic variations in either the genes encoding for pre-pro-ghrelin, GHSR, GLP-1, or its receptor, are associated with AUD and heavy alcohol drinking. While central mechanisms appear to modulate the ability of either ghrelin or GLP-1 to regulate alcohol-related responses the exact mechanisms have not been defined. Taken together these preclinical and clinical data imply that gut–brain peptides participate in the addiction process and should be considered as potential targets for AUD treatment.

International Review of Neurobiology, Volume 178
ISSN 0074-7742, https://doi.org/10.1016/bs.irn.2024.07.006

Abbreviations

AUD	alcohol use disorder
AUDIT	alcohol use disorder identification test
DAG	des-acyl-ghrelin
DPP-IV	dipeptidyl peptidase-IV
Ex4	exendin-4
GHSR	growth hormone secretagogue receptor, ghrelin receptor
GLP-1	glucagon-like peptide-1
GLP-1R	glucagon-like peptide-1 receptor
GOAT	ghrelin *O*-acyltransferase
LDTg	laterodorsal tegmental area
NAc	nucleus accumbens
NTS	nucleus tractus solitarius
SNP	single nucleotide polymorphism
VTA	ventral tegmental area

1. Gut–brain axis

The gut–brain axis constitutes a bidirectional communication between the gut and brain, where feeding and appetite appear as central aspects regulated by this axis. While a vast number of gut–brain peptides exist, both ghrelin and glucagon-like peptide-1 (GLP-1) appear to be central players.

The production of the orexigenic peptide ghrelin is mainly localized to peripheral tissues such as endocrine cells in the gastrointestinal mucosa of the stomach (Kojima et al., 1999). Additionally, ghrelin production may to some extent occur in the brain (Cowley et al., 2003; Lu et al., 2002; Mondal et al., 2005). After its acylation by the enzyme ghrelin *O*-acyltransferase (GOAT), ghrelin ("acyl-ghrelin") exerts its physiological effects after binding to the ghrelin receptor (growth hormone secretagogue receptor). Intriguingly, these ghrelin receptors are expressed in peripheral organs, as well as in the brain (for reviews see, e.g., Morganstern, Barson, & Leibowitz, 2011; Wolfe & Maisto, 2000). The ghrelin receptor is a G-protein coupled receptor characterized by its intrinsic activity and ability to heterodimer with other receptors (for review see, Cornejo et al., 2021). While ghrelin has been attributed to a vast number of physiological functions, the most well-characterized is its ability to enhance feeding (for review see, Cornejo et al., 2021). Specifically, ghrelin augments both the hedonic and homeostatic aspects of feeding (for review see, Cornejo et al., 2021). Although ghrelin and its receptor modulate various physiological properties, clinically available ghrelin receptor ligands do not exist. However, in research, both antagonists of the ghrelin receptor (e.g., JMV2959 or [D-Lys^3]-GHRP-6) and the inverse agonist of the receptor (i.e., PF-5190457) are used. After its release, ghrelin is de-acylated to des-acyl-ghrelin (DAG) (De Vriese et al., 2004), whose physiological properties are less defined (Delhanty et al., 2010; Fernandez et al., 2016; Gauna et al., 2006; Hosoda, Kojima, Matsuo, & Kangawa, 2000; Inoue et al., 2013). The liver-expressed antimicrobial peptide 2 (LEAP2) is a recently identified substrate of the ghrelin pathway as it is an endogenous inverse agonist on ghrelin receptors (Ge et al., 2018; M'Kadmi et al., 2019). While LEAP2 mainly is produced in the small intestine and liver (Krause et al., 2003), central production may exist as the expression of the LEAP2 gene has been detected in the brain (Islam et al., 2020). LEAP2, being an inverse ghrelin receptor agonist, suppresses the intake of regular food (e.g., chow) (Ge et al., 2018), and palatable foods (e.g., high-fat diet intake) (Cornejo et al., 2019). In addition, LEAP2 blocks the ghrelin-induced food intake (Cornejo et al., 2019; Islam et al., 2020).

Whereas ghrelin is well-known for its orexigenic properties, GLP-1 is a crucial orexigenic peptide. GLP-1 production is mainly attributed to L-cells of the intestines (Novak, Wilks, Buell, & McEwen, 1987) and neurons of the nucleus tractus solitarius (NTS) (Alvarez, Roncero, Chowen, Thorens, & Blazquez, 1996). After its release, GLP-1 is degraded by the enzyme dipeptidyl peptidase-IV (DPP-IV), and inhibition of the enzyme (i.e., DPP-IV inhibitors) causes an elevation of circulating GLP-1 levels. The short half-life of GLP-1 has led to the development of GLP-1 receptor (GLP-1R) agonists with improved pharmacokinetic properties. As such, short-acting (Exendin-4; Ex4; exenatide), moderate-acting (liraglutide) and long-acting (including dulaglutide or semaglutide) GLP-1R agonists are to date available. GLP-1R agonists are approved for diabetes type 2 (for review see, Srivastava & Apovian, 2018), as they regulate glucose-dependent insulin secretion (Holst & Seino, 2009). Additionally, their ability to reduce body weight in rodents and humans has led to their approval for obesity (for review see, Srivastava & Apovian, 2018). Preclinical and clinical studies further reveal that GLP-1R agonists suppress the intake of regular and palatable foods (for review see, Srivastava & Apovian, 2018). When it comes to GLP-1Rs, they have been identified both in the periphery as well as in the brain (for reviews see, e.g., Morganstern et al., 2011; Wolfe & Maisto, 2000).

Although both the ghrelin pathway and GLP-1 signaling system are well-known for regulating energy and glucose homeostasis, they are also attributed to an important regulatory role in reward and addiction, which is summarized below.

1.1 Ghrelin and alcohol

As reviewed below, both preclinical and clinical studies reveal that the ghrelin pathway modulates alcohol-related responses.

1.1.1 Preclinical studies

1.1.1.1 Alcohol consumption

One animal model reflecting alcohol use disorder (AUD) does not exist, and therefore alcohol-related responses can be investigated in various animal models that together reflect the AUD pathophysiology. Effects on alcohol intake can be evaluated in the intermittent access model, where rats escalate their intake over prolonged periods reaching stable and high alcohol intake (Simms et al., 2008). Other models measuring alcohol intake include the continuous access paradigm, drinking in the dark, or binge drinking.

The impact of ghrelin and ghrelin receptor antagonists on alcohol intake has been substantially investigated. When it comes to ghrelin, initial studies revealed that central (into the third ventricle) ghrelin infusion increased the intake of alcohol in male mice consuming alcohol for months before treatment (Jerlhag et al., 2009). This escalated alcohol intake was also observed after local infusion of ghrelin into brain areas modulating reward, such as laterodorsal tegmental area (LDTg) and ventral tegmental area (VTA) (Jerlhag et al., 2009). A modulatory role of ghrelin was further supported as the alcohol intake was lower in male mice with an overall ghrelin knockout compared to their wild-type littermates (Bahi et al., 2013). When it comes to systemic injection of ghrelin, the impact of treatment on alcohol intake is less conclusive. Systemic administration of ghrelin elevated alcohol intake in male rats exposed to alcohol for a short period before treatment (Cepko et al., 2014). Additionally, the intake of alcohol was elevated after systemic ghrelin infusion in male rats that had undergone Roux-en-Y gastric bypass (Davis et al., 2012). Another way to modulate circulating ghrelin is to target the formation of acylated ghrelin (i.e., "active" ghrelin). After suppression of this formation a lowered alcohol intake was demonstrated in male mice (Godlewski et al., 2019). In contrast to the above-mentioned articles, neither systemic ghrelin infusion (male mice) (Lyons, Lowery, Sparta, & Thiele, 2008), nor neutralization of circulating ghrelin (male rats) (Jerlhag, Ivanoff, Vater, & Engel, 2013) altered alcohol intake. Supportively, binge drinking was unaffected by the neutralization of circulating ghrelin (anti-ghrelin vaccine) (Richardson et al., 2023). Although reasons for this discrepancy are unknown and remain to be elucidated, they may be associated with animals/strain, time of alcohol exposure, or the drinking model used.

When it comes to ghrelin receptor suppression, the treatment outcome is profound and consequent between studies. The first study demonstrating this association was published in 2009 and showed that both acute systemic and central administration of a ghrelin receptor antagonist lowered the alcohol intake in male mice with prolonged alcohol exposure (Jerlhag et al., 2009). Since then, this initial study has since been verified and extended. Indeed, high alcohol-preferring (either outbred or genetically modified, AA) male rats with long-term alcohol exposure displayed a lower alcohol intake after acute ghrelin receptor antagonist treatment (Landgren et al., 2012; Suchankova, Steensland, Fredriksson, Engel, & Jerlhag, 2013). Intriguingly, this reduction was dose-dependent and greater after long- compared to short-term alcohol exposure (Suchankova et al., 2013).

It should however be noted that a reduction in alcohol drinking after ghrelin receptor suppression was evident in animals exposed to alcohol for shorter periods. As such, acute or repeated treatment with a ghrelin receptor antagonist lowered the alcohol intake in various animal species (male mice/rats/prairie voles) (Bahi et al., 2013; Gomez & Ryabinin, 2014; Gomez et al., 2015; Kaur & Ryabinin, 2010; Stevenson et al., 2015). These initial studies were recently verified in an animal model of binge drinking, as systemic or central (into ventricle) injection of different ghrelin receptor antagonists reduced alcohol intake in male and female mice (Richardson et al., 2023). Moreover, no sex differences in treatment outcomes were demonstrated herein (Richardson et al., 2023). Additionally, suppressing the ghrelin receptor could potentially mitigate the development of AUD, as ghrelin receptor knockout mice display a lower alcohol intake compared to their wild-type counterparts when consuming alcohol in the continuous access model (Jerlhag et al., 2009). Supportively, rats with a ghrelin receptor knockout displayed a lower alcohol intake when exposed to drinking in the dark or operant-self-administration models (Zallar et al., 2019). Additionally, pharmacological suppression of ghrelin receptors reduced the intake of alcohol in obese rats that have undergone a vertical sleeve gastrostomy or Roux-en-Y gastric bypass (Hajnal et al., 2012; Orellana, Jamis, Horvath, & Hajnal, 2018). Taken together these studies revealed a profound reduction in alcohol drinking after ghrelin receptor suppression, and that this decline persists across species, strain, antagonist used, time of exposure, and sex.

1.1.1.2 Motivation to consume alcohol and relapse drinking

While the above-summarized studies show that ghrelin receptor antagonism reduces alcohol intake, additional studies reveal that ghrelin receptors regulate other aspects of the AUD process. One of these is the motivation to consume alcohol in the operant self-administration paradigm, which was reduced after either pharmacological (Gomez et al., 2015; Landgren et al., 2012) or genetic (Zallar et al., 2019) suppression of the ghrelin receptor. On a similar note, local infusion of a ghrelin receptor antagonist into nucleus accumbens (NAc) of female mice reduced the intake of alcohol (Abtahi, Howell, & Currie, 2018). Another aspect of relevance for patients with AUD is the escalated alcohol intake induced by periods of alcohol abstinence (i.e., white period). This relapse drinking in humans is influenced by alcohol cravings, a behavior that in animals can be studied in the alcohol deprivation effect model (Spanagel, 2000). The findings that a ghrelin receptor antagonist prevents relapse drinking in male mice (Jerlhag et al., 2009) and male rats (Suchankova et al., 2013), further suggest that the ghrelin pathway modulates various aspects of AUD.

In a tentative clinical study, the ghrelin receptor antagonist may have some advantages. One of these is that they might reduce alcohol drinking more in patients with severe AUD compared to those with mild AUD as this is seen in rats (Suchankova et al., 2013). Another tentative advantage may be a lack of tolerance as male rats do not display such effects after prolonged treatment (Suchankova et al., 2013). Another possibility is that ghrelin receptor antagonists suppress craving for alcohol during abstinence as this treatment blocks relapse drinking in male rodents (Jerlhag et al., 2009; Suchankova et al., 2013).

1.1.1.3 Circulating ghrelin and alcohol

While ghrelin or ghrelin receptor antagonist was administered exogenously in the above-summarized studies, the role of endogenous ghrelin (in plasma or serum) for alcohol drinking has been studied in additional experiments. In rats, acute alcohol administration reduced ghrelin, possibly due to its caloric content (Deschaine et al., 2021; Szulc et al., 2013; Yoshimoto et al., 2017). On the other hand, rats fed with alcohol chronically show enhanced ghrelin levels in one study (Rasineni et al., 2019). It should, however, be noted that this association was not evident in another study where the ghrelin levels were similar between low- and high-alcohol-consuming rats (Landgren et al., 2012). The inconsistency between these studies on endogenous ghrelin and alcohol may be due to the limited penetration of ghrelin from the blood to the brain (Furness et al., 2011; Grouselle et al., 2008; Pirnik et al., 2011; Sakata et al., 2009; Schaeffer et al., 2013).

1.1.1.4 Alcohol-related behaviors and neurochemistry

In humans and animals, alcohol causes a release of dopamine in NAc (for review see, Jayaram-Lindström, Ericson, Steensland, & Jerlhag, 2016). Importantly, the elevated dopamine levels are associated with alcohol's rewarding properties as well as enhance the risk for an AUD diagnosis later in life (Boileau et al., 2003; King et al., 2021). These rewarding properties can in animals be measured by a release of dopamine in NAc, locomotor stimulation, and a conditioned place preference.

A study conducted in 2009 demonstrated the initial evidence demonstrating the capacity of ghrelin receptors to influence the rewarding aspects of alcohol. Supportively, systemic, or central administration of a ghrelin receptor antagonist attenuates the ability of alcohol to cause locomotor stimulation, release dopamine in NAc, and condition a place preference in male mice (Jerlhag et al., 2009). This interaction was further evidenced as these rewarding properties were attenuated in male mice with an overall genetic suppression of ghrelin receptors (Jerlhag et al., 2009). Further, five days of ghrelin receptor antagonist pre-treatment of male mice attenuated the ability of alcohol to cause locomotor stimulation (Suchankova, Engel, & Jerlhag, 2016a). While these initial studies investigated the role of ghrelin receptors for alcohol reward, subsequent preclinical studies explored the impact of ghrelin on alcohol's rewarding properties. Indeed, male mice with an overall knockout of ghrelin did not display a locomotor stimulation, or dopamine release in NAc after alcohol infusion (Jerlhag, Landgren, Egecioglu, Dickson, & Engel, 2011b). On a similar note, the ability of alcohol to cause reward in the conditioned place preference paradigm is attenuated in ghrelin receptor knockout mice (Bahi et al., 2013; Jerlhag et al., 2011b). As alcohol's rewarding properties modulate alcohol intake and relapse drinking, these data may indicate that ghrelin receptor antagonism lowers alcohol intake due to an attenuation of alcohol's rewarding properties.

1.1.1.5 Tentative mechanisms of action

Although the mechanisms through which ghrelin and its receptor regulate alcohol-related behaviors are unknown, some mechanisms of action have been suggested. One of these is vagal afferents, as a ghrelin receptor antagonist only can reduce alcohol drinking in male mice with intact vagal afferents (Godlewski et al., 2019). Additionally, brain regions central to reward have been suggested as ghrelin receptors expressing brain nodes regulating alcohol drinking. One of these is NAc, as local infusion of a ghrelin receptor antagonist into this region decreases alcohol intake (Abtahi et al., 2018). Additionally, the expression of ghrelin receptors was higher in high- compared to low-alcohol-consuming rats (Suchankova et al., 2013).

NAc is a target of dopaminergic projection from the VTA, another area through which ghrelin may act. Indeed, local infusion of ghrelin into the VTA elevated alcohol consumption both in male mice with prolonged alcohol consumption (Jerlhag et al., 2009) and male rats with short-time alcohol exposure (Colvin et al., 2022). Moreover, rewarding behaviors were observed after ghrelin into the VTA (Cornejo et al., 2018; Jensen et al., 2016; Jerlhag et al., 2006, 2007; Jerlhag, Egecioglu, Dickson, & Engel, 2011a; Suchankova et al., 2016a). Further support for an involvement of VTA in the alcohol-ghrelin connection is provided as the expression of ghrelin receptors was lower in high alcohol-consuming rats compared to those consuming less (Suchankova et al., 2013). Supportively, the reward induced by optogenetic stimulation was modulated by ghrelin receptors expressed on dopaminergic neurons of the VTA (You et al., 2022). Moreover, ghrelin augmented the activity of VTA-dopamine neurons in rats that have undergone Roux-en-Y gastric bypass and consume alcohol (Orellana et al., 2018). Nitric oxide in the VTA may be one downstream target of ghrelin, as attenuation of the nitric oxide pathway prevented ghrelin's ability to cause a locomotor stimulation, release dopamine in NAc, and condition a place preference (Engel, Palsson, Vallof, & Jerlhag, 2023). Somatodendritic release of dopamine in the VTA may be of tentative relevance for ghrelin's actions in the brain. As such, ghrelin caused a release of dopamine in the VTA, and a ghrelin receptor antagonist attenuated this dopamine elevation, and the subsequent dopamine release in NAc (Edvardsson, Vestlund, & Jerlhag, 2021). The finding that ghrelin enhanced the dopamine release in VTA via VTA-located ghrelin receptors was later replicated (Navarro et al., 2022). Further on, the ability of ghrelin to act on the reward systems (Kern, Albarran-Zeckler, Walsh, & Smith, 2012; Kern et al., 2015; Mustafa et al., 2021), specifically this VTA dopamine release (Navarro et al., 2022), may involve heterodimerization of ghrelin receptors with dopamine receptors. The cholinergic input from the LDTg appears to play a central role in both the activity of dopaminergic neurons of the VTA and ghrelin's actions in the brain. Indeed, ghrelin caused a release of acetylcholine in the VTA via local effects in the LDTg (Jerlhag, Janson, Waters, & Engel, 2012), and inhibition of nicotinic acetylcholine receptors in the VTA prevented the ghrelin-induced feeding and reward

(Dickson et al., 2010; Jerlhag, Egecioglu, Dickson, Svensson, & Engel, 2008). On a similar note, the alcohol intake (Jerlhag et al., 2009) and the activity of dopaminergic VTA-NAc projection (Jerlhag et al., 2007) escalated after ghrelin infusion into the LDTg.

Amygdala, a ghrelin receptor expressing brain region (Cruz, Herman, Cote, Ryabinin, & Roberto, 2013; Landgren et al., 2011b) central to addiction (Koob & Volkow, 2010), may participate in ghrelin's ability to regulate alcohol drinking as ghrelin augments GABAergic neurotransmission in rats either naïve or exposed to alcohol (Cruz et al., 2013; Yoshimoto et al., 2017). This is further evidenced as high alcohol intake augmented the protein levels of ghrelin receptors and the immunoreactive ghrelin receptor cells in the amygdala (Yoshimoto et al., 2017). While examined to a lesser extent, both the lateral hypothalamus (Yoshimoto et al., 2017) and Edinger–Westphal nucleus (Kaur & Ryabinin, 2010) may be other areas where ghrelin acts to control alcohol drinking.

1.1.2 Clinical studies

1.1.2.1 Plasma-association studies

While there is a vast number of preclinical studies on alcohol-related responses together with ghrelin or ghrelin receptor antagonists, initial clinical studies explored the association between circulating ghrelin levels and alcohol. In such plasma-association studies of healthy volunteers or social drinkers, the plasma levels were found to be lower after acute oral ingestion of alcohol (Calissendorff, Danielsson, Brismar, & Rojdmark, 2005; Calissendorff, Danielsson, Brismar, & Rojdmark, 2006; Ralevski et al., 2017; Zimmermann, Buchmann, Steffin, Dieterle, & Uhr, 2007). An effect is also demonstrated in heavy-drinking AUD patients (Deschaine et al., 2021). It was further found that an appetite for alcohol did not influence this reduction in ghrelin by alcohol (Calissendorff, Gustafsson, Holst, Brismar, & Rojdmark, 2012). This correlation was further supported by findings from studies on social drinkers in which alcohol was infused intravenously. Indeed, intravenous infusion of alcohol tended to reduce the circulating levels of ghrelin in one study (Leggio, Schwandt, Oot, Dias, & Ramchandani, 2013), and significantly reduced it in another study (Ralevski et al., 2017, 2022). When it comes to prolonged exposure to alcohol, it reduced the plasma levels of ghrelin in both actively drinking AUD patients (Addolorato et al., 2006; Badaoui et al., 2008; Kraus et al., 2005; Olsson et al., 2023) and in those abstaining from alcohol (Kim et al., 2005; Kraus et al., 2005; Wurst et al., 2007). While these initial studies focused on the effects of alcohol on ghrelin, other studies have associated ghrelin with the subjective experiences of alcohol.

Higher ghrelin levels were associated with a more intense subjective response to alcohol in both social drinkers (Ralevski et al., 2017) and heavy alcohol consumers (Ralevski et al., 2022). Additionally, those individuals with higher ghrelin levels were also more sensitive to alcohol's rewarding properties (Ralevski et al., 2018). On a similar note, in abstaining AUD patients elevated ghrelin levels were associated with alcohol craving (Addolorato et al., 2006; Leggio et al., 2012; Wurst et al., 2007). This association was also evident during the early alcohol abstinence phase (Koopmann et al., 2012, 2019; Sha, Dey, Khess, & Khitiz, 2021). A positive correlation between ghrelin and alcohol craving was also found during a cue reactivity task in AUD patients (Bach et al., 2019). Furthermore, ghrelin levels in plasma were positive correlation with impulsivity, observed in healthy individuals, social drinkers, and patients diagnosed with AUD (Agtas-Ertan, Dogan, & Ilhan, 2023; Ralevski et al., 2018). It should however be noted that blood ghrelin levels were not associated with the quantity of daily alcohol intake (Farokhnia et al., 2021).

1.1.2.2 Human genetic studies

Human genetic studies, either single nucleotide polymorphism (SNP) or haplotype, revealed further associations between ghrelin and AUD. In two different population cohorts, high alcohol intake (Landgren et al., 2008), higher scores in the alcohol use disorder identification test (AUDIT) (Suchankova et al., 2016b) and personality traits of AUD patients (Landgren et al., 2011a) were associated with variations in the ghrelin receptor genes. Similarly, within a female cohort, ghrelin receptor polymorphisms were linked to AUD type 2, which is characterized by early onset and hereditary factors (Landgren et al., 2010). Variations within the gene encoding ghrelin, the pre-pro-ghrelin gene, have been associated with various aspects of alcohol drinking and AUD. Specifically, a female cohort study revealed associations between variations in the pre-pro-ghrelin gene paternal alcohol dependence and reports of withdrawal symptoms (Landgren et al., 2010). In another cohort, associations between the pre-pro-ghrelin gene and AUD personality traits were evident (Landgren et al., 2011a). A case-control study revealed further correlations, as genetic variations in the pre-pro-ghrelin gene were linked to AUD diagnosis, decreased average daily alcohol consumption, fewer instances of heavy drinking, and diminished subjective response to alcohol within the intravenous self-administration paradigm among patients with AUD (Suchankova et al., 2017). It should however be noted that these genetic studies are small, and data from these have not been corroborated in genome-wide studies.

1.1.2.3 Exogenous administrations of ghrelin or inverse ghrelin receptor agonist

In an initial study involving patients with AUD, intravenous administration of a high dose of ghrelin intensified alcohol craving (Leggio et al., 2014). Similarly, intravenous infusion of ghrelin not only reduced the time to the first alcohol-containing drink but also increased the number of intravenous alcohol infusions (Farokhnia et al., 2017). Besides enhanced craving and alcohol consumption, intravenous administration of ghrelin reduced the hangover symptoms observed after intravenous alcohol infusions (Farokhnia et al., 2018). While ghrelin augmented alcohol-related responses, an inverse ghrelin receptor agonist decreased the craving for alcohol in a cue reactivity test (Lee et al., 2020b) without affecting the ability of alcohol to cause hangover symptoms (Farokhnia et al., 2018). Effects of intravenous ghrelin in healthy controls have also been studied, in which ghrelin did not affect the monetary reward while attenuating the anticipation of loss (Pietrzak et al., 2023).

1.1.2.4 Tentative mechanisms of action

These human studies have to some extent contributed to the identification of downstream mechanisms through which the ghrelin pathway regulates alcohol-related responses. Like animal studies demonstrating that ghrelin acts in NAc to regulate alcohol intake (Abtahi et al., 2018), the ghrelin-induced alcohol craving involved activation of the ventral striatum (i.e., NAc in rodents) in humans (Koopmann et al., 2019). In addition, the ghrelin-enhanced alcohol craving may involve the amygdala, since the ability of alcohol-associated cues to activate this area was augmented by an intravenous infusion of ghrelin (Farokhnia et al., 2017). Moreover, inflammatory markers (Farokhnia et al., 2020), cortisone, corticosterone, and fatty acids (Haass-Koffler et al., 2019; Karkkainen et al., 2021), rather than endocrine peptides (e.g., GLP-1, amylin, insulin, leptin, GIP) (Lee et al., 2020a), may participate in this interaction. It should however be noted that the expression of ghrelin or its receptor was similar between AUD patients and healthy controls (Deschaine et al., 2021), indicating that such mechanisms do not influence alcohol drinking.

1.2 GLP-1 and alcohol use disorder

1.2.1 Preclinical studies

Since the first preclinical study that demonstrated a suppression in alcohol-related behaviors after GLP-1R activation (Egecioglu et al., 2013), additional experiments have explored this interaction, which is summarized below. In contrast to ghrelin studies in which the names of antagonists were not mentioned, the name of GLP-1R agonists is listed as they are clinically used and thus may be relevant in a clinical situation.

1.2.1.1 Alcohol consumption

Acute systemic administration of Ex4 reduced the alcohol intake in male rats with prolonged exposure to alcohol before treatment (Egecioglu et al., 2013). Supportively, the alcohol consumption was lowered after an Ex4 injection in male rats, especially those who were high-alcohol consuming (Davis et al., 2012; Shirazi, Dickson, & Skibicka, 2013). A finding corroborated in a recent study, in which Ex4 not only reduced alcohol intake but also suppressed the ability of ghrelin to elevate alcohol drinking (Brigande, Darwich, & Currie, 2023). The ability of Ex4 to reduce alcohol intake persists among species as Ex4 lowered the alcohol consumption in group-housed male mice (Thomsen et al., 2017) as well as non-human primates (vervet monkeys) (Thomsen et al., 2019). It should be further noted that Ex4 changed the drinking pattern as it elevated the time to the first drink as well as lowered the number of drinking bouts (Thomsen et al., 2017). Additionally, in male mice, the intravenous self-administration of alcohol was attenuated after Ex4 treatment (Sorensen, Caine, & Thomsen, 2016). Besides Ex4, acute or repeated injections of liraglutide reduced the alcohol intake as well as the preference for alcohol in male rats (Marty et al., 2020; Vallof et al., 2015). Furthermore, both alcohol intake and preference were reduced in male mice after treatment with liraglutide (Liu et al., 2024). On a similar note, alcohol intake was suppressed by liraglutide in non-human primates (vervet monkeys) (Thomsen et al., 2019). While these initial studies explored shorter treatment effects in male rodents, a recent study explored effects of long-term treatment in both male and female rats. This study revealed that 9 or 5 weeks with weekly dulaglutide injections profoundly lowered the consumption of alcohol as well as the preference for alcohol in both male and female rats (Vallof, Kalafateli, & Jerlhag, 2020). On both sexes, the treatment effect persisted after treatment termination (Vallof et al., 2020). It should however be noted that the treatment effect was more profound in male, compared to female rats (Vallof et al., 2020). Semaglutide is another long-acting GLP-1R agonist found to reduce alcohol intake in rodents. One initial report showed that the alcohol intake was lower in male rats treated acutely with semaglutide compared to those treated with vehicle (Marty et al., 2020). This was further studied in male and female rats exposed to alcohol for prolonged periods. Indeed, acute or repeated semaglutide treatment dose-dependently reduced the alcohol intake and lowered the preference thereof (Aranas et al., 2023b). Additionally, this dose-dependent reduction in alcohol intake was more profound in females compared to males (Aranas et al., 2023b). Similarly, binge-like alcohol drinking was dose-dependently suppressed by semaglutide in mice and rats (Chuong et al., 2023). Although semaglutide profoundly reduced the alcohol intake, its combination with anti-smoking cessation agents (bupropion or varenicline) did not synergistically add toward semaglutide's ability to reduce alcohol intake (Aranas, Blid Skoldheden, & Jerlhag, 2023a). Considering that combination treatments potentially could lead to a significant decrease in alcohol intake with fewer adverse effects, future studies should explore different combinations with GLP-1R agonists further. When compared to other GLP-1R agonists, dulaglutide and semaglutide exhibited the most significant reduction in alcohol intake (60%–80% reduction) (Aranas et al., 2023b; Vallof et al., 2020). Additionally, there was no tolerance development towards treatment in these studies, and the treatment effect persisted after treatment combination (Suchankova et al., 2015; Vallof et al., 2015, 2020).

While the summarized preclinical studies indicate that the activation of GLP-1R suppresses alcohol intake, the impact of endogenous GLP-1 has been explored to a lesser extent since it is rapidly metabolized in the body. Supportively, systemic administration of GLP-1 has a short-lasting ability to reduce alcohol intake in male rats (Shirazi et al., 2013). Another way to study the effect of endogenous GLP-1 on alcohol intake is to block the GLP-1R with exendin-9–39. As such, infusion of exendin-9–39 into the VTA resulted in an increase in alcohol intake in male rats, whereas the intake was unaffected after its systemic administration (Marty et al., 2020; Shirazi et al., 2013). Additional evidence suggests that circulating GLP-1 holds diminished significance on alcohol consumption, given that heightened endogenous GLP-1 levels, achieved through increased release (agonists of the G-protein receptor 119) or reduced degradation (DPP-IV inhibitor), did not alter alcohol intake (Marty et al., 2020).

1.2.1.2 Motivation to consume alcohol and relapse drinking

Like ghrelin receptor antagonists, GLP-1R agonists suppress both the motivation to consume alcohol as well as relapse drinking. Specifically, in the operant self-administration model systemic administration of Ex4 reduced the motivation to consume alcohol in male rats (Bornebusch, Fink-Jensen, Wortwein, Seeley, & Thomsen, 2019; Egecioglu et al., 2013), in a dose-dependent manner (Bornebusch et al., 2019). As these initial self-administration studies focus on male rodents, one recent study explored the influence of Ex4 on mice of both sexes. Thus, Ex4 diminished the drive to consume alcohol in both male and female mice, although the efficacy of the treatment appeared to be less pronounced in females (Diaz-Megido & Thomsen, 2023). Additionally, Ex4 attenuated the alcohol-seeking in the progressive ration test of male rats (Egecioglu et al., 2013) and male mice (Diaz-Megido & Thomsen, 2023), whereas this behavior was unaffected by Ex4 in female mice (Diaz-Megido & Thomsen, 2023). A similar outcome was observed in alcohol-preferring rats (sP rats), as repeated injections of liraglutide decreased the motivation to obtain alcohol (Vallof et al., 2015). As mentioned above, relapse drinking during alcohol abstinence (i.e., white period) is a substantial problem for patients with AUD. Intriguingly, this relapse of drinking was prevented by systemic injection of Ex4 to male rats (Vallof et al., 2015) or socially housed male mice (Thomsen et al., 2017). These findings were consistent among different GLP-1R agonists as liraglutide suppressed the relapse drinking in male mice (Liu et al., 2024). Repeated cycles of abstinence escalate relapse drinking further, a drinking behavior attenuated by another GLP-1R agonist (AC3174) (Suchankova et al., 2015). Abstinence symptoms during alcohol withdrawal represent another notable aspect among patients with AUD, an area that has not been explored in relation to the ghrelin pathway. However, these symptoms are influenced by both GLP-1 and its receptor activation. Supportively, liraglutide attenuated the ability of repeated alcohol exposure to induce abstinence symptoms (Sharma, Pise, Sharma, & Shukla, 2014b). Enhanced anxiety-like behaviors during alcohol withdrawal was alleviated by liraglutide in male mice (Liu et al., 2024). A suppressive effect was also obtained after inhibition of DPP-IV causing elevated GLP-1 levels (Sharma, Pise, Sharma, & Shukla, 2014a). Consequently, these findings collectively suggest that GLP-1R may regulate various criteria associated with the AUD diagnosis.

1.2.1.3 Alcohol-related behaviors and neurochemistry

The hypothesis suggesting that ghrelin receptor antagonists suppress alcohol consumption by reducing alcohol-induced reward could potentially extend to GLP-1R agonists as well. Supportively, Ex4 blocked the alcohol-induced locomotor stimulation, accumbal dopamine release, and the expression of reward in the conditioned place preference paradigm male mice (Egecioglu et al., 2013). This finding was further validated by a subsequent study, which confirmed that Ex4 prevented the expression of alcohol reward in the conditioned place preference model (Shirazi et al., 2013). Like Ex4, liraglutide diminished these rewarding properties of alcohol in male mice (Vallof et al., 2015). It should be further noted that in contrast to liraglutide, either Ex4 or GLP-1 suppresses the memory consolidation of alcohol reward in the conditioned place preference test; a behavior of importance for the manifestation of AUD (Egecioglu et al., 2013; Shirazi et al., 2013; Vallof et al., 2015). While liraglutide did not influence the memory of alcohol-induced reward (Vallof et al., 2015), liraglutide augmented the memory impairments induced by long-term alcohol exposure as measured in the Morris water maze (Liu et al., 2024). It is of utter importance that the different GLP-1R agonists share treatment outcomes on alcohol-induced reward, whereas the effect on memory of alcohol reward is diverging. While the reasons for this difference remain to be defined, factors like brain penetration or downstream mechanism may influence this effect (Gu et al., 2013; Salinas et al., 2018).

Although studied to a lesser extent, Ex4 has been found to have other beneficial effects in rodents exposed to alcohol for extended periods. Indeed, Ex4 improves indicators of hepatic steatosis caused by long-term alcohol injections (Mahalingam et al., 2023). As alcohol exposure impairs brain penetration, compromises gut barrier function, disrupts the intestinal mucosa layer, and affects various bodily functions, studies should investigate the potential impact of GLP-1R agonists on such outcomes.

1.2.1.4 Tentative mechanisms of action

The findings that both genetic and pharmacological suppression of central GLP-1R blocked the ability of Ex4 to reduce alcohol-related behaviors indicate that central rather than peripheral GLP-1R regulates the effects of alcohol (Sirohi, Schurdak, Seeley, Benoit, & Davis, 2016; Vallof, Vestlund, & Jerlhag, 2019b). Studies have since then revealed brain regions central to this interaction, where NAc is one potential area of interest. This is suggested as fluorescently tagged Ex4 can be detected on both neurons and astrocytes in NAc after its systemic administration to male rats (Hernandez, O'Donovan, Ortinski, & Schmidt, 2019). Similarly, following systemic administration of fluorescently labeled semaglutide to male and female rats with a history of alcohol consumption, its presence was observable in the NAc (Aranas et al., 2023b). A modulatory role of GLP-1R in NAc was further implied as the alcohol consumption was lower in male and female rats after intra-NAc infusion of Ex4 (Abtahi et al., 2018; Colvin et al., 2020; Vallof, Kalafateli, & Jerlhag, 2019a). Supportively, Ex4 into NAc attenuated the ability of alcohol to cause a locomotor stimulation and memory of alcohol reward in the conditioned place preference paradigm

(Vallof et al., 2019a) as well as reduced operant self-administration of alcohol (Allingbjerg, Hansen, Secher, & Thomsen, 2023). Alcohol consumption patterns were correlated with the GLP-1R expression in the NAc, showing higher levels in male rats with high alcohol intake compared to those with low consumption (Vallof et al., 2019a). Electrophysiological recordings of NAc slices revealed that both Ex4 and liraglutide, but not dulaglutide, suppressed neurotransmission in NAc without affecting the inhibitory neurotransmission in this area (Vestlund, Bergquist, Licheri, Adermark, & Jerlhag, 2020). It should be further noted that the effects of Ex4 on neurotransmission depend on exposure to cocaine, and tentatively alcohol. Indeed, in cocaine-experienced rats Ex4 enhanced the frequency of action potential firing in NAc, whereas no such effect was observed in saline-treated controls (Hernandez et al., 2019). The findings that GLP-1Rs are expressed presynaptically on GABAergic nerve terminals (Babateen et al., 2017; Korol, Jin, & Birnir, 2015; Korol, Jin, Babateen, & Birnir, 2015; Rebosio, Balbi, Passalacqua, Ricciarelli, & Fedele, 2018), indicate that such mechanisms may be involved.

Another area of tentative interest when it comes to the ability of GLP-1R to modulate alcohol-related behaviors is the VTA. Given its heterogeneous nature, it is not surprising that the treatment effect of Ex4 varies depending on the subregion of the VTA. Thus, when infused to the anterior part of VTA, Ex4 neither influenced the alcohol-induced locomotor stimulation, memory of alcohol reward, nor alcohol intake in male rodents (Vallof et al., 2019a). In contrast, the alcohol intake and self-administration were decreased after Ex4 infusion into the posterior part (Colvin et al., 2020; Dixon, McNally, & Ong, 2020; Shirazi et al., 2013). As expected, antagonism of the GLP-1R in posterior VTA elevated the alcohol intake in male rats (Shirazi et al., 2013). The findings that Ex4 into posterior VTA did not alter the alcohol-induced locomotor stimulation, memory of alcohol reward, nor alcohol intake in male rodents (Vallof et al., 2019a), indicate that some but not all alcohol-related behaviors are modulated by GLP-1R in the posterior part.

The discovery that both NAc and VTA are targeted by GLP-1-containing neurons from the NTS (Rinaman, 2010), raises the possibility that GLP-1Rs expressed in the NTS play a regulatory role in alcohol-related responses. Support for this contention is provided as Ex4 into NTS attenuated the ability of alcohol to cause locomotor stimulation, dopamine release in NAc, memory of alcohol reward, and alcohol intake in male rodents (Vallof et al., 2019b).

Given the widespread expression of GLP-1Rs in the brain, it is not surprising that additional brain regions may participate in their ability to control alcohol-related behaviors. Alcohol-related behaviors were attenuated by local infusion of Ex4 into the LDTg (Vallof et al., 2019a), lateral hypothalamus (Colvin et al., 2020), or lateral septum (Allingbjerg et al., 2023). Moreover, local infusion of Ex4 into hippocampus reduced the operant-self administration, (Allingbjerg et al., 2023) and liraglutide prevented the ability of alcohol consumption to lower the dendritic spine density and levels of synaptic transport-related proteins in this region (Liu et al., 2024). On the other hand, areas like the arcuate nucleus, paraventricular nucleus, basolateral amygdala (Colvin et al., 2020) or putamen (Allingbjerg et al., 2023) appear less important for this interaction as local infusion of Ex4 into either area did not alter alcohol-related responses.

The above-summarized studies have revealed some mechanisms of interest, but additional studies are warranted to pinpoint the exact neurocircuits through which the different GLP-1R agonists act.

1.2.2 Clinical studies

In contrast to the vast number of studies exploring the effects of alcohol on plasma ghrelin levels, only one study has so far explored the impact of alcohol on circulating GLP-1 levels. Thus, alcohol exposure reduced the GLP-1 levels in plasma, an effect consistent between the administration route or time of alcohol exposure (Farokhnia et al., 2022a). However, no differences in GLP-1 levels were found between AUD patients and healthy controls (Olsson et al., 2023).

Associations between alcohol and GLP-1R were further found in human genetic studies. An initial study of three separate cohorts demonstrated an association between AUD and polymorphisms in the GLP-1R gene (Suchankova et al., 2015). Moreover, variations in the GLP-1R gene were positively correlated with intravenous self-administration of alcohol and its following enhanced breath alcohol concentrations (Suchankova et al., 2015). Similarly, there was a significant interaction between AUDIT scores and SNPs of the GLP-1R genes (Farokhnia et al., 2022b). While not directly related to alcohol, GLP-1R polymorphisms were correlated to the ability of monetary reward to elevate the BOLD response in the right globus pallidus (Suchankova et al., 2015).

Rat studies have found that GLP-1Rs within the hippocampus regulate alcohol drinking (Allingbjerg et al., 2023). Correspondingly, a human study shows that the expression of GLP-1R in the hippocampus was higher in patients with AUD compared to controls (Farokhnia et al., 2022a).

The first indication of the ability of GLP-1R agonists to regulate alcohol intake in humans was provided by data from a small, preliminary pilot study in which liraglutide lowered alcohol consumption in patients with diabetes type II (Kalra, 2011). Moreover, data from a randomized clinical trial revealed that exenatide reduced alcohol intake in overweight AUD patients (Klausen et al., 2022). It should however be noted that this treatment effect was not observed in AUD patients with regular weight (Klausen et al., 2022). As further evident in patients with AUD, 12 weeks of dulaglutide treatment lowered alcohol consumption by 29%, and the effect was unaffected by smoking status (Probst et al., 2023). Additional studies explored how GLP-1R influences alcohol intake in overweight patients who also are alcohol drinkers. As such, in overweight patients, semaglutide reduced the self-reported intake of alcohol, AUDIT scores, and binge drinking (Quddos et al., 2023). Besides, in this patient group, semaglutide attenuated both the stimulatory and sedative properties of alcohol (Quddos et al., 2023). Moreover, as shown in a register-based study the alcohol-related events were found to be reduced by GLP-1R agonists (Wium-Andersen et al., 2022). Using machine learning, semaglutide was also found to decrease alcohol-related posts (craving, desire, negative effects) in overweight patients who drink alcohol (Quddos et al., 2023). A similar outcome was observed when analyzing the effects of GLP-1R agonists on anonymous online alcohol reports since a decline in intake, desire, and loss of interest was reported (Bremmer & Hendershot, 2024). Similarly, when comparing the AUDIT scores before and after semaglutide treatment in overweight patients with AUD, a reduction was observed by treatment (Richards et al., 2023). The efficacy of GLP-1R agonists in decreasing alcohol consumption among humans seems to be consistently observed across a study involving DPP-IV inhibitors, which are recognized for their role in augmenting circulating GLP-1 levels. Indeed, a register-based study revealed that the number of alcohol-related events was lowered by DPP-IV inhibitors (Wium-Andersen et al., 2022). Together, these studies demonstrate the promising potential of GLP-1R agonists in reducing alcohol intake among individuals with both AUD and obesity, warranting further investigation through larger randomized clinical trials.

2. Summary and future directions

While preclinical and clinical studies demonstrate the modulation of alcohol-related responses by both ghrelin and its receptor, the influence of other components within the ghrelin pathway remains unexplored. One of these is the LEAP2, an endogenous inverse agonist on ghrelin receptors (Ge et al., 2018; M'Kadmi et al., 2019). The production of LEAP2 is mainly located in the small intestine and liver (Krause et al., 2003), but recent studies reveal its expression in the brain (Islam et al., 2020). While LEAP2 reduces both chow and high-fat diet intake (Cornejo et al., 2019; Ge et al., 2018), its effects on alcohol are less studied. As only one limited study reveals that central infusion of LEAP2 reduces alcohol binge drinking in mice that consumed a sweetened alcohol solution (Richardson et al., 2023), upcoming studies should explore this interaction further. DAG is another unexplored substrate of the ghrelin pathway, which previously was considered to be inactive (Delhanty et al., 2010; Fernandez et al., 2016; Gauna et al., 2006; Hosoda et al., 2000; Inoue et al., 2013). While a limited number of studies reveal that DAG reduces food intake (Asakawa et al., 2005), and prevents the effects of ghrelin (Asakawa et al., 2005; Fernandez et al., 2016; Fujimiya et al., 2010; Heppner et al., 2014; Inoue et al., 2013), its effects on alcohol-related responses remains to be determined. Although a case-control study reveals that the expression of GOAT, the enzyme octanoylating ghrelin, is similar between healthy controls and AUD patients (Deschaine et al., 2021), the impact of GOAT inhibitors on alcohol-related responses should be defined. The unexploited domains necessitating exploration within the alcohol and ghrelin pathway include elucidating mechanisms of action, discerning sex-specific variances, and evaluating treatment efficacy in clinical trials involving individuals diagnosed with AUD. These areas should also be addressed when it comes to future studies using GLP-1R agonists. Specifically, mechanisms of action should be defined for each of the clinically GLP-1R agonists as they most likely diverge to some extent.

As summarized elsewhere (Shevchouk, Tufvesson-Alm, & Jerlhag, 2021), ghrelin receptor antagonists and GLP-1R agonists suppress the behavioral and neurochemical responses to addictive drugs and reduce the consumption thereof. Upcoming studies should therefore explore their impact on alcohol and drug taking in rodents and humans with multiple drug use. When it comes to feeding experiments, these drugs reduce the most rewarding ones while leaving less rewarding ones largely unaffected.

It is therefore plausible that ghrelin receptor antagonists and GLP-1R agonists only reduce the most rewarding addictive drug in a choice situation, a subject for future studies.

When it comes to complex and heterogenous disorders like AUD, one approach with potentially beneficial treatment outcomes is to combine different agents that individually reduce alcohol. Supportively, the beneficial effects on alcohol drinking have been found by the combination of bupropion and varenicline (Soderpalm, Danielsson, de Bejczy, Adermark, & Ericson, 2020) as well as naltrexone together with either varenicline or bupropion (Grodin et al., 2021; Zhou, Leri, Low, & Kreek, 2019). However, this approach is unexplored when it comes to the impact of gut–brain peptide combinations on alcohol-related responses and upcoming studies should explore this potential interaction. On a similar note, the combination of the gut–brain peptides GLP-1 and amylin synergistically reduces feeding and body weight (Liberini et al., 2019).

Acknowledgment

EJH has no conflict of interest, and this review is supported by grants from the Swedish Research Council (2023–2600).

References

Abtahi, S., Howell, E., & Currie, P. J. (2018). Accumbal ghrelin and glucagon-like peptide 1 signaling in alcohol reward in female rats. *Neuroreport, 29*(12), 1046–1053. https://doi.org/10.1097/WNR.0000000000001071.

Addolorato, G., Capristo, E., Leggio, L., Ferrulli, A., Abenavoli, L., Malandrino, N., et al. (2006). Relationship between ghrelin levels, alcohol craving, and nutritional status in current alcoholic patients ([Research Support, Non-U.S. Gov't]) *Alcoholism, Clinical and Experimental Research, 30*(11), 1933–1937. https://doi.org/10.1111/j.1530-0277.2006.00238.x.

Agtas-Ertan, E., Dogan, O., & Ilhan, I. (2023). Ghrelin and impulsivity relationship in alcohol-dependent patients and healthy individuals. *Alcohol and Alcoholism (Oxford, Oxfordshire), 58*(5), 497–504. https://doi.org/10.1093/alcalc/agad032.

Allingbjerg, M. L., Hansen, S. N., Secher, A., & Thomsen, M. (2023). Glucagon-like peptide-1 receptors in nucleus accumbens, ventral hippocampus, and lateral septum reduce alcohol reinforcement in mice. *Experimental and Clinical Psychopharmacology, 31*(3), 612–620. https://doi.org/10.1037/pha0000620.

Alvarez, E., Roncero, I., Chowen, J. A., Thorens, B., & Blazquez, E. (1996). Expression of the glucagon-like peptide-1 receptor gene in rat brain [Research Support, Non-U.S. Gov't]. *Journal of Neurochemistry, 66*(3), 920–927. http://www.ncbi.nlm.nih.gov/pubmed/8769850.

Aranas, C., Blid Skoldheden, S., & Jerlhag, E. (2023). Antismoking agents do not contribute synergistically to semaglutide's ability to reduce alcohol intake in rats. *Frontiers in Pharmacology, 14*, 1180512. https://doi.org/10.3389/fphar.2023.1180512.

Aranas, C., Edvardsson, C. E., Shevchouk, O. T., Zhang, Q., Witley, S., Blid Skoldheden, S., et al. (2023). Semaglutide reduces alcohol intake and relapse-like drinking in male and female rats. *EBioMedicine, 93*, 104642. https://doi.org/10.1016/j.ebiom.2023.104642.

Asakawa, A., Inui, A., Fujimiya, M., Sakamaki, R., Shinfuku, N., Ueta, Y., et al. (2005). Stomach regulates energy balance via acylated ghrelin and desacyl ghrelin. *Gut, 54*(1), 18–24. <Go to ISI>://000225659500006.

Babateen, O., Korol, S. V., Jin, Z., Bhandage, A. K., Ahemaiti, A., & Birnir, B. (2017). Liraglutide modulates GABAergic signaling in rat hippocampal CA3 pyramidal neurons predominantly by presynaptic mechanism. *BMC Pharmacology and Toxicology, 18*(1), 83. https://doi.org/10.1186/s40360-017-0191-0.

Bach, P., Bumb, J. M., Schuster, R., Vollstadt-Klein, S., Reinhard, I., Rietschel, M., et al. (2019). Effects of leptin and ghrelin on neural cue-reactivity in alcohol addiction: Two streams merge to one river? *Psychoneuroendocrinology, 100*, 1–9. https://doi.org/10.1016/j.psyneuen.2018.09.026.

Badaoui, A., De Saeger, C., Duchemin, J., Gihousse, D., de Timary, P., & Starkel, P. (2008). Alcohol dependence is associated with reduced plasma and fundic ghrelin levels. *European Journal of Clinical Investigation, 38*(6), 397–403. https://doi.org/10.1111/j.1365-2362.2008.01947.x.

Bahi, A., Tolle, V., Fehrentz, J. A., Brunel, L., Martinez, J., Tomasetto, C. L., et al. (2013). Ghrelin knockout mice show decreased voluntary alcohol consumption and reduced ethanol-induced conditioned place preference. *Peptides, 43*, 48–55. https://doi.org/10.1016/j.peptides.2013.02.008.

Boileau, I., Assaad, J. M., Pihl, R. O., Benkelfat, C., Leyton, M., Diksic, M., et al. (2003). Alcohol promotes dopamine release in the human nucleus accumbens. *Synapse (New York, N. Y.), 49*(4), 226–231. https://doi.org/10.1002/syn.10226.

Bornebusch, A. B., Fink-Jensen, A., Wortwein, G., Seeley, R. J., & Thomsen, M. (2019). Glucagon-like peptide-1 receptor agonist treatment does not reduce abuse-related effects of opioid drugs. *eNeuro, 6*(2), https://doi.org/10.1523/ENEURO.0443-18.2019.

Bremmer, M. P., & Hendershot, C. S. (2024). Social media as pharmacovigilance: The potential for patient reports to inform clinical research on glucagon-like peptide 1 (GLP-1) receptor agonists for substance use disorders. *Journal of Studies on Alcohol and Drugs, 85*(1), 5–11. https://doi.org/10.15288/jsad.23-00318.

Brigande, A. M., Darwich, J. G., & Currie, P. J. (2023). Mesolimbic exendin-4 attenuates reward salience evoked by neuropeptide Y and ghrelin. *Behavioural Brain Research, 440*, 114249. https://doi.org/10.1016/j.bbr.2022.114249.

Calissendorff, J., Danielsson, O., Brismar, K., & Rojdmark, S. (2005). Inhibitory effect of alcohol on ghrelin secretion in normal man [Clinical Trial Controlled Clinical Trial Research Support, Non-U.S. Gov't]. *European Journal of Endocrinology/European Federation of Endocrine Societies, 152*(5), 743–747. https://doi.org/10.1530/eje.1.01905.

Calissendorff, J., Danielsson, O., Brismar, K., & Rojdmark, S. (2006). Alcohol ingestion does not affect serum levels of peptide YY but decreases both total and octanoylated ghrelin levels in healthy subjects. *Metabolism-Clinical and Experimental, 55*(12), 1625–1629. <Go to ISI>://000242856500012.

Calissendorff, J., Gustafsson, T., Holst, J. J., Brismar, K., & Rojdmark, S. (2012). Alcohol intake and its effect on some appetite-regulating hormones in man: Influence of gastroprotection with sucralfate. *Endocrine Research, 37*(3), 154–162. https://doi.org/10.3109/07435800.2012.662662.

Cepko, L. C., Selva, J. A., Merfeld, E. B., Fimmel, A. I., Goldberg, S. A., & Currie, P. J. (2014). Ghrelin alters the stimulatory effect of cocaine on ethanol intake following mesolimbic or systemic administration. *Neuropharmacology, 85*, 224–231. https://doi.org/10.1016/j.neuropharm.2014.05.030.

Chuong, V., Farokhnia, M., Khom, S., Pince, C. L., Elvig, S. K., Vlkolinsky, R., et al. (2023). The glucagon-like peptide-1 (GLP-1) analogue semaglutide reduces alcohol drinking and modulates central GABA neurotransmission. *JCI Insight, 8*(12), https://doi.org/10.1172/jci.insight.170671.

Colvin, K. J., Killen, H. S., Kanter, M. J., Halperin, M. C., Engel, L., & Currie, P. J. (2020). Brain site-specific inhibitory effects of the GLP-1 analogue exendin-4 on alcohol intake and operant responding for palatable food. *International Journal of Molecular Sciences, 21*(24), https://doi.org/10.3390/ijms21249710.

Colvin, K. J., Killen, H. S., Kanter, M. J., Halperin, M. C., Engel, L., Dickinson, M. B., et al. (2022). Differential effects of intra-ventral tegmental area ghrelin and glucagon-like peptide-1 on the stimulatory action of D-amphetamine and cocaine-induced ethanol intake in male Sprague Dawley rats. *Behavioural Brain Research, 421*, 113726. https://doi.org/10.1016/j.bbr.2021.113726.

Cornejo, M. P., Barrile, F., De Francesco, P. N., Portiansky, E. L., Reynaldo, M., & Perello, M. (2018). Ghrelin recruits specific subsets of dopamine and GABA neurons of different ventral tegmental area sub-nuclei. *Neuroscience, 392*, 107–120. https://doi.org/10.1016/j.neuroscience.2018.09.027.

Cornejo, M. P., Castrogiovanni, D., Schioth, H. B., Reynaldo, M., Marie, J., Fehrentz, J. A., et al. (2019). Growth hormone secretagogue receptor signalling affects high-fat intake independently of plasma levels of ghrelin and LEAP2, in a 4-day binge eating model. *Journal of Neuroendocrinology, 31*(10), e12785. https://doi.org/10.1111/jne.12785.

Cornejo, M. P., Mustafa, E. R., Barrile, F., Cassano, D., De Francesco, P. N., Raingo, J., et al. (2021). The intriguing ligand-dependent and ligand-independent actions of the growth hormone secretagogue receptor on reward-related behaviors. *Neuroscience and Biobehavioral Reviews, 120*, 401–416. https://doi.org/10.1016/j.neubiorev.2020.10.017.

Cowley, M. A., Smith, R. G., Diano, S., Tschop, M., Pronchuk, N., Grove, K. L., et al. (2003). The distribution and mechanism of action of ghrelin in the CNS demonstrates a novel hypothalamic circuit regulating energy homeostasis [In Vitro Research Support, Non-U.S. Gov't Research Support, U.S. Gov't, P.H.S]. *Neuron, 37*(4), 649–661. http://www.ncbi.nlm.nih.gov/pubmed/12597862.

Cruz, M. T., Herman, M. A., Cote, D. M., Ryabinin, A. E., & Roberto, M. (2013). Ghrelin increases GABAergic transmission and interacts with ethanol actions in the rat central nucleus of the amygdala. *Neuropsychopharmacology: Official Publication of the American College of Neuropsychopharmacology, 38*(2), 364–375. https://doi.org/10.1038/npp.2012.190.

Davis, J. F., Schurdak, J. D., Magrisso, I. J., Mul, J. D., Grayson, B. E., Pfluger, P. T., et al. (2012). Gastric bypass surgery attenuates ethanol consumption in ethanol-preferring rats. *Biological Psychiatry, 72*(5), 354–360. https://doi.org/S0006-3223(12)00136-9 [pii] https://doi.org/10.1016/j.biopsych.2012.01.035 .

De Vriese, C., Gregoire, F., Lema-Kisoka, R., Waelbroeck, M., Robberecht, P., & Delporte, C. (2004). Ghrelin degradation by serum and tissue homogenates: Identification of the cleavage sites. *Endocrinology, 145*(11), 4997–5005. https://doi.org/10.1210/en.2004-0569.

Delhanty, P. J., Sun, Y., Visser, J. A., van Kerkwijk, A., Huisman, M., van Ijcken, W. F., et al. (2010). Unacylated ghrelin rapidly modulates lipogenic and insulin signaling pathway gene expression in metabolically active tissues of GHSR deleted mice. *PLoS One, 5*(7), e11749. https://doi.org/10.1371/journal.pone.0011749.

Deschaine, S. L., Farokhnia, M., Gregory-Flores, A., Zallar, L. J., You, Z. B., Sun, H., et al. (2021). A closer look at alcohol-induced changes in the ghrelin system: Novel insights from preclinical and clinical data. *Addiction Biology, 27*, e13033. https://doi.org/10.1111/adb.13033.

Diaz-Megido, C., & Thomsen, M. (2023). Sex-dependent divergence in the effects of GLP-1 agonist exendin-4 on alcohol reinforcement and reinstatement in C57BL/6J mice. *Psychopharmacology (Berl), 240*(6), 1287–1298. https://doi.org/10.1007/s00213-023-06367-x.

Dickson, S. L., Hrabovszky, E., Hansson, C., Jerlhag, E., Alvarez-Crespo, M., Skibicka, K. P., et al. (2010). Blockade of central nicotine acetylcholine receptor signaling attenuate ghrelin-induced food intake in rodents. *Neuroscience, 171*(4), 1180–1186. https://doi.org/S0306-4522(10)01335-7 [pii] https://doi.org/10.1016/j.neuroscience.2010.10.005.

Dixon, T. N., McNally, G. P., & Ong, Z. Y. (2020). Glucagon-like peptide-1 receptor signaling in the ventral tegmental area reduces alcohol self-administration in male rats. *Alcoholism, Clinical and Experimental Research, 44*(10), 2118–2129. https://doi.org/10.1111/acer.14437.

Edvardsson, C. E., Vestlund, J., & Jerlhag, E. (2021). A ghrelin receptor antagonist reduces the ability of ghrelin, alcohol or amphetamine to induce a dopamine release in the ventral tegmental area and in nucleus accumbens shell in rats. *European Journal of Pharmacology, 899*, 174039. https://doi.org/10.1016/j.ejphar.2021.174039.

Egecioglu, E., Steensland, P., Fredriksson, I., Feltmann, K., Engel, J. A., & Jerlhag, E. (2013). The glucagon-like peptide 1 analogue exendin-4 attenuates alcohol mediated behaviors in rodents. *Psychoneuroendocrinology, 38*(8), 1259–1270. https://doi.org/10.1016/j.psyneuen.2012.11.009.

Engel, J. A., Palsson, E., Vallof, D., & Jerlhag, E. (2023). Ghrelin activates the mesolimbic dopamine system via nitric oxide associated mechanisms in the ventral tegmental area. *Nitric Oxide: Biology and Chemistry/Official Journal of the Nitric Oxide Society, 131*, 1–7. https://doi.org/10.1016/j.niox.2022.12.001.

Farokhnia, M., Browning, B. D., Crozier, M. E., Sun, H., Akhlaghi, F., & Leggio, L. (2022). The glucagon-like peptide-1 system is modulated by acute and chronic alcohol exposure: Findings from human laboratory experiments and a post-mortem brain study. *Addiction Biology, 27*(5), e13211. https://doi.org/10.1111/adb.13211.

Farokhnia, M., Fede, S. J., Grodin, E. N., Browning, B. D., Crozier, M. E., Schwandt, M. L., et al. (2022b). Differential association between the GLP1R gene variants and brain functional connectivity according to the severity of alcohol use. *Scientific Reports, 12*(1), 13027. https://doi.org/10.1038/s41598-022-17190-3.

Farokhnia, M., Grodin, E. N., Lee, M. R., Oot, E. N., Blackburn, A. N., Stangl, B. L., et al. (2017). Exogenous ghrelin administration increases alcohol self-administration and modulates brain functional activity in heavy-drinking alcohol-dependent individuals. *Molecular Psychiatry*. https://doi.org/10.1038/mp.2017.226.

Farokhnia, M., Lee, M. R., Farinelli, L. A., Ramchandani, V. A., Akhlaghi, F., & Leggio, L. (2018). Pharmacological manipulation of the ghrelin system and alcohol hangover symptoms in heavy drinking individuals: Is there a link? *Pharmacology, Biochemistry, and Behavior, 172*, 39–49. https://doi.org/10.1016/j.pbb.2018.07.004.

Farokhnia, M., Murphy, G., Weinstein, S. J., Shah, N. N., Parisi, D., Albanes, D., et al. (2021). A population-based investigation of the association between alcohol intake and serum total ghrelin concentrations among cigarette-smoking, non-alcohol-dependent male individuals. *Drug and Alcohol Dependence, 226*, 108835. https://doi.org/10.1016/j.drugalcdep.2021.108835.

Farokhnia, M., Portelli, J., Lee, M. R., McDiarmid, G. R., Munjal, V., Abshire, K. M., et al. (2020). Effects of exogenous ghrelin administration and ghrelin receptor blockade, in combination with alcohol, on peripheral inflammatory markers in heavy-drinking individuals: Results from two human laboratory studies. *Brain Research, 1740*, 146851. https://doi.org/10.1016/j.brainres.2020.146851.

Fernandez, G., Cabral, A., Cornejo, M. P., De Francesco, P. N., Garcia-Romero, G., Reynaldo, M., et al. (2016). Des-acyl ghrelin directly targets the arcuate nucleus in a ghrelin-receptor independent manner and impairs the orexigenic effect of ghrelin. *Journal of Neuroendocrinology, 28*(2), 12349. https://doi.org/10.1111/jne.12349.

Fujimiya, M., Asakawa, A., Ataka, K., Chen, C. Y., Kato, I., & Inui, A. (2010). Ghrelin, des-acyl ghrelin, and obestatin: Regulatory roles on the gastrointestinal motility. *International Journal of Peptides, 2010*. https://doi.org/10.1155/2010/305192.

Furness, J. B., Hunne, B., Matsuda, N., Yin, L., Russo, D., Kato, I., et al. (2011). Investigation of the presence of ghrelin in the central nervous system of the rat and mouse. *Neuroscience, 193*, 1–9. https://doi.org/10.1016/j.neuroscience.2011.07.063.

Gauna, C., Delhanty, P. J. D., van Aken, M. O., Janssen, J. A. M. J. L., Themmen, A. P. N., Hofland, L. J., et al. (2006). Unacylated ghrelin is active on the INS-1E rat insulinoma cell line independently of the growth hormone secretagogue receptor type 1a and the corticotropin releasing factor 2 receptor. *Molecular and Cellular Endocrinology, 251*(1–2), 103–111. <Go to ISI>://000238344900012.

Ge, X., Yang, H., Bednarek, M. A., Galon-Tilleman, H., Chen, P., Chen, M., et al. (2018). LEAP2 is an endogenous antagonist of the ghrelin receptor. *Cell Metabolism, 27*(2), 461–469.e6. https://doi.org/10.1016/j.cmet.2017.10.016.

Godlewski, G., Cinar, R., Coffey, N. J., Liu, J., Jourdan, T., Mukhopadhyay, B., et al. (2019). Targeting peripheral CB1 receptors reduces ethanol intake via a gut-brain axis. *Cell Metabolism, 29*(6), 1320–1333.e8. https://doi.org/10.1016/j.cmet.2019.04.012.

Gomez, J. L., Cunningham, C. L., Finn, D. A., Young, E. A., Helpenstell, L. K., Schuette, L. M., et al. (2015). Differential effects of ghrelin antagonists on alcohol drinking and reinforcement in mouse and rat models of alcohol dependence. *Neuropharmacology, 97*, 182–193. https://doi.org/10.1016/j.neuropharm.2015.05.026.

Gomez, J. L., & Ryabinin, A. E. (2014). The effects of ghrelin antagonists [D-Lys(3)]-GHRP-6 or JMV2959 on ethanol, water, and food intake in C57BL/6J mice. *Alcoholism, Clinical and Experimental Research, 38*(9), 2436–2444. https://doi.org/10.1111/acer.12499.

Grodin, E. N., Burnette, E. M., Green, R., Lim, A. C., Miotto, K., & Ray, L. A. (2021). Combined varenicline and naltrexone attenuates alcohol cue-elicited activation in heavy drinking smokers. *Drug and Alcohol Dependence, 225*, 108825. https://doi.org/10.1016/j.drugalcdep.2021.108825.

Grouselle, D., Chaillou, E., Caraty, A., Bluet-Pajot, M. T., Zizzari, P., Tillet, Y., et al. (2008). Pulsatile cerebrospinal fluid and plasma ghrelin in relation to growth hormone secretion and food intake in the sheep. *Journal of Neuroendocrinology, 20*(10), 1138–1146. https://doi.org/10.1111/j.1365-2826.2008.01770.x.

Gu, G., Roland, B., Tomaselli, K., Dolman, C. S., Lowe, C., & Heilig, J. S. (2013). Glucagon-like peptide-1 in the rat brain: Distribution of expression and functional implication. *The Journal of Comparative Neurology, 521*(10), 2235–2261. https://doi.org/10.1002/cne.23282.

Haass-Koffler, C. L., Long, V. M., Farokhnia, M., Magill, M., Kenna, G. A., Swift, R. M., et al. (2019). Intravenous administration of ghrelin increases serum cortisol and aldosterone concentrations in heavy-drinking alcohol-dependent individuals: Results from a double-blind, placebo-controlled human laboratory study. *Neuropharmacology, 158*, 107711. https://doi.org/10.1016/j.neuropharm.2019.107711.

Hajnal, A., Zharikov, A., Polston, J. E., Fields, M. R., Tomasko, J., Rogers, A. M., et al. (2012). Alcohol reward is increased after Roux-en-Y gastric bypass in dietary obese rats with differential effects following ghrelin antagonism. *PLoS One*, 7(11), e49121. https://doi.org/10.1371/journal.pone.0049121.

Heppner, K. M., Piechowski, C. L., Muller, A., Ottaway, N., Sisley, S., Smiley, D. L., et al. (2014). Both acyl and des-acyl ghrelin regulate adiposity and glucose metabolism via central nervous system ghrelin receptors. *Diabetes, 63*(1), 122–131. https://doi.org/10.2337/db13-0414.

Hernandez, N. S., O'Donovan, B., Ortinski, P. I., & Schmidt, H. D. (2019). Activation of glucagon-like peptide-1 receptors in the nucleus accumbens attenuates cocaine seeking in rats. *Addiction Biology, 24*(2), 170–181. https://doi.org/10.1111/adb.12583.

Holst, J. J., & Seino, Y. (2009). GLP-1 receptor agonists: Targeting both hyperglycaemia and disease processes in diabetes. *Diabetes Research and Clinical Practice, 85*(1), 1–3. https://doi.org/10.1016/j.diabres.2009.02.017.

Hosoda, H., Kojima, M., Matsuo, H., & Kangawa, K. (2000). Ghrelin and des-acyl ghrelin: Two major forms of rat ghrelin peptide in gastrointestinal tissue [Research Support, Non-U.S. Gov't]. *Biochemical and Biophysical Research Communications, 279*(3), 909–913. https://doi.org/10.1006/bbrc.2000.4039.

Inoue, Y., Nakahara, K., Maruyama, K., Suzuki, Y., Hayashi, Y., Kangawa, K., et al. (2013). Central and peripheral des-acyl ghrelin regulates body temperature in rats. *Biochemical and Biophysical Research Communications, 430*(1), 278–283. https://doi.org/10.1016/j.bbrc.2012.10.137.

Islam, M. N., Mita, Y., Maruyama, K., Tanida, R., Zhang, W., Sakoda, H., et al. (2020). Liver-expressed antimicrobial peptide 2 antagonizes the effect of ghrelin in rodents. *The Journal of Endocrinology, 244*(1), 13–23. https://doi.org/10.1530/JOE-19-0102.

Jayaram-Lindström, N., Ericson, M., Steensland, P., & Jerlhag, E. (2016). Dopamine and alcohol dependence: From bench to clinic. In W. Meil (Ed.). *Recent advances in drug addiction research and clinical applications* (pp. 81–114). IntechOpen (Vol. https://doi.org/10.5772/63144).

Jensen, M., Ratner, C., Rudenko, O., Christiansen, S. H., Skov, L. J., Hundahl, C., et al. (2016). Anxiolytic-like effects of increased ghrelin receptor signaling in the amygdala. *The International Journal of Neuropsychopharmacology/Official Scientific Journal of the Collegium Internationale Neuropsychopharmacologicum (CINP), 19*(5), https://doi.org/10.1093/ijnp/pyv123.

Jerlhag, E., Egecioglu, E., Dickson, S. L., Andersson, M., Svensson, L., & Engel, J. A. (2006). Ghrelin stimulates locomotor activity and accumbal dopamine-overflow via central cholinergic systems in mice: Implications for its involvement in brain reward. *Addiction Biology, 11*(1), 45–54. https://doi.org/ADB002 [pii] https://doi.org/10.1111/j.1369-1600.2006.00002.x.

Jerlhag, E., Egecioglu, E., Dickson, S. L., Douhan, A., Svensson, L., & Engel, J. A. (2007). Ghrelin administration into tegmental areas stimulates locomotor activity and increases extracellular concentration of dopamine in the nucleus accumbens. *Addiction Biology, 12*(1), 6–16. https://doi.org/ADB041 [pii] https://doi.org/10.1111/j.1369-1600.2006.00041.x.

Jerlhag, E., Egecioglu, E., Dickson, S. L., & Engel, J. A. (2011). Glutamatergic regulation of ghrelin-induced activation of the mesolimbic dopamine system [Research Support, Non-U.S. Gov't]. *Addiction Biology, 16*(1), 82–91. https://doi.org/10.1111/j.1369-1600.2010.00231.x.

Jerlhag, E., Egecioglu, E., Dickson, S. L., Svensson, L., & Engel, J. A. (2008). Alpha-conotoxin MII-sensitive nicotinic acetylcholine receptors are involved in mediating the ghrelin-induced locomotor stimulation and dopamine overflow in nucleus accumbens [Research Support, Non-U.S. Gov't]. *European Neuropsychopharmacology: the Journal of the European College of Neuropsychopharmacology, 18*(7), 508–518. https://doi.org/10.1016/j.euroneuro.2008.02.006.

Jerlhag, E., Egecioglu, E., Landgren, S., Salome, N., Heilig, M., Moechars, D., et al. (2009). Requirement of central ghrelin signaling for alcohol reward [Research Support, Non-U.S. Gov't]. *Proceedings of the National Academy of Sciences of the United States of America, 106*(27), 11318–11323. https://doi.org/10.1073/pnas.0812809106.

Jerlhag, E., Ivanoff, L., Vater, A., & Engel, J. A. (2013). *Peripherally circulating ghrelin does not mediate alcohol-induced reward and alcohol intake in rodents.* ACER, In Press.

Jerlhag, E., Janson, A. C., Waters, S., & Engel, J. A. (2012). Concomitant release of ventral tegmental acetylcholine and accumbal dopamine by ghrelin in rats. *PLoS One, 7*(11), e49557. https://doi.org/10.1371/journal.pone.0049557.

Jerlhag, E., Landgren, S., Egecioglu, E., Dickson, S. L., & Engel, J. A. (2011). The alcohol-induced locomotor stimulation and accumbal dopamine release is suppressed in ghrelin knockout mice [Research Support, Non-U.S. Gov't]. *Alcohol (Fayetteville, N. Y.), 45*(4), 341–347. https://doi.org/10.1016/j.alcohol.2010.10.002.

Kalra, S. (2011). Change in alcohol consumption following liraglutide initiation: A real life experience. In: *71st American Diabetes Association conference, 2019-P.*

Karkkainen, O., Farokhnia, M., Klavus, A., Auriola, S., Lehtonen, M., Deschaine, S. L., et al. (2021). Effect of intravenous ghrelin administration, combined with alcohol, on circulating metabolome in heavy drinking individuals with alcohol use disorder. *Alcoholism, Clinical and Experimental Research, 45*(11), 2207–2216. https://doi.org/10.1111/acer.14719.

Kaur, S., & Ryabinin, A. E. (2010). Ghrelin receptor antagonism decreases alcohol consumption and activation of perioculomotor urocortin-containing neurons [Research Support, N.I.H., Extramural]. *Alcoholism, Clinical and Experimental Research, 34*(9), 1525–1534. https://doi.org/10.1111/j.1530-0277.2010.01237.x.

Kern, A., Albarran-Zeckler, R., Walsh, H. E., & Smith, R. G. (2012). Apo-ghrelin receptor forms heteromers with DRD2 in hypothalamic neurons and is essential for anorexigenic effects of DRD2 agonism. *Neuron, 73*(2), 317–332. https://doi.org/10.1016/j.neuron.2011.10.038.

Kern, A., Mavrikaki, M., Ullrich, C., Albarran-Zeckler, R., Brantley, A. F., & Smith, R. G. (2015). Hippocampal dopamine/DRD1 signaling dependent on the ghrelin receptor. *Cell, 163*(5), 1176–1190. https://doi.org/10.1016/j.cell.2015.10.062.

Kim, D. J., Yoon, S. J., Choi, B., Kim, T. S., Woo, Y. S., Kim, W., et al. (2005). Increased fasting plasma ghrelin levels during alcohol abstinence. *Alcohol and Alcoholism, 40*(1), 76–79. <Go to ISI>://000226189000009.

King, A., Vena, A., Hasin, D. S., deWit, H., O'Connor, S. J., & Cao, D. (2021). Subjective responses to alcohol in the development and maintenance of alcohol use disorder. *The American Journal of Psychiatry, 178*(6), 560–571. https://doi.org/10.1176/appi.ajp.2020.20030247.

Klausen, M. K., Jensen, M. E., Moller, M., le Dous, N., Jensen, A. O., Zeeman, V. A., et al. (2022). Exenatide once weekly for alcohol use disorder investigated in a randomized, placebo-controlled clinical trial. *JCI Insight*. https://doi.org/10.1172/jci.insight.159863.

Kojima, M., Hosoda, H., Date, Y., Nakazato, M., Matsuo, H., & Kangawa, K. (1999). Ghrelin is a growth-hormone-releasing acylated peptide from stomach [Research Support, Non-U.S. Gov't]. *Nature, 402*(6762), 656–660. https://doi.org/10.1038/45230.

Koob, G. F., & Volkow, N. D. (2010). Neurocircuitry of addiction. *Neuropsychopharmacology: Official Publication of the American College of Neuropsychopharmacology, 35*(1), 217–238. https://doi.org/npp2009110 [pii] https://doi.org/10.1038/npp.2009.110.

Koopmann, A., Bach, P., Schuster, R., Bumb, J. M., Vollstadt-Klein, S., Reinhard, I., et al. (2019). Ghrelin modulates mesolimbic reactivity to alcohol cues in alcohol-addicted subjects: A functional imaging study. *Addiction Biology, 24*(5), 1066–1076. https://doi.org/10.1111/adb.12651.

Koopmann, A., von der Goltz, C., Grosshans, M., Dinter, C., Vitale, M., Wiedemann, K., et al. (2012). The association of the appetitive peptide acetylated ghrelin with alcohol craving in early abstinent alcohol dependent individuals. *Psychoneuroendocrinology, 37*(7), 980–986. https://doi.org/10.1016/j.psyneuen.2011.11.005.

Korol, S. V., Jin, Z., Babateen, O., & Birnir, B. (2015). GLP-1 and exendin-4 transiently enhance GABAA receptor-mediated synaptic and tonic currents in rat hippocampal CA3 pyramidal neurons. *Diabetes, 64*(1), 79–89. https://doi.org/10.2337/db14-0668.

Korol, S. V., Jin, Z., & Birnir, B. (2015). The GLP-1 receptor agonist exendin-4 and diazepam differentially regulate GABAA receptor-mediated tonic currents in rat hippocampal CA3 pyramidal neurons. *PLoS One, 10*(4), e0124765. https://doi.org/10.1371/journal.pone.0124765.

Kraus, T., Schanze, A., Groschl, M., Bayerlein, K., Hillemacher, T., Reulbach, U., et al. (2005). Ghrelin levels are increased in alcoholism. *Alcoholism, Clinical and Experimental Research, 29*(12), 2154–2157. http://www.ncbi.nlm.nih.gov/pubmed/16385185.

Krause, A., Sillard, R., Kleemeier, B., Kluver, E., Maronde, E., Conejo-Garcia, J. R., et al. (2003). Isolation and biochemical characterization of LEAP-2, a novel blood peptide expressed in the liver. *Protein Science: a Publication of the Protein Society, 12*(1), 143–152. https://doi.org/10.1110/ps.0213603.

Landgren, S., Landgren, K., Jerlhag, E., Fahlke, C., Balldin, J., Berggren, U., et al. (2011a). Reward-related genes and personality traits in alcohol-dependent individuals: A pilot case control study. *Neuropsychobiology, 64*(1), 38–46. https://doi.org/000324992 [pii] https://doi.org/10.1159/000324992.

Landgren, S., Engel, J. A., Hyytia, P., Zetterberg, H., Blennow, K., & Jerlhag, E. (2011). Expression of the gene encoding the ghrelin receptor in rats selected for differential alcohol preference [Comparative Study Research Support, Non-U.S. Gov't]. *Behavioural Brain Research, 221*(1), 182–188. https://doi.org/10.1016/j.bbr.2011.03.003.

Landgren, S., Jerlhag, E., Hallman, J., Oreland, L., Lissner, L., Strandhagen, E., et al. (2010). Genetic variation of the ghrelin signaling system in females with severe alcohol dependence [Research Support, Non-U.S. Gov't]. *Alcoholism, Clinical and Experimental Research, 34*(9), 1519–1524. https://doi.org/10.1111/j.1530-0277.2010.01236.x.

Landgren, S., Jerlhag, E., Zetterberg, H., Gonzalez-Quintela, A., Campos, J., Olofsson, U., et al. (2008). Association of pro-ghrelin and GHS-R1A gene polymorphisms and haplotypes with heavy alcohol use and body mass [Comparative Study Research Support, Non-U.S. Gov't]. *Alcoholism: Clinical and Experimental Research, 32*(12), 2054–2061. https://doi.org/10.1111/j.1530-0277.2008.00793.x.

Landgren, S., Simms, J. A., Hyytia, P., Engel, J. A., Bartlett, S. E., & Jerlhag, E. (2012). Ghrelin receptor (GHS-R1A) antagonism suppresses both operant alcohol self-administration and high alcohol consumption in rats [Research Support, Non-U.S. Gov't]. *Addiction Biology, 17*(1), 86–94. https://doi.org/10.1111/j.1369-1600.2010.00280.x.

Lee, M. R., Farokhnia, M., Cobbina, E., Saravanakumar, A., Li, X., Battista, J. T., et al. (2020a). Endocrine effects of the novel ghrelin receptor inverse agonist PF-5190457: Results from a placebo-controlled human laboratory alcohol co-administration study in heavy drinkers. *Neuropharmacology, 170*, 107788. https://doi.org/10.1016/j.neuropharm.2019.107788.

Lee, M. R., Tapocik, J. D., Ghareeb, M., Schwandt, M. L., Dias, A. A., Le, A. N., et al. (2020b). The novel ghrelin receptor inverse agonist PF-5190457 administered with alcohol: Preclinical safety experiments and a phase 1b human laboratory study. *Molecular Psychiatry, 25*(2), 461–475. https://doi.org/10.1038/s41380-018-0064-y.

Leggio, L., Ferrulli, A., Cardone, S., Nesci, A., Miceli, A., Malandrino, N., et al. (2012). Ghrelin system in alcohol-dependent subjects: Role of plasma ghrelin levels in alcohol drinking and craving. *Addiction Biology, 17*(2), 452–464. https://doi.org/10.1111/j.1369-1600.2010.00308.x.

Leggio, L., Schwandt, M. L., Oot, E. N., Dias, A. A., & Ramchandani, V. A. (2013). Fasting-induced increase in plasma ghrelin is blunted by intravenous alcohol administration: A within-subject placebo-controlled study. *Psychoneuroendocrinology, 38*(12), 3085–3091. https://doi.org/10.1016/j.psyneuen.2013.09.005.

Leggio, L., Zywiak, W. H., Fricchione, S. R., Edwards, S. M., de la Monte, S. M., Swift, R. M., et al. (2014). Intravenous ghrelin administration increases alcohol craving in alcohol-dependent heavy drinkers: A preliminary investigation. *Biological Psychiatry*. https://doi.org/10.1016/j.biopsych.2014.03.019.

Liberini, C. G., Koch-Laskowski, K., Shaulson, E., McGrath, L. E., Lipsky, R. K., Lhamo, R., et al. (2019). Combined Amylin/GLP-1 pharmacotherapy to promote and sustain long-lasting weight loss. *Scientific Reports, 9*(1), 8447. https://doi.org/10.1038/s41598-019-44591-8.

Liu, W., Wang, Z., Wang, W., Wang, Z., Xing, Y., & Holscher, C. (2024). Liraglutide reduces alcohol consumption, anxiety, memory impairment, and synapse loss in alcohol dependent mice. *Neurochemical Research, 49*(4), 1061–1075. https://doi.org/10.1007/s11064-023-04093-6.

Lu, S., Guan, J. L., Wang, Q. P., Uehara, K., Yamada, S., Goto, N., et al. (2002). Immunocytochemical observation of ghrelin-containing neurons in the rat arcuate nucleus [Research Support, Non-U.S. Gov't]. *Neuroscience Letters, 321*(3), 157–160. http://www.ncbi.nlm.nih.gov/pubmed/11880196.

Lyons, A. M., Lowery, E. G., Sparta, D. R., & Thiele, T. E. (2008). Effects of food availability and administration of orexigenic and anorectic agents on elevated ethanol drinking associated with drinking in the dark procedures. *Alcoholism, Clinical and Experimental Research, 32*(11), 1962–1968. https://doi.org/10.1111/j.1530-0277.2008.00784.x.

M'Kadmi, C., Cabral, A., Barrile, F., Giribaldi, J., Cantel, S., Damian, M., et al. (2019). N-terminal liver-expressed antimicrobial peptide 2 (LEAP2) region exhibits inverse agonist activity toward the ghrelin receptor. *Journal of Medicinal Chemistry, 62*(2), 965–973. https://doi.org/10.1021/acs.jmedchem.8b01644.

Mahalingam, S., Bellamkonda, R., Arumugam, M. K., Perumal, S. K., Yoon, J., Casey, C., et al. (2023). Glucagon-like peptide 1 receptor agonist, exendin-4, reduces alcohol-associated fatty liver disease. *Biochemical Pharmacology, 213*, 115613. https://doi.org/10.1016/j.bcp.2023.115613.

Marty, V. N., Farokhnia, M., Munier, J. J., Mulpuri, Y., Leggio, L., & Spigelman, I. (2020). Long-acting glucagon-like peptide-1 receptor agonists suppress voluntary alcohol intake in male Wistar rats. *Frontiers in Neuroscience, 14*, 599646. https://doi.org/10.3389/fnins.2020.599646.

Mondal, M. S., Date, Y., Yamaguchi, H., Toshinai, K., Tsuruta, T., Kangawa, K., et al. (2005). Identification of ghrelin and its receptor in neurons of the rat arcuate nucleus [Research Support, Non-U.S. Gov't]. *Regulatory Peptides, 126*(1-2), 55–59. https://doi.org/10.1016/j.regpep.2004.08.038.

Morganstern, I., Barson, J. R., & Leibowitz, S. F. (2011). Regulation of drug and palatable food overconsumption by similar peptide systems. *Current Drug Abuse Reviews, 4*(3), 163–173. http://www.ncbi.nlm.nih.gov/pubmed/21999690.

Mustafa, E. R., Cordisco Gonzalez, S., Damian, M., Cantel, S., Denoyelle, S., Wagner, R., et al. (2021). LEAP2 impairs the capability of the growth hormone secretagogue receptor to regulate the dopamine 2 receptor signaling. *Frontiers in Pharmacology, 12*, 712437. https://doi.org/10.3389/fphar.2021.712437.

Navarro, G., Rea, W., Quiroz, C., Moreno, E., Gomez, D., Wenthur, C. J., et al. (2022). Complexes of ghrelin GHS-R1a, GHS-R1b, and dopamine D(1) receptors localized in the ventral tegmental area as main mediators of the dopaminergic effects of ghrelin. *The Journal of Neuroscience, 42*(6), 940–953. https://doi.org/10.1523/JNEUROSCI.1151-21.2021.

Novak, U., Wilks, A., Buell, G., & McEwen, S. (1987). Identical mRNA for preproglucagon in pancreas and gut. *European Journal of Biochemistry/FEBS, 164*(3), 553–558. http://www.ncbi.nlm.nih.gov/pubmed/3569278.

Olsson, Y., Hodzic, K., Wass, C., Lidö, H., Stangl, B. L., O'Connor, S., et al. (2023). Free-access intravenous alcohol self-administration in social drinkers and individuals with alcohol use disorder: Evaluation of relationships with phosphatidylethanol and self-reported alcohol consumption. *Alcohol, Clinical & Experimental Research (Hoboken, N.J.), 47*(8), 1453–1466.

Orellana, E. R., Jamis, C., Horvath, N., & Hajnal, A. (2018). Effect of vertical sleeve gastrectomy on alcohol consumption and preferences in dietary obese rats and mice: A plausible role for altered ghrelin signaling. *Brain Research Bulletin, 138*, 26–36. https://doi.org/10.1016/j.brainresbull.2017.08.004.

Pietrzak, M., Yngve, A., Hamilton, J. P., Kampe, R., Boehme, R., Asratian, A., et al. (2023). A randomized controlled experimental medicine study of ghrelin in value-based decision making. *The Journal of Clinical Investigation, 133*(12), https://doi.org/10.1172/JCI168260.

Pirnik, Z., Bundzikova, J., Holubova, M., Pychova, M., Fehrentz, M., Martinez, J., et al. (2011). Ghrelin agonists impact on Fos protein expression in brain areas related to food intake regulation in male C57BL/6 mice. *Neurochemistry International, 59*(6), 889–895. https://doi.org/10.1016/j.neuint.2011.08.001.

Probst, L., Monnerat, S., Vogt, D. R., Lengsfeld, S., Burkard, T., Meienberg, A., et al. (2023). Effects of dulaglutide on alcohol consumption during smoking cessation. *JCI Insight, 8*(22), https://doi.org/10.1172/jci.insight.170419.

Quddos, F., Hubshman, Z., Tegge, A., Sane, D., Marti, E., Kablinger, A. S., et al. (2023). Semaglutide and tirzepatide reduce alcohol consumption in individuals with obesity. *Scientific Reports, 13*(1), 20998. https://doi.org/10.1038/s41598-023-48267-2.

Ralevski, E., Horvath, T. L., Shanabrough, M., Hayden, R., Newcomb, J., & Petrakis, I. (2017). Ghrelin is supressed by intravenous alcohol and is related to stimulant and sedative effects of alcohol. *Alcohol and Alcoholism (Oxford, Oxfordshire), 52*(4), 431–438. https://doi.org/10.1093/alcalc/agx022.

Ralevski, E., Horvath, T. L., Shanabrough, M., Newcomb, J., Pisani, E., & Petrakis, I. (2022). Ghrelin predicts stimulant and sedative effects of alcohol in heavy drinkers. *Alcohol and Alcoholism (Oxford, Oxfordshire)*. https://doi.org/10.1093/alcalc/agac058.

Ralevski, E., Shanabrough, M., Newcomb, J., Gandelman, E., Hayden, R., Horvath, T. L., et al. (2018). Ghrelin is related to personality differences in reward sensitivity and impulsivity. *Alcohol and Alcoholism (Oxford, Oxfordshire), 53*(1), 52–56. https://doi.org/10.1093/alcalc/agx082.

Rasineni, K., Thomes, P. G., Kubik, J. L., Harris, E. N., Kharbanda, K. K., & Casey, C. A. (2019). Chronic alcohol exposure alters circulating insulin and ghrelin levels: Role of ghrelin in hepatic steatosis. *American Journal of Physiology. Gastrointestinal and Liver Physiology, 316*(4), G453–G461. https://doi.org/10.1152/ajpgi.00334.2018.

Rebosio, C., Balbi, M., Passalacqua, M., Ricciarelli, R., & Fedele, E. (2018). Presynaptic GLP-1 receptors enhance the depolarization-evoked release of glutamate and GABA in the mouse cortex and hippocampus. *Biofactors (Oxford, England), 44*(2), 148–157. https://doi.org/10.1002/biof.1406.

Richards, J. R., Dorand, M. F., Royal, K., Mnajjed, L., Paszkowiak, M., & Simmons, W. K. (2023). Significant decrease in alcohol use disorder symptoms secondary to semaglutide therapy for weight loss: A case series. *The Journal of Clinical Psychiatry, 85*(1), https://doi.org/10.4088/JCP.23m15068.

Richardson, R. S., Sulima, A., Rice, K. C., Kucharczk, J. A., Janda, K. D., Nisbett, K. E., et al. (2023). Pharmacological GHSR (ghrelin receptor) blockade reduces alcohol binge-like drinking in male and female mice. *Neuropharmacology, 238*, 109643. https://doi.org/10.1016/j.neuropharm.2023.109643.

Rinaman, L. (2010). Ascending projections from the caudal visceral nucleus of the solitary tract to brain regions involved in food intake and energy expenditure. *Brain Research, 1350*, 18–34. https://doi.org/10.1016/j.brainres.2010.03.059.

Sakata, I., Nakano, Y., Osborne-Lawrence, S., Rovinsky, S. A., Lee, C. E., Perello, M., et al. (2009). Characterization of a novel ghrelin cell reporter mouse. *Regulatory Peptides, 155*(1–3), 91–98. https://doi.org/10.1016/j.regpep.2009.04.001.

Salinas, C. B. G., Lu, T. T., Gabery, S., Marstal, K., Alanentalo, T., Mercer, A. J., et al. (2018). Integrated brain atlas for unbiased mapping of nervous system effects following liraglutide treatment. *Scientific Reports, 8*(1), 10310. https://doi.org/10.1038/s41598-018-28496-6.

Schaeffer, M., Langlet, F., Lafont, C., Molino, F., Hodson, D. J., Roux, T., et al. (2013). Rapid sensing of circulating ghrelin by hypothalamic appetite-modifying neurons. *Proceedings of the National Academy of Sciences of the United States of America, 110*(4), 1512–1517. https://doi.org/10.1073/pnas.1212137110.

Sha, L., Dey, P., Khess, C. R., & Khitiz, K. K. (2021). The association of plasma acyl ghrelin level with alcohol craving in early abstinent alcohol dependent patients. *Journal of Postgraduate Medicine, 67*(1), 12–17. https://doi.org/10.4103/jpgm.JPGM_1018_20.

Sharma, A. N., Pise, A., Sharma, J. N., & Shukla, P. (2014a). Dipeptidyl-peptidase IV (DPP-IV) inhibitor delays tolerance to anxiolytic effect of ethanol and withdrawal-induced anxiety in rats. *Metabolic Brain Disease*. https://doi.org/10.1007/s11011-014-9603-7.

Sharma, A. N., Pise, A., Sharma, J. N., & Shukla, P. (2014b). Glucagon-like peptide-1 (GLP-1) receptor agonist prevents development of tolerance to anti-anxiety effect of ethanol and withdrawal-induced anxiety in rats. *Metabolic Brain Disease*. https://doi.org/10.1007/s11011-014-9627-z.

Shevchouk, O. T., Tufvesson-Alm, M., & Jerlhag, E. (2021). An overview of appetite-regulatory peptides in addiction processes; from bench to bed side. *Frontiers in Neuroscience, 15*, 774050. https://doi.org/10.3389/fnins.2021.774050.

Shirazi, R. H., Dickson, S. L., & Skibicka, K. P. (2013). Gut peptide GLP-1 and its analogue, exendin-4, decrease alcohol intake and reward. *PLoS One, 8*(4), e61965. https://doi.org/10.1371/journal.pone.0061965.

Simms, J. A., Steensland, P., Medina, B., Abernathy, K. E., Chandler, L. J., Wise, R., et al. (2008). Intermittent access to 20% ethanol induces high ethanol consumption in Long-Evans and Wistar rats. [Research Support, N.I.H., Extramural Research Support, Non-U.S. Gov't Research Support, U.S. Gov't, Non-P.H.S.] *Alcoholism: Clinical and Experimental Research, 32*(10), 1816–1823. https://doi.org/10.1111/j.1530-0277.2008.00753.x.

Sirohi, S., Schurdak, J. D., Seeley, R. J., Benoit, S. C., & Davis, J. F. (2016). Central & peripheral glucagon-like peptide-1 receptor signaling differentially regulate addictive behaviors. *Physiology & Behavior, 161*, 140–144. https://doi.org/10.1016/j.physbeh.2016.04.013.

Soderpalm, B., Danielsson, K., de Bejczy, A., Adermark, L., & Ericson, M. (2020). Combined administration of varenicline and bupropion produces additive effects on accumbal dopamine and abolishes the alcohol deprivation effect in rats. *Addiction Biology, 25*(5), e12807. https://doi.org/10.1111/adb.12807.

Sorensen, G., Caine, S. B., & Thomsen, M. (2016). Effects of the GLP-1 agonist exendin-4 on intravenous ethanol self-administration in mice. *Alcoholism, Clinical and Experimental Research, 40*(10), 2247–2252. https://doi.org/10.1111/acer.13199.

Spanagel, R. (2000). Recent animal models of alcoholism. *Alcohol Research & Health: the Journal of the National Institute on Alcohol Abuse and Alcoholism, 24*(2), 124–131. http://www.ncbi.nlm.nih.gov/entrez/query.fcgi?cmd=Retrieve&db=PubMed&dopt=Citation&list_uids=11199279.

Srivastava, G., & Apovian, C. (2018). Future pharmacotherapy for obesity: New anti-obesity drugs on the horizon. *Current Obesity Reports, 7*(2), 147–161. https://doi.org/10.1007/s13679-018-0300-4.

Stevenson, J. R., Buirkle, J. M., Buckley, L. E., Young, K. A., Albertini, K. M., & Bohidar, A. E. (2015). GHS-R1A antagonism reduces alcohol but not sucrose preference in prairie voles. *Physiology & Behavior, 147*, 23–29. https://doi.org/10.1016/j.physbeh.2015.04.001.

Suchankova, P., Engel, J. A., & Jerlhag, E. (2016). Sub-chronic ghrelin receptor blockade attenuates alcohol- and amphetamine-induced locomotor stimulation in mice. *Alcohol and Alcoholism (Oxford, Oxfordshire), 51*(2), 121–127. https://doi.org/10.1093/alcalc/agv100.

Suchankova, P., Nilsson, S., Von Der Pahlen, B., Santtila, P., Sandnabba, K., Jern, P., et al. (2016b). Genetic variation of the growth hormone secretagogue receptor gene is associated with Alcohol Use Disorders Identification Test scores and smoking. *Addiction Biology, 21*(2) 481.

Suchankova, P., Steensland, P., Fredriksson, I., Engel, J. A., & Jerlhag, E. (2013). Ghrelin receptor (GHS-R1A) antagonism suppresses both alcohol consumption and the alcohol deprivation effect in rats following long-term voluntary alcohol consumption. *PLoS One, 8*(8), e71284. https://doi.org/10.1371/journal.pone.0071284.

Suchankova, P., Yan, J., Schwandt, M. L., Stangl, B. L., Caparelli, E. C., Momenan, R., et al. (2015). The glucagon-like peptide-1 receptor as a potential treatment target in alcohol use disorder: Evidence from human genetic association studies and a mouse model of alcohol dependence. *Translational Psychiatry, 5*, e583.

Suchankova, P., Yan, J., Schwandt, M. L., Stangl, B. L., Jerlhag, E., Engel, J. A., et al. (2017). The Leu72Met polymorphism of the prepro-ghrelin gene is associated with alcohol consumption and subjective responses to alcohol: Preliminary findings. *Alcohol and Alcoholism (Oxford, Oxfordshire), 52*(4), 425–430. https://doi.org/10.1093/alcalc/agx021.

Szulc, M., Mikolajczak, P. L., Geppert, B., Wachowiak, R., Dyr, W., & Bobkiewicz-Kozlowska, T. (2013). Ethanol affects acylated and total ghrelin levels in peripheral blood of alcohol-dependent rats. *Addiction Biology, 18*(4), 689–701. https://doi.org/10.1111/adb.12025.

Thomsen, M., Dencker, D., Wortwein, G., Weikop, P., Egecioglu, E., Jerlhag, E., et al. (2017). The glucagon-like peptide 1 receptor agonist exendin-4 decreases relapse-like drinking in socially housed mice. *Pharmacology, Biochemistry, and Behavior, 160*, 14–20. https://doi.org/10.1016/j.pbb.2017.07.014.

Thomsen, M., Holst, J. J., Molander, A., Linnet, K., Ptito, M., & Fink-Jensen, A. (2019). Effects of glucagon-like peptide 1 analogs on alcohol intake in alcohol-preferring vervet monkeys. *Psychopharmacology (Berl), 236*(2), 603–611. https://doi.org/10.1007/s00213-018-5089-z.

Vallof, D., Kalafateli, A. L., & Jerlhag, E. (2019). Brain region specific glucagon-like peptide-1 receptors regulate alcohol-induced behaviors in rodents. *Psychoneuroendocrinology, 103*, 284–295. https://doi.org/10.1016/j.psyneuen.2019.02.006.

Vallof, D., Kalafateli, A. L., & Jerlhag, E. (2020). Long-term treatment with a glucagon-like peptide-1 receptor agonist reduces ethanol intake in male and female rats. *Translational Psychiatry, 10*(1), 238. https://doi.org/10.1038/s41398-020-00923-1.

Vallof, D., Maccioni, P., Colombo, G., Mandrapa, M., Jornulf, J. W., Egecioglu, E., et al. (2015). The glucagon-like peptide 1 receptor agonist liraglutide attenuates the reinforcing properties of alcohol in rodents. *Addiction Biology*. https://doi.org/10.1111/adb.12295.

Vallof, D., Vestlund, J., & Jerlhag, E. (2019). Glucagon-like peptide-1 receptors within the nucleus of the solitary tract regulate alcohol-mediated behaviors in rodents. *Neuropharmacology, 149*, 124–132. https://doi.org/10.1016/j.neuropharm.2019.02.020.

Vestlund, J., Bergquist, F., Licheri, V., Adermark, L., & Jerlhag, E. (2020). Activation of glucagon-like peptide-1 receptors and skilled reach foraging. *Addiction Biology*, e12953. https://doi.org/10.1111/adb.12953.

Wium-Andersen, I. K., Wium-Andersen, M. K., Fink-Jensen, A., Rungby, J., Jorgensen, M. B., & Osler, M. (2022). Use of GLP-1 receptor agonists and subsequent risk of alcohol-related events. A nationwide register-based cohort and self-controlled case series study. *Basic & Clinical Pharmacology & Toxicology, 131*(5), 372–379. https://doi.org/10.1111/bcpt.13776.

Wolfe, W. L., & Maisto, S. A. (2000). The relationship between eating disorders and substance use: Moving beyond co-prevalence research [Review]. *Clinical Psychology Review, 20*(5), 617–631. http://www.ncbi.nlm.nih.gov/pubmed/10860169.

Wurst, F. M., Graf, I., Ehrenthal, H. D., Klein, S., Backhaus, J., Blank, S., et al. (2007). Gender differences for ghrelin levels in alcohol-dependent patients and differences between alcoholics and healthy controls. *Alcoholism-Clinical and Experimental Research, 31*(12), 2006–2011. <Go to ISI>://000251184800006.

Yoshimoto, K., Nagao, M., Watanabe, Y., Yamaguchi, T., Ueda, S., Kitamura, Y., et al. (2017). Enhanced alcohol-drinking behavior associated with active ghrelinergic and serotoninergic neurons in the lateral hypothalamus and amygdala. *Pharmacology, Biochemistry, and Behavior, 153*, 1–11. https://doi.org/10.1016/j.pbb.2016.12.001.

You, Z. B., Gardner, E. L., Galaj, E., Moore, A. R., Buck, T., Jordan, C. J., et al. (2022). Involvement of the ghrelin system in the maintenance of oxycodone self-administration: Converging evidence from endocrine, pharmacologic and transgenic approaches. *Molecular Psychiatry, 27*(4), 2171–2181. https://doi.org/10.1038/s41380-022-01438-5.

Zallar, L. J., Beurmann, S., Tunstall, B. J., Fraser, C. M., Koob, G. F., Vendruscolo, L. F., et al. (2019). Ghrelin receptor deletion reduces binge-like alcohol drinking in rats. *Journal of Neuroendocrinology, 31*(7), e12663. https://doi.org/10.1111/jne.12663.

Zhou, Y., Leri, F., Low, M. J., & Kreek, M. J. (2019). Sex differences in the effect of bupropion and naltrexone combination on alcohol drinking in mice. *Pharmacology, Biochemistry, and Behavior, 181*, 28–36. https://doi.org/10.1016/j.pbb.2019.04.004.

Zimmermann, U. S., Buchmann, A., Steffin, B., Dieterle, C., & Uhr, M. (2007). Alcohol administration acutely inhibits ghrelin secretion in an experiment involving psychosocial stress [Controlled Clinical Trial Research Support, Non-U.S. Gov't]. *Addiction Biology, 12*(1), 17–21. https://doi.org/10.1111/j.1369-1600.2006.00026.x.